ISBN 978-0-266-53883-7
PIBN 10862904

Journal of Mathematics.

SIMON NEWCOMB, Editor.
THOMAS CRAIG, Associate Editor.

PUBLISHED UNDER THE AUSPICES OF THE

JOHNS HOPKINS UNIVERSITY.

Πραγμάτων ἔλεγχος οὐ βλεπομένων.

VOLUME VII.

BALTIMORE: PRESS OF ISAAC FRIEDENWALD.

AGENTS:

B. WESTERMANN & CO., *New York.*
D. VAN NOSTRAND, *New York.*
E. STEIGER & CO., *New York.*
JANSEN, MCCLURG & CO., *Chicago.*
CUSHINGS & BAILEY, *Baltimore.*
CUPPLES, UPHAM & CO., *Boston.*

TRÜBNER & CO., *London.*
A. HERMANN, *Paris.*
GAUTHIER-VILLARS, *Paris.*
MAYER & MÜLLER, *Berlin.*
ULRICO HOEPLI, *Milan.*
KARL J. TRÜBNER, *Strassburg.*

1885.

INDEX.

ERRATA.

VOL. VI.

Page 65, Contents 2. For "the particular cognate forms" read "the unequal particular cognate forms."

" 70. The fourth line should be

$$f(\omega_1, \theta_1, \text{etc.}) = h_1 \omega_1^{n-1} + h_2 \omega_1^{n-2} + \ldots + h_n.$$

" 101, line 13. For "by Cor. Prop. XIV" read "by Cor. Prop. XIII."

VOL. VII.

Page 4. In the third and fourth lines from top, for 126 read 136.

" 210. Tenth line from bottom, for "on verrait que y étant" read "on verrait que y étant."

" 211. First line below the three equations, for "reconnaitrait" read "reconnaîtrait."

" 219. Line above equation (2), for "on" read "ou."

" 222. Seventh line from top, for "on" read "ou."

" 228. First line above formulæ near middle of the page, for "différe" read "diffère."

" 244. First line above equation (5), for "qui" read "que."

" 248. First line above formula $Q_i =$ &c., for "la coëfficient" read "le coefficient."

" 255. Ninth line from bottom, for "considerons" read "considérons."

" 296. In the three first equations of 2, for X read χ.

A Memoir on Seminvariants.

By Professor Cayley.

INTRODUCTORY. Art. Nos. 1 to 8.

1. A very remarkable discovery in the Theory of Seminvariants has been recently made by Capt. MacMahon, viz., considering the equation

$$0 = 1 + b\frac{x}{1} + c\frac{x^2}{1.2} + d\frac{x^3}{1.2.3} + \text{etc.}$$

and its roots α, β, γ . . . , as defined by the identity

$$1 + b\frac{x}{1} + c\frac{x^2}{1.2} + \text{etc.} = 1 - \alpha x.1 - \beta x.1 - \gamma x \ldots$$

then, any symmetrical function of the roots being represented by a partition symbol in the usual manner, $1 = \Sigma\alpha$, $2 = \Sigma\alpha^2$, 11 or $1^2 = \Sigma\alpha\beta$, etc., the theorem is that any symmetric function represented by a non-unitary symbol (or symbol not containing a 1), say for shortness any non-unitary symmetric function, is a seminvariant in regard to the coefficients 1, b, c, d, e, etc.

We have for instance

$$\begin{aligned} 2 &= -\ (c - b^2), \\ 3 &= -\tfrac{1}{2}(d - 3bc + 2b^3), \\ 4 &= \ \ \tfrac{1}{6}(-e + 4bd + 3c^2 - 12bc + 6b^4), \end{aligned}$$

(where to verify that this is a seminvariant, observe that the value may be written $= \frac{1}{6}\{-(e - 4bd + 3c^2) + 6(c - b^2)^2\}$)

$$22 = \tfrac{1}{12}(e - 4bd + 3c^2),$$

etc.

and observe further that the forms 2, 4 and 22, are connected by the identical relation $$2.2 = 4 + 2.22.$$

2. We conclude that the theory of seminvariants is a part of that of symmetric functions. I take the opportunity of remarking that (the subject of the memoir being seminvariants) I use in general non-unitary symbols, even in cases where the restriction is unnecessary, and the symbols might have contained 1's: thus instead of $2.2 = 4 + 2.22$, the equation $1.1 = 2 + 2.11$ would have served equally well as an instance of an identical relation between symmetric functions; and so in general, in formulæ relating to symmetric functions, the symbols are not restricted to be non-unitary. I remark also that, for instance, instead of 4443322222, or $4^3 3^2 2^5$, I usually write 444332^5, introducing the index only for the 2; the reason is only that the 2 is often repeated a large number of times, so that the abbreviation, which I dispense with for the higher numbers, becomes convenient for the 2.

3. Reckoning the coefficients $1, b, c, d, e \ldots$ as being each of them of the degree 1, and of the weights $0, 1, 2, 3, 4 \ldots$ respectively, then any symmetric function is of a degree which is equal to the highest number, and of a weight which is equal to the sum of all the numbers, in the partition-symbol. And we frequently speak of the deg. weight: thus for the function 22 the deg. weight is $= 2.4$.

MULTIPLICATION OF TWO SYMMETRIC FUNCTIONS. ART. NOS. 4 TO 17.

4. We require a theory for the multiplication of two symmetric functions. We have for instance $3.2 = 5 + 32$: for 3 denoting $\Sigma\alpha^3$, and 2 denoting $\Sigma\alpha^2$, the product contains the term α^5, and the term $\alpha^3\beta^2$, and it is thus $= \Sigma\alpha^5 + \Sigma\alpha^3\beta^2$, which is $= 5 + 32$. But multiplying for instance 2 by itself, the product contains the term α^4, and the term $\alpha^2\beta^2$ twice, and it is thus $= \Sigma\alpha^4 + 2\Sigma\alpha^2\beta^2$, and we thus have the before-mentioned formula $2.2 = 4 + 2.22$.

And so, l, m being different

$$l.m = (l + m) + lm,$$

but when $m = l$,
$$l.l = (2l) + 2.ll,$$

and in general for any symmetric function $lmnpqr \ldots$, where the numbers are all of them different, if any two of the m's become equal, we must multiply the term by 2; if any three of them become equal, we must multiply the term by 6; and so in other cases, viz., if the term becomes $l^\alpha m^\beta n^\gamma \ldots$, we must multiply it by $[\alpha]^\alpha [\beta]^\beta [\gamma]^\gamma \ldots$

5. We may, taking in the first instance the numbers $l, m, n, p, q \ldots$ to be all of them different, develop an algorithm as follows:

$$\begin{array}{l|l} l & m \\ \hline m & \\ & m \end{array} = \begin{array}{l}(l+m)\\ lm\end{array} \qquad \text{that is} \quad l.m = \begin{array}{l}(l+m)\\ +\, lm\end{array}$$

$$\begin{array}{r|l} lm & n \\ \hline n & \\ n & \\ & n \end{array} = \begin{array}{l}(l+n)\,m\\ l\,(m+n)\\ lmn\end{array} \qquad \text{that is } lm.n = \begin{array}{l}\,(l+n)\,m\\ +\, l\,(m+n)\\ +\, lmn,\end{array}$$ and so in other cases;

$$\begin{array}{r|l} lmn & p \\ \hline p\;\;\; & \\ p\;\; & \\ p & \\ & p \end{array} = \begin{array}{l}(l+p)\,mn\\ l\,(m+p)\,n\\ lm\,(n+p)\\ lmnp\end{array} \qquad \begin{array}{r|l} lm & np \\ \hline np & \\ pn & \\ n\; & p \\ p\; & n \\ n & p \\ p & n \\ & np \end{array} = \begin{array}{l}(l+n)(m+p)\\ (l+p)(m+n)\\ (l+n)\,mp\\ (l+p)\,mn\\ l\,(m+n)\,p\\ l\,(m+p)\,n\\ lmnp\end{array}$$

$$\begin{array}{r|l} lmn & pq \\ \hline pq\;\;\; & \\ qp\;\;\; & \\ p\;\;q\;\; & \\ q\;\;p\;\; & \\ pq & \\ qp & \\ p\;\;\;\;\;\; & q \\ q\;\;\;\;\;\; & p \\ p\;\; & q \\ q\;\; & p \\ p & q \\ q & p \\ & pq \end{array} = \begin{array}{l}(l+p)(m+q)\,n\\ (l+q)(m+p)\,n\\ (l+p)\,m\,(n+q)\\ (l+q)\,m\,(n+p)\\ l\,(m+p)(n+q)\\ l\,(m+q)(n+p)\\ (l+p)\,mnq\\ (l+q)\,mnp\\ l\,(m+p)\,q\\ l\,(m+q)\,p\\ lm\,(n+p)\,q\\ lm\,(n+q)\,p\\ lmnpq\end{array}$$ and so on.

6. Observe that if the two factors contain i numbers and j numbers respectively, $i >$ or $= j$, then in the product we have

$[i]^j$ terms each containing i numbers,

$\dfrac{[j]^1}{[1]^1}[i]^{j-1}$ " " $i+1$ "

$\dfrac{[j]^2}{[2]^2}[i]^{j-2}$ " " $i+2$ "

$\vdots$

$\dfrac{[j]^j}{[j]^j}[i]^0, = 1$, term containing $i+j$ "

so that the whole number of terms in the product is

$$\{i, j\}, = [i]^j + \frac{j}{1}[i]^{j-1} + \ldots + 1.$$

We may, if we please, take the smaller number first, and we then have

$$\{j, i\} = \{i, j\};$$

the $\{j, i\}$ series in fact begins with zero terms, but following these we have terms which are identical each to each with those of the $\{i, j\}$ series. Thus

$$\{5,3\}= \qquad [5]^3+3[5]^2+3[5]^1+[5]^0= \qquad 60+3.20+3.5+1, =126,$$
$$\{3,5\}=[3]^5+5[3]^4+10[3]^3+10[3]^2+5[3]^1+[3]^0=0+0+10.6+10.6+5.3+1, =126.$$

and it is easy to see that the general theorem can be verified in like manner.

In particular we have Putting $i=1$ these give

$$\{i,1\}=[i]^1+1 \qquad =i+1 \qquad , \{1,1\}=2,$$
$$\{i,2\}=[i]^2+2[i]^1+1 \qquad =i^2+i+1 \qquad , \{1,2\}=3,$$
$$\{i,3\}=[i]^3+3[i]^2+3[i]^1+1 \qquad =i^3+2i+1 \qquad , \{1,3\}=4,$$
$$\{i,4\}=[i]^4+4[i]^3+6[i]^2+4[i]^1+1 \qquad =i^4-2i^3+5i^2+1 \qquad , \{1,4\}=5,$$
$$\{i,5\}=[i]^5+5[i]^4+10[i]^3+10[i]^2+5[i]^1+1=i^5-5i^4+15i^3-15i^2+9i+1, \ \{1,5\}=6,$$

which agree with $\{i, 1\}=i+1$.

Hence also the values of

$\{1,1\}$,	are	2,
$\{1,2\}$, $\{2,2\}$,		3, 7,
$\{1,3\}$, $\{2,3\}$, $\{3,3\}$,		4, 13, 34,
$\{1,4\}$, $\{2,4\}$, $\{3,4\}$, $\{4,4\}$,		5, 21, 73, 209,
$\{1,5\}$, $\{2,5\}$, $\{3,5\}$, $\{4,5\}$, $\{5,5\}$,		6, 31, 136, 501, 1546.

7. In forming a product $lmn \ldots pqr \ldots$, we may have equalities among the numbers $l, m, n \ldots$ of the first symbol, and also between the numbers $p, q, r \ldots$ of the second symbol: moreover (whether there are or are not any such equalities), we may have equalities presenting themselves between the numbers $l+p$, $m+q, \ldots$ of any symbol $(l+p)(m+q) \ldots$ on the right-hand side of the equation: and the process must be performed so as to take account of all these equalities. The actual process is best shown by an example: say we require the product 3332.322, which is of the deg. weight 6.18.

3332	322			+12	
322		6	6552	2	1
32 2		12	6543	1	1
22 3		6	5553	6	3
32	2	12	65322	2	2
3 2	2	6	64332	2	1
2 3	2	6	55332	4	2 }
22	3	6	55332	4	2 }
2 2	3	6	54333	6	3
3	22	3	633222	12	3
3	22	1	533322	12	1 }
2	32	6	533322	12	6 }
2	32	2	433332	24	4
	322	1	3333222	144	12
		73			

8. Observe first that $\{4, 3\} = 24 + 36 + 12 + 1, = 73$. In placing all or any of the numbers of the 322 under those of the 3332, we do this in all the really distinct ways, inserting a numerical coefficient for the frequency of each way. Thus when the whole of the 322 is thus placed, the 3 may be under a 3, and the two 2's may then be under 33 or under 32: or else the 3 may be under a 2, and the two 2's must then be under 33: there are thus three ways, and these have the frequencies 6, 12, 6 respectively. For as to the first way, the 3 may be under any one of the three 3's, and for each such position of the 3, the 22 can be placed in either of two orders under the other two 3's; the frequency is thus $3.2, = 6$. And similarly the other two frequencies are 12 and 6; the sum $6 + 12 + 6, = 24$, is, the first term of $\{4, 3\}$. In like manner when two of the numbers of the 322 are placed under the 3332 there are five ways having the frequencies 12, 6, 6, 6, 6 respectively, the sum of these frequencies is 36, which is the second term of $\{4, 3\}$. And so when one number of the 322 is placed under the 3332 there are four ways having the frequencies 3, 1, 6, 2 respectively: the sum of these frequencies is 12, which is the third term of $\{4, 3\}$. Finally, when no number of the 322 is placed under the 3332, there is a single way, having the frequency 1, which is the last term of $\{4, 3\}$. These agreements are a useful verification for the frequencies: the sum of all the frequencies is of course $= 73$, the value of $\{4, 3\}$.

9. We next form the column of symmetric functions by adding to each line the 3332: to avoid accidental errors of addition, it is proper to verify for each of these that the weight is the sum, $= 18$, of the weights 11 and 7 of the two factors respectively. It is to be observed that the same symmetric function may present itself more than once: thus we have the functions 55332, and 533322, each of them twice. We then form a column of multiplicities: in 6552 the two 5's give the multiplicity 2; in 5553, the three 5's give 6: in 633222, the two 3's and the three 2's give $2.6, = 12$, and so on. There is in like manner for the factor 3332 a multiplicity 6, and for the factor 322 a multiplicity 2; and these combined together give $6.2, = 12$, viz., this is the heading $\div 12$ of the column. And then forming the products 6.2, 12.1, 6.6, etc. of the corresponding frequencies and multiplicities and dividing in each case by 12, we have the right-hand column of numerical coefficients.

10. The result is to be read

$$\begin{aligned} 3332.322 = \; & 1.6552 \\ & + 1.6543 \\ & + 3.5553 \\ & + \text{etc.,} \end{aligned}$$

the coefficients 2, 2 and 1, 6 of the repeated terms being of course united together, so that we have

$$\begin{aligned} &+{}_4.55322\\ &+{}_7.533322. \end{aligned}$$

With a little practice the operation is performed without difficulty and with very small risk of error.

11. We may apply the process to obtain analytical formulæ for certain forms of products. Consider, for instance, the product $2^\alpha.\,2^\beta$, where $\alpha >$ or $= \beta$: and in the coefficients write for shortness α, instead of $[\alpha]^\alpha$ or $\Pi\alpha$, to denote $1.2.3\ldots\alpha$. We have

$$\begin{array}{c|c|l|l} 2^\alpha & 2^\beta & \qquad\qquad\qquad\qquad\qquad \div\,\alpha.\beta & \\ \hline 2^A & 2^{\beta-A} & \dfrac{\alpha}{A.\alpha-A}\,\dfrac{\beta}{\beta-A}\cdot 4^A\,2^{\alpha+\beta-2A}.A.\alpha+\beta-2A & \dfrac{\alpha+\beta-2A}{\alpha-A.\beta-A}, \end{array}$$

viz. the formula is

$$2^\alpha.\,2^\beta = \Sigma\,\frac{\alpha+\beta-2A}{\alpha-A.\beta-A}\cdot 4^A\,2^{\alpha+\beta-2A},$$

where A has any integer value from β to 0; the first term is $= 1.4^\beta\,2^{\alpha-\beta}$, and the last term is $\dfrac{\alpha+\beta}{\alpha.\beta}\,2^{\alpha+\beta}$. The weight of any term is $4A + 2(\alpha+\beta-2A)$, $= 2\alpha + 2\beta$, as it should be.

In explanation as to the frequency, observe that out of the β 2's we take any A 2's, and place them in any order under any A of the α 2's. The number of combinations taken is $\dfrac{\beta}{A.\beta-A}$, which (in A orders) gives $\dfrac{\beta}{\beta-A}$ sets to be placed under $\dfrac{\alpha}{A.\alpha-A}$ sets out of the α 2's: we thus have the foregoing coefficient of frequency $\dfrac{\alpha}{A.\alpha-A}\cdot\dfrac{\beta}{\beta-A}$, where as mentioned above α etc. are written to denote $[\alpha]^\alpha$, etc.

12. We may in like manner find a formula for the product $3^\alpha 2^\beta.3^\gamma 2^\delta$. Taking $\gamma, = x + y + z$, any partition of γ, and $\delta, = p + q + r$, any partition of δ, and writing also $A = x$, $B = y + p$, $C = q$, we have

$$\begin{array}{r|l|} 3^\alpha 2^\beta & 3^\gamma 2^\delta \qquad\qquad\qquad\qquad\qquad\qquad \div\,\alpha.\beta.\gamma.\delta \\ \hline 3^x 2^p 3^y 2^q & 3^z 2^r\ \ M.6^A 5^B 4^C 3^{\alpha+\gamma-2A-B}\,2^{\beta+\delta-B-2C}.N \end{array}$$

where for the frequency, M, we have

$$\begin{aligned} &\text{No. of terms } 3^x, = \frac{\gamma}{x.\gamma-x}; && 3^y, = \frac{\gamma-x}{y.z},\\ &\text{"}\qquad\text{"}\qquad 2^p, = \frac{\delta}{p.\delta-p}; && 2^q, = \frac{\delta-p}{q.r}, \end{aligned}$$

to be placed under

$$\begin{aligned} &\text{sets of terms } 3^\alpha, = \frac{\alpha}{x+p.\alpha-x-p}; && 2^\beta, = \frac{\beta}{y+q.\beta-y-q},\\ &\qquad\text{in orders, } = x+p; && \qquad = y+q, \end{aligned}$$

whence, multiplying, we have for the frequency

$$M=\frac{\alpha.\beta.\gamma.\delta}{x.y.z.p.q.r.\alpha-x-p.\beta-y-q};$$

and for the multiplicity, N, we have

$$N=A.B.C.\alpha+\gamma-2A-B.\beta+\delta-B-2C,$$
$$=x.B.q.\alpha+\gamma-2A-B.\beta+\delta-B-2C.$$

13. The coefficient is thus

$$\frac{M.N}{\alpha.\beta.\gamma.\delta},=\frac{B.\alpha+\gamma-2A-B.\beta+\delta-B-2C}{y.z.p.r.\alpha-x-p.\beta-y-q},$$
$$=\frac{B.\alpha+\gamma-2A-B.\beta+\delta-B-2C}{y.p.\gamma-A-y.\alpha-A-p.\beta-C-y.\delta-C-p},$$

or putting also $D=\alpha+\gamma-2A-B,$

$$E=\beta+\delta\qquad -B-2C,$$

whence $6A+5B+4C+3D+2E=3(\alpha+\gamma)+2(\beta+\delta),$

the formula finally is $3^\alpha 2^\beta.3^\gamma 2^\delta=\Sigma\Lambda.6^A 5^B 4^C 3^D 2^E,$

where

$$\Lambda=\frac{B\ .\qquad D\qquad .\qquad E}{y.p.\gamma-A-y.\alpha-A-p.\beta-C-y.\delta-C-p},$$

or as this may also be written

$$=\frac{B}{y.B-y}\cdot\frac{D}{z.D-z}\cdot\frac{E}{r.E-r},$$

so that the coefficient Λ is in fact the product of three binomial coefficients: it must, however, be recollected that the same term $6^A 5^B 4^C 3^D 2^E$ may occur more than once, with different coefficients Λ, so that in the final result, when these terms are united together, the numerical coefficients are not each of them of the form in question.

14. The limits of the summation are conveniently defined by means of the diagram

A	p	$\alpha-A-p$	α
y	C	$\beta-C-y$	β
z	r		
γ	δ		

$D=\alpha+\gamma-2A-B,$

$E=\beta+\delta\qquad -B-2C,$

ut suprà,

viz., here the sums of the first and second lines are α and β respectively, and the sums of the first and second columns are γ and δ respectively; we have $B=y+p$, a partition of B: but the values of y, p must be such as not to render negative any one of the four terms z, r, $\alpha-A-p$, $\beta-C-y$ of the diagram (for if any one of these numbers were negative, the corresponding factorial in the denominator of Λ would be infinite and we should have $\Lambda = 0$). For any given term $6^A 5^B 4^C 3^D 2^E$ of the proper weight $3(\alpha+\gamma)+2(\beta+\delta)$, there may be no suitable values of y, p, and the coefficient is then $=0$: there may be a single pair of values, and the coefficient is then (as remarked above) = a product of three binomial coefficients: or there may be more than a single pair of values, and the entire coefficient of the term has not in this case a like simple form.

15. To exhibit the working of the formula, I apply it to the recalculation of the foregoing product 3332.322 ($\alpha, \beta, \gamma, \delta = 3, 1, 1, 2$). Properly the whole series of symbols 666, 6642, 6633, etc. of the symmetric functions of the weight 18 should be written down, and the coefficient be calculated for each of them: but I write down only those which have coefficients not $=0$: for each of these, I take *all* the partitions $B = y + p$, several of these giving, as will be seen, zero values, and the others giving the values already obtained for the coefficients of the several terms.

							1			2			3					
	A	B	C	D	E	B	y	p	D	$\gamma-A-y$	$\epsilon-A-p$	E	$\beta-C-y$	$\delta-C-p$	1	2	3	
6552	1	2	0	0	1	2	2 1 0	0 1 2	0	−2 −1 0	2 1 0	1	−1 0 1	2 1 0	1 2 1	0 0 1	0 1 1	1.6552
6543	1	1	1	1	0	1	1 0	0 1	1	−1 0	2 1	0	−1 0	1 0	1 1	0 1	0 1	1.6543
6532^2	1	1	0	1	2	1	1 0	0 1	1	−1 0	2 1	2	0 1	2 1	1 1	0 1	1 2	2.6532^2
64332	1	0	1	2	1	0	0	0	2	0	2	1	0	1	1	1	1	1.64332
6332^3	1	0	0	2	3	0	0	0	2	0	2	3	1	2	1	1	3	3.6332^3
5553	0	3	0	1	0	3	3 2 1 0	0 1 2 3	1	−2 −1 0 1	3 2 1 0	0	−2 −1 0 1	2 1 0 −1	1 3 3 1	0 0 1 1	0 0 1 0	3.5553
55332	0	2	0	2	1	2	2 1 0	0 1 2	2	−1 0 1	3 2 1	1	−1 0 1	2 1 0	1 2 1	0 1 2	0 1 1	2 2 } 4.55332
54333	0	1	1	3	0	1	1 0	0 1	3	0 1	3 2	0	−1 0	1 0	1 1	1 3	0 1	3.54333
53332^2	0	1	0	3	2	1	1 0	0 1	3	0 1	3 2	2	0 2	1 1	1 1	1 3	1 2	1 6 } 7.53332^2
433332	0	0	1	4	1	0	0	0	4	1	3	1	0	1				

It is quite possible that abbreviations and verifications might be introduced, but the process as it stands seems to be at once less expeditious and less safe than the one first made use of.

16. Particular cases of the general formula are

$$2^\beta . 2^\delta = \Sigma \Lambda 4^C 2^E; \quad E = \beta + \delta - 2C, \text{ and thence } 4C + 2E = 2(\beta + \delta)$$

$$\Lambda = \frac{E}{\beta - C . \delta - C},$$

which agrees with a result before obtained.

$$3^\alpha 2^\beta . 2^\delta = \Sigma \Lambda . 5^B 4^C 3^D 2^E; \quad D = \alpha - B,$$
$$E = \beta + \delta - B - 2C,$$

and thence $5B + 4C + 3D + 2E = 3\alpha + 2\beta + 2\delta$,

$$\Lambda = \frac{E}{\beta - C . \delta - B - C}.$$

17. We may in like manner with the formula for $3^\alpha 2^\beta . 3^\gamma 2^\delta$, obtain the following:

$$4^\theta 3^\alpha 2^\beta . 2^\delta = \Sigma \Lambda . 6^A 5^B 4^C 3^D 2^E,$$

where

$$D = \alpha - B,$$
$$E = 2\theta + \beta + \delta - 3A - B - 2C,$$

and thence $6A + 5B + 4C + 3D + 2E = 4\theta + 3\alpha + 2\beta + 2\delta$,

and the value of the coefficient Λ is

$$\Lambda = \frac{C \quad . \quad E}{\theta - A . C - \theta + A . \theta + \beta - A - C . \theta + \delta - 2A - B - C},$$

viz. Λ is the product of two binomial coefficients: and since here a given term $6^A 5^B 4^C 3^D 2^E$ occurs once only, each numerical coefficient is actually of this form.

In the particular case $\theta = 0$, we must have $A = 0$ (this appears *à posteriori* from the denominator factor $-A$, a factorial which is infinite for any positive value of A); and we thus obtain the first given formula for $3^\alpha 2^\beta . 2^\delta$.

CAPITATION AND DECAPITATION. ART. NOS. 18 TO 21.

18. I explain the converse processes of Decapitation and Capitation. In any symmetric function, for instance 6552 of the degree 6, the whole coefficient of α^6 is 552, this symbol referring in the first instance to the series of remaining roots $\beta, \gamma, \delta \ldots$, but as the series of roots is unlimited we may ultimately replace these by $\alpha, \beta, \gamma, \ldots$, and so use 552 in its original sense. Similarly in 6652 the whole coefficient of α^6 is $= 652$, the only difference being that, while in the former case the degree is reduced to 5, in the present case it remains $= 6$. In every case we decapitate the symbol by striking out the highest number—in

the case of two or more equal numbers, one only of these being struck out. Observe that by decapitation we always diminish the weight, but we do or do not diminish the degree. In a product such as 3332.322 we obtain in like manner the whole coefficient of a^6 by the decapitation of each factor, viz. the coefficient is $= 332.22$, and in any equation such as that obtained above for the product 3332.322, the whole coefficient of the highest power of a must be $= 0$, viz. we can by decapitation obtain a new equation of lower weight: thus from the equation in question of weight 18, we obtain the new equation of weight 12,

$$\begin{aligned} 332.22 = {} & 1.552 \\ & + 1.543 \\ & + 2.5322 \\ & + 1.4332 \\ & + 3.33222 \end{aligned}$$

where observe that the terms of a degree lower than 6 in the original equation give no term in the new equation. The new equation of the deg. weight 5.12 might of course be obtained independently in like manner with the original equation.

19. We capitate a symbol by prefixing to it a number which is not less than the highest number contained in it: thus 552 may be capitated into 5552, 6552, etc.: and so a product 332.22 may be capitated into 3332.222, 3332.322, etc.; moreover a single symbol may be capitated into a product, 552 into 5552.4: in fact the capitation may be any operation such that by decapitation we reproduce the original symbol. The increase of weight may be any number not less than the degree of the original symbol: but it is usually taken to be a given number: thus for any symbol of a degree not exceeding 6, we may capitate so as to increase the weight by 6. The capitation does or does not increase the degree.

20. An identical equation may be capitated in a variety of ways, but instead of an identity we obtain only a congruence, that is an equation which requires to be completed by the adjunction to it of proper terms of lower degrees. Thus from the above equation of the deg. weight 5.12 we may obtain

$$\begin{aligned} 3332.322 \equiv {} & 1.6552 \\ & + 1.6543 \\ & + 2.65322 \\ & + 1.64332 \\ & + 3.633222 \end{aligned}$$

Imagine here all the terms brought to the same side so that the form is $\Omega \equiv 0$: Ω is a function not containing a^6, for the whole coefficient of a^6 therein is precisely that function which by the equation of the deg. weight 5.12 is expressed to be $= 0$: and hence Ω, *quâ* symmetric function cannot contain $\beta^6, \gamma^6 \ldots$; viz. Ω is a symmetric function of the degree 5 at most: the congruence $\Omega \equiv 0$ thus means, $\Omega =$ a properly determined function of the degree 5 at most. Obviously by development of the term 3332.322, that is, by substituting for this term its value as given by the equation of the deg. weight 6.18, the function of the degree 5 at most would be found to be $=$ the sum of those terms in the expression of 3332.322 which are of a degree inferior to 6: and the congruence $\Omega \equiv 0$ thus completed would be nothing else than the equation of the deg. weight 6.18.

21. We might have capitated in a different manner: for instance

$$\begin{aligned} 4332.222 = \; & {}_1.6552 \\ & + {}_1.6543 \\ & + {}_2.65322 \\ & + {}_1.44332.2 \\ & + {}_3.33222.3, \end{aligned}$$

there are here three products requiring to be developed: on replacing them by their values there would be found for Ω a value which would be a determinate function of a degree less than 6: and putting $\Omega =$ this value, the congruence $\Omega \equiv 0$ would be completed into an equation.

SEMINVARIANTS OF A GIVEN DEGREE; PERPETUANTS, ETC. ART. NOS. 22 TO 38.

22. We consider now seminvariants according to their degrees; in particular those of the degrees 2, 3, 4, 5 and 6, or say quadric, cubic, quartic, quintic and sextic seminvariants: the forms of these are $2^{\alpha+1}$, $3^{\alpha+1}2^{\beta}$, $4^{\alpha+1}3^{\beta}2^{\gamma}$, $5^{\alpha+1}4^{\beta}3^{\gamma}2^{\delta}$, $6^{\alpha+1}5^{\beta}4^{\gamma}3^{\delta}2^{\varepsilon}$, where each exponent $\alpha, \beta, \gamma, \delta, \varepsilon$ is a positive integer not excluding zero: the exponent of the highest number is in each case written $\alpha + 1$, and has thus the value 1 at least, for otherwise the form would not be of the proper degree. The several weights are $2(\alpha + 1)$, $3(\alpha + 1) + 2\beta$, $4(\alpha + 1) + 3\beta + 2\gamma$, etc., or what is the same thing, we have for the several degrees respectively

$$\begin{aligned} w - 2 &= 2\alpha, \\ w - 3 &= 3\alpha + 2\beta, \\ w - 4 &= 4\alpha + 3\beta + 2\gamma, \\ w - 5 &= 5\alpha + 4\beta + 3\gamma + 2\delta, \\ w - 6 &= 6\alpha + 5\beta + 4\gamma + 3\delta + 2\varepsilon, \end{aligned}$$

and we have for a given degree and weight as many seminvariants as there are systems of exponents satisfying the corresponding equation.

23. These numbers are at once expressible by means of a series of Generating Functions (G. F.), viz. writing for shortness 2, 3, 4, 5, 6 to denote $1-x^2$, $1-x^3$, $1-x^4$, $1-x^5$, $1-x^6$, the G. F.'s are

$$\frac{x^2}{2},\ \frac{x^3}{2.3},\ \frac{x^4}{2.3.4},\ \frac{x^5}{2.3.4.5},\ \frac{x^6}{2.3.4.5.6}.$$

In fact the number of seminvariants of a given weight is = coefficient of x^w in the corresponding G. F.; for the quadric seminvariants (or those of deg. weight $2.w$) in $\frac{x^2}{2}$; for the cubic seminvariants (or those of deg. weight $3.w$) in $\frac{x^3}{2.3}$; and so for the others.

24. A seminvariant of a given degree may be a sum of products (of that degree) of seminvariants of lower degrees, and of seminvariants of lower degrees: and it is in this case said to be reducible: a seminvariant which is not reducible is said to be irreducible, or otherwise to be a perpetuant. This notion of a perpetuant is due to Sylvester, see his Memoir "On Subinvariants, *i. e.* Semi-Invariants to Binary Quantics of an Unlimited Order," *American Math. Journal*, V (1882–83), pp. 76–137 (§4 Perpetuants, pp. 105–118). In speaking of the number of perpetuants of a given deg. weight, we assume throughout that these are independent perpetuants, not connected by any linear relation.

25. Since the seminvariants used for the reduction of a given reducible seminvariant can themselves be expressed in terms of perpetuants, we may say more definitely that a seminvariant of a given degree, which is a sum of products (of that degree) of perpetuants of lower degrees, and of perpetuants of lower degrees, is reducible. The words "of that degree" are essential to the definition: a seminvariant may be expressible as a sum of products (of a higher degree) of perpetuants of lower degrees, and of perpetuants of lower degrees, and it is not on this account reducible: a seminvariant so expressible is said to be a "syzygant"; but as to this, see No. 49.

26. Every quadric or cubic seminvariant is obviously a perpetuant: the quadric and cubic perpetuants have thus the before-mentioned G. F.'s $\frac{x^2}{2}$ and $\frac{x^3}{2.3}$ respectively.

27. A reducible quartic seminvariant can only be a sum of products (2.2) of two quadric perpetuants, and of quadric perpetuants, and it is clear that no

quartic seminvariant the symbol of which contains a 3 is thus expressible. If the symbol does not contain a 3, viz. when the form is $4^{\alpha+1}2^{\gamma}$, the seminvariant is reducible: we have for instance

$$4 = 2.2 - 2.22,$$
$$42 = 22.2 - 3.222, \text{ etc.,}$$

and to show that this is so in general, observe that any symmetric function $2^{\alpha+1}1^{\gamma}$, *quâ* symmetric function can be expressed as a rational and integral function of the degree $2(\alpha+1)+\gamma$, of the coefficients 1, 1^2, 1^3, etc.: instead of the roots considering their squares, we have thence an expression for the quartic seminvariant $4^{\alpha+1}2^{\gamma}$ in terms of the quadric perpetuants 2, 2^2, 2^3, etc., and such expression will be of the same degree $4(\alpha+1)+2\gamma$, as the quartic seminvariant.

It thus appears, as regards the quartic seminvariants, that whenever the symbol contains a 3, and in this case only, the seminvariant is a perpetuant: or what is the same thing, the form of a quartic perpetuant is $4^{\alpha+1}3^{\beta+1}2^{\gamma}$: for the weight w, the number is equal to that of the sets of values α, β, γ, such that $w-7=4\alpha+3\beta+2\gamma$: or what is the same thing the G. F. of the quartic perpetuants is $=\dfrac{x^7}{2\cdot3\cdot4}$.

28. Sylvester, in the memoir referred to, obtained this result in a different manner: the quartic seminvariants of a given weight are the quartic perpetuants of that weight and also the products (of that weight) of two quadric perpetuants, the same or different: say (4) is the G. F. for the perpetuants, and (2, 2) for the products: then the G. F. for the quartic seminvariants being as already mentioned $\dfrac{x^4}{2\cdot3\cdot4}$, we have his equation

$$(4)+(2,\ 2)=\frac{x^4}{2\cdot3\cdot4}.$$

He deduces (2, 2) from the G. F. $=\dfrac{x^2}{2}$ of the quadric perpetuants and thence obtains (4), $=\dfrac{x^7}{2\cdot3\cdot4}$, as above.

29. Write for a moment ϕx to represent the G. F. $=\dfrac{x^2}{2}$ of the quadric perpetuants, and A, B, C, ... to represent these quadric perpetuants: we have, in an algorithm which will be readily understood,

$$\phi x \;= (A+B+C\ldots)$$
$$(\phi x)^2 = (A+B+C\ldots)^2, = A^2+2AB+\ldots$$
$$\phi x^2 = \qquad\qquad\qquad A^2+\text{etc.,}$$

and thence $\tfrac{1}{2}\{(\phi x)^2+\phi x^2\} = \qquad A^2+AB+\text{etc.}$

viz. the G. F. (2, 2), is

$$\tfrac{1}{2}((\phi x)^2+\phi x^2),\ =\tfrac{1}{2}\left(\frac{x^4}{2\cdot 2}+\frac{x^4}{4}\right),\ =\tfrac{1}{2}\left(\frac{x^4(1+x^2)}{2\cdot 4}+\frac{x^4(1-x^2)}{2\cdot 4}\right),\ =\frac{x^4}{2\cdot 4},$$

and we thence have

$$(4)+\frac{x^4}{2\cdot 4}=\frac{x^4}{2\cdot 3\cdot 4},$$

that is $(4)=\frac{x^7}{2\cdot 3\cdot 4}$, the same result as was found above by independent considerations.

30. Sylvester established in like manner (but without the terms S which will be presently explained) the equations

$$(5)+(3,\ 2)=\frac{x^5}{2\cdot 3\cdot 4\cdot 5}+S_5,$$

$$(6)+(4,\ 2)+(3,\ 3)+(2,\ 2,\ 2)=\frac{x^6}{2\cdot 3\cdot 4\cdot 5\cdot 6}+S_6,$$

viz. here (5) is the G. F. for the quintic perpetuants, (2, 3) that for the products of the quadric and the cubic perpetuants: and similarly (6) is the G. F. for the sextic perpetuants, (4, 2) that for the products of the quadric and the quartic perpetuants, (3, 3) for the products of two cubic perpetuants, the same or different: and (2, 2, 2) for the product of three quadric perpetuants, the same or different. We have at once $(3,\ 2)=(3).(2)$; $(4,\ 2)=(4).(2)$; $(3,\ 3)$ is found by the same process as was used for finding (2, 2), substituting only $\frac{x^3}{3}$ for $\frac{x^2}{2}$; and (2, 2, 2) is found by a like process, viz. the G. F. for A^3+A^2B+ABC is $\tfrac{1}{6}\{(\phi x)^3+3\phi x.\phi x^2+2\phi x^3\}$, $\phi x=\frac{x^2}{2}$ as before, viz. this is $\tfrac{1}{6}\left\{\frac{x^6}{2\cdot 2\cdot 2}+\frac{3x^6}{2\cdot 4}+\frac{2x^6}{6}\right\}$: reducing to the common denominator $2\cdot 4\cdot 6$, the numerator is

$$=\tfrac{1}{6}x^6\{(1+x^2)(1+x^2+x^4)+3(1-x^6)+2(1-x^2)(1-x^4)\},$$

viz. this is $=x^6$. The several functions thus are

$$(3,\ 2)=\frac{x^5}{2\cdot 2\cdot 3},\qquad (4,\ 2)=\frac{x^9}{2\cdot 2\cdot 3\cdot 4},$$

$$(3,\ 3)=\frac{x^6}{3\cdot 6},\qquad (2,\ 2,\ 2)=\frac{x^6}{2\cdot 4\cdot 6}.$$

31. Mr. Hammond, in regard to the equation for the quintic perpetuants, made the very important observation (see his paper "On the Solution of the Differential Equation of Sources," *Amer. Math. Jour.*, IV (1882), pp. 218–227) that the products (3, 2) of a cubic perpetuant and a quadric perpetuant were not independent: we have between them syzygies such as $32.2-3.22\equiv 0$ (viz. the difference of the two products contains no term of the degree 5; the actual

value is $= 43 + 322$: Hammond's equation (12), p. 322); hence the necessity in the equation of a term S_5 referring to these syzygies, and he moreover obtained the expression, $S_5 = \frac{x^7}{2.4}$ of the G. F. for these syzygies.

The equation gives
$$(5) = \frac{-x^7 + x^{10} + x^{12}}{2.3.4.5} + S_5,$$
where of course the first term is the value of (5) which would be given by the equation without the term S_5; and substituting herein for S_5 the foregoing value, viz.
$$S_5 = \frac{x^7}{2.4}, = \frac{x^7(1-x^3)(1-x^5)}{2.3.4.5},$$
we find
$$(5) = \frac{x^{15}}{2.3.4.5},$$
which is the correct value of the G. F. for the quintic perpetuants: the lowest quintic perpetuant is thus of the weight 15.

32. The equation for the sextic perpetuants gives
$$(6) = \frac{-x^6 - x^{13} + 2x^{16} + x^{18}}{2.3.4.5.6} + S_6,$$
which is an equation connecting (6) and S_6, the G. F.'s for the sextic perpetuants, and the sextic syzygies respectively. I have, in the investigation of the value of S_6, met with a difficulty which I have not been able to overcome: but I find that
$$S_6 = \frac{x^6 + x^{13} - 2x^{16} - x^{18} + \omega(x)}{2.3.4.5.6},$$
where $\omega(x)$ is possibly the monomial function x^{31}, but this result (which Capt. MacMahon believes to be true) is not yet completely established; it is a function containing no term lower than x^{31}. We have therefore
$$(6) = \frac{\omega(x)}{2.3.4.5.6},$$
and there is, it would appear, no sextic perpetuant of a weight lower than 31.

33. But before entering on the investigation it is proper to further develop the theory of the quintic perpetuants. We have quintic seminvariants: 5 for weight 5; 52 for weight 7; 53 for weight 8; 54, 522 for weight 9; 55, 532 for weight 10; 542, 533, 5222 for weight 11; and so on. These are reduced by means of the products 3.2 of a cubic perpetuant and a quadric perpetuant; viz. for any given weight we have all the products $3^\alpha 2^\beta . 2^\gamma$ of that weight. Thus for weight 5 there is only the product 3.2 and this in fact serves to reduce the seminvariant 5: we have $3.2 = 5 + 32$, and therefore $5 = 3.2 - 32$.

For the weight 7; there is only the seminvariant 52, and there are the two products 32.2 and 3.22: either of these would serve for the reduction: we have

$$32.2 = 52, + 43, + 2.322, \qquad 3.22 = 52, + 322,$$

and these two equations imply the before-mentioned syzygy, $32.2 - 3.22 \equiv 0$, in virtue of this, the two reductions become equivalent; or say there remains a single equation serving for the reduction: the most simple form is $52 = 3.22 - 322$.

Weight 8: there is only the seminvariant 53, and the product 33.2: this gives the reduction.

Weight 9: seminvariants 54, 522: products $32^2.2$, 32.2^2, 3.2^3; there is between the first and last of these a syzygy; and this being satisfied there remain two equations for the reductions.

Weight 10: seminvariants 55, 532: products 332.2, 33.22, these give the reductions.

Weight 11: seminvariants 533; 542, 5222: products 333.2; $32^3.2$, $32^2.2^2$, 32.2^3, 32^4.

34. Observe that the seminvariants, and in like manner the products, form two classes, according as the symbols contain three odd numbers or a single odd number: these correspond separately to each other, for the development of any product will contain only seminvariants having each of them as many odd numbers as there are odd numbers in the product. Hence for the weight 11 just referred to, 333.2 serves for the reduction of 533; $32^3.2$, $32^2.2^2$, 32.2^3, 3.2^4 are connected the first and fourth of them by a syzygy, and the second and third by a syzygy; and there remain two equations serving for the reduction of the two seminvariants 542, 5222.

It is easy to show in this manner that there is no quintic perpetuant for any weight under 15; and that there is a single quintic perpetuant for the weight 15.

35. Generally the syzygies exist only for an odd weight $w = 2\beta + 3$ between the products $32^{\beta-1}.2$, $32^{\beta-2}.2^2$, ... $32.2^{\beta-1}$, 3.2^β; viz. there is a syzygy between the first and last terms: a syzygy between the second and last but one terms; and so on. The existence of these syzygies at once appears from the principle of decapitation: decapitating $32^{\beta-1}.2 - 3.2^\beta$, we have $2^{\beta-1} - 2^{\beta-1}$, which is identically $= 0$, hence the function contains no term of the degree 5, that is $32^{\beta-1}.2 - 3.2^\beta \equiv 0$; and similarly for the other pairs of terms.

There are β terms: hence in the case β even, we have $\frac{1}{2}\beta$ pairs of terms and therefore $\frac{1}{2}\beta$ syzygies: in the case β odd, there is a middle term, not connected by a syzygy with any other term, and the number of syzygies is thus $= \frac{1}{2}(\beta - 1)$: writing for β its value $\frac{1}{2}(w-3)$, the number is $= \frac{1}{4}(w-3)$, and $\frac{1}{4}(w-5)$ in the two cases respectively: and it thus appears that the G. F. is $= \frac{x^7}{2.4}$ as already mentioned.

36. In the case w an odd number we have seminvariants, and in like manner products, containing respectively one odd number, three odd numbers, five odd numbers, and so on: thus $w = 15$ we have

SEMINVARIANTS.	PRODUCTS.	
555	$3332^2.2$	3 equations
5532	3332.2^2	
5433	333.2^3	
5332^2		
5442	$32^5.2$	$\frac{1}{2}6$, = 3 equations
542^3	$32^4.2^2$	
52^5	$32^3.2^3$	
	$32^2.2^4$	
	32.2^5	
	3.2^6	

hence the seminvariants 555, 5532, 5433, 5332^2 with three odd numbers are not reducible, but they can be linearly expressed in terms of the 3 like products $3332^2.2$, 3332.2^2, 333.2^3, and of $4 - 3$, $= 1$ arbitrary quantity (observe, however, that this must not be the seminvariant 5332^2, for this is in fact reducible): the seminvariants 5442, 542^3, 52^5, with one odd number are reducible. And the like as regards any other odd value of w.

37. In the case w an even number, we have seminvariants, and in like manner products, containing respectively two odd numbers, four odd numbers, six odd numbers and so on. Thus $w = 16$, we have

SEMINVARIANTS.	PRODUCTS.	
5533	33332.2	2 equations
53332	3333.2^2	
5542	$332^4.2$	5 equations
552^3	$332^3.2^2$	
5443	$332^2.2^3$	
5432^2	332.2^4	
532^4	33.2^5	

and thus the seminvariants with four odd numbers, and those with two odd numbers, are each set reducible.

38. I give in the case $w = 19$ the following results: the expression for the G. F. shows that there are 2 quintic perpetuants; viz. two forms X, Y such that every quintic seminvariant of the weight 19 is expressible as a linear function of these, of products (3.2) of a cubic and a quadric perpetuant, and of forms of a degree inferior to 5, that is, quartic, cubic and quadric perpetuants. Attending only to the terms in X, Y, the actual values are:

$$\begin{aligned}
5554 &= X, \\
5552^3 &= Y, \\
55432 &= -X, \\
55333 &= 0, \\
5532^3 &= -Y, \\
54442 &= 0, \\
54433 &= X, \\
5442^3 &= 0, \\
54332^3 &= Y, \\
542^5 &= 0, \\
533332 &= 0, \\
5332^4 &= 0, \\
52^7 &= 0,
\end{aligned}$$

viz. of the 13 quintic seminvariants of the weight in question there are 7, which as not containing either X or Y are each of them reducible; while the remaining 6 can only be expressed as linear functions of X and Y. It would be allowable to select $5554\,(=X)$ and $5552^3\,(=Y)$ as the two representative perpetuants, but there is no particular advantage in this.

SEXTIC PERPETUANTS AND SEXTIC SYZYGIES: SYZYGANTS. ART. NOS. 39 TO 51.

39. Returning now to the sextic seminvariants: these are weight 6; 6: weight 8; 62: weight 9; 63: weight 10; 64, 62^2: and so on. And they are reducible by means of the products (4, 2), (3, 3) and (2, 2, 2), that is of a quartic perpetuant and a quadric perpetuant, of two cubic perpetuants, and of three quadric perpetuants: this last form of product existing only in the case of an even weight.

40. For weight 6, we have seminvariant 6, and the two products 3.3 and 2.2.2; this implies a syzygy $3.3 - 2.2.2 \equiv 0$; and there then remains a single

equation for the reduction of the seminvariant. The formulæ are

$$3.3 = 6 + s.33 \qquad 2.2.2 = 6 + s.22.2 - s.222$$

so that the complete syzygy, and the most simple reduction are

$$3.3 - 2.2.2 - s.33 + s.22.2 - s.222 = 0, \qquad 6 = 3.3 - s.33$$

and we might in this way verify that for the successive weights 8, 9, 10, etc., there are no sextic perpetuants; and find for these weights respectively, the number of the sextic syzygies. But such direct investigation becomes soon impracticable.

41. I endeavor to determine the number of sextic syzygies for the weight w; and for this purpose I establish the following relation:

$$(S_6) = ((0)) + ((2))' + ((3))' + ((2,\ 2))' + ((3,\ 2))' + (S_5)' + (S_6)' - ((5))' + ((\theta))',$$

where (S_6) is the number of sextic syzygies for the weight w, or what is the same thing, it is the coefficient of x^w in the function S_6, which is the G. F. for these syzygies: $((0))$ has the value 1 for $w = 6$, and the value 0 in all other cases. The accented symbols refer to the weight $w - 6$; $(S_6)'$ is thus the number of sextic syzygies for this weight: and for the same weight $w - 6$, $((2))'$ denotes the number of quadric perpetuants, or coefficient of x^{w-6} in the function (2) which is the G. F. for these perpetuants: $((3))'$ the number of cubic perpetuants, $((2,\ 2))'$ the number of products of two quadric perpetuants, the same or different, $((3,\ 2))'$ the number of products of a cubic perpetuant and a quadric perpetuant, $(S_5)'$ the number of quintic syzygies, $((5))'$ the number of quintic perpetuants, and $((\theta))'$ a term of unascertained form which will be explained further on. Transposing the term $(S_6)'$ to the left-hand side, and passing to the generating functions, we have

$$(1 - x^6)\, S_6 = (0) + (2)' + (3)' + (2,\ 2)' + (3,\ 2)' + S_5' - (5)' + (\theta)',$$

or as this may be written

$$= x^6 \{1 + (2) + (3) + (2,\ 2) + (3,\ 2) + (S_5) - (5) + (\theta)\},$$

where (2), (3), etc. have the values already obtained for these G. F.'s respectively: viz. writing $x^6(\theta) = \frac{\omega x}{2 \cdot 3 \cdot 4 \cdot 5}$, the equation is

$$6 \cdot S_6 = x^6 + \frac{x^8}{2} + \frac{x^9}{3} + \frac{x^{10}}{2 \cdot 4} + \frac{x^{11}}{2 \cdot 2 \cdot 3} + \frac{x^{13}}{2 \cdot 4} - \frac{x^{21}}{2 \cdot 3 \cdot 4 \cdot 5} + \frac{\omega(x)}{2 \cdot 3 \cdot 4 \cdot 5} .$$

42. Reducing on the right-hand side the known terms to the common denominator $2 \cdot 3 \cdot 4 \cdot 5$, the numerator is

$$\begin{array}{lllllll}
x^6 . & 1-x^2.1-x^3.1-x^4.1-x^5 = & x^6-x^8-x^9-x^{10} & +x^{12}+2x^{13}+x^{14} & -x^{16}-x^{17}-x^{18}+x^{20} & & \\
+x^8 . & 1-x^3.1-x^4.1-x^5 & +x^8 & -x^{11}-x^{12}-x^{13} & +x^{15}+x^{16}+x^{17} & -x^{20} & \\
+x^9 . & 1-x^4.1-x^5 & +x^9 & -x^{13}-x^{14} & & +x^{18} & \\
+x^{10} . & 1-x^3.1-x^5 & +x^{10} & -x^{13} & -x^{15} & +x^{18} & \\
+x^{11} . & 1+x^2.1-x^5 & +x^{11} & +x^{13} & -x^{16} & -x^{18} & \\
+x^{13} . & 1-x^3.1-x^5 & & +2^{13} & -x^{16} & -x^{18} & +x^{21} \\
-x^{21} . & & & & & & -x^{21} \\
\hline
 & & x^6 & +x^{13} & -2x^{16} & -x^{18} &
\end{array}$$

whence, dividing by 6, we have the before-mentioned formula

$$S_6 = \frac{x^6 + x^{13} - 2x^{16} - x^{18} + \omega(x)}{2 \cdot 3 \cdot 4 \cdot 5 \cdot 6} .$$

43. We have to prove the formula for (S_6): this symbol denotes for the weight w, the number of syzygies between the products (4, 2), (3, 3) and (2, 2, 2): we have to consider separately the cases w odd, and w even. First, if w be odd, there are no products (2, 2, 2) and the only forms are (4, 2) and (3, 3).

I consider a particular value of w, say $w = 15$. The whole series of seminvariants weight 15 is

663	555	4443	33333
654	5532	$44 32^2$	3332^3
652^3	5442	43332	32^6
6432	5433	432^4	
6333	542^3		
632^3	5332^2		
	52^5		

and from the quartic and cubic forms we obtain the forms of the products (4, 2) and (3, 3), viz. these are

4432 . 2	3333 . 3
$443 . 2^2$	333 . 33
4333 . 2	$332^2 . 3$
$432^2 . 2$	$332^2 . 32$
$432^2 . 2^2$	$332 . 32^2$
$432 . 2^3$	$33 . 32^3$
$43 . 2^4$	

(S_9) will denote the number of syzygies between these products. Now from any such syzygy, we obtain by decapitation, it may be an identity, but if not an identity, then a syzygy, of the weight $15-6, =9$: and from these lower identities or syzygies we can pass back to the syzygies of the weight 15. To show how this is, I decapitate the several products, thus obtaining the forms

432	333 ;	viz. distinct		432 ; and of these	333	occur each of
43.2	33.3	forms are		43.2	32^3	them twice.
333	32^3			333	$32^2.2$	
32^3	$32^2.2$		33.3	32^3	32.2^2	
$32^2.2$	32.2^2			$32^2.2$	3.2^3	
32.2^2	3.2^3			32.2^2		
3.2^3				3.2^3		

The forms occurring each twice are 333, 32^3 (viz. these are the forms (3), or cubic perpetuants of the weight 9) and $32^2.2$, 32.2^2, 3.2^3 (viz. these are the forms (3, 2) or products of a cubic perpetuant and a quadric perpetuant for the weight 9); and any form thus occurring twice gives a syzygy of the weight 15: thus 333, we capitate it with 4.2 or with 3.3, and so obtain the syzygy $4333.2 - 3333.3 \equiv 0$; and in like manner for each of the other forms 32^3, $32^2.2$, 32.2^2, 3.2^3. And so for any other odd weight: (S_6) contains the terms $((3))'$ and $((3, 2))'$, and for an odd value of w we may assume that S_6 contains also the terms $((2))'$ and $((2, 2))'$: for these, it is clear, vanish for any odd value of w.

44. When w is even it appears by a similar investigation that (S_6) contains the terms $((2))'$, and $((2, 2))'$ (which in this case do not vanish), and also the before-mentioned terms $((3))'$ and $((3, 2))'$: so that whether w be even or odd, (S_6) contains the terms $((2))'$, $((3))'$, $((2, 2))'$, $((3, 2))'$.

In the particular case $w=6$, there is the sextic syzygy $3.3-2.2.2\equiv 0$, obtained by capitation from the identity $1-1=0$; and by reason hereof, we introduce into the formula the term $((0)), =1$ for $w=6$, and $=0$ in every other case.

In what immediately follows I revert to the instance $w=19$, but this now represents indifferently an odd or an even value of w, there being no distinction between the two cases.

45. Attending next to the remaining distinct terms, these are 333, 32^3 of the degree 3; 432 of the degree 4; $32^2.2$, 32.2^2, 3.2^3 of the degree 5; and 43.2, 33.3 of the degree 6. For the degrees 3 and 4, there are no syzygies:

but for the degree 5 we have a syzygy: this, written as a congruence is $32^2.2 - 3.2^3 \equiv 0$, and *quâ* quintic syzygy, it will, when completed, not contain any 5; the completed form in fact is

$$\begin{array}{rl} & 32^2.2 \\ - & 3.2^3 \\ - & 432 \\ - & {}_2.32^3 = 0. \end{array}$$

We can capitate this, each term with 4.2 or else 3.3, or it may be indifferently either with 4.2 or 3.3; and so obtain therefrom a syzygy of the weight 15; such a syzygy (in the congruence form) is

$$\begin{array}{rl} & 432^2.2^2 \\ - & 43.2^4 \\ - & 4432.2 \\ - & {}_2.432^3.2 \equiv 0; \end{array}$$

and it is to be observed that it is quite indifferent how the capitation is performed: if for instance the first term had been capitated into $332^2.32$, then in virtue of the before-obtained syzygy $432^2.2^2 - 332^2.32 \equiv 0$, the new form would be equivalent to the old one. (It is I think convenient to capitate, when this can be done, with 4.2; and only the other terms with 3.3.) Clearly the case is the same with any other odd weight, and we thus see that (S_6) contains the term $(S_5)'$.

46. But further we have between the terms 43.2 and 33.3 of the degree 6, a syzygy, $43.2 - 33.3 \equiv 0$. Completing this, there will be a term containing a 5, viz. the syzygy is

$$\begin{array}{rl} & 43.2 \\ - & 33.3 \\ - & 54 \\ - & 432 \\ - & {}_3.333 = 0; \end{array}$$

and in this form we cannot capitate it, for the quintic term 54 is not to be capitated either with 4.2 or with 3.3. But 54 is not a perpetuant, we have

$$54 = \begin{array}{rl} & 32.2^2 \\ - & {}_2.3.2^3 \\ - & 432 \\ - & 32^3 \end{array} \quad \text{and thence syzygy is} \quad \begin{array}{rl} & 43.2 \\ - & 33.3 \\ - & 32.2^2 \\ + & {}_2.3.2^3 \\ + & {}_3.333 \\ + & 32^3 = 0 \end{array}$$

and it can be capitated, for instance, into

$$
\begin{array}{rl}
 & 443.2^3 \\
- & 333.33 \\
- & 432.2^3 \\
+ & 2.43.2^4 \\
+ & 3.4333.2 \\
+ & 432^3.2 \equiv 0.
\end{array}
$$

the form of capitation being (for the reason mentioned above) quite immaterial. Observe that in every case where the sextic syzygy contains in the first instance any quintic seminvariants, it is assumed that each of these is expressed in terms of quintic perpetuants, as shown in No. 38; and this being done, the sextic syzygy exhibits itself as a syzygy containing, or else not containing, a quintic perpetuant or perpetuants.

47. The conclusion is that from any sextic syzygy of the weight $w-6$, *which does not contain a quintic perpetuant*, we can obtain by capitation a sextic syzygy of the weight w. The number of sextic syzygies of the weight $w-6$ is $(S_6)'$, and the number of quintic perpetuants of the same weight is $((5))'$: the former of these is (for not too large values of w) the greater; and at first sight it would appear that we can, by elimination of the quintic perpetuants, obtain from the $(S_6)'$ syzygies, $(S_6)'-((5))'$ syzygies which do not contain a quintic perpetuant: if this was always the case, we should have in (S_6) the term $(S_6)'-((5))'$, completing the series of terms, and the formula would be

$$(S_6) = ((0)) + ((2))' + ((3))' + ((2,\ 2))' + ((3,\ 2))' + (S_5)' + (S_6)' - ((5))',$$

leading to

$$S_6 = \frac{x^6 + x^{13} - 2x^{16} - x^{18}}{2.3.4.5.6}.$$

48. But this result is on the face of it wrong, for as remarked by Sylvester in the memoir referred to, from the mere fact that the sum $1+1-2-1$ of the numerator coefficients is negative, it follows that the coefficients of the development ultimately become negative; and the actual calculation showing when this happens is given by him. And it is further to be noticed that not only the formula cannot be correct beyond the point at which the coefficients become negative, but it cannot be correct beyond the point for which $(S_6)'-((5))'$ becomes negative: the sextic syzygies of the weight $w-6$ may add nothing to, but they cannot take anything away from, the number of the sextic syzygies of the weight w.

49. If for a moment we further consider these syzygies of the weight $w-6$; so long as the number of these is greater than the number of quintic perpetuants of the same weight, we can by means of them presumably express each of the quintic perpetuants in terms of sextic products, viz., in the language of Capt. MacMahon, express each quintic perpetuant as a "Sextic Syzygant." The syzygy of the weight 9, above obtained, will serve as an example: 54 is not a quintic perpetuant, but ignoring this, it is by the syzygy in question expressible in the form

$$\begin{aligned} 54 = &\ \ 43.2 \\ - &\ \ 33.3 \\ - &\ \ 432 \\ - &\ s.333 \end{aligned}$$

viz., as a Sextic Syzygant, inasmuch as on the right-hand side we have terms 43.2 and 33.3, of the degree 6, which exceeds the degree 5 of the seminvariant 54 in question. Referring back to the definition of reduction, No. 25, observe that this is *not* a reduction of the seminvariant 54. It may be remarked that for the weight 19 we have 15 sextic syzygies: the number of quintic perpetuants is $=3$: so that while it is conceivable that the 15 equations might be such that they would fail to determine the 3 perpetuants, it is *primâ facie* very unlikely that this should be so. I have in fact ascertained that the equations are sufficient for the determination; that is, that (weight 19) each of the three quintic perpetuants is a sextic syzygant. So in the case $w=23$, the number of the sextic syzygies is $=28$, and that of the quintic perpetuants is $=5$; here also the 28 equations are sufficient to determine the 5 perpetuants, viz. (weight 23) each of the 5 quintic perpetuants is a sextic syzygant.

50. Supposing that for any given weight $w-6$, each of the quintic perpetuants *is* a sextic syzygant: this implies that the number of sextic syzygies $(S_6)'$ is at least equal to the number $((5))'$ of quintic perpetuants (for each expression of a quintic perpetuant as a sextic syzygant is in fact a sextic syzygy): and not only so, but it further implies that the number of the sextic syzygies which do not contain a quintic perpetuant, is precisely equal $(S_6)'-((5))'$: for if besides the equations which serve to express the perpetuants as syzygants, we have any other sextic syzygy, then either this does not contain a quintic perpetuant, or it can (by substituting therein for every quintic perpetuant its value as a sextic syzygant) be reduced to a syzygy which does not contain any quintic perpetuant.

51. In the general case we have $(S_6)'$ sextic syzygies of the weight $w-6$, and $((5))'$ quintic perpetuants of this weight: but it may happen that certain of

the quintic perpetuants do not enter into any of the sextic syzygies; and those which enter, may do so in definite combinations: by elimination of these combinations of perpetuants we obtain (it may be) α sextic syzygies not containing any quintic perpetuant; and the remaining $(S_6)' - \alpha$ equations will then serve to express each of them a quintic perpetuant, or combination of quintic perpetuants, as a sextic syzygant. The number α is at most $= ((5))'$, or taking it to be $= ((5))' - ((\theta))'$, the number of sextic syzygies not containing any quintic perpetuant will be $= (S_6)' - ((5))' + ((\theta))'$, that is the number of sextic syzygies not containing any quintic perpetuant will be equal to the whole number $(S_6)'$ of sextic syzygies diminished by some number $((5))' - ((\theta))'$, which is less than or at most equal to the whole number $((5))'$ of quintic perpetuants of the weight in question $w - 6$. But as already mentioned I have not been able to obtain the expression of the function (θ), $= \frac{\omega(x)}{2.3.4.5}$, which is the G. F. of the number $((\theta))'$.

CAMBRIDGE, ENGLAND, 17th *March*, 1884.

On Perpetuants.

By Captain P. A. MacMahon, R. A.

Professor Cayley has recently treated the subject of Perpetuants in the *American Journal of Mathematics*, and has largely developed the theory of the syzygies existing between them. I here make a few remarks upon the general subject, and then proceed to carry on the theory from the point where it was left by Professor Cayley.

The particular result that I obtain is the discovery that the simplest sextic perpetuant is of weight thirty-one; this result is, as will be seen, a somewhat remarkable one, since the *prima facie* probability was much against there being one of so low a weight. Postponing all explanations of the language made use of, the way in which this result comes out is as follows: for the weight $31 - 6$, $= 25$, the number of exemplar quintic perpetuants $\left(\text{coeff. of } x^{25} \text{ in } \frac{x^{15}}{\mathbf{2.3.4.5}}\right)$ is $= 7$, viz. these are $\widetilde{112}$, $\widetilde{114}$, $\widetilde{122}$, $\widetilde{124}$, $\widetilde{132}$, $\widetilde{213}$, $\widetilde{312}$,
but it appears from the discussion of the syzygies of the third kind, that these are not each of them a sextic syzygant, but that only the combinations

$$\widetilde{112},\ \widetilde{114},\ \widetilde{122},\ \widetilde{132},\quad \widetilde{213} + 2.\widetilde{124},\ \widetilde{312} - 2.\widetilde{124},$$

are each of them a sextic syzygant, viz. the number of these is $= 6$. Hence for the weight 31 the number of sextic perpetuants is $7 - 6$, $= 1$.

Of a certain degree λ and weight w, there exist in general perpetuant forms of two kinds, which may be called *exemplar* and *non-exemplar;* certain linear relations subsist between them, so that the non-exemplar forms are reducible by aid of the exemplar forms; of the second, third and fourth degrees, every form is exemplar, the simplest forms being symbolized as is well known by the partitions 2, 3, 43, respectively; for the fifth degree, we have for the lowest weight fifteen, one exemplar form 543^2, two non-exemplar $5^2 32$, 5^3, the exemplar form being the simplest of the three; but there is another reason for choosing this form as the representative, for it will be observed that the symbol 543^3 contains in itself the symbol 43 of the simplest quartic perpetuant, and it will be proved that it is proper to take for the exemplar symbol of a perpetuant of any given degree, that one which contains the exemplar symbol of the degree next below, and that such forms are in fact the only ones that it is necessary to consider, the remainder being certainly non-exemplar.

For suppose the form $\lambda\mu\nu\pi\ldots$ of degree λ to be not a perpetuant, that is to be reducible by aid of compound forms and exemplar forms of lower degrees; then by the process, named by Prof. Cayley 'capitation,' it is at once obvious that $\varkappa \geqq \lambda$,

$$\varkappa\lambda\mu\nu\pi\ldots$$

is reducible, at the worst, by aid of $\varkappa^{\text{ic}}$ perpetuants, involving lower exemplar symbols; for instance consider the form 532; we have

$$532 = 3^2 2\cdot 2 - 43^2 - 2(3^2 2^2),$$

whence $5^2 32 = 3^3 2\cdot 2^2 - 543^2 - 2(3^3 2^2 . 2) +$ terms of lower degree; this equation may be operated upon in the same manner any number of times by a capitation of any degree $\nless 5$, proving that any form thus obtained on the left-hand side is reducible by means of exemplar forms, simply because 532 is not a perpetuant.

It follows that every sextic exemplar form must contain in itself the symbol 543^2 and that it is unnecessary to consider other forms.

Another useful principle is, that if a form $\lambda\mu\nu\pi\ldots$ of degree λ is reducible, then, by the reverse process of decapitation, it follows that the form $\mu\nu\pi\ldots$ must be expressible as a sextic syzygant: note that we must have $\lambda = \mu + 1$, as otherwise $\lambda\mu\nu\pi\ldots$ would be certainly reducible.

Thus to every perpetuant of degree μ, that is not a $(\mu+1)^{\text{ic}}$ syzygant, will certainly correspond a $(\mu+1)^{\text{ic}}$ perpetuant, by simply capitating the μ^{ic} symbol with the part $\mu + 1$; *ex. gr.* we know that the generating function for quintic syzygies is $\dfrac{x^7}{1-x^3.1-x^4}$, and that each syzygy involves a quartic perpetuant whose symbol contains one and only one part 3, the number of syzygies being exactly equal to the number of such quartic forms; consequently, no quartic form containing more than one part 3 can possibly be a quintic syzygant, and therefore 543^3 must be the simplest exemplar form of the fifth degree. The question is then in reality, not to determine the total generating function for the syzygies, but rather to find it for those syzygies which involve the exemplar perpetuant forms of the degree next below; thus, representing $1 - x^\mu$ by μ for brevity, the generating function for μ^{ic} perpetuants is $\dfrac{x^\mu}{\mu}$ {G. F. for $(\mu - 1)^{\text{ic}}$

perpetuants — G. F. for μ^{ic} syzygies which involve $(\mu - 1)^{\text{ic}}$ perpetuants}; *ex. gr.* for degree 5, we arrive at the generating function

$$\frac{x^5}{5}\left(\frac{x^7}{2.3.4} - \frac{x^7}{2.4}\right) = \frac{x^{15}}{2.3.4.5},$$

the well-known result; and for degree 6, it will be $\frac{x^6}{6}\left(\frac{x^{15}}{2.3.4.5} - \omega(x)\right)$, wherein $\omega(x)$ is to be determined. There appears to be another way of regarding the question which it may be useful to mention.

It is a remarkable fact that there is an exact correspondence between the reducible quartic forms and the quintic syzygies which involve quartic perpetuants (which are in this case the whole of the syzygies); and what is still more strange is that presuming this same correspondence to hold between the reducible quintic forms and the sextic syzygies, a result is reached which exactly accords with that rigorously obtained in the sequel. This is not all, for the quintic syzygies can be derived from the reduced expressions of the non-perpetuant quartic forms by a direct operation. I have (Proc. L. M. S. Vol. XV, p. 31) explained the laws of a series of inverse operators, and it is by means of these that the derivation is effected; if $d_0, d_1, \ldots d_\lambda$ be Mr. Hammond's operators, then the type of the operators referred to is

$$V_{-\kappa} = (\kappa)\, d_0 - (\kappa + 1)\, d_1 + (\kappa + 2)\, d_2 - \ldots (-)^\lambda (\kappa + \lambda)\, d_\lambda + \ldots$$

$(\kappa + \lambda)$ being symbolic of a symmetric function; consider now the reduction of the quartic form (4), viz. $(4) = (2)^2 - 2\,(2^2)$,

and operating on both sides of this with V_{-3}, we get at once

$$(43) + (32^2) = 32.2 - 3.2^2,$$

which is the simplest quintic syzygy; in like manner the operator V_{-3} gives the whole of the quintic syzygies, and each involves of necessity a quartic perpetuant; this follows of course from the occurrence of the number 3 in the operator V_{-3}.

It follows that G. F. for quintic perpetuants $= \frac{x^5}{5} \cdot x^3 \cdot \frac{x^7}{2.3.4} = \frac{x^{15}}{2.3.4.5}$.

Now, just as by superposing any quartic form with the symbolic number 3 (3 being the simplest cubic form which is not a quartic syzygant) we obtain a quartic perpetuant, so by superposing any quintic form with the symbolic numbers 43^2 (43^2 being the simplest quartic form which is not a quintic syzygant) we obtain a quintic perpetuant; consider the reduction of the simplest quintic form, viz. $(5) = 3.2 - 32$,

and the operator $V_{-43^2} = (43^2)d_0 - (53^2)d_1 + (63^2)d_2 - \ldots;$

operating we find $\quad 543^2 = 43^3.2 - 3^4.3 - 43^3 2 + 5(3^5)$,
that is 543^2 expressed as a sextic syzygant; now, assuming that this operator will have a similar effect in the case of every reducible quintic form, and not in the case of an irreducible form, it will follow that the generating function for sextic perpetuants is $\frac{x^6}{6} \cdot x^{10} \cdot \frac{x^{15}}{2.3.4.5} = \frac{x^{31}}{2.3.4.5.6}$; the actual result, so far as I have obtained it rigorously, agrees with this.

It seems worth observing that if this principle is sound, the G. F. for n^{ic} perpetuants is $(n > 2)$

$$\frac{x^{2^{n-1}-1}}{2.3.4\ldots n}.$$

Before attacking the general subject of the sextic syzygies it is absolutely necessary to inquire more minutely into the theory of the quintic perpetuants; the main object is to discover the law of occurrence of the exemplar forms in the reduction of the non-exemplar forms. For weight 15, these latter forms are

$$5^3 \quad ,$$
$$5^2 32,$$

which arise from the fact that the forms 5^3, 532 involve the form 43^2 in their reduction, 43^2 not being a quintic syzygant; in general, every μ^{ic} form will be a non-exemplar perpetuant, which, being decapitated, requires for its reduction a $(\mu - 1)^{ic}$ perpetuant which is not a μ^{ic} syzygant.

The characteristic of the forms is that their symbols contain each three odd numbers, and it is obvious that every quintic form containing in itself either 5^3 or $5^2 32$ will be a non-exemplar perpetuant, so long as it does not also contain 43^2, when of course it would be exemplar.

Consider now the form of weight w,

$$5^{\varkappa} 4^{\lambda} 3^{\mu} 2^{\frac{1}{2}(w - 5\varkappa - 4\lambda - 3\mu)},$$

and for brevity denote it by $\widetilde{\varkappa\lambda\mu}$; $\widetilde{\varkappa\lambda\mu}$ is exemplar if $\lambda \not< 1$ and $\mu \not< 2$; the forms to be considered are three, namely

$$\widetilde{\varkappa 0 \mu}\ (\varkappa > 1), \quad \widetilde{\varkappa\lambda 1}\ (\varkappa > 1), \quad \widetilde{\varkappa\lambda 0}\ (\varkappa > 2);$$

Prof. Cayley has given the law for compounding two symmetric functions into a series of monomials and illustrated it by examples; by thus developing the quintic compounds, the non-exemplar forms we are considering are found to involve, in their reduction, the exemplar forms according to very beautiful laws connected with the binomial coefficients; it will have been observed in Prof. Cayley's paper that he finds that, in the general case, the coefficient of a monomial arising in the expansion is the sum of a number of terms, each of which is a binomial coefficient; the simplicity which is found in the case of perpetuants is therefore rather remarkable. I do not attempt the laborious proof of the following laws in the general case; every number however occurring in the tabulations has been verified for high values of κ, λ and μ.

In what follows, those terms involving forms of lower degree are omitted, so that such a result as

$$\overbrace{411} \equiv 2\ \overbrace{134} - \overbrace{144} + \overbrace{223} - \overbrace{233} - \overbrace{322},$$

will not be misunderstood.

All the forms are of the general weight, a number of two's being always supposed present, so as to bring up the weight to w; all the forms of course will not be present when the weight is low, those terms being absent in which the index of the number 2 becomes negative.

The annexed table is formed

	$\{\theta.1.\kappa+\mu-\theta$	$\{\theta.2.\kappa+\mu-\theta$	$\{\theta.3.\kappa+\mu-\theta$	$\{\theta.4.\kappa+\mu-\theta$	$\{\theta.5.\kappa+\mu-\theta$	$\{\theta.6.\kappa+\mu-\theta$	$\{\theta.7.\kappa+\mu-\theta$	$\{\theta.8.\kappa+\mu-\theta$	$\{\theta.9.\kappa+\mu-\theta$	$\{\theta.10.\kappa+\mu-\theta$	$\{\theta.11.\kappa+\mu-\theta$	. . .
Row 1	−1											.
" 2	+1	−1										.
" 3		+2	−1									.
" 4		−1	+3	−1								.
" 5			−3	+4	− 1							.
" 6			+1	−6	+ 5	− 1						.
" 7				+4	−10	+ 6	− 1					.
" 8				−1	+10	−15	+ 7	− 1				.
" 9					− 5	+20	−21	+ 8	− 1			.
" 10					+ 1	−15	+35	−28	+ 9	− 1		.
" 11	. . .	. . .	. . .	. . .	. . .	+ 6	−35	+56	−36	+10	−1	.

The law is obvious, the column headed $\overbrace{\theta.t.x+\mu-\theta}$ involving the coefficients in the expansion of $(x+1)^t$, with signs alternately negative and positive. Then $\widetilde{x0\mu} \equiv$ Row 1 $(\theta = x-1)$ + Row 2 $(\theta = x-2) + \ldots +$ Row $x-1$ $(\theta = 1)$, or more concisely

$$\widetilde{x0\mu} \equiv \sum_{t=x-1}^{t=1} \text{Row } t\,(\theta = x - t).$$

As an instance $\widetilde{504} \equiv -\widetilde{415} + \widetilde{316} - \widetilde{326} + 2\,\widetilde{227} - \widetilde{237} - \widetilde{128} + 3\,\widetilde{138} - \widetilde{148}$.

Again forming a table, viz.

	$\theta.\lambda+1.2$	$\theta.\lambda+1.3$	$\theta.\lambda+2.3$	$\theta.\lambda+2.4$	$\theta.\lambda+2.5$	$\theta.\lambda+3.4$	$\theta.\lambda+3.5$	$\theta.\lambda+3.6$	$\theta.\lambda+3.7$	$\theta.\lambda+4.5$	$\theta.\lambda+4.6$	$\theta.\lambda+4.7$	$\theta.\lambda+4.8$	$\theta.\lambda+4.9$	$\theta.\lambda+5.6$	$\theta.\lambda+5.7$	$\theta.\lambda+5.8$	$\theta.\lambda+5.9$	$\theta.\lambda+5.10$	$\theta.\lambda+5.11$	$\theta.\lambda+6.7$	. . .
Row 1	−1																					.
" 2		+1	−1																			.
" 3				+2		−1																.
" 4					−1		+3			−1												.
" 5								−3			+4				−1							.
" 6									+1			−6				+5					−1	.
. . .																						

the law being again clear, then

$$\widetilde{x\lambda 1} \equiv \sum_{t=x-1}^{t=1} \text{Row } t\,(\theta = x - t).$$

And again from the table

	$\theta.\lambda+1.2$	$\theta.\lambda+2.3$	$\theta.\lambda+2.4$	$\theta.\lambda+3.4$	$\theta.\lambda+3.5$	$\theta.\lambda+3.6$	$\theta.\lambda+4.5$	$\theta.\lambda+4.6$	$\theta.\lambda+4.7$	$\theta.\lambda+4.8$	$\theta.\lambda+5.6$	. . .
Row 1	+1											.
" 2		+1										.
" 3			−1	+1								.
" 4					−2		+1					.
" 5						+1		−3			+1	.
. . .												

which is to be continued on the same principle as the foregoing one to which it is very similar, we have:

$$\widetilde{x\lambda 0} \equiv \sum_{t=x-2}^{t=1} \text{Row } t(\theta = x - t - 1).$$

Thus $$\widetilde{420} \equiv \widetilde{232} + \widetilde{143}.$$

This result may be verified from the previous table by operating with Mr. Hammond's operator d_3 upon the value of $\widetilde{421}$.

Thus	$\widetilde{421} \equiv - \widetilde{332} + \widetilde{233} - \widetilde{243} + 2\ \widetilde{144} - \widetilde{154}$
Whence	$\widetilde{420} \equiv - \widetilde{331} + \widetilde{232} - \widetilde{242} + 2\ \widetilde{143} - \widetilde{153}$
and	$\widetilde{331} \equiv - \widetilde{242} + \widetilde{143} - \widetilde{153}$
Therefore	$\widetilde{420} \equiv \widetilde{232} + \widetilde{143}$,

as we before found.

It was absolutely necessary to obtain these theorems in order to form certain sextic syzygies of the general weight as will be done presently.

The sextic syzygies of any weight w are of three kinds: firstly, those formed by capitation of simple identities, which will be alluded to hereafter as simple syzygies; secondly, those formed by capitating the quintic syzygies; thirdly, those formed by capitating the capitatable syzygies of weight $w - 6$.

The first kind, with which for the present we are mostly concerned, arise from any forms, simple or composite, that it is possible to submit to both a 4.2 and a 3.3 capitation; such must be, it is clear, of the form $3^x 2^\lambda . 2^\mu$, where x, λ and μ may have any positive, including zero values, consistent with the relation

$$3x + 2(\lambda + \mu) = w - 6;$$

thus from the identity $$3^x 2^\lambda . 2^\mu = 3^x 2^\lambda . 2^\mu,$$

we get the syzygy $$43^x 2^\lambda . 2^{\mu+1} - 3^{x+1} 2^\lambda . 32^\mu \equiv ,$$

wherein if x is zero, the resulting quartic form is to be expressed, as it can be at once, in terms of quadric perpetuants.

It will now be proved by considering these simple syzygies in conjunction with another special set of even weight, derived by a second capitation of certain of them, that there can be no sextic perpetuant below weight 31.

It will be convenient to call those syzygies which are derived by a first, second, etc., capitation of simple identities, of the first, second, etc., class.

The special set above referred to will then be of the second class; the syzygies of the first class naturally arrange themselves into groups according to the number of threes occurring in the components of the identities from which they are derived; as will be seen there is only one group in the second class syzygies here discussed. In what follows $\equiv 0$ indicates that the dexter of the syzygy can be expressed in terms of perpetuants of a degree lower than 5, and $\equiv \widetilde{112}$ indicates $\equiv \widetilde{112}$ + lower forms and so on. Thus the first of the following syzygies in its complete form is

$$\begin{aligned} 432^{\kappa}.2 &= 542^{\kappa} \\ -\ 3^2 2^{\kappa}.3 \quad & -\ 53^2 2^{\kappa-1} \\ & +\ 2.(4^2 32^{\kappa-1}) \\ & +\ (\kappa+1)\,432^{\kappa+1} \\ & -\ 3.(3^3 2^{\kappa}), \end{aligned}$$

where on the right-hand side everything is to be rejected. The following values have been thoroughly verified and I think may be relied upon; the syzygies are denoted by the capital letters with suffixes, the suffix denoting the greatest number of threes in any term of the syzygy.

SEXTIC SYZYGIES.

Class 1. Group 1. $w = 2\kappa + 9$.

A_3 $432^{\kappa}.2 \quad - 3^2 2^{\kappa}.3 \quad \equiv 0,$

B_3 $432^{\kappa-1}.2^2 - 3^2 2^{\kappa-1}.32 \equiv \widetilde{112},$

C_3 $432^{\kappa-2}.2^3 - 3^2 2^{\kappa-2}.32^2 \equiv (\kappa-5)\,\widetilde{112} + \widetilde{122},$

D_3 $432^{\kappa-3}.2^4 - 3^2 2^{\kappa-3}.32^3 \equiv \tfrac{1}{2}(\kappa^2 - 11\kappa + 26)\,\widetilde{112} + (\kappa-7)\,\widetilde{122} + \widetilde{132},$

E_3 $432^{\kappa-4}.2^5 - 3^2 2^{\kappa-4}.32^4 \equiv \tfrac{1}{6}(\kappa-3)(\kappa^2 - 15\kappa + 50)\,\widetilde{112} + \tfrac{1}{2}(\kappa^2 - 15\kappa + 52)\,\widetilde{122}$
$\qquad + (\kappa-9)\,\widetilde{132} + \widetilde{142},$

etc. $\equiv$ etc.,

whence it is easy to infer that every form $\widetilde{1\kappa 2}$ is a sextic syzygant.

Class 1. Group 2. $w = 2\kappa + 12$.

A_4 $43^2 2^{\kappa}.2 \quad - 3^3 2^{\kappa}.3 \quad \equiv 0,$

B_4 $43^2 2^{\kappa-1}.2^2 - 3^3 2^{\kappa-1}.32 \equiv \widetilde{113},$

C_4 $43^2 2^{\kappa-2}.2^3 - 3^3 2^{\kappa-2}.32^2 \equiv (\kappa-5)\,\widetilde{113} + 2\,\widetilde{123} + \widetilde{212},$

D_4 $43^2 2^{\kappa-3}.2^4 - 3^3 2^{\kappa-3}.32^3 \equiv \tfrac{1}{2}(\kappa^2 - 11\kappa + 26)\,\widetilde{113} + (2\kappa - 13)\,\widetilde{123}$
$\qquad + 2\ \widetilde{133} + (\kappa-6)\,\widetilde{212} + \widetilde{222},$

E_4 $43^2 2^{\kappa-4}.2^5 - 3^3 2^{\kappa-4}.32^4 \equiv \tfrac{1}{6}(\kappa-3)(\kappa^2 - 15\kappa + 50)\,\widetilde{113} + (\kappa^2 - 14\kappa + 45)\,\widetilde{123}$
$\qquad + (2\kappa - 17)\,\widetilde{133} + 2\,\widetilde{143} + \tfrac{1}{2}(\kappa^2 - 13\kappa + 38)\,\widetilde{212} + (\kappa-8)\,\widetilde{222} + \widetilde{232},$

etc. $\equiv$ etc.

Class 1. Group 3. $w = 2x + 9$.

A_5 $43^3 2^{x-3}.2 - 3^4 2^{x-3}.3 \equiv \widehat{112}$,

B_5 $43^3 2^{x-4}.2^2 - 3^4 2^{x-4}.32 \equiv (x-3)\widehat{112} + \widehat{114} + \widehat{122}$,

C_5 $43^3 2^{x-5}.2^3 - 3^4 2^{x-5}.32^2 \equiv \tfrac{1}{2}(x-3)(x-4)\widehat{112} + (x-8)\widehat{114} + (x-5)\widehat{122}$
$+ 2\ \widehat{124} + \widehat{132} + \widehat{213}$,

D_5 $43^3 2^{x-6}.2^4 - 3^4 2^{x-6}.32^3 \equiv \tfrac{1}{6}(x-3)(x-4)(x-5)\widehat{112} + \tfrac{1}{2}(x^2 - 17x + 68)\widehat{114}$
$+ \tfrac{1}{2}(x-5)(x-6)\widehat{122} + 2(x-10)\widehat{124} + (x-7)\widehat{132} + 3\ \widehat{134} + \widehat{142}$
$+ (x-9)\widehat{213} + 2\ \widehat{223} + \widehat{312}$,

E_5 $43^3 2^{x-7}.2^5 - 3^4 2^{x-7}.32^4 \equiv \tfrac{1}{24}(x-3)(x-4)(x-5)(x-6)\widehat{112}$
$+ \tfrac{1}{6}(x-6)(x-8)(x-13)\widehat{114} + \tfrac{1}{6}(x-5)(x-6)(x-7)\widehat{122} + (x^2 - 21x + 106)\widehat{124}$
$+ \tfrac{1}{2}(x-7)(x-8)\widehat{132} + (3x-34)\widehat{134} + (x-9)\widehat{142} + 3\ \widehat{144} + \widehat{152}$
$+ \tfrac{1}{2}(x^2 - 19x + 86)\widehat{213} + (2x-21)\widehat{223} + 2\ \widehat{233} + (x-10)\widehat{312} + \widehat{322}$

etc. $\equiv$ etc.

Class 1. Group 4. $w = 2x + 12$.

A_6 $43^4 2^{x-3}.2 - 3^5 2^{x-3}.3 \equiv \widehat{113}$,

B_6 $43^4 2^{x-4}.2^2 - 3^5 2^{x-4}.32 \equiv (x-3)\widehat{113} + \widehat{115} + 2\ \widehat{123} + \widehat{212}$.

C_6 $43^4 2^{x-5}.2^3 - 3^5 2^{x-5}.32^2 \equiv \tfrac{1}{2}(x-3)(x-4)\widehat{113} + (x-8)\widehat{115} + (2x-9)\widehat{123}$
$+ 2\ \widehat{125} + 2\ \widehat{133} + (x-4)\widehat{212} + \widehat{214} + \widehat{222}$,

D_6 $43^4 2^{x-6}.2^4 - 3^5 2^{x-6}.32^3 \equiv \tfrac{1}{6}(x-3)(x-4)(x-5)\widehat{113} + \tfrac{1}{2}(x^2 - 17x + 68)\widehat{115}$
$+ (x-5)^2\widehat{123} + 2(x-10)\widehat{125} + (2x-13)\widehat{133} + 3\ \widehat{135} + 2\ \widehat{143}$
$+ \tfrac{1}{2}(x-4)(x-5)\widehat{212} + (x-9)\widehat{214} + (x-6)\widehat{222} + 2\ \widehat{224} + \widehat{232} + \widehat{313}$,

E_6 $43^4 2^{x-7}.2^5 - 3^5 2^{x-7}.32^4 \equiv \tfrac{1}{24}(x-3)(x-4)(x-5)(x-6)\widehat{113}$
$+ \tfrac{1}{6}(x-6)(x-8)(x-13)\widehat{115} + \tfrac{1}{6}(x-5)(x-6)(2x-11)\widehat{123}$
$+ (x^2 - 21x + 106)\widehat{125} + (x-7)^2\widehat{133} + 3(x-12)\widehat{135} + (2x-17)\widehat{143}$
$+ 4\ \widehat{145} + 2\ \widehat{153} + \tfrac{1}{6}(x-4)(x-5)(x-6)\widehat{212} + \tfrac{1}{2}(x^2 - 19x + 86)\widehat{214}$
$+ \tfrac{1}{2}(x-6)(x-7)\widehat{222} + 2(x-11)\widehat{224} + (x-8)\widehat{232} + 3\ \widehat{234} + \widehat{242}$
$+ (x-10)\widehat{313} + 2\ \widehat{323} + \widehat{412}$,

etc. $\equiv$ etc.

Class 1. Group 5. $w = 2x + 9$.

A_7 $43^5 2^{x-6}.2 - 3^6 2^{x-6}.3 \equiv \widehat{114}$,

B_7 $43^5 2^{x-7}.2^2 - 3^6 2^{x-7}.32 \equiv (x-6)\widehat{114} + \widehat{116} + 2\ \widehat{124} + \widehat{213}$,

C_7 $43^5 2^{x-8}.2^3 - 3^6 2^{x-8}.32^2 \equiv \tfrac{1}{2}(x-6)(x-7)\widehat{114} + (x-11)\widehat{116} + 2(x-8)\widehat{124}$
$+ 2\ \widehat{126} + 3\ \widehat{134} + (x-7)\widehat{213} + \widehat{215} + 2\ \widehat{223} + \widehat{312}$,

$D_7 \;\; 43^5 2^{\kappa-9}.2^4 - 3^6 2^{\kappa-9}.32^3 \equiv \frac{1}{6}(x-6)(x-7)(x-8)\overbrace{114} + \frac{1}{2}(x^2-23x+128)\overbrace{116}$
$+ (x-8)(x-9)\overbrace{124} + 2(x-13)\overbrace{126} + (3x-28)\overbrace{134} + 3\;\overbrace{136}$
$+ 3\;\overbrace{144} + \frac{1}{2}(x-7)(x-8)\overbrace{213} + (x-12)\overbrace{215} + (2x-17)\overbrace{223}$
$+ 2\;\overbrace{225} + 2\;\overbrace{233} + (x-8)\overbrace{312} + \overbrace{314} + \overbrace{322},$

$E_7 \;\; 43^5 2^{\kappa-10}.2^5 - 3^6 2^{\kappa-10}.32^4 \equiv \frac{1}{24}(x-6)(x-7)(x-8)(x-9)\overbrace{114}$
$+ \frac{1}{6}(x-9)(x-11)(x-16)\overbrace{116} + \frac{1}{3}(x-8)(x-9)(x-10)\overbrace{124}$
$+ (x^2-27x+178)\overbrace{126} + \{\frac{3}{2}(x-10)(x-11) + 2x-19\}\overbrace{134}$
$+ 3(x-15)\overbrace{136} + (3x-34)\overbrace{144} + 4\;\overbrace{146} + 3\;\overbrace{154}$
$+ \frac{1}{6}(x-7)(x-8)(x-9)\overbrace{213} + \frac{1}{2}(x^2-25x+152)\overbrace{215} + (x-9)^2\overbrace{223}$
$+ 2(x-14)\overbrace{225} + (2x-21)\overbrace{233} + 3\;\overbrace{235} + 2\;\overbrace{243} + \frac{1}{2}(x-8)(x-9)\overbrace{312}$
$+ (x-13)\overbrace{314} + (x-10)\overbrace{322} + 2\;\overbrace{324} + \overbrace{332} + \overbrace{413},$

etc. $\equiv$ etc.

Class 1. Group 6. $w = 2x + 12$.

$A_8 \;\; 43^6 2^{\kappa-6}.2 - 3^7 2^{\kappa-6}.3 \equiv \overbrace{115},$

$B_8 \;\; 43^6 2^{\kappa-7}.2^2 - 3^7 2^{\kappa-7}.32 \equiv (x-6)\overbrace{115} + \overbrace{117} + 2\;\overbrace{125} + \overbrace{214},$

$C_8 \;\; 43^6 2^{\kappa-8}.2^3 - 3^7 2^{\kappa-8}.32^2 \equiv \frac{1}{2}(x-6)(x-7)\overbrace{115} + (x-11)\overbrace{117} + 2(x-8)\overbrace{125}$
$+ 2\;\overbrace{127} + 3\;\overbrace{135} + (x-7)\overbrace{214} + \overbrace{216} + 2\;\overbrace{224} + \overbrace{313},$

$D_8 \;\; 43^6 2^{\kappa-9}.2^4 - 3^7 2^{\kappa-9}.32^3 \equiv \frac{1}{6}(x-6)(x-7)(x-8)\overbrace{115} + \frac{1}{2}(x^2-23x+128)\overbrace{117}$
$+ (x-8)(x-9)\overbrace{125} + 2(x-13)\overbrace{127} + 3(x-10)\overbrace{135} + 3\;\overbrace{137}$
$+ 4\;\overbrace{145} + \frac{1}{2}(x-7)(x-8)\overbrace{214} + (x-12)\overbrace{216} + 2(x-9)\overbrace{224}$
$+ 2\;\overbrace{226} + 3\;\overbrace{234} + (x-8)\overbrace{313} + \overbrace{315} + 2\;\overbrace{323} + \overbrace{412},$

$E_8 \;\; 43^6 2^{\kappa-10}.2^5 - 3^7 2^{\kappa-10}.32^4 \equiv \frac{1}{24}(x-6)(x-7)(x-8)(x-9)\overbrace{115}$
$+ \frac{1}{6}(x-9)(x-11)(x-16)\overbrace{117} + \frac{1}{3}(x-8)(x-9)(x-10)\overbrace{125}$
$+ (x^2-27x+178)\overbrace{127} + \{\frac{3}{2}(x-10)(x-11) - 1\}\overbrace{135} + 3(x-15)\overbrace{137}$
$+ (4x-45)\overbrace{145} + 4\;\overbrace{147} + 4\;\overbrace{155} + \frac{1}{6}(x-7)(x-8)(x-9)\overbrace{214}$
$+ \frac{1}{2}(x^2-25x+152)\overbrace{216} + (x-9)(x-10)\overbrace{224} + 2(x-14)\overbrace{226}$
$+ (3x-31)\overbrace{234} + 2\;\overbrace{236} + 3\;\overbrace{244} + \frac{1}{2}(x-8)(x-9)\overbrace{313} + (x-13)\overbrace{315}$
$+ (2x-19)\overbrace{323} + 2\;\overbrace{325} + 2\;\overbrace{333} + (x-9)\overbrace{412} + \overbrace{414} + \overbrace{422},$

etc. $\equiv$ etc.

Class 1. Group 7. $w = 2x + 9$.

$$A_9\ 43^7 2^{x-9}.2 - 3^8 2^{x-9}.3 \equiv \widetilde{116},$$

$$B_9\ 43^7 2^{x-10}.2^2 - 3^8 2^{x-10}.32 \equiv (x-9)\widetilde{116} + \widetilde{118} + 2\ \widetilde{126} + \widetilde{215},$$

$$C_9\ 43^7 2^{x-11}.2^3 - 3^8 2^{x-11}.32^2 \equiv \tfrac{1}{2}(x-9)(x-10)\widetilde{116} + (x-14)\widetilde{118}$$
$$+ 2(x-11)\widetilde{126} + 2\ \widetilde{128} + 3\ \widetilde{136} + (x-10)\widetilde{215} + \widetilde{217} + 2\ \widetilde{225} + \widetilde{314},$$

$$D_9\ 43^7 2^{x-12}.2^4 - 3^8 2^{x-12}.32^3 \equiv \tfrac{1}{6}(x-9)(x-10)(x-11)\widetilde{116}$$
$$+ \tfrac{1}{2}(x^2 - 29x + 206)\widetilde{118} + (x-11)(x-12)\widetilde{126} + 2(x-16)\widetilde{128}$$
$$+ 3(x-13)\widetilde{136} + 3\ \widetilde{138} + 4\ \widetilde{146} + \tfrac{1}{2}(x-10)(x-11)\widetilde{215}$$
$$+ (x-15)\widetilde{217} + 2(x-12)\widetilde{225} + 2\ \widetilde{227} + 3\ \widetilde{235} + (x-11)\widetilde{314}$$
$$+ \widetilde{316} + 2\ \widetilde{324} + \widetilde{413},$$

$$E_9\ 43^7 2^{x-13}.2^5 - 3^8 2^{x-13}.32^4 \equiv \tfrac{1}{24}(x-9)(x-10)(x-11)(x-12)\widetilde{116}$$
$$+ \tfrac{1}{6}(x-12)(x-14)(x-19)\widetilde{118} + \tfrac{1}{3}(x-11)(x-12)(x-13)\widetilde{126}$$
$$+ (x^2 - 33x + 268)\widetilde{128} + \tfrac{3}{2}(x-13)(x-14)\widetilde{136} + 3(x-18)\widetilde{138}$$
$$+ 4(x-15)\widetilde{146} + 4\ \widetilde{148} + 5\ \widetilde{156} + \tfrac{1}{6}(x-10)(x-11)(x-12)\widetilde{215}$$
$$+ \tfrac{1}{2}(x^2 - 31x + 236)\widetilde{217} + (x-12)(x-13)\widetilde{225} + 2(x-17)\widetilde{227}$$
$$+ 3(x-14)\widetilde{235} + 3\ \widetilde{237} + 4\ \widetilde{245} + \tfrac{1}{2}(x-11)(x-12)\widetilde{314}$$
$$+ (x-16)\widetilde{316} + 2(x-13)\widetilde{324} + 2\ \widetilde{326} + 3\ \widetilde{334} + (x-12)\widetilde{413}$$
$$+ \widetilde{415} + 2\ \widetilde{423} + \widetilde{512},$$

etc. $\equiv$ etc.

It will be observed that the A, B, C, D, E syzygies attain their final forms in groups 3, 4, 5, 6, 7 respectively, and generally the N syzygies reach their final form in group $n + 2$; a general formula can be obtained for the N syzygies for group $n + 2$ and succeeding groups; it is evident on inspection that the number of terms in a final form is a sum of odd numbers, proceeding regularly from unity; thus in the case of the C syzygies there is one term of the form $\widetilde{3\lambda\mu}$, three of $\widetilde{2\lambda\mu}$, five of $\widetilde{1\lambda\mu}$, giving a total of nine terms; the numbers of terms in the final forms of the $A, B, C \ldots$ syzygies are therefore the successive square numbers $1, 4, 9 \ldots$. After the final forms have been reached, the syzygies $A_x, B_x, C_x, D_x, E_x \ldots$ are derived from the syzygies $A_{x-3}, B_{x-3}, C_{x-3}, D_{x-3}, E_{x-3}, \ldots$ by substituting therein $x - 3$ for x and increasing the third number under the symbol $\frown$ by unity.

In Group 7 all the syzygies written down are in their final forms; what is apparently the law of the coefficients in the A, B, C, D, E syzygies may be shown by the following scheme. Consider a group λ, of weight $3\lambda + 2\mu + 6$, commencing with the syzygy

$$A_{\lambda+2}\ 43^\lambda 2^\mu.2 - 3^{\lambda+1} 2^\mu.3 \equiv \widetilde{1.1.\lambda - 1},$$

any number in the table represents the coefficient of the perpetuant in the same row and in the left-hand column, in the syzygy denoted by the capital letter with suffix, at the head of its column.

	$A_{\lambda+2}$	$B_{\lambda+2}$	$C_{\lambda+2}$	$D_{\lambda+2}$	$E_{\lambda+2}$
$\overbrace{1.1.\lambda-1}$	1	μ	$\frac{1}{2}\mu(\mu-1)$	$\frac{1}{6}\mu(\mu-1)(\mu-2)$	$\frac{1}{24}\mu(\mu-1)(\mu-2)(\mu-3)$
$\overbrace{1.2.\lambda-1}$		2	$2(\mu-2)$	$\frac{2}{2}(\mu-2)(\mu-3)$	$\frac{2}{6}(\mu-2)(\mu-3)(\mu-4)$
$\overbrace{1.3.\lambda-1}$			3	$3(\mu-4)$	$\frac{3}{2}(\mu-4)(\mu-5)$
$\overbrace{1.4.\lambda-1}$				4	$4(\mu-6)$
$\overbrace{1.5.\lambda-1}$					5
$\overbrace{1.1.\lambda+1}$		1	$\mu-5$	$\frac{1}{2}(\mu-5)(\mu-6)-2$	$\frac{1}{6}(\mu-5)(\mu-6)(\mu-7)-2(\mu-5)$
$\overbrace{1.2.\lambda+1}$			2	$2(\mu-7)$	$\frac{2}{2}(\mu-7)(\mu-8)-4$
$\overbrace{1.3.\lambda+1}$				3	$3(\mu-9)$
$\overbrace{1.4.\lambda+1}$					4
$\overbrace{2.1.\lambda-2}$		1	$\mu-1$	$\frac{1}{2}(\mu-1)(\mu-2)$	$\frac{1}{6}(\mu-1)(\mu-2)(\mu-3)$
$\overbrace{2.2.\lambda-2}$			2	$2(\mu-3)$	$\frac{2}{2}(\mu-3)(\mu-4)$
$\overbrace{2.3.\lambda-2}$				3	$3(\mu-5)$
$\overbrace{2.4.\lambda-2}$					4
$\overbrace{2.1.\lambda}$			1	$\mu-6$	$\frac{1}{2}(\mu-6)(\mu-7)-2$
$\overbrace{2.2.\lambda}$				2	$2(\mu-8)$
$\overbrace{2.3.\lambda}$					3
$\overbrace{3.1.\lambda-3}$			1	$\mu-2$	$\frac{1}{2}(\mu-2)(\mu-3)$
$\overbrace{3.2.\lambda-3}$				2	$2(\mu-4)$
$\overbrace{3.3.\lambda-3}$					3
$\overbrace{3.1.\lambda-1}$				1	$\mu-7$
$\overbrace{3.2.\lambda-1}$					2
$\overbrace{4.1.\lambda-4}$				1	$\mu-3$
$\overbrace{4.2.\lambda-4}$					2
$\overbrace{4.1.\lambda-2}$					1
$\overbrace{5.1.\lambda-5}$					1

From this it appears that the coefficients in any block are derived from those in the preceding one, by shifting them one column to the right and writing $\mu - 1$ for μ; further in any one block, the coefficients in the second part of the block are derived from those in the first part by shifting them one column to the right and writing $\mu - 5$ for μ; at the same time subtracting from them twice the column to the left, next to it but one, with $\mu - 5$ written for μ therein.

It only remains to examine the first part of the first block; the law of the first row therein is evident; the tth row is formed from the first row by multiplying it by t, writing $\mu - 2(t-1)$ for μ and shifting it $t - 1$ columns to the right.

The complete law thus appears to be defined, but I have attempted no proof of it.

Passing now to the special group of syzygies of Class 2, before referred to; they arise from the simple identity $2^{\kappa-\lambda}.2^{\lambda} = 2^{\kappa-\lambda}.2^{\lambda}$; a first capitation gives the Class 1 syzygies included in $32^{\kappa-\lambda}.32^{\lambda} - 42^{\kappa-\lambda}.2^{\lambda+1} \equiv$; in which the component $42^{\kappa-\lambda}$ is to be at once expressed in terms of quadric perpetuants; the dexter of this syzygy cannot involve quintic perpetuants, because the sinister contains no term with three odd symbolic numbers; the whole of these syzygies can therefore be at once again capitated as they are; and since the dexter of the above written syzygy consists entirely of quintic compounds with two odd numbers, quartic and cubic perpetuants with two odd numbers, together with terms consisting wholly of even numbers, and since the first three species of terms can be submitted to a 4.2 capitation, an operation which does not increase the number of odd numbers which they contain, it easily follows that the only term which need be considered after the second capitation is $3^2 2^{\kappa-\lambda}.3^2 2^{\lambda}$; this is part of a theory of abbreviation which is more fully entered upon afterwards. We have then:

Class 2. Special Group. $w = 2\kappa + 12$.

$A'_4 \quad 3^2 2^{\kappa}.3^2 \quad - \ldots \equiv -, \ \widetilde{113}$

$B'_4 \quad 3^2 2^{\kappa-1}.3^2 2 - \ldots \equiv -(\kappa - 4)\,\widetilde{113} - 3\ \widetilde{123} - 2\ \widetilde{212},$

$C'_4 \quad 3^2 2^{\kappa-2}.3^2 2^2 - \ldots \equiv -\{\tfrac{1}{2}(\kappa-4)(\kappa-5) - 2\}\widetilde{113} - 3(\kappa-6)\,\widetilde{123} - 3\ \widetilde{133}$
$\qquad - 2(\kappa-6)\,\widetilde{212} - 2\ \widetilde{222},$

$D'_4 \quad 3^2 2^{\kappa-3}.3^2 2^3 - \ldots \equiv -\{\tfrac{1}{6}(\kappa-4)(\kappa-5)(\kappa-6) - 2(\kappa-4)\}\widetilde{113}$
$\qquad - 3\{\tfrac{1}{2}(\kappa-6)(\kappa-7) - 2\}\,\widetilde{123} - 3(\kappa-8)\,\widetilde{133} - 3\ \widetilde{143}$
$\qquad - 2\{\tfrac{1}{2}(\kappa-6)(\kappa-7) - 2\}\,\widetilde{212} - 2(\kappa-8)\,\widetilde{222} - 2\ \widetilde{232}.$

etc. $\equiv$ etc.

Whence forming a scheme as before,

	A'_4	B'_4	C'_4	D'_4
$\widetilde{113}$	-1	$-(\kappa-4)$	$-\{\frac{1}{2}(\kappa-4)(\kappa-5)-2\}$	$-\{\frac{1}{6}(\kappa-4)(\kappa-5)(\kappa-6)-2(\kappa-4)\}$
$\widetilde{123}$		-3	$-3(\kappa-6)$	$-3\{\frac{1}{2}(\kappa-6)(\kappa-7)-2\}$
$\widetilde{133}$			-3	$-3(\kappa-8)$
$\widetilde{143}$				-3
$\widetilde{212}$		-2	$-2(\kappa-6)$	$-2\{\frac{1}{2}(\kappa-6)(\kappa-7)-2\}$
$\widetilde{222}$			-2	$-2(\kappa-8)$
$\widetilde{232}$				-2

and the law is apparent.

Proceeding to examine the groups of syzygies, it is seen that $\widetilde{11\mu}$ is a sextic syzygant, and as remarked before, Group 1 shows that $\widetilde{1\lambda 2}$ is so also; from C_4 and B'_4 we see that so are $\widetilde{123}$ and $\widetilde{212}$; from D_4 and C'_4, $\widetilde{133}$ and $\widetilde{222}$ are so also, and so on, combining the members of the second and special groups in pairs it is seen that every quintic perpetuant with four odd numbers is a sextic syzygant; those containing a lesser number have been shown to be so also. If the attempt is now made to express either of the forms $\widetilde{312}$, $\widetilde{213}$, $\widetilde{124}$ in a similar manner, it is found that the coincidence of the coefficients involved presents an insuperable obstruction; by making a list of the forms occurring for different weights, commencing with the weight 15, it will be seen to what point an advance has been made. We have

$w=$							
15	$\widetilde{112}$.						
17	$\widetilde{112}$.						
18	$\widetilde{113}$.						
19	$\widetilde{112}$,	$\widetilde{122}$.					
20	$\widetilde{113}$,	$\widetilde{212}$.					
21	$\widetilde{112}$,	$\widetilde{114}$,	$\widetilde{122}$.				
22	$\widetilde{113}$,	$\widetilde{123}$,	$\widetilde{212}$.				
23	$\widetilde{112}$,	$\widetilde{114}$,	$\widetilde{122}$,	$\widetilde{132}$,	$\widetilde{213}$.		
24	$\widetilde{113}$,	$\widetilde{115}$,	$\widetilde{123}$,	$\widetilde{212}$,	$\widetilde{222}$.		
25	$\widetilde{112}$,	$\widetilde{114}$,	$\widetilde{122}$,	$\widetilde{132}$,	$\widetilde{124}$,	$\widetilde{213}$,	$\widetilde{312}$.
26	$\widetilde{113}$,	$\widetilde{115}$,	$\widetilde{123}$,	$\widetilde{133}$,	$\widetilde{212}$,	$\widetilde{222}$,	$\widetilde{214}$.

so that certainly every form of a lower weight than 23 is a sextic syzygant; since the form $\widetilde{124}$ does not occur for weight 23, it is evident from B_7 or C_5 that for weight 23, $\widetilde{213}$ is a syzygant; all forms of weight 24 are certainly syzygants and from B_8 so are those of weight 26.

The conclusion arrived at is that the only possible weight, below weight 33, for which there can be a sextic perpetuant is 31.

From the groups of simple syzygies of Class 1 are derivable two series of capitatable syzygies; thus for an odd weight:

$$A_3 \equiv 0,$$
$$B_3 - A_5 \equiv 0,$$
$$2B_3 + C_3 - B_5 + A_7 \equiv 0,$$
$$3B_3 + 2C_3 + D_3 - C_5 - 2A_7 + B_7 - A_9 \equiv 0,$$
$$4B_3 + 3C_3 + 2D_3 + E_3 - D_5 + A_7 - 2B_7 + C_7 + 4A_9 - B_9 + A_{11} \equiv 0,$$
$$\cdots\cdots\cdots\cdots\cdots\cdots$$

These may also be written

$$A_3 \equiv 0,$$
$$B_3 - A_5 \equiv 0,$$
$$2B_3 + C_3 - (B_5 - A_7) \equiv 0,$$
$$3B_3 + 2C_3 + D_3 - \{2B_5 + C_5 - (B_7 - A_9)\} + 2(B_5 - A_7) \equiv 0,$$
$$4B_3 + 3C_3 + 2D_3 + E_3 - [3B_5 + 2C_5 + D_5 - \{2B_7 + C_7 - (B_9 - A_{11})\} + 2(B_7 - A_9)]$$
$$+ 2\{2B_5 + C_5 - (B_7 - A_9)\} - (B_5 - A_7) \equiv 0,$$
$$\cdots\cdots\cdots\cdots\cdots\cdots$$

but the law does not seem to be clear.

Making a unit increase of suffixes throughout, a similar series is obtained for each even weight.

It is now necessary to examine the remaining syzygies of weight 25, in order to discover if the forms $\widetilde{312}$, $\widetilde{213}$, $\widetilde{124}$ of that weight are separately expressible as sextic syzygants; for this purpose, there is no need to attend to those syzygies which are derived by a sextic capitation of quintic syzygies; for consider the syzygy $32.2 - 3.2^3 = 43 + 32^3$; this is capable of 4.2 capitation throughout, and it is plain that it can be operated upon an infinite number of times in a similar way, without there being any necessity to introduce another odd number into any term; similarly the whole of the quintic syzygies admit of infinite 4.2 capitation and can never involve a quintic perpetuant. It is only requisite to consider then those that are derived from the sextic syzygies of weight 19. We first require the syzygies of weight 13; these are

$3^2 2.32$	$432^2.2$	432.2^2	43.2^3
$-3^2.32^3$	$-3^2 2^2.3$	$-3^3 2.32$	$-3^2.32^3$
$-4^2 3.2$	$-32^3.2^2$	$-2(32^2.2^3)$	$-32^3.2^2$
$-3^2 2^2.3$	$+4(32^4.2)$	$+2(32^3.2^2)$	$+4(32^4.2)$
$-3^3.2^2$	$+3^3 2.2$	$+4(32^4.2)$	$+3^3.2^2$
$+3^3 2.2$	$-4^2 32$	$+2(3^3.2^2)$	$+3^3 2.2$
$+32^2.2^3$	-43^3	$+3^3 2.2$	$+4^2 32$
$-2(32^3.2^2)$	$-4(432^3)$	$-2(4^2 32)$	-43^3
$+2(32^4.2)$	$+3^3 2^2$	$-7(432^3)$	$-2(432^3)$
$+2(4^2 32)$	$-10(32^5)=0$	$+2(43^3)$	$-10(32^5)=0$
$-4(43^3)$		$+2(3^3 2^2)$	
$+432^3$		$-20(32^5)=0$	
$-3^3 2^2=0$			

Capitating these, and placing the simple syzygies first, we have the following fifteen syzygies of weight 19; the $\varkappa$ forms being those derived from quintic syzygies.

$A_3 = 432^5.2$	$B_3 = 432^4.2^2$	$C_3 = 432^3.2^3$	$D_3 = 432^2.2^4$
$-3^2 2^5.3$	$-3^2 2^4.32$	$-3^2 2^3.32^2$	$-3^2 2^2.32^3$
$\equiv 0$	$\equiv \widetilde{112}$	$\equiv \widetilde{122}$	$\equiv -2\,\widetilde{112} - 2\,\widetilde{122}$
$E_3 = 432.2^5$	$F_3 = 43.2^6$		
$-3^2 2.32^4$	$-3^2.32^5$		
$\equiv \widetilde{122}$	$\equiv \widetilde{112}$		
$A_5 = 43^3 2^2.2$	$B_5 = 43^3 2.2^2$	$C_5 = 43^3.2^3$	
$-3^4 2^2.3$	$-3^4 2.32$	$-3^4.32^2$	
$\equiv \widetilde{112}$	$\equiv \widetilde{122} + 2\,\widetilde{112}$	$\equiv \widetilde{112}$	
$\varkappa_1 = 432^4.2^2$	$\varkappa'_1 = 432^3.2^3$		
-43.2^6	-432.2^5		
$-4^2 32^3.2$	$-4^3 32.2$		
$-4(432^5.2)$	$-2(4^2 32^3.2)$		
$\equiv 0$	$-5(432^5.2)$		
	$\equiv 0$		

$$\begin{aligned}
A'_5 \equiv\; & 3^32 . 3^22 \\
& - 3^3 . 3^22^2 \\
& - 4^23 . 2^2 \\
& - 3^32^2 . 3^2 \\
& - 43^3 . 2^3 \\
& + 43^32 . 2^2 \\
& + 432^2 . 2^4 \\
& - 2(432^3 . 2^3) \\
& + 2(432^4 . 2^2) \\
& + 2(4^332 . 2) \\
& - 4(4^23^3 . 2) \\
& + 4^232^3 . 2 \\
& - 43^32^2 . 2 \\
& \equiv - 3\; \overbrace{122}
\end{aligned}
\qquad
\begin{aligned}
B'_5 \equiv\; & 4^232^2 . 2^2 \\
& - 3^32^2 . 3^2 \\
& - 432^3 . 2^3 \\
& + 4(432^4 . 2^2) \\
& + 43^32 . 2^2 \\
& - 4^332 . 2 \\
& - 4^23^3 . 2 \\
& - 4(4^232^3 . 2) \\
& + 43^32^2 . 2 \\
& - 10(432^5 . 2) \\
& \equiv 3\; \overbrace{112}
\end{aligned}$$

$$\begin{aligned}
C'_5 \equiv\; & 4^232 . 2^3 \\
& - 3^32 . 3^22 \\
& - 2(4\quad . 2^4 \\
& + 2(4^{32^3} . \\
& + 4(\;^{2^0}\; . 2^2) \\
& + 2(43^22^2\,2^3)) \\
& + 43^22 . 2^3 \\
& - 2(4^332 . 2) \\
& - 7(4^232^3 . 2) \\
& + 2(4^23^3 . 2) \\
& + 2(43^32^2\; 2) \\
& - 20(4\quad .. 2) \\
& \equiv 3\; \overbrace{122}^{32^5} + 6\; \overbrace{112}
\end{aligned}
\qquad
\begin{aligned}
D'_5 \equiv\; & 4^23 . 2^4 \\
& - 3^3 . 3^22^2 \\
& - 432^3 . 2^3 \\
& + 4(432^4 . 2^2) \\
& + 43^3 . 2^3 \\
& + 43^32 . 2^2 \\
& + 4^332 . 2 \\
& - 4^23^3 . 2 \\
& - 2(4^232^3 . 2) \\
& - 10(432^5 . 2) \\
& \equiv 3\; \overbrace{112}
\end{aligned}$$

which may be verified by the relations:

$$4^23^3 . 2 \equiv - 3^22^3 . 32^2 \equiv - 3^22 . 32^4 \equiv \overbrace{122},$$

$$43^32^2 . 2 \equiv 43^3 . 2^3 \equiv - 3^22^4 . 32 \equiv - 3^3 . 32^5 \equiv \overbrace{112},$$

$$43^32 . 2^2 \equiv 2\; \overbrace{112} + \overbrace{122},$$

$$3^22^2 . 32^3 \equiv 2\; \overbrace{112} + 2\; \overbrace{122},$$

every other being a null form.

From the foregoing is obtained the capitatable series of thirteen syzygies, viz:

$$\begin{array}{llll}
A_3 \equiv 0, & A_5 - B_3 \equiv 0, & \varkappa_1 \equiv 0, & A'_5 + 3C_3 \equiv 0, \\
D_3 + 2C_3 + 2B_3 \equiv 0, & B_5 - 2B_3 - C_3 \equiv 0, & \varkappa'_1 \equiv 0, & B'_5 - 3B_3 \equiv 0, \\
E_3 - C_3 \equiv 0, & C_5 - B_3 \equiv 0, & & C'_5 - 3C_3 - 6B_3 \equiv 0, \\
F_3 - B_3 \equiv 0, & & & D'_5 - 3B_3 \equiv 0.
\end{array}$$

Before proceeding it is convenient to make a few remarks in order to shorten the remaining work.

Reflection will show that we can at this stage neglect a large number of terms as having no influence on the present investigation; the capitatable syzygies above written down are more strictly what Prof. Cayley terms Congruences, which have to be completed into full syzygies prior to capitation; now it is clear that any term of the congruence, which contains but one part three, will only give rise to such terms when it is completed, and that it and its dependent terms are capable of 4.2 capitation, which will produce no term that can give rise to a quintic perpetuant; for the present purpose therefore all terms containing but one three are at once neglected.

Bearing in mind that we are only concerned with the forms $\widetilde{312}$, $\widetilde{213}$, $\widetilde{124}$, which contain five odd numbers, it having been shown that the remainder are sextic syzygants, another great abbreviation may be employed; for any term containing three threes will give rise to terms containing three threes capable of 4.2 capitation; an operation which cannot produce perpetuants with five odd numbers; it follows therefore that every 4.2 form containing three threes may be at once neglected, and that every 3.3 form containing three threes, at most, may be capitated as it stands, *i. e.* without completion.

The only terms then that need completion are the 3.3 terms containing five threes. We have

$$3^4 2^2 \cdot 3 \equiv 5 3^4 2 + 5(3^5 2^2),$$

$$3^5 2 \cdot 2 = 5 3^4 2 + 4 3^5 + 2(3^5 2^2),$$

thus

$$3^4 2^2 \cdot 3 \equiv 3^5 2 \cdot 2 - 4 3^5 + 3(3^5 2^2);$$

and

$$3^4 2 . 3 2 \equiv 2(3^5 . 2^2) + 3(3^5 2 . 2) + 2(4 3^5) + 2(3^5 2^2),$$

$$3^4 . 3 2^2 \equiv 3(3^5 . 2^2) + 3^5 2 \cdot 2 - 4 3^5,$$

$$3^3 . 3^2 2^2 \equiv 3(3^5 . 2^2) + 3(3^5 2 . 2) - 3(4 3^5) + 3^5 2^2,$$

$$3^3 2 . 3^2 2 \equiv 6(3^5 . 2^2) + 4(3^5 2 . 2) + 6(4 3^5) + 6(3^5 2^2),$$

$$3^3 2^2 . 3^2 \equiv 3^5 . 2^2 + 3(3^5 2 . 2) - 3(4 3^5) + 3(3^5 2^2).$$

Using these results in completing and then capitating, we obtain the following thirteen syzygies, in which only those terms which are essential to our purpose are retained.

No. 1.	No. 2.	No. 3.	No. 4.	No. 5.
⋮	⋮	⋮	⋮	⋮
$+ 3^3 2^5 . 3^2$	$- 3^3 2^2 . 3^2 2^3$	$- 3^3 2 . 3^2 2^4$	$- 3^3 . 3^2 2^5$	$- 3^5 2^2 . 3^2$
$\equiv$	$- 2(3^3 2^3 . 3^2 2^2)$	$+ 3^3 2^3 . 3^2 2^2$	$+ 3^3 2^4 . 3^2 2$	$+ 4 3^5 2 . 2^2$
	$- 2(3^3 2^4 . 3^2 2)$	$\equiv$	$\equiv$	$- 4^2 3^5 . 2$
	$\equiv$			$+ 3(4 3^5 2^2 . 2)$
				$+ 3^3 2^4 . 3^2 2$

No. 6.

$$\vdots$$

$$\begin{aligned}&- 3^52.3^22\\&+ 2(43^5.2^3)\\&+ 3(43^52.2^2)\\&+ 2(4^23^5.2)\\&+ 2(43^52^2.2)\\&+ 2(3^32^4.3^22)\\&+ 3^32^3.3^22^2\\&\equiv\end{aligned}$$

No. 7.

$$\vdots$$

$$\begin{aligned}&- 3^5.3^22^2\\&+ 3(43^5.2^3)\\&+ 43^52.2^2\\&- 4^23^5.2\\&+ 3^32^4.3^22\\&\equiv\end{aligned}$$

No. 8.

$$\vdots$$

$$\equiv 0$$

No. 9.

$$\vdots$$

$$\equiv 0$$

No. 10.

$$\vdots$$

$$\begin{aligned}&3^42.3^32\\&- 2(43^5.2^3)\\&+ 2(43^52.2^2)\\&- 12(4^23^5.2)\\&- 2(43^52^2.2)\\&- 3(3^32^3.3^22^2)\\&\equiv\end{aligned}$$

No. 11.

$$\vdots$$

$$\begin{aligned}&- 3^42^2.3^3\\&+ 43^5.2^3\\&+ 3(43^52.2^2)\\&- 3(4^23^5.2)\\&+ 3(43^52^2.2)\\&+ 3(3^32^4.3^22)\\&\equiv\end{aligned}$$

No. 12.

$$\vdots$$

$$\begin{aligned}&- 3^42.3^32\\&+ 6(43^5.2^3)\\&+ 4(43^52.2^2)\\&+ 6(4^23^5.2)\\&+ 6(43^52^2.2)\\&+ 3(3^32^3.3^22^2)\\&+ 6(3^32^4.3^22)\\&\equiv\end{aligned}$$

No. 13.

$$\vdots$$

$$\begin{aligned}&- 3^4.3^32^2\\&+ 3(43^5.2^3)\\&+ 3(43^52.2^2)\\&- 3(4^23^5.2)\\&+ 43^52^2.2\\&+ 3(3^32^4.3^22)\\&\equiv\end{aligned}$$

The end of this long investigation has now been nearly reached, and, by reason of the extreme peculiarity of the result presently obtained, great care has been taken to afford a means of checking the work at each stage.

The following values are now required; only the terms $\widetilde{312}$, $\widetilde{213}$, $\widetilde{124}$, being retained.

$$\begin{aligned}
3^32^5.3^2 &\equiv 0, & 43^52^2.2 &\equiv 0,\\
3^32^4.3^22 &\equiv - 2\ \widetilde{213} - 3\ \widetilde{124}, & 43^52.2^2 &\equiv + \widetilde{213} + 2\ \widetilde{124},\\
3^32^3.3^22^2 &\equiv - 3\ \widetilde{312} + \widetilde{213} + 4\ \widetilde{124}, & 43^5.2^3 &\equiv + \widetilde{312} + \widetilde{213},\\
3^32^2.3^22^3 &\equiv + 6\ \widetilde{312} + 3\ \widetilde{213}, & 3^42^2.3^3 &\equiv 0,\\
3^32.3^22^4 &\equiv - 3\ \widetilde{312} - \widetilde{213}, & 3^42.3^32 &\equiv 0,\\
3^3.3^22^5 &\equiv - \widetilde{213} - \widetilde{124}, & 3^42^2.3^32^2 &\equiv 0,\\
4^23^5.2 &\equiv + \widetilde{124}.
\end{aligned}$$

Substituting these values in the above written syzygies, we find:

$$
\begin{aligned}
&\text{No. } 1 \equiv 0,\\
&\text{No. } 2 \equiv -(\widetilde{213} + 2\ \widetilde{124}),\\
&\text{No. } 3 \equiv +2\ (\widetilde{213} + 2\ \widetilde{124}),\\
&\text{No. } 4 \equiv -(\widetilde{213} + 2\ \widetilde{124}),\\
&\text{No. } 5 \equiv -(\widetilde{213} + 2\ \widetilde{124}),\\
&\text{No. } 6 \equiv -(\widetilde{312} - 2\ \widetilde{124}) + 2(\widetilde{213} + 2\ \widetilde{124}),\\
&\text{No. } 7 \equiv +3(\widetilde{312} - 2\ \widetilde{124}) + 2(\widetilde{213} + 2\ \widetilde{124}),\\
&\text{No. } 8 \equiv 0,\\
&\text{No. } 9 \equiv 0,\\
&\text{No. } 10 \equiv +7(\widetilde{312} - 2\ \widetilde{124}) - 3(\widetilde{213} + 2\ \widetilde{124}),\\
&\text{No. } 11 \equiv +(\widetilde{312} - 2\ \widetilde{124}) - 2(\widetilde{213} + 2\ \widetilde{124}),\\
&\text{No. } 12 \equiv -3(\widetilde{312} - 2\ \widetilde{124}) + (\widetilde{213} + 2\ \widetilde{124}),\\
&\text{No. } 13 \equiv +3(\widetilde{312} - 2\ \widetilde{124}),
\end{aligned}
$$

from which it appears that, although we have

$$
\begin{aligned}
\widetilde{312} - 2\ \widetilde{124} &= \text{sextic syzygant},\\
\widetilde{213} + 2\ \widetilde{124} &= \text{sextic syzygant},
\end{aligned}
$$

yet it is impossible to express $\widetilde{312}$, $\widetilde{213}$, $\widetilde{124}$ each separately as a sextic syzygant. This means that there is a sextic perpetuant of the weight 31.

The extraordinary character of this result consists in the fact that there are altogether no fewer than sixteen syzygies involving two or all of the three forms, and that yet they are so locked together that the elimination of two of them is impossible.

Interpreting the symbols, it is found that there are three sextic perpetuants, so far of weight 31, viz.: $65^3 43^2$, $65^2 43^3 2$, $654^2 3^4$, of which the last is obviously the exemplar form, the first two being non-exemplar.

The only other possible non-exemplar forms are $6^2 5^2 3^3$, $6^2 53^4 2$, $6^2 43^5$, $6^2 3^5 2^2$, the first three of which are not, when decapitated, reducible without the aid of the form $54^2 3^4$. The last is so.

We have therefore for weight 31

$$
\begin{aligned}
&1 \text{ exemplar form } 654^2 3^4\\
&5 \text{ non-exemplar forms } \begin{cases} 6^2 5^2 3^3\\ 6^2 53^4 2\\ 6^2 43^5\\ 65^3 43^2\\ 65^2 43^3 2 \end{cases}
\end{aligned}
$$

connected by the five relations

$$6^2 5^2 3^3 \equiv - 654^2 3^4,$$
$$6^2 53^4 2 \equiv + 654^2 3^4,$$
$$6^2 43^5 \equiv - 654^2 3^4,$$
$$65^3 43^2 \equiv + 2\,(654^2 3^4),$$
$$65^2 43^3 2 \equiv - 2\,(654^2 3^4).$$

So far therefore it has been proved that the generating function for sextic perpetuants is

$$\frac{x^{31} + 0x^{32} + \ldots}{2.3.4.5.6},$$

and for sextic syzygies (*vide* Prof. Cayley's paper)

$$\frac{x^6 + x^{13} - 2x^{16} - x^{18} + x^{31} + 0x^{32} + \ldots}{2.3.4.5.6}.$$

There seems to be a very great probability that the true form of the numerator of the G. F. for perpetuants is monomial, *i. e.* simply x^{31}, but the way to show this does not seem clear; the question seems to be: 'Do any quintic forms exist of a weight superior to 25, which, not containing in their symbols the symbol $\overbrace{124}$, are not connected, by forms containing the symbol $\overbrace{124}$, with sextic syzygies?'

The possible forms are $\overbrace{\varkappa 1 \mu}$, $(\varkappa > 1\ \mu > 1)$,

$\overbrace{\varkappa \lambda 3}$, $(\varkappa > 1)$,

$\overbrace{\varkappa \lambda 2}$, $(\varkappa > 2)$,

and it is easily seen from a consideration of the lettered syzygies '*ante*' that $\mu > 2$, the form $\overbrace{\varkappa 1 \mu}$ is impossible, but as regards the other forms there seem to be difficulties, and it is likely that the simple syzygies will have to be capitated, a work of great labor, in order to settle this point.

Royal Military Academy, Woolwich, March 18, 1884.

Tables of the Symmetric Functions of the Roots, to the Degree 10, for the form

$$1 + bx + \frac{cx^2}{1.2} + \ldots = (1 - \alpha x)(1 - \beta x)(1 - \gamma x) \ldots$$

BY PROFESSOR CAYLEY.

The tables are derived from the tables (b) of my "Memoir on the Symmetric Functions of the Roots of an Equation," *Phil. Trans.*, Vol. 147 (1857), pp. 489–496. These refer in effect to the form $1 + bx + cx^2 + \ldots$, and we have consequently to change $b, c, d, \ldots$ into $\frac{b}{1}, \frac{c}{1.2}, \frac{d}{1.2.3}, \ldots$ respectively. Thus in the heading of the original table $V(b)$, we must

instead of $f,\quad be,\quad cd,\quad b^2d,\quad bc^2,\quad b^3c,\quad b^5,$

write $\frac{f}{120},\quad \frac{be}{24},\quad \frac{cd}{12},\quad \frac{b^2d}{6},\quad \frac{bc^2}{4},\quad \frac{b^3c}{2},\quad \frac{b^5}{1}$

$$= \frac{1}{120}(f,\quad 5be,\quad 10cd,\quad 20b^2d,\quad 30bc^2,\quad 60b^3c,\quad 150b^5),$$

the several columns of the original table are then multiplied by 1, 5, 10, 20, 30, 60, 120, and we thus obtain the new table with the heading

$$\frac{1}{120}(f,\quad be,\quad cd,\quad b^2d,\quad bc^2,\quad b^3c,\quad b^5).$$

In the original tables, there is a remarkable property (very easily proved) in regard to the sums of the numbers in a *column*. Thus for the table $V(b)$ these sums are $-1,\quad +2,\quad +2,\quad -3,\quad -3,\quad +4,\quad -1.$ where the sign is + or — according as the heading is the product of an even or an odd number of letters; and the numerical value depends only on the indices in the heading: these indices are

1 11 11 21 21 31 5

and they give the foregoing values

1, 2, 2, 3, 3, 4, 1,

viz., $b^3c, = 31$ gives the value $\Pi 4 \div \Pi 3 . \Pi 1, = 4$; b^2d, bc^2, each $= 21$, give the value $\Pi 3 \div \Pi 2 . \Pi 1, = 3$; and so in other cases.

In the new tables we have a property in regard to the sums of the numbers in a *line*: viz., except for the last line of each table, where there is only a single number $+1$ or -1, this sum is always $=0$. I have given in the several tables on the right-hand of each line, the sums for the positive and negative coefficients separately: thus $V(b)$, line 1, the number ± 375 means that these sums are $+375$ and -375 respectively, the sum of all the coefficients being of course $=0$. The property is an important verification as well of the original tables (b) as of the new tables derived from them, and I had the pleasure of thus ascertaining that there was not a single inaccuracy in the original tables (b).

The symbols in the left-hand outside column of each table denote symmetric functions of the roots $\alpha, \beta, \gamma, \ldots$; $5 = \Sigma\alpha^5$, $41 = \Sigma\alpha^4\beta$, etc.: and the tables are read according to the lines: thus in table $V(b)$,

$$5(=\Sigma\alpha^5) = \frac{1}{120}(5f + 25be + 50cd - 100b^2d - 150bc^2 + 300\,b^3c - 120b^5),$$

$$41(=\Sigma\alpha^4\beta) = \frac{1}{120}(5f - 5be - 50cd + 20b^2d + 90bc^2 - 60b^3c), \text{ etc.}$$

I (b)

	$= b$	
1	-1	-1

II (b)

$\div 2$

	$= c$	b^2	
2	-2	$+2$	± 2
1^2	$+1$	$+1$	

III (b)

$\div 6$

	$= d$	bc	b^3	
3	-3	$+9$	-6	± 9
21	$+3$	-3	± 3	
1^3	-1	-1		

IV (b)

$\div 24$

	$= e$	bd	c^2	b^2c	b^4	
4	-4	$+16$	$+12$	-48	$+24$	± 52
31	$+4$	-4	-12	$+12$	± 16	
2^2	$+2$	-8	$+6$	± 8		
21^2	-4	$+4$	± 4			
1^4	$+1$	$+1$				

V (b)

$\div 120$

	f	be	cd	b^2d	bc^2	b^3c	b^5	
5	-5	$+25$	$+50$	-100	-150	$+300$	-120	± 375
41	$+5$	-5	-50	$+20$	$+90$	-60	± 115	
32	$+5$	-25	$+10$	$+40$	-30	± 55		
31^2	-5	$+5$	$+20$	-20	± 25			
2^21	-5	$+15$	-10	± 15				
21^3	$+5$	-5	± 5					
1^5	-1	-1						

VI (*b*)

÷ 720	g	bf	ce	b^2e	d^2	bcd	b^3d	c^3	b^2c^2	b^4c	b^6	
6	− 6	+36	+90	−180	+60	−720	+720	−180	+1620	−2160	+720	±3246
51	+ 6	− 6	−90	+ 30	−60	+420	−120	+180	− 720	+ 360	±996	
42	+ 6	−36	+30	+ 60	−60	+240	−240	−180	+ 180	±516		
3^2	+ 3	−18	−45	+ 90	+60	−180	0	+ 90	±243			
41^2	− 6	+ 6	+30	− 30	+60	−180	+120	±216				
321	−12	+42	+60	− 90	−60	+ 60	±163					
2^3	− 2	+12	−30	0	+20	±32						
31^3	+ 6	− 6	−30	+ 30	±36							
2^21^2	+ 9	−24	+15	±15								
21^4	− 6	+ 6	±6									
1^6	+ 1	±1										

VII (*b*)

÷ 5040	h	bg	cf	b^2f	de	bce	b^3e	bd^2	c^2d	b^2cd	b^4d	bc^3	b^3c^2	b^5c	
7	− 7	+49	+147	−294	+245	+1470	+1470	−980	−1470	+8820	−5880	+4410	−17640	+17640	—
61	+ 7	− 7	−147	+ 42	−245	+ 840	− 210	+560	+1470	−3780	+ 840	−3150	+ 6300	− 2520	±
52	+ 7	−49	+ 63	+ 84	−245	+ 420	− 420	+980	− 630	−2520	+1680	+1890	− 1200	±5124	
43	+ 7	−49	−147	+294	+175	+ 210	− 630	−700	+ 210	+1260	0	− 630	±2156		
51^2	− 7	+ 7	+ 42	− 42	+245	− 315	+ 210	−560	− 420	+1680	− 840	±2184			
421	−14	+56	+ 84	−126	+ 70	− 840	+ 630	+140	+ 420	− 420	±1400				
3^21	− 7	+28	+147	−168	−175	+ 105	0	+280	− 210	±560					
32^2	− 7	+49	− 63	− 84	+ 35	+ 210	0	−140	±294						
41^3	+ 7	− 7	− 42	+ 42	−105	+ 315	− 210	±364							
321^2	+21	−63	−126	+168	+105	− 105	±294								
2^31	+ 7	−35	+ 63	0	− 35	±70									
31^4	− 7	+ 7	+ 42	− 42	±49										
2^21^3	−14	+35	− 21	±35											
21^5	+ 7	− 7	±7												
1^7	− 1	−1													

VIII (*b*)

÷ 40320

	i	bh	cg	b^2g	df	bcf	b^3f	e^2	bde	c^2e	b^2ce	b^4e	cd^2
8	− 8	+ 64	+224	−448	+448	−2688	+2688	+280	−4480	−3360	+20160	−13440	−4480
71	+ 8	− 8	−224	+ 56	−448	+1512	− 336	−280	+2520	+3360	− 8400	+ 1680	+4480
62	+ 8	− 64	+112	+112	−448	+ 672	− 672	−280	+4480	−1680	− 5040	+ 3360	+1120
53	+ 8	− 64	−224	+448	+392	+ 168	−1008	−280	+ 280	+3360	− 7560	+ 5040	−3920
4^2	+ 4	− 32	−112	+224	−224	+1344	−1344	+420	−2240	−1680	+ 3360	0	+2240
61^2	− 8	+ 8	+ 56	− 56	+448	− 504	+ 336	+280	−2520	− 840	+ 3360	− 1680	−2800
521	−16	+ 72	+112	−168	+ 56	−1354	+1008	+560	−2800	−1680	+ 9240	− 5040	+2800
431	−16	+ 72	+448	−504	+ 56	−1680	+1344	−560	+2800	0	− 840	0	− 560
42^2	− 8	+ 64	−112	−112	+448	− 672	+ 672	−280	0	+1680	− 1680	0	−1120
3^22	− 8	+ 64	+ 56	−280	−392	+ 840	0	+280	− 280	− 840	0	0	+ 560
51^3	+ 8	− 8	− 56	+ 56	−168	+ 504	− 336	−280	+1120	+ 840	− 3360	+ 1680	±4208
421^2	+24	− 80	−168	+224	−504	+1848	−1344	+280	− 280	− 840	+ 840	±3216	
3^21^2	+12	− 40	−252	+280	+168	− 168	0	+140	− 560	+ 420	±1020		
32^21	+24	−136	0	+280	+336	− 504	0	−280	+ 280	±920			
2^4	+ 2	− 16	+ 56	0	−112	0	0	+ 70	±128				
41	− 8	+ 8	+ 56	56	+168	− 504	+ 336	±568					
321^3	32	+ 88	+224	−280	−168	+ 168	±480						
2^31^2	−16	+ 72	−112	0	+ 56	±128							
31^5	+ 8	− 8	− 56	+ 56	±64								
2^21^4	+20	− 48	+ 28	±48									
21^6	8	+ 8	±8										
1^8	+ 1	+1											

Concluded infrà.

÷ 40320

	b^2d^2	bc^2d	b^3cd	b^5d	c^4	b^2c^3	b^4c^2	b^6c	b^8	
8	+13440	+40320	−107520	+53760	+5040	−80640	+201600	−161280	+40320	±377344
71	− 5600	−28560	+ 36960	− 6720	−5040	+45360	− 60480	+ 20160	±116096	
62	−10080	0	+ 26880	−13440	+5040	−20160	+ 10080	±51864		
53	+ 3360	+10080	− 10080	0	−5040	+ 5040	±28176			
4^2	+ 2240	− 6720	0	0	+2520	±12352				
61^2	+ 5600	+ 8400	− 16800	+ 6720	±25208					
521	− 1120	− 5040	+ 3360	±17208						
431	− 2240	+ 1680	±6400							
42^2	+ 1120	±3984								
	±1800									

IX (b)

÷ 362880

	j	bi	ch	b^2h	dg	bcg	b^3g	ef	bdf	c^2f	b^2cf	b^4f	be^2
9	− 9	+ 81	+324	−648	+ 756	−4536	+4536	+1134	−9072	−6804	+40824	−27216	−5670
81	+ 9	− 9	−324	+ 72	− 756	+2520	− 504	−1134	+5040	+6804	−16632	+ 3024	+3150
72	+ 9	− 81	+180	+144	− 756	+1008	−1008	−1134	+9072	−3780	− 9072	+ 6048	+5670
63	+ 9	− 81	−324	+648	+ 756	0	−1512	−1134	0	+6804	−13608	+ 9072	+5670
54	+ 9	− 81	−324	+648	− 756	+4536	−4536	+1386	−1008	− 756	−10584	+12096	−6930
71^2	− 9	+ 9	+ 72	− 72	+ 756	− 756	+ 504	+1134	−5040	−1512	+ 6048	− 3024	−3150
621	−18	+ 90	+144	−216	0	−2016	+1512	+2268	−5040	−3024	+16632	− 9072	−8820
531	−18	+ 90	+648	−720	0	−2520	+2016	− 252	−2520	−6048	+22680	−12096	+3780
4^21	− 9	+ 45	+324	−360	+ 756	−3528	+2520	−1386	+3024	+ 756	− 1512	0	+3150
52^2	− 9	+ 81	−180	−144	+ 756	−1008	+1008	− 126	−4032	0	+ 9072	− 6048	+ 630
432	−18	+162	+144	−792	0	−1008	+2520	− 252	+1008	+4536	− 7560	0	+1260
3^3	− 3	+ 27	+108	−216	− 504	+ 756	0	+ 378	+1512	−2268	0	0	−1890
61^3	+ 9	− 9	− 72	+ 72	− 252	+ 756	− 504	−1124	+2016	+1512	− 6048	+ 3024	+3150
521^2	+27	− 99	−216	+288	− 756	+2772	−2016	− 882	+7560	+4536	−22680	+12096	− 630
431^2	+27	− 99	−720	+792	− 756	+3276	−2520	+1638	−2520	0	+ 1512	0	−3150
42^21	+27	−171	+ 36	+360	− 756	+3024	−2520	+ 378	−1008	−4536	+ 4536	0	+ 630
3^221	+27	−171	−468	+864	+1512	−1764	0	− 882	−2016	+2268	0	0	+1890
32^3	+ 9	− 81	+180	+144	− 252	− 504	0	+ 126	+1008	0	0	0	− 630
51^4	− 9	+ 9	+ 72	− 72	+ 252	− 756	+ 504	+ 504	−2016	−1512	+ 6048	− 3024	±7889
421^3	−36	+108	+288	−360	+1008	−3528	+2520	− 504	+ 504	+1512	− 1512	±5940	
3^21^3	−18	+ 54	+396	−432	− 252	+ 252	0	− 252	+1008	− 756	±1710		
32^21^2	−54	+270	+180	−648	− 756	+1008	0	+ 504	− 504	±1962			
2^41	− 9	+ 63	−180	0	+ 252	0	0	− 126	±315				
41^5	+ 9	− 9	− 72	+ 72	− 252	+ 756	− 504	±887					
321^4	+45	−117	−360	+432	+ 252	− 252	±729						
2^31^3	+30	−126	+180	0	− 84	±210							
31^6	− 9	+ 9	+ 72	− 72	±81								
2^21^5	−27	+ 63	− 36	±63									
21^7	+ 9	− 9	±9										
1^9	− 1	−1											

Continued next page.

IX (*b*)

	cde	b^2de	bc^2e	b^3ce	b^5e	d^3	bcd^2	b^3d^2	c^3d	b^2c^2d
9	−22680	+68040	+102060	−272160	+136080	− 5040	+136080	−181440	+68040	−816480
81	+22680	−27720	− 71820	+ 90720	− 15120	+ 5040	− 95760	+ 60480	−68040	+453600
72	+ 5040	−50400	+ 3780	+ 60480	− 30240	+ 5040	− 65520	+110880	+37800	+ 75600
63	0	−22680	− 34020	+ 90720	− 45360	−10080	+ 90720	− 30240	−22680	−136080
54	− 2520	+32760	+ 11340	− 30240	0	+ 5040	− 35280	− 20160	+ 7560	+ 60480
71^2	−13860	+27720	+ 18900	− 37800	+ 15120	− 5040	+ 60480	− 60480	+15120	−136080
621	− 5040	+32760	+ 45360	−105840	+ 45360	+ 5040	− 35280	+ 10080	−15120	+ 60480
531	+ 2520	−12600	− 7560	+ 7560	0	+ 5040	− 20160	+ 20160	+15120	− 15120
4^21	+ 2520	−12600	+ 3780	0	0	− 5040	+ 15120	0	− 7560	±31995
52^2	+ 7560	0	− 22680	+ 15120	0	− 5040	+ 15120	− 10080	±29347	
432	−10080	+ 2520	+ 7560	0	0	+ 5040	− 5040	±24750		
3^3	+ 3780	0	0	0	0	− 1680	±6561			
61^3	+ 6300	−12600	− 18900	+ 37800	− 15120	±54369				
521^2	− 6300	+ 2520	+ 11340	− 7560	±41139					
431^2	+ 1260	+ 5040	− 3780	±13545						
42^21	+ 2520	− 2520	±11511							
3^221	− 1260	±6561								
	±1467									

÷ 40320

	b^4cd	b^6d	bc^4	b^3c^3	b^5c^2	b^7c	b^9	
9	+1360800	−544320	−204120	+1360800	−2449440	+1632960	−362880	±4912515
81	− 393120	+ 60480	+158760	− 635040	+ 635040	− 181440	±1507419	
72	− 302400	+120960	−113400	+ 226800	− 90720	±668511		
63	+ 90720	0	+ 68040	− 45360	±363159			
54	0	0	− 22680	±135855				
71^2	+ 181440	− 60480	±327303					
621	− 30240	±219726						
	±79164							

Concluded infra.

X (*b*).

÷ 3628800

		k	*bj*	*ci*	b^2i	*dh*	*bch*	b^3h	*eg*	*bdg*	c^2g	b^2cg	b^4g
1	10	−10	+100	+450	−900	+1200	−7200	+7200	+2100	−16800	−12600	+75600	−50400
2	91	+10	−10	−450	+90	−1200	+3960	−720	−2100	+9240	+12600	−30240	+5040
3	82	+10	−100	+270	+180	−1200	+1440	−1440	−2100	+16800	−7560	−15120	+10080
4	73	+10	−100	−450	+900	+1320	−360	−2160	−2100	−840	+12600	−22680	+15120
5	64	+10	−100	−450	+900	−1200	+7200	−7200	+2940	−3360	−2520	−15120	+20160
6	5^2	+5	−50	−225	+450	−600	+3600	−3600	−1050	+8400	+6300	−37800	+25200
7	81^2	−10	+10	+90	−90	+1200	−1080	+720	+2100	−9240	−2520	+10080	−5040
8	721	−20	+110	+180	−270	−120	−2880	+2160	+4200	−8400	−5040	+27720	−15120
9	631	−20	+110	+900	−990	−120	−3600	+2880	−840	−3360	−10080	+37800	−20160
10	541	−20	+110	+900	−990	+2400	−11160	+7920	−840	−5880	−10080	+45360	−25200
11	62^2	−10	+100	−270	−180	+1200	−1440	+1440	−420	−6720	.	+15120	−10080
12	532	−20	+200	+180	−1080	−120	−1080	+3600	+4200	−15960	−5040	+37800	−25200
13	4^22	−10	+100	+90	−540	+1200	−4320	+4320	−2940	+3360	+12600	−15120	.
14	43^2	−10	+100	+450	−900	−1320	+360	+2160	−420	+10920	−5040	−7560	.
15	71^3	+10	−10	−90	+90	−360	+1080	−720	−2100	+3360	+2520	−10080	+5040
16	621^2	+30	−120	−270	+360	−1080	+3960	−2880	−1260	+12600	+7560	−37800	+20160
17	531^2	+30	−120	−990	+1080	−1080	+4680	−3600	−1260	+12600	+12600	−47880	+25200
18	4^21^2	+15	−60	−495	+540	−1800	+6120	−4320	+1890	−2520	−1260	+2520	.
19	52^21	+30	−210	+90	+450	−1080	+4320	−3600	−3780	+15120	+5040	−42840	+25200
20	4321	+60	−420	−1260	+2340	+360	+7560	−8640	+2520	−12600	−10080	+17640	.
21	3^31	+10	−70	−450	+630	+1320	−1440	.	+420	−5880	+5040	.	.
22	42^3	+10	−100	+270	+180	−1200	+1440	−1440	+2100	.	−5040	+5040	.
23	3^22^2	+15	−150	+45	+630	+720	−2520	.	−1890	+2520	+2520	.	.
24	61^4	−10	+10	+90	−90	+360	−1080	+720	+840	−3360	−2520	+10080	−5040
25	521^3	−40	+130	+360	−450	+1440	−5040	+3600	+3360	−15960	−10080	+47880	−25200
26	431^3	−40	+130	+1080	−1170	+1440	−5760	+4320	−1680	+4200	.	−2520	.
27	42^21^2	−60	+330	+180	−810	+2160	−8280	+6480	−2520	+2520	+10080	−10080	.
28	3^221^2	−60	+330	+1260	−1890	−2880	+3240	.	.	+5040	−5040	.	.
29	32^31	−40	+310	−360	−630	−240	+1800	.	+1680	−2520	.	.	.
30	2^5	−2	+20	−90	.	+240	.	.	−420	.	.	.	.

Continued next page.

X (*b*).

		f^2	bef	cdf	b^2df	bc^2f	b^3cf	b^5f	ce^2	b^2e^2	d^2e
1	10	+1260	−25200	−50400	+151200	+226800	−604800	+302400	− 31500	+ 94500	−42000
2	91	−1260	+13860	+50400	− 60480	−158760	+196560	− 30240	+ 31500	− 37800	+42000
3	82	−1260	+25200	+10080	−110880	+ 15120	+120960	− 60480	+ 6300	− 69300	+42000
4	73	−1260	+25200	− 2520	− 45360	− 68040	+181440	− 90720	+ 31500	− 94500	−46200
5	64	−1260	− 5040	+50400	− 30240	−136080	+241920	−120960	− 44100	+ 56700	− 8400
6	5^2	+2520	−18900	−37800	+ 50400	+ 75600	− 75600	.	+ 15750	+ 31500	+21000
7	81^2	+1260	−13860	−30240	+ 60480	+ 37800	− 75600	+ 30240	− 18900	+ 37800	−42000
8	721	+2520	−39060	− 7560	+ 65520	+ 90720	−211680	+ 90720	− 37800	+107100	+ 4200
9	631	+2520	− 8820	−47880	+ 75600	+136080	−287280	+120960	+ 12600	− 18900	+54600
10	541	−3780	+28980	+25200	− 60480	− 7560	+ 15120	.	+ 12600	− 50400	−33600
11	62^2	+1260	−10080	−10080	+ 50400	+ 30240	−120960	+ 60480	+ 31500	− 6300	−16800
12	532	−3780	+12600	+42840	− 20160	− 98280	+ 75600	.	− 37800	+ 6300	+ 4200
13	4^22	+1260	+ 5040	−30240	+ 10080	+ 15120	.	.	+ 6300	− 18900	+ 8400
14	43^2	+1260	−10080	+ 2520	− 15120	+ 22680	.	.	+ 6300	+ 18900	− 4200
15	71^3	−1260	+13860	+12600	− 25200	− 37800	+ 75600	− 30240	+ 18900	− 37800	+12600
16	621^2	−3780	+22680	+37800	− 95760	−128520	+287280	−120960	− 18900	+ 6300	−12600
17	531^2	+2520	−15120	−10080	+ 15120	+ 15120	− 15120	.	+ 6300	+ 12600	−12600
18	4^21^2	+1260	−10080	+ 2520	+ 10080	− 7560	.	.	− 9450	+ 18900	+12600
19	52^21	+2520	− 1260	−32760	+ 10080	+ 68040	− 45360	.	+ 6300	− 6300	+12600
20	4321	−1260	.	+12600	+ 20160	− 22680	.	.	+ 12600	− 18900	−12600
21	3^31	−1260	+ 6300	− 2520	.	.	.	.	− 6300	.	+ 4200
22	42^3	−1260	.	+10080	− 10080	.	.	.	− 6300	+ 6300	±25420
23	3^22^2	+1260	− 1260	− 5040	.	.	.	.	+ 3150	±10860	
24	61^4	+1260	− 6300	−12600	+ 25200	+ 37800	− 75600	+ 30240	±106600		
25	521^3	−1260	+ 1260	+12600	− 5040	− 22680	+ 15120	±35750			
26	431^3	−1260	+ 6300	− 2520	− 10080	+ 7560	±25030				
27	42^21^2	+1260	− 1260	− 5040	+ 5040	±28050					
28	3^221^2	+1260	− 3780	+ 2520	±13650						
29	32^31	−1260	+ 1260	±5050							
30	2^5	+ 252	±512								

Continued next page.

X (*b*).

		$bcde$	b^3de	c^3e	b^2c^2e	b^4ce	b^6e	bd^3	c^2d^2	b^2cd^2	b^4d^2
1	10	+756000	−1008000	+189000	−2268000	+3780000	−1512000	+168000	+378000	−3024000	+2520000
2	91	−529200	+ 327600	−189000	+1247400	−1058400	+ 151200	−117600	−378000	+1663200	− 705600
3	82	−352800	+ 604800	+113400	+ 151200	− 756000	+ 302400	−168000	+ 25200	+1411200	−1310400
4	73	+ 37800	+ 302400	−189000	+ 680400	−1134000	+ 453600	+184800	+151200	−1209600	+ 302400
5	64	+151200	− 201600	+189000	− 453600	+ 302400	·	+ 33600	−226800	·	+ 201600
6	5^2	− 63000	− 126000	− 94500	+ 189000	·	·	− 84000	+126000	+ 252000	·
7	81^2	+327600	− 327600	+ 37800	− 340200	+ 453600	− 151200	+117600	+176400	−1058400	+ 705600
8	721	+264600	− 403200	+ 75600	− 869400	+1285200	− 453600	− 67200	−176400	+ 453600	− 100800
9	631	−189000	+ 126000	·	+ 113400	− 75600	·	−117600	+ 75600	+ 302400	− 201600
10	541	·	+ 126000	·	− 37800	·	·	+ 84000	− 25200	− 151200	·
11	62^2	−100800	·	− 75600	+ 302400	− 151200	·	+ 67200	+ 50400	− 201600	+ 100800
12	532	+ 63000	− 25200	+ 75600	− 75600	·	·	− 16800	− 50400	+ 50400	±376520
13	4^22	+ 50400	·	− 37800	·	·	·	− 33600	+ 25200	±143470	
14	43^2	− 37800	·	·	·	·	·	+ 16800	±82450		
15	71^3	−151200	+ 151200	− 37800	+ 340200	− 453600	+ 151200	±788260			
16	621^2	+ 88200	− 25200	+ 37800	− 151200	+ 75600	±600330				
17	531^2	+ 50400	− 50400	− 37800	+ 37800	±196050					
18	4^21^2	− 37800	·	+ 18900	±75345						
19	52^21	− 37800	+ 25200	±174990							
20	4321	+ 12600	±88440								
		±17920									

Continued next page.

X (*b*)

		bc^3d	b^3c^2d	b^5cd	b^7d	c^5	b^2c^4	b^4c^3	b^6c^2	b^8c	
1	10	−3024000	+15120000	−18144000	+6048000	−226800	+5670000	−22680000	+31752000	−18144000	
2	91	+2343600	− 6955200	+ 4536000	− 604800	+226800	−3628800	+ 9072000	− 7257600	+ 1814400	±2
3	82	− 604800	− 1814400	+ 3628800	−1209600	−226800	+2041200	− 2721600	+ 907200	±9433840	
4	73	− 151200	+ 1814400	− 907200	.	+226800	− 907200	+ 453600	±4875490		
5	64	+ 604800	− 604800	.	.	−226800	+ 226800	±2089630			
6	5^2	− 378000	.	.	.	+113400	±921125				
7	81^2	− 529200	+ 2116800	− 2116800	+ 604800	±4721980					
8	721	+ 378000	− 756000	+ 302400	±3154550						
9	631	− 226800	+ 151200	±1212650							
10	541	+ 75600	±424190								
		±712540									

		k	bf	ci	b^2i	dh	bch	b^3h	eg	bdg	c^2g	b^2cg	b^4
31	51^5	+ 10	− 10	− 90	+ 90	− 360	+1080	− 720	−840	+3360	+2520	−10080	
32	421^4	+ 50	−140	−450	+ 540	−1800	+6126	−4320	+840	− 840	−2520	+ 2520	±1
33	3^21^4	+ 25	− 70	−585	+ 630	+ 360	− 360	.	+420	−1680	+1260	±2695	
34	32^21^3	+100	−460	−540	+1260	+1440	−1800	.	−840	+ 840	±3640		
35	241^2	+ 25	−160	+405	.	− 480	.	.	+210	±640			
36	41^6	− 10	+ 10	+ 90	− 90	+ 360	−1080	+ 720	±1180				
37	321^5	− 60	+150	+540	− 630	− 360	+ 360	±1050					
38	2^31^4	− 50	+200	−270	.	+ 120	±320						
39	31^7	+ 10	− 10	− 90	+ 90	±100							
40	2^21^6	+ 35	− 80	+ 45	±80								
41	21^8	− 10	+ 10	±10									
42	1^{10}	+ 1	+1										

By Professor Cayley.

In the theory of Seminvariants we are concerned with the non-unitary partitions of a number, that is, the number of ways of making up the number with the parts 2, 3, 4, . . . ; or what is the same writing, writing $\mathbf{2} = 1 - x^2$, $\mathbf{3} = 1 - x^3$, etc. with the Generating Functions having in their denominators the factors $\mathbf{2}$, $\mathbf{3}$, $\mathbf{4}$, etc. In the present short paper, I give the developments up to x^{100} of the functions $1 \div \mathbf{2}$, $\mathbf{2}.\mathbf{3}$, $\mathbf{2}.\mathbf{3}.\mathbf{4}$, $\mathbf{2}.\mathbf{3}.\mathbf{4}.\mathbf{5}$, $\mathbf{2}.\mathbf{3}.\mathbf{4}.\mathbf{5}.\mathbf{6}$, respectively: and also of the function $x^6 + x^{13} - 2x^{16} - x^{18} + x^{31} \div \mathbf{2}.\mathbf{3}.\mathbf{4}.\mathbf{5}.\mathbf{6}$, which function is (there is strong reason to believe) the G. F. for the number of sextic syzygies of a given weight: the same function without the term x^{31} occurs p. 115 in Professor Sylvester's paper "On Subinvariants, *i. e.* Seminvariants to Binary Quantics of an Unlimited Order," *A. M. J.* t. V (1882), pp. 79–136.

In the tables X is written to denote $x^6 + x^{13} - 2x^{16} - x^{18} + x^{31}$.

Ind. x	1÷ 2.3	2.3.4	2.3.4.5	2.3.4.5.6	X÷ 2.3.4.5.6	Ind. x	1÷ 2.3	2.3.4	2.3.4.5	2.3.4.5.6
0	1	1	1	1		50	9	65	258	750
1	0	0	0	0		51	9	61	268	783
2	1	1	1	1		52	9	70	286	854
3	1	1	1	1		53	9	65	297	891
4	1	2	2	2		54	10	75	316	972
5	1	1	2	2		55	9	70	328	1010
6	2	3	3	4	1	56	10	80	348	1098
7	1	2	3	3	0	57	10	75	361	1144
8	2	4	5	6	1	58	10	85	382	1236
9	2	3	5	6	1	59	10	80	396	1287
10	2	5	7	9	2	60	11	91	419	1391
11	2	4	7	9	2	61	10	85	433	1443
12	3	7	10	14	4	62	11	96	457	1555
13	2	5	10	13	4	63	11	91	473	1617
14	3	8	13	19	6	64	11	102	598	1734
15	3	7	14	20	7	65	11	96	515	1802
16	3	10	17	26	8	66	12	108	541	1932
17	3	8	18	27	11	67	11	102	559	2002
18	4	12	22	36	13	68	12	114	587	2142
19	3	10	23	36	15	69	12	108	606	2223
20	4	14	28	47	17	70	12	120	635	2369
21	4	12	29	49	21	71	12	114	655	2457
22	4	16	34	60	22	72	13	127	686	2618
23	4	14	36	63	28	73	12	120	707	2709
24	5	19	42	78	29	74	13	133	739	2881
25	4	16	44	80	35	75	13	127	762	2985
26	5	21	50	97	36	76	13	140	795	3164
27	5	19	53	102	44	77	13	133	819	3276
28	5	24	60	120	43	78	14	147	854	3472
29	5	21	63	126	54	79	13	140	879	3588
30	6	27	71	149	53	80	14	154	916	3797
31	5	24	74	154	64	81	14	147	942	3927
32	6	30	83	180	62	82	14	161	980	4144
33	6	27	87	189	78	83	14	154	1008	4284
34	6	33	96	216	72	84	15	169	1048	4520
35	6	30	101	227	89	85	14	161	1077	4665
36	7	37	111	260	84	86	15	176	1118	4915
37	6	33	116	270	102	87	15	169	1149	5076
38	7	40	127	307	96	88	15	184	1192	5336
39	7	37	133	322	117	89	15	176	1224	5508
40	7	44	145	361	108	90	16	192	1269	[illegible]
41	7	40	151	378	133	91	15	184	1302	
42	8	48	164	424	123	92	16	200	1349	
43	7	44	171	441	149	93	16			
44	8	52	185	492	137	94	16			
45	8	48	193	515	167	95	16			
46	8	56	207	568	152	96	17			
47	8	52	216	594	186	97	16			
48	9	61	232	656	169	98	17			
49	8	56	241	682	205	99	17			

Seminvariant Tables.

By Professor Cayley.

The present tables are not, I think, superseded by the tables A, pp. 149–163, contained in Capt. MacMahon's paper, "Seminvariants and Symmetric Functions," *A. M. J.* t. VI (1883), pp. 131–163. His order of the terms, though a very ingenious one, and giving rise to a most remarkable symmetry in the form of the tables, seems to me too artificial—and I cannot satisfy myself that it ought to be adopted in preference to the more simple one which I use: I attach also considerable importance to the employment of the simple letters b, c, d, e, etc. in place of the suffixed ones a_1, a_2, a_3, a_4, etc. There is, moreover, the question of the identification of the seminvariants with their expressions as non-unitary symmetric functions of the roots of the equation $1 + bx + \frac{cx^2}{1.2} + \text{etc.} = 0$, which requires to be considered.

As to the form in which the tables present themselves, I remark that every seminvariant is a rational and integral function of the fundamental seminvariants

$$\mathrm{c} = (1,\ b,\ c \between -b,\ 1)^2$$
$$\mathrm{d} = (1,\ b,\ c,\ d \between -b,\ 1)^3$$
$$\mathrm{e} = (1,\ b,\ c,\ d,\ e \between -b,\ 1)^4,\ \text{etc.},$$

viz. up to g these are

c =	d =	e =	f =	g =	ce =	d² =	c³ =
$c+1$	$d\ +1$	$e\ \ +1$	$f\ \ +\ 1$	$g\ \ +\ 1$			
b^2-1	$bc-3$	$bd\ \ -4$	$be\ \ -\ 5$	$bf\ \ -\ 6$			
	b^3+2	c^2	cd	ce	$+1$		
		$b^2c\ \ +6$	b^2d+10	d^2		$+\ 1$	
		$b^4\ \ -3$	bc^2	b^2e+15	-1		
			b^3c-10	bcd	-4	$-\ 6$	
			$b^5\ \ +\ 4$	c^3			$+1$
				b^3d-20	$+4$	$+\ 4$	
				b^2c^2	$+6$	$+\ 9$	-3
				b^4c+15	-9	-12	$+3$
				$b^6\ \ -\ 5$	$+3$	$+\ 4$	-1

and if to the value of g, which is of the weight 6, we join those of the products ce, d^2, c^3 of this same weight 6, we have as just written down, what is in effect a table of the assyzygetic seminvariants of the weight 6 and which I call the Crude Table. But we do not, in this way, obtain immediately the seminvariants of the lowest degrees: in fact, the only seminvariant containing g is given by the first column as a function $g \ldots - 5b^6$ of the degree 6, whereas, there is the seminvariant $g - 6bf + 15ce - 10d^2$ of the degree 2: to obtain this, we have to form a linear combination of the columns: the proper combination is $g + 15ce - 10d^2$, giving rise to the column g of the table $(g = 6)$. And similarly each other column of the same table is a linear combination of columns of the Crude Table: and so in every case. The process would be a very laborious one, and the tables were not, in fact, thus calculated; but we see very clearly in this manner, the origin and meaning of the tables.

The mere inspection of the tables gives rise to several remarks. We see that each column begins with a non-unitary term (term without the letter b), and that it ends with a power-ending term (product wherein the last letter enters as a power)—thus, weight 8

initial terms are	i,	cg,	df,	e^2,	c^2e,	cd^2,	c^4
finals are	e^2,	cd^2,	b^2d^2,	c^4,	b^2c^3,	b^4c^2,	b^8,

and it will be observed further that in this case the initial terms are all the non-unitary terms taken in order, and the corresponding final terms are all the power-ending terms taken also in order. The arrangement of the columns *inter se* is of course arbitrary, and they are, in fact, arranged so that the initial terms are the non-unitary terms taken in order—and this being so, then for each weight up to the weight 9, the final terms are the power-ending terms taken in order: but for each of the weights 10 and 11, there is a single deviation from this order; and for the weight 12, there are a great many deviations from the order.

The initial terms being in order, the broken line which bounds the tops of the columns forms a series of continually descending steps; and when the final terms are also in order, the case is the same with the broken line bounding the bottom of the columns: any deviation in the order of the final terms is shown by an ascending step or steps in the broken line bounding the bottom of the columns: thus in the table $(k = 10)$ the column cdf is longer than the next following one ce^2, and there is an ascending step accordingly.

It is to be remarked that any ascending step gives rise to a certain indeterminateness in a preceding column or columns: thus in the case just referred to, the column cdf might be replaced by any linear combination of itself with the column ce^2, it would still have the original initial and final terms cdf and b^4d^2 respectively. It would be possible to fix a standard form; we might, for instance, say that the column cdf should be that combination $cdf + 2ce^2$, which does not contain the leading term ce^2 of the ce^2-column: but I have not thought it worth while to attend to this.

It will be observed that except in the case of an ascending step or steps, each column is completely determinate: we cannot with any column combine a preceding column, for this would give it a higher initial term: nor can we with it combine a succeeding column, for this would give it a lower final term. The numbers in the column may be taken to be without any common divisor, for any such divisor, if it existed, might be divided out: and the leading coefficient of the column may be taken to be positive.

I add certain subsidiary tables to enable the expression of any column in terms of the non-unitary symmetric functions of the roots of the equation $0 = 1 + bx + \frac{cx^2}{1.2} + \text{etc.}$ These consist of left-hand tables and right-hand tables: the left-hand table for any weight is the original table for that weight, writing therein $b = 0$ and converting the columns into lines: thus weight $= 6$, we have

$$\begin{aligned}
\text{col. } g &= g + 15ce - 10d^2,\\
\text{col. } ce &= \quad ce - d^2 - c^3,\\
\text{col. } d^2 &= \quad d^2 + 4c^3,\\
\text{col. } c^3 &= \quad c^3,
\end{aligned}$$

viz. these are the values of the original columns writing therein $b = 0$.

The right-hand table is the table for the same weight taken from my paper "Tables of the Symmetric Functions of the Roots, to the Degree 10, for the form $1 + bx + \frac{cx^2}{1.2} + \ldots = (1 - \alpha x)(1 - \beta x)(1 - \gamma x)\ldots$" *ante* p. 47, writing therein $b = 0$, and giving only those lines of the table which relate to the non-unitary symmetric functions. Thus weight 6, we have

$$\begin{aligned}
6\ (= \Sigma\alpha^6) &= \tfrac{1}{720}(-6g + 90ce + 60d^2 - 180c^3),\\
42 (= \Sigma\alpha^4\beta^2) &= \tfrac{1}{720}(\ \ 6g + 30ce - 60d^2 - 180c^3),\\
3^2\ (= \Sigma\alpha^3\beta^3) &= \tfrac{1}{720}(\ \ 3g - 45ce + 60d^2 + 90c^3)\\
2^3\ (= \Sigma\alpha^2\beta^2\gamma^2) &= \tfrac{1}{720}(-2g - 30ce + 20d^2),
\end{aligned}$$

we thus have on the one side col. g, col. ce, col. d^2 and col. c^3, and on the other side the symmetric functions 6, 42, 3^2 and 2^3, each of them expressed as a linear function of g, ce, d^2 and c^3. It follows that each of the columns can be expressed as a linear function of the symmetric functions: and conversely each of the symmetric functions as a linear function of the columns: and this being done, each of the columns is to be regarded as having its complete value as a function of b and the other letters: for the columns *quâ* seminvariants *are* linear functions of the symmetric functions: and assuming them to be so, they can only be the linear functions determined by the foregoing process of writing $b=0$.

The left-hand tables are carried up to $m=12$; the right-hand only up to $k=10$, the limit of the tables in the memoir last referred to.

SEMINVARIANT TABLES UP TO ($m=12$).

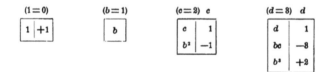

(1=0)	
1	+1

($b=1$)
b

($c=2$)	c
c	1
b^2	−1

($d=3$)	d
d	1
bc	−3
b^3	+2

($e=4$)	e	c^2
e	1	
bd	−4	
c^2	+3	1
b^2c		−2
b^4		+1

($f=5$)	f	cd
f	1	
be	−5	
cd	+2	+1
b^2d	+8	−1
bc^2	−6	−3
b^3c		+5
b^5		−2

($g=6$)	g	ce	d^2	c^3
g	+ 1			
bf	− 6			
ce	+15	+1		
d^2	−10	−1	+1	
b^2e		−1		
bcd		+2	−6	
c^3		−1	+4	+1
b^3d			+4	
b^2c^2			−3	−3
b^4c				+3
b^6				−1

$(h=7)$	h	cf	de	c^2d
h	+1			
bg	−7			
cf	+9	+1		
de	−5	−1	+1	
b^2f	+12	−1		
bce	−30	−2	−8	
bd^2	+20	+4	−4	
c^2d		−1	+8	+1
b^3e		+3	+2	
b^2cd		−6	+12	−2
bc^3		+3	−9	−3
b^4d			−8	+1
b^3c^2			+6	+8
b^5c				−7
b^7				+2

$(i=8)$	i	cg	df	e^2	c^2e	cd^2	c^4
i	+1						
bh	−8						
cg	+28	+1					
df	−56	−8	+3				
e^2	+35	+2	−2	+1			
b^2g		−1					
bcf		+8	−9				
bde		−1	+1	−8			
c^2e		−3	+18	+6	+1		
cd^2		+2	−12		−1	+1	
b^3f			+6				
b^2ce			−15		−2		
b^2d^2			+10	+16	+1	−1	
bc^2d				−24	+2	−6	
c^4				+9	−1	+4	+1
b^4e					+1		
b^3cd					−2	+10	
b^2c^3					+1	−7	−4
b^5d						−4	
b^4c^2						+3	+6
b^6c							−4
b^8							+1

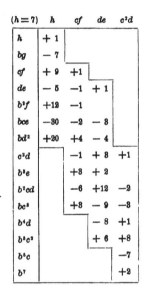

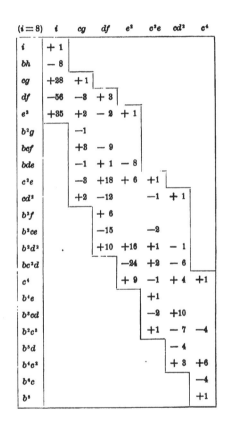

$(j=9)$	j	ch	dg	ef	c^2f	cde	d^3	c^3d
j	+ 1							
bi	− 9							
ch	+ 20	+ 2						
dg	− 28	− 7	+ 1					
ef	+ 14	+ 5	− 1	+ 1				
b^2h	+ 16	− 2						
bcg	− 56	+ 7	− 8					
bdf	+112	+22	− 2	− 4				
be^2	− 70	−25	+ 5	− 5				
c^2f		−27	+ 9	− 9	+ 2			
cde		+45	−17	+32	− 5	+1		
d^3		−20	+ 8	−18	+ 3	−1	+ 1	
b^3g			+ 2					
b^2cf			− 6	+24	− 4			
b^2de			+ 2	− 2	+ 5	−1		
bc^2e			+ 6	−51	+ 5	−8		
bcd^2			− 4	+34	− 7	+5	− 9	
c^3d					+ 1	−1	+ 4	+ 1
b^4f				−12	+ 2			
b^3ce				+30	− 5	+5		
b^3d^2				−20	− 2	−2	+ 6	
b^2c^2d					+ 8	−6	+15	− 3
bc^4					− 8	+8	−12	− 3
b^5e						−2		
b^4cd						+4	−24	+ 8
b^3c^3						−2	+17	+11
b^6d							+ 8	− 1
b^5c^2							− 6	−15
b^7c								+ 9
b^9								− 2

$(k=10)$	k	ci	dh	eg	f^2	c^2g	cdf	ce^2	d^2e	c^3e	c^2d^2	c^5
k	+ 1											
bj	− 10											
ci	+ 45	+1										
dh	−120	−4	+ 4									
eg	+210	+8	− 8	+ 16								
f^2	−126	−5	+ 5	+ 15	+ 1							
b^2i		−1										
bch		+4	−12									
bdg		−4	+ 4	− 64								
bef		+2	− 2	+ 54	−10							
c^2g		−4	+32	+ 48		+1						
cdf		+8	−64	− 60	+ 4	−3	+ 3					
ce^2		−5	+40		+16	+2	− 2	+ 1				
d^2e				+ 20	−12			− 1	+ 1			
b^3h			+ 8									
b^2cg			−28			−2						
b^2df			+56	+144	+16	+8	− 3					
b^2e^2			−35	−135	+ 9	−2	+ 2	− 1				
bc^2f				−108	−12	+3	− 9					
$bcde$				+180	−76	−1	+ 1	− 2	− 6			
bd^3				− 80	+48			+ 4	− 4			
					+48	−3	+18	+ 2	+ 4	+ 1		
					−32	+2	−12	− 3	+ 3	− 1	+ 1	
						+1						
						−3	+15					
						+1	− 1	+ 4	+ 4			
						+3	−33	− 3	− 3	− 3		
						−2	+22	− 3	+24	+ 2	− 2	
								+10	−34	+ 2	− 6	
								− 3	+12	− 1	+ 4	+ 1
							− 6					
							+15			+ 3		
							−10		−16	− 1	+ 1	
									+24	− 4	+16	
									− 9	+ 2	−11	
										− 1		− 5
										+ 2	−14	
										− 1	+10	+10

$(l=11)$	l	cj	di	eh	fg	c^2h	cdg	cef	d^2f	de^2	c^3f	c^2de	cd^3	c^4d
l	+ 1													
bk	− 11													
cj	+ 35	+ 2												
di	− 75	− 9	+ 1											
eh	+ 90	+14	− 2	+ 1										
fg	− 42	− 7	+ 1	− 1	+ 1									
b^2j	+ 20	− 2												
bci	− 90	+ 9	− 8											
bdh	+240	+16		− 4										
beg	−420	−63	+ 9	− 2	− 5									
bf^2	+252	+42	− 6	+ 6	− 6									
c^2h		−30	+10	+ 3	− 16	+ 2								
cdg		+70	−26	− 2	+ 58	− 7	+ 1							
cef		−21	+ 7	− 6	+ 5	+ 5	− 3	+ 1						
d^2f		−56	+24	+10	−100		+ 6	− 8	+ 1					
de^2		+35	−15	− 5	+ 60		− 4	+ 2	− 1	+ 1				
b^3i			+ 2											
b^2ch			− 8		+ 32	− 4								
b^2dg			+ 8	+20	− 48	+ 7	− 1							
b^2ef			− 4	−18	+ 40	− 5	+ 3	− 1						
bc^2g			+ 8	−15	− 62	+ 7	− 8							
$bcdf$			−16	−24	+232	+22	−30	+14	− 6					
bce^2			+10	+45	−205	−25	+27	−11	+ 8	− 8				
bd^2e				−10	+ 20		+ 2	− 1	+ 8	− 8				
c^3f				+27	− 54	−27	+27	− 9	+ 4	−12	+ 2			
c^2de				−45	+ 90	+45	−45	+14	− 6	+36	− 5	+ 1		
cd^3				+20	− 40	−20	+20	− 6	+ 2	−18	+ 8	− 1	+ 1	
b^4h					− 16	+ 2								

Continued next page.

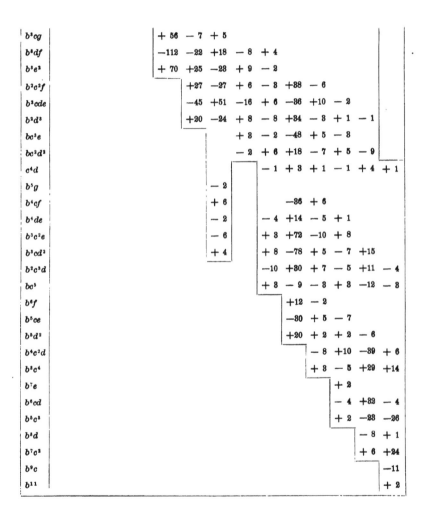

b^3cg	+56	−7	+5							
b^3df	−112	−22	+18	−8	+4					
b^3e^2	+70	+25	−23	+9	−2					
b^2c^2f		+27	−27	+6	−3	+38	−6			
b^2cde		−45	+51	−16	+6	−36	+10	−2		
b^2d^3		+20	−24	+8	−8	+34	−3	+1	−1	
bc^3e				+3	−2	−48	+5	−3		
bc^2d^2				−2	+6	+18	−7	+5	−9	
c^4d					−1	+3	+1	−1	+4	+1
b^5g			−2							
b^4cf			+6			−36	+6			
b^4de			−2		−4	+14	−5	+1		
b^3c^2e			−6		+3	+72	−10	+8		
b^3cd^2			+4		+8	−78	+5	−7	+15	
b^2c^3d					−10	+30	+7	−5	+11	−4
bc^5					+3	−9	−3	+3	−12	−3
b^6f						+12	−2			
b^5ce						−30	+5	−7		
b^5d^2						+20	+2	+2	−6	
b^4c^2d							−8	+10	−39	+6
b^3c^4							+3	−5	+29	+14
b^7e								+2		
b^6cd								−4	+32	−4
b^5c^3								+2	−23	−26
b^8d									−8	+1
b^7c^2									+6	+24
b^9c										−11
b^{11}										+2

$(m=12)$	m	ck	dj	ei	fh	g^2	c^2i	cdh	ceg	cf^2	d^2g	def	e^3	c^3g	c^2df
	+ 1														
	− 12														
	+ 66	+ 3													
	−220	−15	+ 15												
	+495	+40	− 40	+ 1											
	−792	−70	+ 70	− 4	+ 25										
	+462	+42	− 42	+ 3	− 24	+ 1									
		− 3													
		+15	− 45												
		−25	+ 25	− 4											
		+30	− 30	+12	−125										
		−14	+ 18	− 3	+113	− 12									
		−15	+150	+ 3			+ 1								
		+40	−400	− 3	+ 50		− 4	+ 4							
		−70	+700	−22	+630	− 70	+ 3	− 3	+ 1						
		+42	−420	+24	−675	+100	− 5	+ 5	− 1	+ 2					
				+24	−570	+ 30			− 1	+ 5	+ 1				
				−36	+925	−200			+ 2	−19	− 3	+ 18			
				+15	−400	+100			− 1	+12	+ 2	− 17	+ 1		
	. . . . .														
			+ 30												
			−135				− 2								
			+360		+200		+ 4	− 4							
			−630		−525	+100	− 3	+ 3	− 1						
			+378		+336	− 64	+ 5	− 5	+ 1	− 2					
					−150		+ 4	− 12							
					+350	−200	− 4	+ 4	+ 2	−30	− 6				
					−105	+ 20	+ 2	− 2	− 2	+37	+ 9	− 54			
					−280	+320			− 2	+46	+ 6	− 72			
					+175	−200			+ 2	−49	− 9	+114	− 12		
						+100	− 4	+ 32	− 1	+20	+ 4			+ 1	
						−200	+ 3	− 64	+ 2	−49	− 9	+ 54		− 3	+ 3
						+125	− 5	+ 40	+ 1	−32	−12	+162	− 18		
									− 3	+91	+27	−342	+ 54	+ 4	
									+ 1	−32	−10	+135	− 27	− 2	− 2
	. . . . .	. . .	. . .	. . .	. . .	. . .									
										+20	+ 4				
										−18	− 6	+ 36			

− 8	+120	−24		+216		+ 6							
+ 5	− 75	+45	+15	−360	+ 54	+ 2		− 2					
		−10	− 6	+ 66	− 6	− 6		+ 2	− 1				
		+27	+ 3	−162		+ 3	− 9						
		−45	−21	−486	−180	− 8	−15	+ 4	− 6				
		+20	+12	−252	+108	+ 7	+24	− 4	+ 8	−12			
			+ 6	− 81	+ 81	+ 1	+30	− 2	+ 4		+ 1		
			− 4	+ 54	− 54	− 2	−28	+ 2	− 5	+ 8	− 1	+ 1	
	− 8												
	+ 28					+ 3							
	− 56			−144		− 3	+ 3						
	+ 35			+135	− 27			+ 1					
				+108		− 6	+24						
				−180	+108	+ 9	+30	− 4	+10				
				+ 80	− 64	+ 1	−16		− 4	+ 8			
					− 54	+ 2	−75	+ 2	− 7		− 4		
					+ 36	−12	+ 6	+ 4	− 9	+30	+ 3	− 3	
						+ 8	+48	− 4	+14	−48	+ 2	− 6	
						− 2	+ 2	+ 1		+16	− 1	+ 4	+ 1
						− 1	−16		− 4				
						+ 3	−21						
						− 1	−15		− 4				
						− 3	+60		+ 3		+ 6		
						+ 2			+ 8	−48	− 3	+ 3	
							−40		−10	+68	− 6	+22	
							+12		+ 3	−24	+ 3	−15	− 6
							+ 6						
							−15				− 4		
							+10			+16	−23	− 1	
										−24	+30	−30	
										+ 9	− 3	+21	+15
											+ 1		
											+ 2	+14	−20
											− 3	− 9	
												− 4	
												+ 3	+15
													− 6
													+ 1

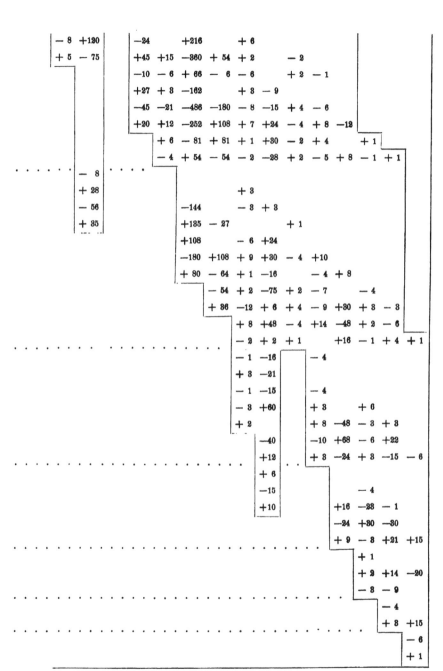

SUBSIDIARY TABLES $b = 0$.

Left-hand up to ($m = 12$).

Col.	c
c	+1

Col.	d
d	+1

Col.	e	c^2
e	+1	+3
c^2		+1

Col.	f	cd
f	+1	+2
cd		+1

Col.	g	ce	d^2	c^3
g	+1	+15	−10	
ce		+1	−1	−1
d^2			+1	+4
c^3				+1

Col.	h	cf	de	c^2d
h	+1	+9	−5	
cf		+1	−1	−1
de			+1	+3
c^2d				+1

Col.	i	cg	df	e^2	c^2e	cd^2	c^4
i	+1	+28	−56	+35			
cg		+1	−3	+2	−3	+2	
df			+3	−2	+18	−12	
e^2				+1	+6	·	+9
c^2e					+1	−1	−1
cd^2						+1	+4
c^4							+1

Col.	j	ch	dg	ef	c^2f	cde	d^3	c^3d
j	+1	+20	−28	+14				
ch		+2	−7	+5	−27	+45	−20	
dg			+1	−1	+9	−17	+3	
ef				+1	−9	+32	−18	
c^2f					+2	−5	+3	+1
cde						+1	−1	−1
d^3							+1	+4
c^3d								+1

Right-hand up to ($k = 10$).

÷2

	c
2	−2

÷6

	d
3	−3

÷24

	e	c^2
4	−4	+12
2^2	+2	+6

÷120

	f	cd
5	−5	+50
32	+5	+10

÷720

	g	ce	d^2	c^3
6	−6	+90	+60	−180
42	+6	+30	−60	−180
3^2	+3	−45	+60	+90
2^3	−2	−30	+20	·

÷5040

	h	cf	de	c^2d
7	−7	+147	+245	−1470
52	+7	+63	−245	−630
43	+7	−147	+175	+210
32^2	−7	−63	+35	·

÷40320

	i	cg	df	e^2	c^2e	cd^2	c^4
8	−8	+224	+448	+280	−3360	−4480	+5040
62	+8	+112	−448	−280	−1680	+1120	+5040
53	+8	−224	+392	−280	+3360	−3920	−5040
4^2	+4	−112	−224	+420	−1680	+2240	+2520
42^2	−8	−112	+448	−280	+1680	−1120	·
3^22	−8	+56	−392	+280	−840	+560	
2^4	+2	+56	−112	+70			

÷362880

	j	ch	dg	ef	c^2f	cde	d^3	c^3d
9	−9	+324	+756	+1134	−6804	−22680	−5040	+68040
72	+9	+180	−756	−1134	−3780	+5040	+5040	+37800
63	+9	−324	+756	−1134	+6804	·	−10080	−22680
54	+9	−324	−756	+1386	−756	−2520	+5040	+7560
52^2	−9	−180	+756	−126	·	+7560	−5040	·
432	−18	+144	·	−252	+4536	−10080	+5040	·
3^3	−3	+108	−504	+378	−2268	+3780	−1680	·
32^3	+9	+180	−252	+126	·	·	·	·

Left-hand. Col.	k	ci	dh	eg	f^2	c^2g	cdf	ce^2	d^2e	c^3e	c^2d^2	c^5
k	+1	+45	−120	+210	−126							
ci		+1	−4	+8	−5	−4	+8	−5				
dh			+4	−8	+5	+32	−64	+40				
eg				+16	−15	+48	−60	·	+20			
f^2					+1	·	+4	+16	−12	+48	−32	
c^2g						+1	−3	+2	·	−3	+2	
cdf							+3	−2	·	+18	−12	
ce^2								+1	−1	+2	−3	−3
d^2e									+1	+4	+3	+12
c^3e										+1	−1	−1
c^2d^2											+1	+4
c^5												+1

÷ 3628800

Right-hand.	k	ci	dh	eg	f^2	c^2g	cdf	ce^2	d^2e	c^3e	c^2d^2	c^5
10	−10	+450	+1320	+2100	+1260	−12600	−50400	−31500	−42000	+189000	+378000	−226800
82	+10	+270	−1320	−2100	−1260	−7560	+10080	+6300	+42000	+113400	+25200	−226800
73	+10	−450	+1320	−2100	−1260	+12600	−2520	+31500	−46200	−189000	+151200	+226800
64	+10	−450	−1200	+2940	−1260	−2520	+50400	−44100	−8400	+189000	−226800	−226800
5^2	+5	−225	−600	−1050	+2520	+6300	−37800	+15700	+21000	−94500	+126000	+113400
62^2	−10	−270	+1200	−420	+1260	·	−10080	+31500	−16800	−75600	+50400	
532	−20	+180	−120	+4200	−3780	−5040	+42840	−37800	+4200	+75600	−50400	
4^22	−10	+90	+1200	2940	+1260	+12600	−30240	+6300	+8400	−37800	+25200	
43^2	−10	+450	−1320	−420	+1260	−5040	+2520	+6300	−4200	·		
42^3	+10	+270	−1200	+2100	−1260	−5040	+10080	−6300				
3^22^2	+15	+45	+720	−1890	+1260	+2520	−5040	+3150				
2^5	−2	−90	+240	−420	+252							

Left-hand, no right-hand.

Col	l	cj	di	eh	fg	c^2h	cdg	cef	d^2f	de^2	c^3f	c^2de	cd^3	c^4d
l	+1	+35	−75	+90	−42									
cj		+ 2	− 9	+14	− 7	−30	+70	−21	− 56	+35				
di			+ 1	− 2	+ 1	+10	−26	+ 7	+ 24	−15				
eh				+ 1	− 1	+ 3	− 2	− 6	+ 10	− 5	+27	−45	+20	
fg					+ 1	−16	+58	+ 5	−100	+60	−54	+90	−40	
c^2h						+ 2	− 7	+ 5	·	·	−27	+45	−20	
cdg							+ 1	− 3	+ 6	− 4	+27	−45	+20	
cef								+ 1	− 3	+ 2	− 9	+14	− 6	
d^2f									+ 1	− 1	+ 4	− 6	+ 2	−1
de^2										+ 1	−12	+36	−18	+3
c^3f											+ 2	− 5	+ 3	+1
c^2de												+ 1	− 1	−1
cd^3													+ 1	+4
c^4d														+1

ck	dj	ei	fh	g^2	c^2i	cdh	ceg	cf^2	d^2g	def	e^3	c^3g	c^2df	c^2e^2	cd^2e	d^4	c^4e	
+66	−220	+495	−792	+462														
+ 3	− 15	+ 40	− 70	+ 42	− 15	+ 40	− 70	+ 42										
	+ 15	− 40	+ 70	− 42	+150	−400	+700	−420										
		+ 1	− 4	+ 3	+ 3	− 8	− 22	+ 24	+ 24	− 36	+ 15							
			+ 25	− 24		+ 50	+680	−675	−570	+925	−400							
				+ 1			− 70	+100	+ 80	−200	+100	− 4	+ 8	− 5				
					+ 1	− 4	+ 8	− 5				− 4	+ 8	− 5				
						4	− 8	+				+32	−64	+ 40				
							+ 1	− 1	− 1	+ 2	− 1	− 1	+ 2	+ 1	− 3	+ 1		
								+ 2	+ 5	− 19	+ 12	+20	−49	− 32	+ 91	− 32		
									+ 1	− 3	+ 2	+ 4	− 9	− 12	+ 27	− 10	+ 6	−
										+ 18	− 17		+54	+162	−342	+135	−81	+
											+ 1			− 18	+ 54	− 27	+81	−
												+ 1	− 3				+ 1	−
													+ 3			− 2	+30	−
														+ 1	− 2	+ 1	− 2	+
															+ 1	− 1	+ 4	−
																1		+
																	+ 1	−

By Morgan Jenkins, M. A.

After reading Professor Sylvester's exposition (Act III, Vol. V, No. 3, p. 286, and No. 4 to p. 296) of the mode of using the bends of a regularized graph for the construction of combined partitions, I thought it would be useful to show how to examine the bends of a graph, having given large numbers as elements without the inconvenience of actually constructing the graph. In pursuing this course one additional result has been obtained giving a set of progressions having the common difference two, instead of a set of sequences.

Let $$m_1 a_{(1)} + m_2 a_{(2)} + \ldots \qquad + m_j a_{(j)} = P$$
and $$\mu_1 \alpha_{(1)} + \mu_2 \alpha_{(2)} + \ldots \qquad + \mu_j \alpha_{(j)} = P$$
represent two conjugate partitions of a number P; $a_{(1)}, a_{(2)} \ldots a_{(j)}$ being the j different elements of one partition, arranged in descending order of magnitude; $m_1, m_2 \ldots$ et cetera, the numbers of times those elements are taken; and let $a_1, a_2, a_3, \ldots$ et cetera, represent the elements of the same partition, taken in order, counting repetitions; that is $a_1, a_2, \ldots a_{m_1}$, each equal to $a_{(1)}$, and so on; also let the μ's and the α's represent corresponding quantities in the conjugate partition. Then we have

$$\alpha_{(1)} = \Sigma m_j = m_1 + m_2 + \ldots + m_j;\ \alpha_{(2)} = \Sigma m_{j-1};\ \ldots\ \alpha_{(j)} = m_1,$$
$$\mu_1 = a_{(j)};\ \mu_2 = a_{(j-1)} - a_{(j)};\ \mu_3 = a_{(j-2)} - a_{(j-1)} \ldots \mu_j = a_{(1)} - a_{(2)},$$
and reciprocally $$a_{(1)} = \Sigma \mu_j;\quad m_1 = \alpha_{(j)},\ \text{and so on.}$$

The two partitions are of the same extent j, as well as of the same content P; and any process which alters the extent of one partition will alter in like manner the extent of the conjugate partition.

The selection of suitable examples will be facilitated if we notice that, instead of supposing the coefficients and elements of one partition to be given, we may take any j different numbers as the elements of one partition, and any other j different numbers (not necessarily differing from the first set) as the elements of the second: then the coefficients of either partition are equal to the differences of consecutive elements of the other partition, the coefficient of the highest element in one partition being equal to the lowest element in the other.

The content of each partition is

$$a_{(j)}\alpha_{(1)} + [a_{(j-1)} - a_j]\,\alpha_{(2)} + [a_{(j-2)} - a_{(j-1)}]\,\alpha_{(3)} + \dots [a_{(1)} - a_{(2)}]\,\alpha_{(j)}$$

which equals

$$\{a_{(j)}\alpha_{(1)} + a_{(j-1)}\alpha_{(2)} + \dots + a_{(1)}\alpha_{(j)}\} - \{a_{(j)}\alpha_{(2)} + a_{(j-1)}\alpha_{(3)} + \dots a_{(2)}\alpha_{(j)}\}$$

a result which is unaltered when the a's and α's are interchanged.

In a regularized graph it will be seen that the different elements are the coordinates of the 'out corners' of the outline of the graph: thus at T, in Fig. 1, the number of nodes in the row and column passing through T may be denoted by $a_{(f)}$ and $\alpha_{(j+1-f)}$ respectively, the sum of the subscripts being $j+1$. At an 'in-corner,' as at S, the number of nodes in the row and column passing through S, reckoned up to and including S, but not beyond, are the same as at the adjacent out-corners R and T respectively: hence the coordinates of an in-corner may be taken to be $a_{(f+1)}$ and $\alpha_{(j+1-f)}$, the sum of the subscripts being $j+2$ instead of $j+1$.

The diagonal of the Durfee-square terminates either at an 'out-corner' (Fig. 1) or at an 'in-corner' (Fig. 2), or it divides a column (Fig. 3) or a row (Fig. 4). If we use i to denote the number of bends in the graph, in each of the four cases i is equal to one of the elements of the graph, either $a_{(h)}$ or $\alpha_{(h)}$.

FIG. 1.

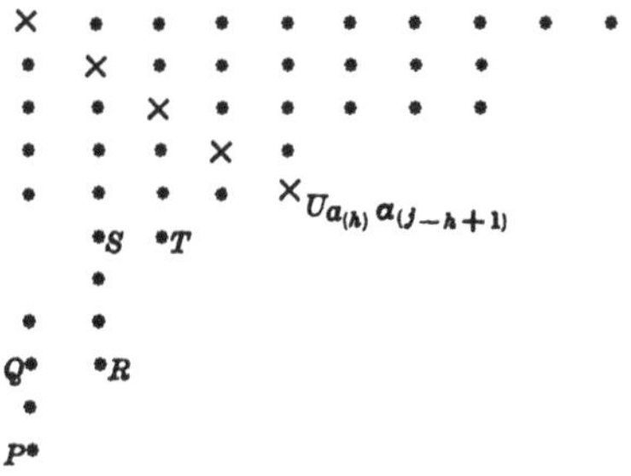

FIG. 2.

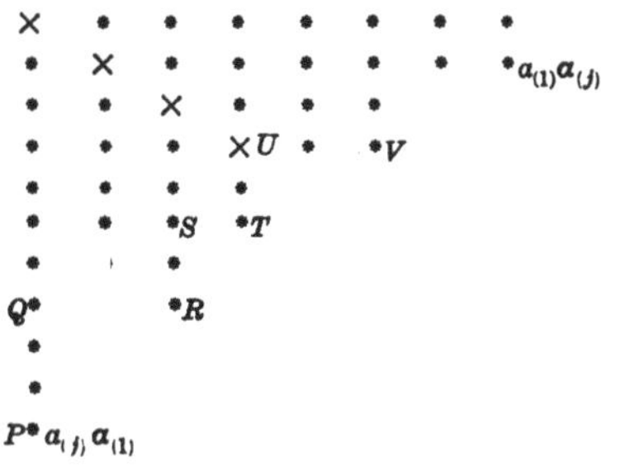

FIG. 3.

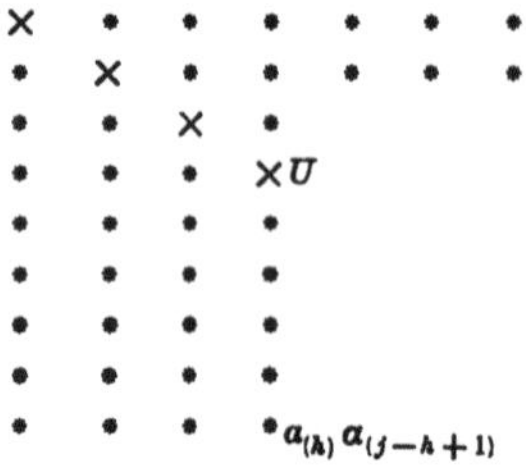

FIG. 4.

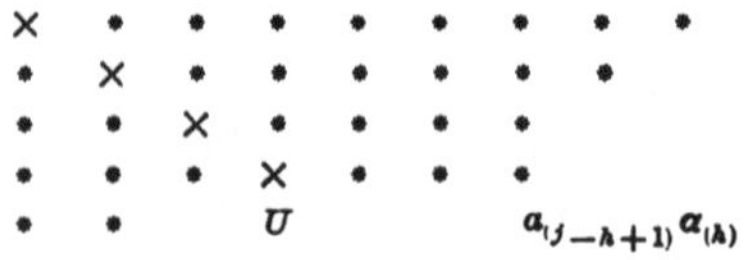

FIG. 5.

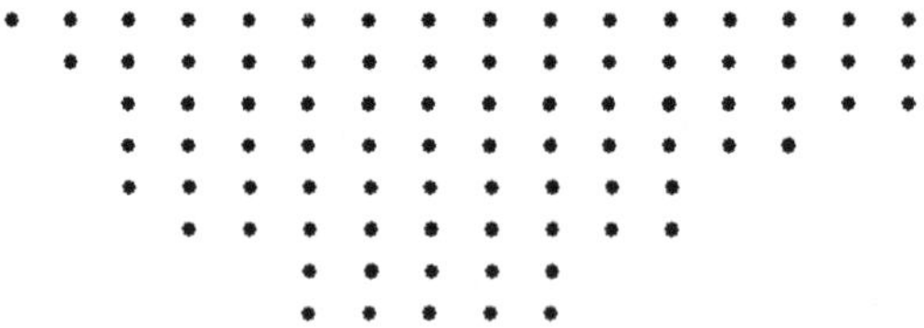

In Fig. 1 $a_{(h)} = a_{(j-h+1)} = i$; $a_i = a_i = i$.
In Fig. 2 $a_{(h)} = a_{(j-h+2)} = i$,
$a_i = a_{(h-1)}$, which is greater than i,
$a_i = a_{(j-h+1)}$, which is greater than i.
In Fig. 3 $a_{(h)}$ is less than $a_{(j+h+1)}$ and greater than $a_{(j-h+2)}$; $a_i = a_{(h)} = i$,
$a_i = a_{(j-h+1)}$ which is greater than i.
In Fig. 4 $a_{(h)}$ is less than $a_{(j-h+1)}$ and greater than $a_{(j-h+2)}$
$a_i = a_{(j-h+1)}$, which is greater than i,
$a_i = a_{(h)} = i$.

Thus in every case we can find an element of one partition, which is between or not outside the pair of consecutive elements of the other partition, which is formed of the corresponding element and the one next below it; and the first named element gives the number of bends in the graph which represents the partition.

Let us now consider the effect of adding two regularized graphs together, row to row, in order. If we suppose them to be placed side by side (Fig. 5) and then all the nodes in every row to be pushed close to the left-hand side, the same effect is produced as if the columns were picked out and rearranged in order of magnitude. The number of different columns in the regularized graph so formed is unaltered by the rearrangement.

Hence if j be the extent, that is the number of different elements in either set of conjugate elements, of one partition, the conjugate elements being denoted by $\dot{a}$'s and α's, and k, b and β be corresponding letters for another partition, then the extent of the combined partition $a_1 + b_1$, $a_2 + b_2$, et cetera, is equal to the number of different elements in $\alpha_{(1)}, \alpha_{(2)}, \ldots \alpha_{(j)}, \beta_{(1)}, \beta_{(2)} \ldots \beta_{(k)}$, considered as a single series, that is, if l be the number of α elements which are equal to β elements, the extent of the combined partition is $j + k - l$. This theorem may be applied to the bends of a graph. The numbers of nodes in these bends are

$$a_1 + \alpha_1 - 1,\ a_2 + \alpha_2 - 3,\ a_3 + \alpha_3 - 5 \ldots a_i + \alpha_i - (2i - 1).$$

If we omit the negative terms we have a regularized graph which is the sum of two regularized graphs, and which exceeds the original graph in content by $1 + 3 + 5 + \ldots + (2i - 1)$ or by i^2, which is the content of the Durfee-square. The second of the two partial graphs may be supposed to be turned through a right angle, so that α_1 is in a line with a_1, α_2 with a_2, ... and α_i with a_i. The extent of the new graph, that is the number of different elements in either set of its conjugate elements will be found, in every case, to be $j - g$, where g is the number of elements in either set of elements of the original graph which are less than the corresponding conjugate element, but equal to some other conjugate element.

Thus in the combined graph formed by the addition of the parts of graph (Fig. 1), if we cut off all the rows below the Durfee-square, the different columns in the remaining graph are $\alpha_{(j)}, \alpha_{(j-1)} \ldots \alpha_{(j-h+1)}$: if we cut off all the columns to the right of the Durfee-square and turn the remaining graph round through a right angle, the different columns in this remaining graph are now $\alpha_{(j)}, \alpha_{(j-1)} \ldots \alpha_{(h)}$. In these two series $\alpha_{(h)} = \alpha_{(j-h+1)}$, and is not included in the g equalities between the elements of the original graph, the remaining equalities are g in number according to the above-stated definition of g. Therefore the number of elements in the two series, including repetitions, is $h + (j - h + 1)$ or $j + 1$, and the number excluding repetitions is $(j + 1) - (g + 1)$ or $j - g$. In the case of Fig. 2, the columns of one graph are $\alpha_{(j)}, \alpha_{(j-1)} \ldots \alpha_{(j-h+2)}$, and the columns, formed out of rows, of the other graph are $\alpha_{(j)}, \alpha_{(j-1)} \ldots \alpha_{(h)}$; $\alpha_{(h)}$ being equal to $\alpha_{(j-h+2)}$ and less than $\alpha_{(j-h+1)}$ is reckoned in the g equalities. Therefore the number of elements in the two series is, including repetitions, $(h-1) + (j-h+1)$ or j, and is, excluding repetitions, $j-g$. In the case of Fig. 3, the columns of one graph are $\alpha_{(j)}, \alpha_{(j-1)} \ldots \alpha_{(j-h+2)} a_{(h)}$, $a_{(h)}$ taking the place of an α element at the end in consequence of the diagonal of the Durfee-square falling between two corners of the original graph; the columns, formed out of rows, of the other partial graph are $a_{(j)}, a_{(j-1)} \ldots a_{(h)}$. $a_{(h)} = \alpha_{(h)}$ is to be counted in the equalities of these two series, and is not one of the g equalities of the original graph. Therefore the number of elements in the two series is, including repetitions, $j+1$, and is, excluding repetitions, $(j+1)-(g+1)$ or $j-g$. The case of Fig. 4 is exactly like that of Fig. 3, interchanging a's and α's.

If we subtract the terms of a single ascending progression having the common difference r from a constant quantity, the remainders will form a single descending progression having the common difference r. If for the constant quantity we substitute the terms of a descending series, containing repetitions, every break in the repetitions produces a break in the progressions arising from the subtraction. Hence the number of distinct progressions produced must be the same as the number of different numbers in the series from which we subtract; also the last number of one progression must exceed the first number of the next progression by a number greater than r.

In applying the theorem with regard to the subtraction of the terms of a single progression from the terms of another series containing repetitions, we should have to notice whether any of the remainders were zero or negative. In this case all the remainders are positive, because they represent the bends of the original graph, and the last term of the series is not zero.

It follows that the numbers of nodes in the bends of a regularized graph when placed in descending order of magnitude contain $j-g$ distinct progressions, having the common difference 2, where j is the number of different elements in either set of conjugate elements of the graph, and g is the number of elements in either set which are less than their corresponding element but equal to some other element of the conjugate set: also the excess of the last number of one progression over the first number of the next progression is greater than two.

EXAMPLE 1. Let

$$\begin{matrix} 11 & 10 & 7 & 5 & 3 & 2 \\ 4 & 5 & 7 & 9 & 12 & 13 \end{matrix}$$

be the elements of two conjugate partitions written one underneath the other in contrary order of magnitude. The extent (j) is equal to 6, the content is equal to 89, and the coefficients are as given by the equations

$$4.11 + 1.10 + 2.7 + 2.5 + 3.3 + 1.2 = 89$$
$$2.13 + 1.12 + 2.9 + 2.7 + 3.5 + 1.4 = 89$$

The numbers of nodes in the bends of the graph which represents these conjugate partitions are found as follows:

1 is less than 4 in lower line, so we take 11 from upper line,
" " 2 " " " " 13 " " "
and the number of nodes in the first bend is 11 + 13 — 1.

2 is less than 4 in lower line, so we take 11 from upper line,
2 is equal to 2 " " " 13 " " "
and the number of nodes in the second bend is 11 + 13 — 3.

3 is less than 4 in lower line, so we take 11 from upper line,
3 is equal to 3 " " " " 12 " " "
and the number of nodes in the third bend is 11 + 12 — 5, and so on. Thus we have for the bends

$$\begin{aligned} 11 + 13 - 1 &= 23 \\ 11 + 13 - 3 &= \underline{21} \\ 11 + 12 - 5 &= \underline{18} \\ 11 + 9 - 7 &= \underline{13} \\ 10 + 9 - 9 &= \underline{10} \\ 7 + 7 - 11 &= 3 \\ 7 + 7 - 13 &= \underline{1} \\ &\ 89 \end{aligned}$$

Since two corresponding elements 7 are equal to each other, 7 is the number of bends. There is one pair of non-corresponding equal elements which do not exceed 7, viz. 5, 5; therefore $g = 1$; $j = 6$ and 6 — 1 or 5 is the number of distinct progressions having the common difference 2. The number of elements which appear in the formation of the bends are $j + 1$ or 7, viz. 11, 10 and 7 in one set and 13, 12, 9, 7 in the other set.

In the next example only the distinctive points will be noticed.

EXAMPLE 2.

16	13	12	9	4	3	1
2	4	9	11	13	14	16

$$2.16 + 2.13 + 5.12 + 2.9 + 2.4 + 1.3 + 2.1 = 149$$
$$1.16 + 2.14 + 1.13 + 5.11 + 3.9 + 1.4 + 3.2 = 149$$

$j = 7$. Two conjugate elements, 9, which are only one place removed from each other are equal, therefore 9 is the number of bends. The number of pairs of non-corresponding equal elements which do not exceed 9 is 2, viz. 4, 4 and 9, 9. Therefore $g = 2$.

BENDS.

$$\begin{aligned} 16 + 16 - 1 &= \underline{31} \\ 16 + 14 - 3 &= \underline{27} \\ 13 + 14 - 5 &= \underline{22} \\ 13 + 13 - 7 &= \underline{19} \\ 12 + 11 - 9 &= 14 \\ 12 + 11 - 11 &= 12 \\ 12 + 11 - 13 &= 10 \\ 12 + 11 - 15 &= 8 \\ 12 + 11 - 17 &= \underline{6} \\ &\ 149 \end{aligned}$$

Also 7 — 2 or 5 is the number of distinct progressions having the common difference 2.

In this case the number of elements from the two sets appearing in the formation of the bends is j or 7 and not $j+1$, viz. 16, 13, 12 from one set and 16, 14, 13, 11 from the other set.

EXAMPLE 3.

$$\begin{array}{ccccccc} 13 & 12 & 11 & 6 & 5 & 3 & 2 \\ 2 & 3 & 5 & 7 & 8 & 9 & 10 \end{array}$$

$$2.13 + 1.12 + 2.11 + 2.6 + 1.5 + 1.3 + 1.2 = 82$$
$$2.10 + 1.9 + 2.8 + 1.7 + 5.5 + 1.3 + 1.2 = 82$$

$j = 7$; $g = 3$ from the equal pairs 2, 2; 3, 3 and 5, 5.

BENDS.

$$\begin{aligned} 13 + 10 - 1 &= 22 \\ 13 + 10 - 3 &= 20 \\ 12 + 9 - 5 &= 16 \\ 11 + 8 - 7 &= 12 \\ 11 + 8 - 9 &= 10 \\ 6 + 7 - 11 &= 2 \\ & \overline{82} \end{aligned}$$

EXAMPLE 4.

$$\begin{array}{ccc} 12 & 10 & 9 \\ 2 & 5 & 6 \end{array}$$

$$2.12 + 3.10 + 1.9 = 63$$
$$9.6 + 1.5 + 2.2 = 63$$

$j = 3$; $g = 0$. Number of bends $= 6$, because 6 is between the corresponding element 9 and 0. $j - g = 3$, and there are 3 distinct progressions.

BENDS.

$$\begin{aligned} 12 + 6 - 1 &= 17 \\ 12 + 6 - 3 &= 15 \\ 10 + 6 - 5 &= 11 \\ 10 + 6 - 7 &= 9 \\ 10 + 6 - 9 &= 7 \\ 9 + 6 - 11 &= 4 \\ & \overline{63} \end{aligned}$$

The method here applied to the bends of the original graph may be used to establish the theorem enunciated and proved by Professor Sylvester, namely, that if we interpolate with the bends of the original graph the bends of the graph obtained by cutting off the highest column from the original graph, the series so obtained will contain j sequences.

The series is $a_1 + a_1 - 1$, $a_1 + a_2 - 2$, $a_2 + a_2 - 3$, $a_2 + a_3 - 4$, ... $a_i + a_i - (2i - 1)$, $a_i - i$.

The last term is $a_i - i$ (as stated in Vol. V, Number 3, p. 288), and it may be written $a_i + i - 2i$. If we omit the negative terms we have the sum of two partial graphs which are formed thus: Double the number of terms of every element from a_1 to a_i inclusive; also double the number of terms of every element from a_1 to a_i inclusive, but cut off the first element a_1 from the beginning and attach an element i at the end.

For each of the three cases (Fig. 1, Fig. 2 and Fig. 4) where i is equal to an a element, say $a_{(h)}$, the conjugate elements, whose differences give the number of times $a_{(1)}$, $a_{(2)}$, et cetera, are taken, are $2a_{(j)}$, $2a_{(j-1)} \dots 2a_{(h)}$, all even numbers.

The conjugate elements whose differences give the number of times $a_{(1)}$, $a_{(2)}$ $\dots$, et cetera, are taken, are $2a_{(j)} - 1$, $2a_{(j-1)} - 1$, $\dots$ $2a_{(j-h+2)} - 1$, $2a_{(h)}$, all odd numbers except the last. The cutting off of the first element to form the second partial graph explains why all the conjugate elements but the last are odd: the reason why the last is even is that one element, i, has been added to the $2\{a_{(h)} - a_{(j-h+2)}\}$ elements, each equal to $a_{(h)}$; and i being equal to $a_{(h)}$ in the three cases of Fig. 1, Fig. 2 and Fig. 4, the conjugate element becomes $2a_{(h)}$ instead of $2a_{(h)} - 1$. The number of different elements in the two sets of conjugate elements taken as a single series is $j - h + 1 + h - 1$ or j, the element $2a_{(h)}$ which is the only element common to the two series being only counted once. When we subtract the terms of the single sequence $1, 2, 3, \dots 2i$, we shall have the remainders forming j distinct sequences. In the case of Fig. 1, and Example 1, the last term is 0, $a_i - i$ being equal to 0; but it can be shown that this 0 is preceded by 1, a_i being also equal to i in this case. Therefore if we cut off this 0 we shall have j sequences left. If we examine the case of Fig. 3 and Example 3, we shall find there are $j + 1$ sequences, including the final term, which is 0; but the penultimate term must be greater than 1.

Hence if we cut off the final zero term we have j sequences left.

Third Note on Weierstrass' Theory of Elliptic Functions.

By A. L. Daniels, *Johns Hopkins University.*

The Sigma-Quotients.

As long as the argument and the quasi-periods 2ω, $2\omega'$, remain the same, we may omit them, and write σ, σ_1, σ_2, σ_3, $\frac{\sigma}{\sigma_1}$, etc. The functions σ_1, σ_2, σ_3 are then thus defined,

$$\sigma_1 u = \frac{e^{-\eta u}\sigma(\omega + u)}{\sigma\omega} = \frac{e^{\eta u}\sigma(\omega - u)}{\sigma\omega},$$

$$\sigma_2 u = \frac{e^{-\eta'' u}\sigma(\omega'' + u)}{\sigma\omega''} = \frac{e^{\eta'' u}\sigma(\omega'' - u)}{\sigma\omega''},$$

$$\sigma_3 u = \frac{e^{-\eta' u}\sigma(\omega' + u)}{\sigma\omega'} = \frac{e^{\eta' u}\sigma(\omega' - u)}{\sigma\omega'},$$

which are seen to be even functions. These apparently arbitrary definitions flow naturally from considerations connected with the "pocket edition,"

$$\wp u - \wp v = \frac{\sigma(u+v)\,\sigma(u-v)}{\sigma^2 u\,\sigma^2 v}.$$

Since $\wp\omega = e_1$, $\wp(\omega + \omega') = e_2$, $\wp\omega' = e_3$, we have, using the general mark α, and writing $v = \omega_\alpha$

$$\wp u - e_\alpha = \frac{\sigma(u + \omega_\alpha)\,\sigma(u - \omega_\alpha)}{\sigma^2 u \,.\, \sigma^2\omega_\alpha}.$$

But $\wp u$ is a truly periodic function: it remains therefore to examine the periodicity of σu. In the second note, p. 261, I have shown that

$$\sigma u = u\Pi'_w\left(1 - \frac{u}{w}\right)e^{\frac{u}{w} + \frac{1}{2}\frac{u^2}{w^2}}$$

$$w = m \,.\, 2\omega + m'2\omega'$$

degenerates into the sine when $m' = 0$, or

$$\lim\,(\sigma u)_{m'=0} = \frac{2\omega}{\pi}\,e^{\frac{1}{6}\left(\frac{u\pi}{2\omega}\right)^2}.\sin\frac{u\pi}{2\omega}.$$

It was also shown that

$$\frac{\sin \pi x}{\pi} = x \prod_{n=-\infty}^{+\infty}{}' \left[\left(1-\frac{x}{n}\right)e^{\frac{x}{n}}\right] = s(x).$$

The following definitions are introduced as convenient:

$$\frac{d}{dx}\log s(x) = \frac{s'(x)}{s(x)} = s_1 x = \frac{1}{x} + \sum_{n=-\infty}^{+\infty}{}' \left(\frac{1}{x-n} + \frac{1}{n}\right)$$

$$-\frac{d}{dx}.s_1 x = s_2 x = \sum_{n=-\infty}^{+\infty} \frac{1}{(x-n)^2}$$

in which last series the value $n = 0$ is included. One sees that $s_2(x+1) = s_2(x)$, and $s_2 x$ is periodic. Incidentally it may be remarked that on comparing the developments for $\sin \pi x$ and $s(x)$,

$$\sin \pi x = \pi x - \frac{\pi^3 x^3}{3!} + \frac{\pi^5 x^5}{5!} - \ldots$$

$$\frac{\sin \pi x}{\pi} = s(x) = x\left(1-\left(\frac{x}{1}\right)^2\right)\left(1-\left(\frac{x}{2}\right)^2\right)\left(1-\left(\frac{x}{3}\right)^2\right)\cdots$$

$$= x - \left(1 + \frac{1}{2^2} + \frac{1}{3^2} + \frac{1}{4^2} + \ldots\right)x^3$$

$$+ \left(\frac{1}{2^2} + \frac{1}{3^2} + \ldots\right)x^5$$

$$- \ \ldots\ldots\ldots\ldots$$

or, $$s(x) = x - x^3 \sum_{1}^{+\infty} \frac{1}{n^2} + x^5 \sum_{m=1}^{+\infty} \sum_{n=2}^{+\infty} \frac{1}{m^2 n^2} - x^7 \sum_{m=1}^{\infty} \sum_{n=2}^{\infty} \sum_{p=3}^{\infty} \frac{1}{m^2.n^2.p^2} + \ldots$$
$$m < n \qquad\qquad m < n < p$$

whence $$\Sigma \frac{1}{n^2} = \frac{\pi^2}{3!}, \quad \Sigma\Sigma \frac{1}{m^2 n^2} = \frac{\pi^4}{5!}, \text{ etc.}$$

We can now in $$\sigma u = u\Pi'\left(1 - \frac{u}{w}\right)e^{\frac{u}{w} + \frac{1}{2}\frac{u^2}{w^2}},$$

give to m' a constant value and carry out the multiplication with respect to m. For $m' = 0$, we have

$$\sigma u = u\Pi'\left(1 - \frac{u}{m2\omega}\right)e^{\frac{u}{m2\omega} + \frac{1}{2}\left(\frac{u}{m2\omega}\right)^2}$$

$$= u\Pi'\left(1 - \frac{u}{m2\omega}\right)e^{\frac{u}{m2\omega}}.\Pi e^{\frac{1}{2}\left(\frac{u}{m2\omega}\right)^2}.$$

But $$\Pi e^{\frac{1}{8}\cdot\frac{1}{m^2}\cdot\frac{u^2}{\omega^2}} = e^{\frac{1}{8}\frac{\pi^2}{6}\cdot\frac{u^2}{\omega^2}},$$

and $$u\Pi'\left[\left(1 - \frac{u}{m2\omega}\right)e^{\frac{u}{m2\omega}}\right] = \frac{2\omega}{\pi}\sin\frac{u\pi}{2\omega},$$

which two factors furnish that part of σu corresponding to those values of m and m' represented in the plane of complex number by points on the real axis distant from each other by 2ω; in other words to the numbers

$$0,\ \pm 1.2\omega,\ \pm 2.2\omega,\ \pm 3.2\omega, \ldots$$

Employing again

$$su = u\Pi'\left(1-\frac{u}{n}\right)e^{\frac{u}{n}},$$

we have

$$\frac{s(u-a)}{s(-a)} = \frac{u-a}{-a}\Pi'\left(1-\frac{u}{n+a}\right)e^{\frac{u}{n}}.$$

Remarking now the identity

$$\frac{u}{n} = \frac{u}{n+a} + \frac{u}{n} + \frac{u}{-n-a}$$

and also

$$\frac{s'(-a)}{s(-a)} = -\frac{1}{a} + \Sigma'\left(\frac{1}{-n-a}+\frac{1}{n}\right)$$

there appears

$$\frac{s(u-a)}{s(-a)} = \left(1-\frac{u}{a}\right)\Pi'\left(1-\frac{u}{n+a}\right)e^{\frac{u}{n+a}+\frac{u}{n}+\frac{u}{-n-a}}.$$

But

$$\frac{s'(-a)}{s(-a)} = s_1(-a)$$

therefore

$$\frac{s(u-a)}{s(-a)} = \left(1-\frac{u}{a}\right)\Pi'\left[\left(1-\frac{u}{n+a}\right)e^{\frac{u}{n+a}}\right].e^{us_1(-a)+\frac{u}{a}}$$

or, taking up $\left(1-\frac{u}{a}\right)e^{\frac{u}{a}}$ into the product as the value of $\left(1-\frac{u}{n+a}\right)e^{\frac{u}{n+a}}$ for $n=0$, we can drop the accent of the product sign and write

$$\frac{s(u-a)}{s(-a)} = \Pi\left(1-\frac{u}{n+a}\right)e^{\frac{u}{n+a}}.e^{us_1(-a)},$$

whereby the only restriction as to a is that it must not be an integer. Transposing,

$$\Pi\left(1-\frac{u}{n+a}\right)e^{\frac{u}{n+a}} = \frac{s(u-a)}{s(-a)}.e^{-u.s_1(-a)}.$$

We are now ready to decompose the sigma-product

$$\Pi\left(1-\frac{u}{m.2\omega+m'2\omega'}\right)e^{\frac{u}{m2\omega+m'2\omega'}+\frac{1}{2}\left(\frac{u}{m2\omega+m'2\omega'}\right)^2},$$

$$m = 0,\ \pm 1,\ \pm 2, \ldots \pm\infty,$$

$$m' = \quad \pm 1,\ \pm 2, \quad \pm\infty,$$

where $m'=0$ is omitted, as already accounted for, and consequently the accent on the product sign is dropped. Dividing by $2\omega_1$, the formula becomes

$$\Pi\left(1-\frac{\frac{u}{2\omega}}{m+m'\frac{\omega'}{\omega}}\right)e^{\frac{\frac{u}{2\omega}}{m+m'\frac{\omega'}{\omega}}}.e^{\frac{1}{2}\left(\frac{\frac{u}{2\omega}}{m+m'\frac{\omega'}{\omega}}\right)^2}.$$

Writing as before $$s_2(x) = -\frac{d}{dx}s_1 x = \Sigma \frac{1}{(n-x)^2},$$
the second exponential factor becomes
$$e^{\frac{1}{2}\left(\frac{u}{2\omega}\right)^2 \cdot \Sigma \frac{1}{\left(m+m'\frac{\omega'}{\omega}\right)^2}} = e^{\frac{1}{2}\left(\frac{u}{2\omega}\right)^2 \cdot s_2\left(-m'\frac{\omega'}{\omega}\right)}$$
for each particular value of m', the summation being taken with respect to m alone. The rest of the product is
$$\Pi\left(1 - \frac{\frac{u}{2\omega}}{m+m'\frac{\omega'}{\omega}}\right)e^{\frac{\frac{u}{2\omega}}{m+m'\frac{\omega'}{\omega}}} = \frac{s\left(\frac{u}{2\omega} - m'\frac{\omega'}{\omega}\right)}{s\left(-m'\frac{\omega'}{\omega}\right)} \cdot e^{-\frac{u}{2\omega}\cdot s_1\left(-m'\frac{\omega'}{\omega}\right)}$$
by the formula above deduced. Collecting the four factors, we have
$$\sigma u = e^{\frac{1}{8}\cdot\frac{\pi^2}{6}\cdot\frac{u^2}{\omega^2}} s\left(\frac{u}{2\omega}\right) \cdot \prod_{m'=-\infty}^{+\infty}\left[e^{\frac{1}{2}\left(\frac{u}{2\omega}\right)^2 s_2\left(-m'\frac{\omega'}{\omega}\right)} \cdot e^{\frac{-u}{2\omega}\cdot s_1\left(-m'\frac{\omega'}{\omega}\right)} \cdot \frac{s\left(\frac{u}{2\omega} - m'\frac{\omega'}{\omega}\right)}{s\left(-m'\frac{\omega'}{\omega}\right)}\right].$$
This formula can however be simplified in form by multiplying together the factors in pairs and taking the product from 1 to ∞, instead of from $-\infty$ to $+\infty$. For we had
$$s(x) = x\prod_{-\infty}^{+\infty}{}'\left(1 - \frac{x}{n}\right)e^{\frac{x}{n}},$$
$$s_1(-x) = -\frac{1}{x} + \sum_{n=-\infty}^{+\infty}{}'\left(\frac{1}{-x-n} + \frac{1}{n}\right),$$
$$s_2(-x) = \sum_{-\infty}^{+\infty}\frac{1}{(x+n)^2},$$
whence it appears that $s_1(-x) = \sum_{n=1}^{+\infty}\frac{-2x_2}{x^2-n}$, or is an odd function, consequently
$$\sum_{m'=-\infty}^{+\infty} s_1(-m') = \sum_{m'=1}^{+\infty} s_1(-m') + \sum_{m'=1}^{+\infty} s_1(+m') = 0,$$
and one exponential factor disappears, and the formula now reads
$$\sigma u = 2\omega e^{\frac{1}{8}\cdot\frac{\pi^2}{6}\cdot\frac{u^2}{\omega^2}} s\frac{u}{2\omega} \cdot e^{\frac{1}{2}\left(\frac{u}{2\omega}\right)^2 \Sigma\left[s_2\left(-m'\frac{\omega'}{\omega}\right) + s_2\left(+m'\frac{\omega'}{\omega}\right)\right]} \prod_{m'=1}^{+\infty}\frac{s\left(\frac{u}{2\omega} - m'\frac{\omega'}{\omega}\right)s\left(\frac{u}{2\omega} + m'\frac{\omega'}{\omega}\right)}{s\left(-m'\frac{\omega'}{\omega}\right) \cdot s\left(m'\frac{\omega'}{\omega}\right)}$$
$$= 2\omega e^{\frac{1}{8}\frac{\pi^2}{6}\frac{u^2}{\omega^2}} s\frac{u}{2\omega} \cdot e^{\frac{1}{8}\frac{\pi^2}{6}\frac{u^2}{\omega^2} + \left(\frac{u}{2\omega}\right)^2 \sum_{m'=1}^{+\infty} s_2\left(m'\frac{\omega'}{\omega}\right)} \prod_{m'=1}^{+\infty}\frac{s\left(\frac{u}{2\omega} - m'\frac{\omega'}{\omega}\right)s\left(\frac{u}{2\omega} + m'\frac{\omega'}{\omega}\right)}{s\left(-m'\frac{\omega'}{\omega}\right)s\left(m'\frac{\omega'}{\omega}\right)}$$
$$= 2\omega s\frac{u}{2\omega} \cdot e^{\left(\frac{u}{2\omega}\right)^2\left[\frac{1}{2}\frac{\pi^2}{6} + \Sigma s_2\left(m'\frac{\omega'}{\omega}\right)\right]} \Pi \frac{s\left(\frac{u}{2\omega} - m'\frac{\omega'}{\omega}\right)s\left(\frac{u}{2\omega} + m'\frac{\omega'}{\omega}\right)}{s\left(-m'\frac{\omega'}{\omega}\right)s\left(m'\frac{\omega'}{\omega}\right)}$$

and, on passing from the s and s_3 to the sine,

$$\sigma u = \frac{2\omega}{\pi} \sin \frac{u\pi}{2\omega} \cdot e^{\frac{1}{6}\left(\frac{\pi u}{2\omega}\right)^2} \Pi \frac{\sin \frac{\pi}{2\omega}(2m'\omega' - u) \sin \frac{\pi}{2\omega}(2m'\omega' + u)}{\sin^2 \frac{m'\omega'\pi}{\omega}} \cdot e^{\frac{\left(\frac{u\pi}{2\omega}\right)^2}{\sin^2 \frac{m'\omega'\pi}{\omega}}}.$$

Professor Schwarz writes

$$\eta = \frac{\pi^2}{2\omega} \left\{ \frac{1}{6} + \Sigma_{m'} \frac{1}{\sin \frac{m'\omega'\pi}{\omega}} \right\},$$

whereupon the sigma-function is thus represented as a singly infinite product of sines

$$\sigma u = e^{\frac{\eta u^2}{2\omega}} \cdot \frac{2\omega}{\pi} \cdot \sin \frac{u\pi}{2\omega} \Pi_{m'} \left(1 - \frac{\sin^2\left(\frac{u\pi}{2\omega}\right)}{\sin^2\left(\frac{m'\omega'\pi}{\omega}\right)} \right).$$

On substituting $u + 2\omega$ for u the expression becomes

$$\sigma(u + 2\omega) = - e^{2\eta(u+\omega)} \sigma u,$$

from which by logarithmic differentiation and writing $u = -\omega$, we find

$$\eta = \frac{\sigma'\omega}{\sigma\omega},$$

Recurring now to the pocket edition

$$\wp u - e_a = \frac{\sigma(u + \omega_a)\sigma(u - \omega_a)}{\sigma^2 u \sigma^2 \omega_a}.$$

From the definitions at the beginning of this paper

$$\sigma_a^2 u = e^{-2\eta_a u} \frac{\sigma^2(\omega_a - u)}{\sigma^2 \omega_a},$$

so that

$$\wp u - e_a = \left(\frac{\sigma_a u}{\sigma u}\right)^2;$$

which is the simplest form of a doubly periodic function. In the second note was deduced the equation $(\wp' u)^2 = 4(\wp u - e_1)(\wp u - e_2)(\wp u - e_3)$, and on comparison with the above

$$\wp' u = 2 \frac{\sigma_1 u . \sigma_2 u . \sigma_3 u}{\sigma u . \sigma u . \sigma u}.$$

The sigma-quotients have not the same pair of fundamental periods as the sigma-function itself. But while

σu	has the quasi-periods	2ω, $2\omega'$
$\frac{\sigma_1 u}{\sigma u}$	has the periods	2ω, $4\omega'$
$\frac{\sigma_3 u}{\sigma u}$	"	4ω, $2\omega'$
$\frac{\sigma_2 u}{\sigma u}$		4ω, $4\omega'$.

This is shown in the following manner. It will be noticed that aside from exponential and constant factors, the function σ_1, σ_2, σ_3, are formed from σu by increasing the argument u by the half-periods ω, $\omega'' = \omega + \omega'$, ω', respectively. If we write $\quad \bar{w} = r\omega + r'\omega', \quad \bar{\eta} = r\eta + r'\eta'$

instead of $\quad w = m2\omega + m'2\omega'$,

it is evident that w and w will only then be equivalent when both r and r' are even. We have then

$$\sigma(u + 2w) = \varepsilon\sigma u . e^{2\bar{\eta}(u+\bar{w})},$$

$$\sigma(u + w) = \varepsilon\sigma(u - w)e^{2\bar{\eta}u} = -\varepsilon\sigma(w - u)e^{2\bar{\eta}u},$$

and, writing $u = 0$, $\sigma\bar{w} = -\varepsilon\sigma(w)$, or $\varepsilon = -1$ when either r or r' is odd. If both are even, then $\bar{w} \equiv w$ and $\sigma w = 0$. To determine the value of ε in this case, develop both sides according to powers of u

$$u . \sigma' w + u^2 + \ldots = \varepsilon . u . \sigma' \bar{w} + \ldots$$

and $\varepsilon = +1$ when both r and r' are even. Now the formula

$$(r + 1)(r' + 1) - 1 = rr' + r + r'$$

is only even when both r and r' are even; we can write therefore,

$$\sigma(u + 2\bar{w}) = (-1)^{rr'+r+r'} . \sigma u . e^{2\bar{\eta}(u+\bar{w})}$$

$$\sigma(u + \omega_a + 2\bar{w}) = (-1)^{rr'+r+r'} . \sigma(u + \omega_a)e^{2\bar{\eta}(u+\omega_a+\bar{w})};$$

but, from the definition $\quad \sigma(u + \omega_a) = e^{\eta_a u} . \sigma_a u . \sigma\omega_a$,

whence, writing for u, $u + 2\bar{w}$

$$\sigma(u + \omega_a + 2\bar{w}) = e^{\eta_a(u+2w)}\sigma_a(u + 2\bar{w}) . \sigma\omega_a,$$

and, equating the right-hand members,

$$\sigma_a(u + 2w)\sigma\omega_a = (-1)^{rr'+r+r'} . \sigma(u + \omega_a)e^{2\bar{\eta}(u+\omega_a+\bar{w}) - \eta_a(u+2\bar{w})}$$

or, writing $u - \bar{w}$ for u

$$\sigma_a(u + \bar{w})\sigma\omega_a = (-1)^{rr'+r+r'} . \sigma_a(u - \bar{w})e^{2(\bar{\eta}\omega_a - \eta_a\bar{w}) + 2\bar{\eta}u},$$

$$\frac{\sigma_a(u + \bar{w})}{\sigma_a(u - \bar{w})} = (-1)^{rr'+r+r'} e^{2(\bar{\eta}\omega_a - \eta_a\bar{w}) + 2\bar{\eta}u},$$

and for $u = 0$, since $\sigma_a(-w) = \sigma_a(+\bar{w})$ we have for the determination of r and r',

$$1 = e^{2(\bar{\eta}w_a - \eta_a w) + (rr'+r+r')\pi i}.$$

For the case $a = 1$, we shall have $\bar{\eta} = r\eta + r'\eta'$, $\bar{w} = r\omega + r'\omega'$, $\eta_a = \eta$, $\omega_a = \omega$, and

$$1 = e^{2r'(\eta'\omega - \eta\omega') + (rr'+r+r')\pi i}.$$

But

$$\eta'\omega - \eta\omega' = \pm\frac{\pi i}{2},$$

whence $\quad \sigma_1(u + 2\bar{w}) = (-1)^{rr'+r} . \sigma_1 u . e^{2\eta}(u + \bar{w})$.

For the case $a = 3$, we shall have

$$\eta_a = \eta', \ \omega_a = \omega', \ 2(\bar{\eta}\omega_a - \eta_a\bar{w}) = 2r(\eta\omega' - \eta'\omega),$$

and $$\sigma_3(u+2\bar{w}) = (-1)^{rr'+r'}\sigma_3 u \,.\, e^{2\bar{\eta}(u+\bar{w})},$$
and likewise $$\sigma_2(u+2\bar{w}) = (-1)^{rr'}\sigma_2 u e^{2\eta(u+\bar{w})};$$
so that
$$\frac{\sigma_1(u+2\bar{w})}{\sigma(u+2\bar{w})} = -(-1)^{rr'+r}\,.\,\frac{\sigma_1 u}{\sigma u},$$
$$\frac{\sigma_3(u+2\bar{w})}{\sigma(u+2\bar{w})} = -(-1)^{rr'+r'}\,.\,\frac{\sigma_3 u}{\sigma u},$$
$$\frac{\sigma_2(u+2\bar{w})}{\sigma(u+2\bar{w})} = -(-1)^{rr'}\,.\,\frac{\sigma_2 u}{\sigma u}.$$
In order therefore that $2\bar{w} = 2(r\omega + r'\omega')$ may be a period of $\frac{\sigma_1 u}{\sigma u}$, we must have $(-1)^{rr'+r+1} = 1$, or $rr'+r+1 = \text{even}$, $r(r'+1) = \text{odd}$, that is $r = \text{odd}$, $r' = \text{even}$, so that $2\bar{w} = 2m\omega + 4m'\omega'$, where m and m' are integers. In like manner, for $\alpha = 3$ we shall have $2\bar{w} = 4m\omega + 2m'\omega'$, and for $\alpha = 2$, $2\bar{w} = 4m\omega + 4m'\omega'$.

The relation will now be shown between the sigma-quotients on the one hand, and the notation of Jacobi and Abel on the other. The Jacobian differential equation is $$\left(\frac{dx}{du}\right)^2 = (1-x^2)(1-k^2x^2).$$
In the second note, p. 267, we had
$$(\wp' u)^2 = 4(\wp u - e_1)(\wp u - e_2)(\wp u - e_3),$$
or, since
$$\wp u - e_\lambda = \left(\frac{\sigma_\lambda u}{\sigma u}\right)^2;\ \lambda = 1,\ 2,\ 3,$$
$$\wp' u = -2\frac{\sigma_\lambda u \,.\, \sigma_\mu u \,.\, \sigma_\nu u}{\sigma u \,.\, \sigma u \,.\, \sigma u}.$$
Writing now for convenience $\frac{\sigma u}{\sigma_\lambda u} = \xi_{0\lambda}$, $\frac{\sigma_\mu u}{\sigma_\nu u} = \xi_{\mu\nu}$, etc., the last equation becomes
$$\frac{d\xi_{0\lambda}}{du} = \xi_{\mu\lambda}\,.\,\xi_{\nu\lambda},\quad \frac{d\xi_{\mu\nu}}{du} = -(e_\mu - e_\nu)\xi_{\lambda\nu}\,.\,\xi_{0\nu},\quad \frac{d\xi_{\lambda 0}}{du} = -\xi_{\mu 0}\xi_{\nu 0}.$$
For $u = 0$ these functions ξ satisfy the conditions
$$\xi_{0\lambda} = 0,\quad \xi_{\mu\nu} = 1,\quad \xi_{\lambda 0} = \infty.$$
From
$$\wp u - e_\lambda = \left(\frac{\sigma_\lambda u}{\sigma u}\right)^2,\ \lambda = 1,\ 2,\ 3,\ = \lambda,\ \mu,\ \nu,$$
we obtain
$$\sigma_\mu^2 u - \sigma_\nu^2 u + (e_\mu - e_\nu)\sigma^2 u = 0,$$
$$\sigma_\nu^2 u - \sigma_\lambda^2 u + (e_\nu - e_\lambda)\sigma^2 u = 0,$$
$$\sigma_\lambda^2 u - \sigma_\mu^2 u + (e_\lambda - e_\mu)\sigma^2 u = 0,$$
$$(e_\mu - e_\nu)\sigma_\lambda u + (e_\nu - e_\lambda)\sigma_\mu u + (e_\lambda - e_\mu)\sigma_\nu u = 0.$$
The differential equations are then thus transformed
$$\left(\frac{d\xi_{0\lambda}}{du}\right)^2 = \left(\frac{d}{du}\frac{\sigma}{\sigma_\lambda}\right)^2 = \xi_{\mu\lambda}^2\,.\,\xi_{\nu\lambda}^2 \equiv \frac{\sigma_\mu^2\sigma_\nu^2}{\sigma_\lambda^2\sigma_\lambda^2}$$
$$= \frac{[\sigma_\lambda^2 + (e_\lambda - e_\mu)\sigma^2][\sigma_\lambda^2 - (e_\nu - e_\lambda)\sigma^2]}{\sigma_\lambda^2\,.\,\sigma_\lambda^2},$$

or
$$\left(\frac{d}{du}\xi_{0\lambda}\right)^2=\left[1-(e_\mu-e_\lambda)\left(\frac{\sigma}{\sigma_\lambda}\right)^2\right]\left[1-(e_\nu-e_\lambda)\left(\frac{\sigma}{\sigma_\lambda}\right)^2\right],$$
and similarly
$$\left(\frac{d}{du}\cdot\xi_{\mu\nu}\right)^2=[1-\xi_{\mu\nu}^2][e_\mu-e_\lambda+(e_\lambda-e_\nu)\xi_{\mu\nu}^2].$$
$$\left(\frac{d}{du}\cdot\xi_{\lambda0}\right)^2=[\xi_{\lambda0}^2-e_\lambda-e_\mu][\xi_{\lambda0}^2+e_\lambda-e_\nu],$$
and, in general, the four functions
$$\frac{\sigma u}{\sigma_\lambda u},\quad \frac{1}{\sqrt{e_\mu-e_\lambda}}\cdot\frac{\sigma_\mu u}{\sigma_\nu u},\quad \frac{1}{\sqrt{e_\nu-e_\lambda}}\cdot\frac{\sigma_\nu u}{\sigma_\mu u},\quad \frac{1}{\sqrt{e_\mu-e_\nu}\sqrt{e_\nu-e_\lambda}}\cdot\frac{\sigma_\lambda u}{\sigma u},$$
satisfy the same differential equation
$$\left(\frac{d\xi}{du}\right)^2=(1-(e_\mu-e_\lambda)\xi^2)(1-(e_\nu-e_\lambda)\xi^2).$$
In order to compare these with the Jacobian differential equation, we have only to write
$$\sqrt{e_\lambda-e_\mu}\,\xi_{0\lambda}=\xi,\quad u_1=\sqrt{e_\lambda-e_\mu}.u,\quad \frac{e_\nu-e_\lambda}{e_\mu-e_\lambda}=k^2$$
whereupon
$$\frac{\xi}{\sqrt{e_\lambda-e_\mu}}=\xi_{0\lambda}=\frac{\sigma u}{\sigma_\lambda u}=\frac{\operatorname{sn} u_1}{\sqrt{e_\lambda-e_\mu}}=\frac{\operatorname{sn}(\sqrt{e_\lambda-e_\mu}.u,\,k)}{\sqrt{e_\lambda-e_\mu}},$$
and in a similar manner all the twelve sigma-quotients are produced,

$$\frac{\sigma u}{\sigma_3 u}=\frac{1}{\sqrt{e_1-e_3}}\operatorname{sn}(\sqrt{e_1-e_3}.u,\,k)$$
$$\frac{\sigma_1 u}{\sigma_3 u}=\operatorname{cn}(\sqrt{e_1-e_3}.u,\,k)$$
$$\frac{\sigma_2 u}{\sigma_3 u}=\operatorname{dn}(\sqrt{e_1-e_3}.u,\,k)$$

$$\frac{\sigma_1 u}{\sigma u}=\sqrt{e_1-e_3}\,\frac{\operatorname{cn}(\sqrt{e_1-e_3}.u,\,k)}{\operatorname{sn}(\sqrt{e_1-e_3}.u,\,k)}$$
$$\frac{\sigma_2 u}{\sigma u}=\sqrt{e_1-e_3}\,\frac{\operatorname{dn}(\sqrt{e_1-e_3}.u,\,k)}{\operatorname{sn}(\sqrt{e_1-e_3}.u,\,k)}$$
$$\frac{\sigma_3 u}{\sigma u}=\sqrt{e_1-e_3}\,\frac{1}{\operatorname{sn}(\sqrt{e_1-e_3}.u,\,k)}$$

$$\frac{\sigma_1 u}{\sigma_2 u}=\operatorname{sn}\operatorname{coam}(\sqrt{e_1-e_3}.u,\,k)$$
$$\frac{\sigma_1 u}{\sigma_2 u}=\frac{1}{\sqrt{e_1-e_2}}\cos\operatorname{coam}(\sqrt{e_1-e_3}.u,\,k)$$
$$\frac{\sigma_3 u}{\sigma_2 u}=\frac{\sqrt{e_1-e_3}}{\sqrt{e_1-e_2}}\Delta\operatorname{coam}(\sqrt{e_1-e_3}.u,\,k)$$

$$\frac{\sigma u}{\sigma_1 u}=\frac{1}{\sqrt{e-e}}\operatorname{tn}(\sqrt{e_1-e_3}.u,\,k)$$
$$\frac{\sigma_2 u}{\sigma_1 u}=\frac{1}{\sin\operatorname{coam}(\sqrt{e_1-e_3}.u,\,k)}$$
$$\frac{\sigma_3 u}{\sigma_1 u}=\frac{1}{\operatorname{cn}(\sqrt{e_1-e_3}.u,\,k)}$$

$$\operatorname{coam}(\sqrt{e_1-e_3}.u,\,k)=\operatorname{am}(K-\sqrt{e_1-e_3}.u,\,k).$$
Abel writes (Oeuvres, t. I, p. 265, nouvelle édition),
$$u=\int_0\frac{dx}{\sqrt{(1-c^2x^2)(1+e^2x^2)}},$$
$$x=\phi u,\quad \sqrt{1-c^2x^2}=fu,\quad \sqrt{1+e^2x^2}=Fu,$$

comparing which with the Weierstrassian notation,

$$x = \phi u = \frac{\sigma u}{\sigma_2 u}, \quad fu = \frac{\sigma_1 u}{\sigma_2 u}, \quad Fu = \frac{\sigma_3 u}{\sigma_2 u},$$

if only $\quad e_1 - e_2 = -c^2, \quad -(e_3 - e_2) = c^2.$

As regards the analogues of Jacobi's K and K', it is to be noticed that, as usually defined by the equations

$$K = \int_0^1 \frac{dt}{\sqrt{1-t^2}.\sqrt{1-k^2t^2}}, \quad K' = \int_0^1 \frac{dt}{\sqrt{1-t^2}.\sqrt{1-k'^2t^2}},$$

the values are only unambiguous when the path of integration is fixed, it being generally understood that the path of integration is the straight line from 0 to 1. Corresponding to this we have, *e. g.*,

$$K = \sqrt{e_1 - e_3}(\omega + 4p\omega + 2q\omega'),$$

where the determination of the path of integration corresponds to the freedom of choice of p and q. Commonly we have $p = q = 0$, and

$$K = \omega\sqrt{e_1 - e_3}, \quad K'i = \omega'\sqrt{e_1 - e_3},$$

and then 2ω, $2\omega'$ form a primitive period-pair for the function $\wp(u, g_1, g_2)$, and, if we write as before $\omega + \omega' = \omega''$, or $\omega_1 + \omega_3 = \omega_2$, then is $\wp\omega_1 = e_1$, $\wp\omega_2 = e_2$, $\wp\omega_3 = e_3$.

The functions $\sigma_1 u$, $\sigma_2 u$, $\sigma_3 u$, can be represented as an infinite product of the same form as that for σu by writing

$$w_1 = (2\mu + 1)\omega + 2\mu'\omega', \quad w_2 = (2\mu + 1)\omega + (2\mu' + 1)\omega', \quad w_3 = 2\mu\omega + (2\mu + 1)\omega',$$

where $\mu, \mu' = 0, \pm 1, \pm 2, \ldots \pm \infty$; namely,

$$\sigma_\lambda u = e^{-\frac{1}{2}e_\lambda u^2} \Pi_{w_\lambda}\left(1 - \frac{u}{w_\lambda}\right) e^{\frac{u}{w_\lambda} + \frac{1}{2}\frac{u^2}{w_\lambda^2}}.$$

But these functions are also representible in the form of singly infinite products. As an aid in transforming, Professor Schwarz makes use of the following table. When the argument u assumes the values $u + \omega$, $u + \omega'$, $u + \omega''$, then the magnitudes $v = \frac{u}{2\omega}$, $z = e^{v\pi i}$, $2\eta\omega v^2$, $e^{2\eta\omega v^2}$, assume the values in the table, where $\tau = \frac{\omega'}{\omega}$, $h = e^{\tau\pi i}$,

u	$u + \omega$	$u + \omega'$	$u + \omega''$
v	$v + \frac{1}{2}$	$v + \frac{1}{2}\tau$	$v + \frac{1}{2} + \frac{1}{2}\tau$
z	iz	$h^{\frac{1}{2}}.z$	$i.h^{\frac{1}{2}}.z$
$2\eta\omega v^2$	$2\eta\omega v^2 + \eta u + \frac{1}{2}\eta\omega$	$2\eta\omega v^2 + \eta' u + \frac{1}{2}\eta'\omega' + \frac{1}{4}\tau\pi i + v\pi i$	$2\eta\omega v^2 + \eta'' u + \frac{1}{2}\eta''\omega'' + \frac{1}{4}\pi i + \frac{1}{4}\tau\pi i + v\pi i$
$e^{2\eta\omega v^2}$	$e^{2\eta\omega v^2}.e^{\eta u}.e^{\frac{1}{2}\eta\omega}$	$e^{2\eta\omega v^2}.e^{\eta' u}.e^{\frac{1}{2}\eta'\omega'}.h^{\frac{1}{4}}.z$	$e^{2\eta\omega v^2}.e^{\eta'' u}.e^{\frac{1}{2}\eta''\omega''}.\sqrt{i}.h^{\frac{1}{4}}.z$

With the help of this table the infinite product for σu on page 261, Vol. VI of this Journal, can be transformed as follows: Developing

$$(1-h^{2n}.z^2)(1-h^{2n}z^{-2}) = 1-2h^{2n}\cos\frac{u\pi}{\omega}+h^{4n}$$

since

$$1-\cos\frac{u\pi}{\omega} = 2\sin^2\frac{u\pi}{2\omega}$$

$$\begin{aligned}(1-h^{2n}z^2)(1-h^{2n}z^{-2}) &= 1-2h^{2n}+4h^{2n}.\sin^2\frac{u\pi}{\omega}+h^{4n}\\ &= (1-h^{2n})^2+4h^{2n}.\sin^2\frac{u\pi}{2\omega}\\ &= (1-h^{2n})^2\left\{1+\left(\frac{2}{h^n-h^{-n}}\right)^2.\sin^2\frac{u\pi}{2\omega}\right\},\end{aligned}$$

but $\frac{h^n-h^{-n}}{2} = i\sin n\frac{\pi\omega'}{\omega}$, consequently

$$(1-h^{2n}z^2)(1-h^{2n}z^{-2}) = (1-h^{2n})^2\left\{1-\frac{\sin^2\frac{u\pi}{2\omega}}{\sin^2 n\frac{\pi\omega'}{\omega}}\right\}$$

and

$$\sigma u = \frac{2\omega}{\pi}\cdot e^{\frac{\eta u^2}{2\omega}}.\frac{z-z^{-1}}{2i}\Pi_n\frac{1-h^{2n}z^2}{1-h^{2n}}\cdot\frac{1-h^{2n}z^{-2}}{1-h^{2n}},$$

which is the desired expression for σu. Since further

$$\sigma_1 u = e^{-\eta u}.\frac{\sigma(u+\omega)}{\sigma\omega},$$

we obtain by the assistance of the table the analogous expressions for σ_1, σ_2, σ_3; namely

$$\begin{aligned}\sigma_1 u &= e^{2\eta\omega v^2}.\cos v\pi\,\Pi_n\frac{\cos(n\pi-v)\pi}{\cos n\tau\pi}.e^{-v\pi i}.\Pi_n\frac{\cos(n\tau+v)\pi}{\cos n\tau\pi}.e^{v\pi i}\\ &= e^{2\eta\omega v^2}.\frac{z+z^{-1}}{2}.\Pi_n\frac{1+h^{2n}z^{-2}}{1+h^{2n}}.\Pi_n\frac{1+h^{2n}z^2}{1+h^{2n}}\\ &= e^{2\eta\omega v^2}.\cos v\pi.\Pi_n\frac{1+2h^{2n}\cos 2v\pi+h^{4n}}{(1+h^{2n})^2},\end{aligned}$$

$$\begin{aligned}\sigma_2 u &= e^{2\eta\omega v^2}.\Pi_n\frac{\cos((n-\frac{1}{2})\tau-v)\pi}{\cos(n-\frac{1}{2})\tau\pi}.e^{-v\pi i}.\Pi_n\frac{\cos((n-\frac{1}{2})\tau+v)\pi}{\cos(n-\frac{1}{2})\tau\pi}.e^{v\pi i}\\ &= e^{2\eta\omega v^2}.\Pi_n\frac{1+h^{2n-1}.z^{-2}}{1-h^{2n-1}}.\Pi_n\frac{1+h^{2n-1}.z^2}{1+h^{2n-1}}\\ &= e^{2\eta\omega v^2}.\Pi_n\frac{1+2h^{2n-1}.\cos 2v\pi+h^{4n-2}}{(1+h^{2n-1})^2},\end{aligned}$$

$$\begin{aligned}\sigma_3 u &= e^{2\eta\omega v^2}.\Pi_n\frac{\sin((n-\frac{1}{2})\tau-v)\pi}{\sin(n-\frac{1}{2})\tau\pi}.e^{-v\pi i}.\Pi_n\frac{\sin((n-\frac{1}{2})\tau+v)\pi}{\sin(n-\frac{1}{2})\tau\pi}.e^{v\pi i}\\ &= e^{2\eta\omega v^2}.\Pi_n\frac{1-h^{2n-1}z^{-2}}{1-h^{2n-1}}.\Pi_n\frac{1-h^{2n-1}z^2}{1-h^{2n-1}}\\ &= e^{2\eta\omega v^2}.\Pi_n\frac{1-2h^{2n-1}.\cos 2v\pi+h^{4n-2}}{(1-h^{2n-1})^2}.\end{aligned}$$

Analogous to the expression for σu at the bottom of p. 261, we have

$$\sigma_1 u = e^{\frac{\eta u^2}{2\omega}} . \cos\frac{u\pi}{2\omega} . \Pi_n \left(1 - \frac{\sin^2 \frac{u\pi}{2\omega}}{\cos^2 n \frac{\omega' \pi}{\omega}} \right)$$

$$\sigma_2 u = e^{\frac{\eta u^2}{2\omega}} . \Pi_n \left(1 - \frac{\sin^2 \frac{u\pi}{2\omega}}{\cos^2 (n - \frac{1}{2}) \frac{\omega' \pi}{\omega}} \right)$$

$$\sigma_3 u = e^{\frac{\eta u^2}{2\omega}} . \Pi_n \left(1 - \frac{\sin^2 \frac{u\pi}{2\omega}}{\sin^2 (n - \frac{1}{2}) \frac{\omega' \pi}{\omega}} \right).$$

In the normal case we shall have ω real and $\frac{\omega'}{\omega}$ imaginary, and therefore none of the quotients under the product sign can assume the value unity. The functions σu and $\sigma_1 u$ disappear accordingly only when $\sin\frac{u\pi}{2\omega}$ and $\cos\frac{u\pi}{2\omega}$ respectively vanish. The Jacobian functions $\frac{\sigma}{\sigma_3}$, $\frac{\sigma_1}{\sigma_3}$, $\frac{\sigma_2}{\sigma_3}$, are analogous, the first to the sine, the second to the cosine, while the third remains positive for real values of u. The Abelian forms $\frac{\sigma}{\sigma_2}$, $\frac{\sigma_1}{\sigma_2}$, $\frac{\sigma_3}{\sigma_2}$, are analogous, the first to the tangent, the second and third to the secant.

The expressions for the root-differences and the connection with the ϑ-functions are obtained in the following manner. Defining as above $h = e^{\frac{\omega'}{\omega}\pi i} = e^{\tau\pi i}$, and writing

$$h_0 = \prod_{n=1}^{\infty}(1 - h^{2n}),\quad h_1 = \prod_{n=1}^{\infty}(1 + h^{2n}),\quad h_2 = \prod_{n=1}^{\infty}(1 + h^{2n-1}),\quad h_3 = \prod_{n=1}^{\infty}(1 - h^{2n-1}),$$

then is

$$h_0 = h_0 . h_1 . h_2 . h_3;\quad h_1 . h_2 . h_3 = 1.$$

For

$$\begin{aligned} h_0 &= (1 - h^2)(1 - h^4)(1 - h^6) \ldots \\ &= (1 + h)(1 + h^2)(1 + h^3) \ldots \\ &\quad . (1 - h)(1 - h^2)(1 - h^3) \ldots \end{aligned}$$

and the proof is apparent. With the aid of these facts the relation between the periods and the root-differences is easily discovered. Starting again from the "pocket-edition"

$$\wp u - \wp v = \frac{\sigma(u+v)\,\sigma(v-u)}{\sigma^2 u\ \sigma^2 v},$$

and writing $u = \omega$, $v = \omega'$, $\omega'' = \omega + \omega'$, the equation becomes

$$e_1 - e_3 = \frac{\sigma\omega'' . \sigma(\omega' - \omega)}{\sigma^2\omega\ \sigma^2\omega'}.$$

But noticing that $\omega' - \omega = \omega'' - 2\omega$, $\omega'' - \omega = \omega'$, and that

$$\sigma(\omega'' - 2\omega) = -\sigma(-\omega'' + 2\omega) = -\sigma\omega'' . e^{-2\eta(\omega' - \omega)},$$
$$\sigma(\omega' - \omega) = \sigma(\omega'' - 2\omega) = -\sigma\omega'' . e^{-2\eta(\omega'' - \omega)},$$

the equation becomes

$$e_1 - e_3 = -\left(\frac{\sigma\omega''}{\sigma\omega\,\sigma\omega'}\right)^2 . e^{-2\eta\omega'}.$$

And in a similar manner, writing $u = \omega$, $v = \omega + \omega'$, we have,

$$e_1 - e_2 = \frac{\sigma(\omega' + 2\omega)\,\sigma(\omega')}{\sigma^2\omega\,\sigma^2\omega''} = -\left(\frac{\sigma\omega'}{\sigma\omega\,\sigma\omega''}\right)^2 . e^{2\eta\omega''},$$

and for $u = \omega''$, $v = \omega'$,

$$e_2 - e_3 = \frac{\sigma(\omega + 2\omega'')\,\sigma(\omega)}{\sigma^2\omega''\,\sigma^2\omega'} = \left(\frac{\sigma\omega'}{\sigma\omega\,\sigma\omega''}\right)^2 . e^{2\eta'\omega''}.$$

But these formulæ can be still further simplified. From

$$\sigma u = \frac{2\omega}{\pi} e^{\frac{\eta u^2}{2\omega}} . \mathrm{sn}\,\frac{u\pi'}{2\omega} . \Pi \frac{1 - h^{2n}z^2}{1 - h^{2n}} \cdot \frac{1 - h^{2n}z^{-2}}{1 - h^{2n}},$$

we have for $u = \omega$

$$\sigma\omega = \frac{2\omega}{\pi} e^{\frac{\eta\omega}{2}} . \Pi \frac{1 + h^{2n}}{1 - h^{2n}} \cdot \frac{1 + h^{2n}}{1 - h^{2n}} = \frac{2\omega}{\pi} e^{\frac{\eta\omega}{2}} . \frac{h_1^2}{h_0^2}.$$

And similarly

$$\sigma\omega' = e^{\eta\frac{\omega'^2}{2\omega}} . \frac{1}{1 - h} . h^{-\frac{1}{4}} . \frac{h^{-1}}{2i} \cdot \frac{h_3^2}{h_0},$$

$$\sigma\omega'' = \frac{2\omega}{\pi} e^{\frac{(\eta + \eta')\omega''}{2}} \sqrt{\frac{i}{2}} . h^{-\frac{1}{4}} . \frac{h_2^2}{h_0^2}.$$

The expressions for the root-differences become then

$$\left(\frac{2\omega}{\pi}\right)^2 (e_1 - e_3) = \left(\frac{h_2 h_0}{h_1 h_3}\right)^4 = h_0^4 h_2^8,$$

$$\left(\frac{2\omega}{\pi}\right)^2 (e_1 - e_2) = \left(\frac{h_3^2 h_0^2}{h_2^2 h_1^2}\right)^2 = h_0^4 h_3^8,$$

$$\left(\frac{2\omega}{\pi^2}\right)^2 (e_2 - e_3) = \qquad 16 . h . h_0^4 h_1^8,$$

where $h = e^{\frac{\omega'}{\omega}\pi i}$, and h_0, h_1, h_2, h_3, are defined above. In accordance also with previous definitions for the k and k' of Jacobi,

$$k^2 = \frac{e_2 - e_3}{e_1 - e_3} = 16h \left\{\frac{(1 + h^2)(1 + h^4)(1 + h^6) \ldots}{(1 + h)(1 + h^3)(1 + h^5) \ldots}\right\}^8$$

$$k'^2 = \frac{e_1 - e_2}{e_1 - e_3} = \qquad \left\{\frac{(1 - h)(1 - h^3)(1 - h^5) \ldots}{(1 + h)(1 + h^3)(1 + h^5) \ldots}\right\}^8.$$

The four sigma-functions are now expressed through the functions ϑ and Θ, as follows. The infinite product $F(z) = \Pi_n (1 - h^{2n})(1 + h^{2n-1}z^{-2})(1 + h^{2n-1}z^{+2})$ can be expressed as a power series of z^2 which converges for all values of z except

$z=0$, so long as $h<1$. This is plain from the following identity,

$$\Pi_n(1-h^{2n})(1+h^{2n-1}z^{-2})(1+h^{2n-1}z^2)$$
$$=1+h(z^2+z^{-2})+h^4(z^4+z^{-4})+h^9(z^6+z^{-6})+\ldots$$

The development of the sigma-function follows at once, since

$$h_2^2 . h_0 . \sigma_2 u = e^{\frac{\eta u^2}{2\omega}} . F(z) \text{ and } \sqrt{\frac{2\omega}{\pi}}\sqrt[4]{e_1-e_3} = h_0 . h_2^2.$$

We have then

$$\sqrt{\frac{2\omega}{\pi}}\sqrt[4]{e_1-e_3} . \sigma_3 u = e^{\frac{\eta u^2}{2\omega}} F(z) = e^{\frac{\eta u^2}{2\omega}} . \sum_n h^{n^2} z^{2n}$$

$$\sqrt{\frac{2\omega}{\pi}}\sqrt[4]{e_1-e_2} . \sigma_3 u = e^{\frac{\eta u^2}{2\omega}} F(zi) = e^{\frac{\eta u^2}{2\omega}} . \sum_n (-1)^n h^{n^2} z^{2n}$$

$$\sqrt{\frac{2\omega}{\pi}}\sqrt[4]{e_2-e_3} . \sigma_1 u = e^{\frac{\eta u^2}{2\omega}} . h^{\frac{1}{4}} . z . F(zh^{\frac{1}{2}}) = e^{\frac{\eta u^2}{2\omega}} . \sum_n h^{\frac{1}{4}(2n+1)^2} z^{2n+1}$$

$$\sqrt{\frac{2\omega}{\pi}}\sqrt[4]{e_1-e_3}\sqrt[4]{e_1-e_2} . \sqrt[4]{e_2-e_3} . \sigma u = \frac{1}{i} e^{\frac{\eta u^2}{2\omega}} h^{\frac{1}{4}} z . F(izh^{\frac{1}{2}}).$$

It will be remembered that e_1, e_2, e_3 are the roots of the equation

$$4x^3-g2x-g_3=0.$$

The discriminant of which, squared

$$\{(e_1-e_2)(e_2-e_3)(e_3-e_1)\}^2 = \frac{g_2^3-27g_3^2}{16} = G,$$

so that
$$\sqrt{\frac{2\omega}{\pi}} \cdot \sqrt[8]{G} = \frac{1}{i} e^{\frac{\eta u^2}{2\omega}} h^{\frac{1}{4}} . z . F(h^{\frac{1}{2}}z) = e^{\frac{\eta u^2}{2\omega}} \frac{1}{i} \sum_n (-1)^n h^{\frac{1}{4}(2n+1)^2} z^{2n+1}$$
$$n=0, \quad \pm 1, \quad \pm 2, \ldots \pm \infty,$$
$$z=e^{\frac{u\pi i}{2\omega}}, \quad h=e^{\frac{\omega'}{\omega}\pi i}.$$

The expression for Fz becomes, since

$$z^{2n}+z^{-2n} = 2\cos\frac{nu\pi i}{\omega}.$$

$$F(z) = 1+2h\cos\frac{u\pi i}{\omega}+2h^4\cos 2\frac{u\pi i}{\omega}+2h^9\cos 3\frac{u\pi i}{\omega}+\ldots$$

which is the ϑ series of Jacobi (*Werke*, Bd. I, p. 501). Weierstrass defines the ϑ functions as follows:

$$\frac{1}{i}\sum_n(-1)^n h^{\frac{1}{4}(2n+1)^2} z^{2n+1} = 2h^{\frac{1}{4}}\sin v\pi - 2h^{\frac{9}{4}}\sin 3v\pi + 2h^{\frac{25}{4}}\sin 5v\pi = \ldots = \vartheta_1(v),$$

$$\sum_n h^{\frac{1}{4}(2n+1)^2} z^{2n+1} = 2h^{\frac{1}{4}}\cos v\pi + 2h^{\frac{9}{4}}\cos 3v\pi + 2h^{\frac{25}{4}}\cos 5v\pi + \ldots = \vartheta_2(v),$$

$$\sum_n h^{n^2} z^{2n} = 1+2h\cos 2v\pi + 2h^4\cos 4v\pi + 2h^9\cos 6v\pi + \ldots = \vartheta_3(v),$$

$$\sum_n (-1)^n h^{n^2} z^{2n} = 1-2h\cos 2v\pi + 2h^4\cos 4v\pi - 2h^9\cos 6v\pi + \ldots = \vartheta_0(v),$$

which agree with Jacobi's notation when $v\pi = x$; $h = q$. Hermite writes

$$\sum_m (-1)^{m\nu} e^{i\pi[(2m+\mu)x + \frac{1}{4}\omega(2m+\mu)^2]} = \theta_{\mu,\nu}(x),$$

$$m = 0, \quad \pm 1, \quad \pm 2, \ldots \pm \infty.$$

If now we define the function Θ by the equation

$$\Theta_\nu(u) = e^{2\eta\omega v^2}.\vartheta_\nu\left(v, \frac{\omega'}{\omega}\right); \; u = 2\omega v,$$

then the four sigma-functions are thus expressed through Θ,

$$\sqrt{\frac{2\omega}{\pi}}\sqrt[4]{G}.\sigma u \qquad = e^{2\eta\omega v^2}\vartheta_1\left(v, \frac{\omega'}{\omega}\right) = \Theta_1(u, \omega, \omega'),$$

$$\sqrt{\frac{2\omega}{\pi}}\sqrt[4]{e_\mu - e_\nu}.\sigma_\lambda u = e^{2\eta\omega v^2}\vartheta_2\left(v, \frac{\omega'}{\omega}\right) = \Theta_2(u, \omega, \omega'),$$

$$\sqrt{\frac{2\omega}{\pi}}\sqrt[4]{e_\lambda - e_\nu}.\sigma_\mu u = e^{2\eta\omega v^2}\vartheta_3\left(v, \frac{\omega'}{\omega}\right) = \Theta_3(u, \omega, \omega'),$$

$$\sqrt{\frac{2\omega}{\pi}}\sqrt[4]{e_\lambda - e_\mu}.\sigma_\nu u = e^{2\eta\omega v^2}\vartheta_0\left(v, \frac{\omega'}{\omega}\right) = \Theta_0(u, \omega, \omega'),$$

where again $v = \frac{u}{2\omega}$. By developing according to powers of v, and comparing we have

$$\sqrt{\frac{2\omega}{\pi}}\sqrt[4]{G} = \frac{1}{2\omega}\vartheta'_1(0) = \frac{\pi}{\omega}h^{\frac{1}{4}}(1 - 3h^{1.2} + 5h^{2.3} - 7h^{3.4} + \ldots)$$

$$\sqrt{\frac{2\omega}{\pi}}\sqrt[4]{e_\mu - e_\nu} = \vartheta_2(0) \qquad = 2h^{\frac{1}{4}}(1 + h^{1.2} + h^{2.3} + h^{3.4} + \ldots)$$

$$\sqrt{\frac{2\omega}{\pi}}\sqrt[4]{e_\lambda - e_\nu} = \vartheta_3(0) \qquad = 1 + 2h + 2h^4 + 2h^9 + \ldots$$

$$\sqrt{\frac{2\omega}{\pi}}\sqrt[4]{e_\lambda - e_\mu} = \vartheta_0(0) \qquad = 1 - 2h + 2h^4 - 2h^9 + \ldots$$

From these spring the following equations which become useful in computation:

$$\vartheta'_1(0) = \pi\vartheta_0(0).\vartheta_2(0).\vartheta_3(0), \quad \vartheta_0^4(0) + \vartheta_2^4(0) = \vartheta_3^4(0),$$

$$e_\lambda = \frac{1}{3}\left(\frac{\pi}{2\omega}\right)^2(\vartheta_3^4(0) + \vartheta_0^4(0)), \quad e_\mu = \frac{1}{3}\left(\frac{\pi}{2\omega}\right)^2(\vartheta_2^4(0) - \vartheta_0^4(0)),$$

$$e_\nu = -\frac{1}{3}\left(\frac{\pi}{2\omega}\right)^2(\vartheta_2^4(0) + \vartheta_3^4(0)),$$

$$\sqrt{\frac{2\omega}{\pi}} = \frac{2h^{\frac{1}{4}} + 2h^{\frac{9}{4}} + 2h^{\frac{25}{4}} + \ldots}{\sqrt[4]{e_\mu - e_\nu}} = \frac{2}{\sqrt[4]{e_\lambda - e_\nu} - \sqrt[4]{e_\lambda - e_\mu}}(2h + 2h^9 + 2h^{25} + \ldots)$$

$$\sqrt{\frac{2\omega}{\pi}} = \frac{1 + 2h + 2h^4 + 2h^9 + \ldots}{\sqrt[4]{e_\lambda - e_\nu}} = \frac{2}{\sqrt[4]{e_\lambda - e_\nu} + \sqrt[4]{e_\lambda - e_\mu}}(1 + 2h^4 + 2h^{16} + \ldots)$$

$$\sqrt{k} = \frac{\sqrt[4]{e_\mu - e_\nu}}{\sqrt[4]{e_\lambda - e_\nu}} = \frac{\vartheta_2\left(0, \frac{\omega'}{\omega}\right)}{\vartheta_3\left(0, \frac{\omega'}{\omega}\right)} = \frac{2h^{\frac{1}{4}} + 2h^{\frac{9}{4}} + 2h^{\frac{25}{4}} + \ldots}{1 + 2h + 2h^4 + 2h^9 + \ldots}$$

$$\sqrt{k'} = \frac{\sqrt[4]{e_\lambda - e_\mu}}{\sqrt[4]{e_\lambda - e_\nu}} = \frac{\vartheta_0\left(0, \frac{\omega'}{\omega}\right)}{\vartheta_3\left(0, \frac{\omega'}{\omega}\right)} = \frac{1 - 2h + 2h^4 - 2h^9 + \ldots}{1 + 2h + 2h^4 + 2h^9 + \ldots}.$$

We define

$$l = \frac{1 - \sqrt{k'}}{1 + \sqrt{k'}} = \frac{\sqrt[4]{e_\lambda - e_\nu} - \sqrt[4]{e_\lambda - e_\mu}}{\sqrt[4]{e_\lambda - e_\nu} + \sqrt[4]{e_\lambda - e_\mu}} = \frac{2h + 2h^9 + \ldots}{1 + 2h^4 + 2h^{16} + \ldots} = \frac{\vartheta_2\left(0, 4\frac{\omega'}{\omega}\right)}{\vartheta_3\left(0, 4\frac{\omega'}{\omega}\right)}$$

which is identically satisfied by writing

$$h = q = \frac{l}{2} + 2\left(\frac{l}{2}\right)^5 + 15\left(\frac{l}{2}\right)^9 + 150\left(\frac{l}{2}\right)^{13} + \ldots$$

These expressions for l and h make the computation of the period 2ω or of Jacobi's K very easy.

In explaining more at length the methods employed for computing, I cannot do better than to give them with scarcely any variation from the words of that most genial expounder of Weierstrass' theories, Prof. Schwarz. When the three roots e_1, e_2, e_3, are once known, we can, by the aid of the formulæ just given, not only compute with the greatest ease the two periods 2ω, $2\omega'$, but we can also express the sigma-quotients through such ϑ series that the argument h shall have the smallest possible value, and the series converge most rapidly. This last end is brought about by so choosing the order of magnitude of e_1, e_2, e_3, that

$$l = \frac{\sqrt[4]{e_\lambda - e_\nu} - \sqrt[4]{e_\lambda - e_\mu}}{\sqrt[4]{e_\lambda - e_\nu} + \sqrt[4]{e_\lambda - e_\mu}}$$

which is used in the computation of h shall be as small as possible. Of the several cases which present themselves according as the invariants g_2, g_3 and the roots e_1, e_2. e_3, are real or imaginary, I shall discuss here but one, where all are real. The roots will be real when the discriminant $G = \frac{1}{16}(g_2^3 - 27g_3^2)$ of the cubic equation $4s^3 - g_2 s - g_3 = 0$ is positive and g_2 and g_3 real. We then assume $e_1 > e_2 > e_3$ and all the radicals positive; farther $\lambda = 1$, $\mu = 2$, $\nu = 3$, whereupon

$$k^2 = \frac{e_2 - e_3}{e_1 - e_3}, \quad k'^2 = \frac{e_1 - e_2}{e_1 - e_3}, \quad \omega_1 = \frac{K}{\sqrt{e_1 - e_3}}, \quad \omega_3 = \frac{K'i}{\sqrt{e_1 - e_3}},$$

$$\frac{\sigma'\omega_1}{\sigma\omega_1} = \eta_1, \quad \frac{\sigma'\omega_3}{\sigma\omega_3} = \eta_3, \quad \tau = \frac{\omega_3}{\omega_1}, \quad h = e^{\tau\pi i}, \quad h_1 = h' = e^{-\frac{\pi i}{\tau}},$$

$$v = \frac{u}{2\omega_1}, \quad v_1 = \frac{ui}{2\omega_3}, \quad \text{where } \omega_1, \ \frac{\omega_3}{i}, \ \frac{\tau}{i}, \ h_1 h, \text{ are positive.}$$

For the computation of the periods the following system of equations is used:

$$l = \frac{\sqrt[4]{e_1 - e_3} - \sqrt[4]{e_1 - e_2}}{\sqrt[4]{e_1 - e_3} + \sqrt[4]{e_1 - e_2}}, \quad h = \frac{l}{2} + 2\left(\frac{l}{2}\right)^5 + 15\left(\frac{l}{2}\right)^9 + 150\left(\frac{l}{2}\right)^{13} + \ldots$$

$$\sqrt{\frac{2\omega_1}{\pi}} = \frac{2}{\sqrt[4]{e_1-e_3}+\sqrt[4]{e_1-e_2}}(1+2h^4+2h^{16}+\ldots), \quad \omega_3 = \frac{\omega_1 i}{\pi}\,\text{log. nat.}\,\frac{1}{h},$$

$$2\eta_1\omega_1 = \frac{\pi^2}{6}\,\frac{1-3^3h^2+5^3h^6-7^3h^{12}+\ldots}{1-3h^2+5h^6-7h^{12}+\ldots}, \quad \eta_1\omega_3-\omega_1\eta_3=\tfrac{1}{2}\pi i,$$

$$\sqrt{\frac{2\omega_1}{\pi}}\,\sqrt[4]{e_2-e_3} = 2h^{\frac{1}{4}}(1+h^2+h^6+h^{12}+\ldots)$$

$$\sqrt{\frac{2\omega_1}{\pi}}\,\sqrt[4]{e_1-e_3} = 1+2h+2h^4+2h^9+\ldots$$

$$\sqrt{\frac{2\omega_1}{\pi}}\,\sqrt[4]{e_1-e_2} = 1-2h+2h^4-2h^9+\ldots$$

$$\sqrt{\frac{2\omega_1}{\pi}}\,\sqrt[4]{G} = \frac{\pi}{\omega_1}\cdot h^{\frac{1}{4}}(1-3h^2+5h^6-7h^9+\ldots)$$

For the calculation of the sigma-functions we shall have

$$\sqrt{\frac{2\omega_1}{\pi}}\,\sqrt[4]{G}\,.\,\sigma u = e^{2\eta_1\omega_1 v^2}\,\vartheta_1(v,\tau)$$

$$\sqrt{\frac{2\omega_1}{\pi}}\,\sqrt[4]{e_2-e_3}\,.\,\sigma_1 u = e^{2\eta_1\omega_1 v^2}\,\vartheta_2(v,\tau)$$

$$\sqrt{\frac{2\omega_1}{\pi}}\,\sqrt[4]{e_1-e_3}\,.\,\sigma_2 u = e^{2\eta_1\omega_1 v^2}\,\vartheta_3(v,\tau)$$

$$\sqrt{\frac{2\omega_1}{\pi}}\,\sqrt[4]{e_1-e_2}\,.\,\sigma_3 u = e^{2\eta_1\omega_1 v^2}\,\vartheta_0(v,\tau)$$

$$\vartheta_0(v,\tau) = 1-2h\cos 2v\pi+2h^4\cos 4v\pi-2h^9\cos 6v\pi+\ldots$$
$$\vartheta_1(v,\tau) = 2h^{\frac{1}{4}}\sin v\pi-2h^{\frac{9}{4}}\sin 3v\pi+2h^{\frac{25}{4}}\sin 5v\pi-\ldots$$
$$\vartheta_2(v,\tau) = 2h^{\frac{1}{4}}\cos v\pi+2h^{\frac{9}{4}}\cos 3v\pi+2h^{\frac{25}{4}}\cos 5v\pi+\ldots$$
$$\vartheta_3(v,\tau) = 1+2h\cos 2v\pi+2h^4\cos 4v\pi+2h^9\cos 6v\pi+\ldots$$

If, however, we have to choose $\lambda=3$, $\mu=2$, $\nu=1$, the equations become

$$l_1 = \frac{\sqrt[4]{e_1-e_3}-\sqrt[4]{e_2-e_3}}{\sqrt[4]{e_1-e_3}+\sqrt[4]{e_2-e_3}}, \quad h_1 = \frac{l_1}{2}+2\left(\frac{l_1}{2}\right)^5+15\left(\frac{l}{2}\right)^9+150\left(\frac{l}{2}\right)^{13}+\ldots$$

$$\sqrt{\frac{2\omega_3}{\pi i}} = \frac{2}{\sqrt[4]{e_1-e_3}+\sqrt[4]{e_2-e_3}}(1+2h_1^4+2h_1^{16}+\ldots) \quad \omega_1 = \frac{\omega_3}{\pi i}\,\text{log. nat.}\left(\frac{1}{h_1}\right)$$

$$\sqrt{\frac{2\omega_3}{\pi i}}\,.\,\sqrt[4]{e_1-e_2} = 2h_1^{\frac{1}{4}}(1+h_1^2+h_1^6+h_1^{12}+\ldots)$$

$$\sqrt{\frac{2\omega_3}{\pi i}}\,.\,\sqrt[4]{e_1-e_3} = 1+2h_1+2h_1^4+2h_1^9+\ldots$$

$$\sqrt{\frac{2\omega_3}{\pi i}}\,.\,\sqrt[4]{e_2-e_3} = 1-2h_1+2h_1^4-2h_1^9+\ldots$$

$$\sqrt{\frac{2\omega_3}{\pi i}}\,.\,\sqrt[4]{G} = \frac{\pi i}{\omega_3}\,h_1^{\frac{1}{4}}(1-3h_1^2+5h_1^6-7h_1^{12}+\ldots)$$

$$\sqrt{\frac{2\omega_3}{\pi i}}\sqrt[4]{G}.\,\sigma u = \frac{1}{i}\,e^{-2\eta_3\omega_3 v_1^2}\,\vartheta_1\left(v_1 i,\ \frac{-1}{\tau}\right),$$

$$\sqrt{\frac{2\omega_3}{\pi i}}\sqrt[4]{e_2-e_3}.\,\sigma_1 u = e^{-2\eta_3\omega_3 v_1^2}\,\vartheta_0\left(v_1 i,\ \frac{-1}{\tau}\right),$$

$$\sqrt{\frac{2\omega_3}{\pi i}}\sqrt[4]{e_1-e_3}.\,\sigma_2 u = e^{-2\eta_3\omega_3 v_1^2}\,\vartheta_3\left(v_1 i,\ \frac{-1}{\tau}\right),$$

$$\sqrt{\frac{2\omega_3}{\pi i}}\sqrt[4]{e_1-e_2}.\,\sigma_3 u = e^{-2\eta_3\omega_3 v_1^2}\,\vartheta_2\left(v_1 i,\ \frac{-1}{\tau}\right).$$

$$\vartheta_0\left(v_1 i,\ \frac{-1}{\tau}\right) = 1 - h_1(e^{2v_1\pi} + e^{-2v_1\pi}) + h_1^4(e^{4v_1\pi} + e^{-4v_1\pi}) - h_1^9 e^{6v_1\pi} + e^{-6v_1\pi}) + \ldots$$

$$\frac{1}{i}\vartheta_1\left(v_1 i,\ \frac{-1}{\tau}\right) = h_1^{\frac{1}{4}}(e^{v_1\pi} - e^{-v_1\pi}) - h_1^{\frac{9}{4}}(e^{3v_1\pi} - e^{3v_1\pi}) + h_1^{\frac{25}{4}}(e^{5v_1\pi} - e^{-5v_1\pi}) - \ldots$$

$$\vartheta_2\left(v_1 i,\ \frac{-1}{\tau}\right) = h_1^{\frac{1}{4}}(e^{v_1\pi} + e^{-v_1\pi}) + h_1^{\frac{9}{4}}(e^{3v_1\pi} + e^{3v_1\pi}) + h_1^{\frac{25}{4}}(e^{5v_1\pi} + e^{-5v_1\pi}) + \ldots$$

$$\vartheta_3\left(v_1 i,\ \frac{-1}{\tau}\right) = 1 + h_1(e^{2v_1\pi} + e^{-2v_1\pi}) + h_1^4(e^{4v_1\pi} + e^{-4v_1\pi}) + h_1^9(e^{6v_1\pi} + e^{-6v_1\pi}) + \ldots$$

When $e_2 - e_3 \lesseqgtr e_1 - e_2$, that is, $e_2 \lesseqgtr 0$, then is

$$l_1 \gtreqless \frac{\sqrt[4]{2}-1}{\sqrt[4]{2}+1},\quad h_1 \gtreqless e^{-\pi} \qquad \text{and } l \lesseqgtr \frac{\sqrt[4]{2}-1}{\sqrt[4]{2}+1},\quad h \lesseqgtr e^{-\pi}$$

When g_3 is positive, that is $e_2 < 0$, it is advisable to use the formulæ for h and l, otherwise those for h_1, l_1 will be found best, because in the first case we have $h < h_1$, in the second $h_1 < h$.

For the calculation of the elliptic integral of the first kind in the ordinary form, Professor Schwarz throws the necessary formulæ into the following shape "Among the values of u for which $\wp u = s$, there are, in consequence of the equation

$$\frac{\sigma_3(u \pm 2\omega')}{\sigma_2(u \pm 2\omega')} = -\frac{\sigma_3 u}{\sigma_2 u},$$

always such for which the real component of $\dfrac{\sqrt[4]{e_1-e_2}}{\sqrt[4]{e_1-e_3}}$ is not negative and consequently the modulus of

$$\frac{\sqrt[4]{e_1-e_3}.\,\sigma_2 u - \sqrt[4]{e_1-e_2}.\,\sigma_3 u}{\sqrt[4]{e_1-e_3}.\,\sigma_2 u + \sqrt[4]{e_1-e_3}.\,\sigma_3 u} = \frac{\sqrt[4]{e_1-e_3} - \sqrt[4]{e_1-e_2}}{\sqrt[4]{e_1-e_3} + \sqrt[4]{e_1-e_2}} \cdot \frac{\sigma_1(2u,\ \omega,\ 4\omega')}{\sigma_2(2u,\ \omega,\ 4\omega')},$$

is not greater than unity. The value of $\sqrt{s-e_2}$ can be chosen at pleasure, after which the value of $\sqrt{s-e_3}$ can be so taken that the real component of

$$\frac{\sqrt[4]{e_1-e_2}.\,\sqrt{s-e_2}}{\sqrt[4]{e_1-e_3}.\,\sqrt{s-e_3}}$$

shall not be negative. We then write

$$\frac{\sqrt[4]{e_1-e_3}-\sqrt[4]{e_1-e_2}}{\sqrt[4]{e_1-e_3}+\sqrt[4]{e_1-e_2}}=l,\quad \frac{\sqrt[4]{e_1-e_3}\cdot\sqrt{s-e_2}-\sqrt[4]{e_1-e_2}\cdot\sqrt{s-e_3}}{\sqrt[4]{e_1-e_2}\cdot\sqrt{s-e_2}+\sqrt[4]{e_1-e_2}\cdot\sqrt{s-e_3}}=lt,$$

$$L_0=1,$$

$$L_1=1+\left(\frac{1}{2}\right)^2 l^4,$$

$$L_2=1+\left(\frac{1}{2}\right)^2 l^4+\left(\frac{1.3}{2.4}\right)^2 l^8,$$

. . . .

. . . .

$$L=1+\left(\frac{1}{2}\right)^2 l^4+\left(\frac{1.3}{2.4}\right)^2 l^8+\left(\frac{1.3.5}{2.4.6}\right)^2 l^{12}+\dots$$

when the equation

$$u=\frac{2}{(\sqrt[4]{e_1-e_3}+\sqrt[4]{e_1-e_2})^2}\int_t^1\frac{dt}{\sqrt{(1-t^2)(1-l^4t^2)}}$$

$$=\frac{2}{(\sqrt[4]{e_1-e_3}+\sqrt[4]{e_1-e_2})^2}\left\{L\cdot\frac{1}{i}\ \text{log. nat.}\ (t+i\sqrt{1-t^2})+\sqrt{1-t^2}\left[\frac{L-L_0}{l}(lt)\right.\right.$$

$$\left.\left.+\frac{2}{3}\,\frac{L-L_1}{l^3}(lt)^3+\frac{2.4}{3.5}\cdot\frac{L-L_2}{l^5}(lt)^5+\dots\right]\right\},$$

determines such a value for u as satisfies the equation $\wp u=s$, when to $\sqrt{1-t^2}$ is given either one of its two values, and to the log. nat. any one of its infinite values."

It is to be hoped that notwithstanding a few gaps in the demonstrations, this sketch has been elaborate enough to give mathematical students a clear idea of these theories in themselves and in relation to the older nomenclature of Jacobi.

A Memoir on the Abelian and Theta Functions.

BY PROFESSOR CAYLEY.

CONTINUED FROM VOL. V. (1882), pp. 137-179.

CHAPTER IV. THE MAJOR FUNCTION $(x, y, z)_{12}^{n-2}$, CONTINUED.

The Conversion, Fixed Curve a Quartic, Continued. Art. Nos. 74 to 82.

74. I resume the question considered *ante* Nos. 66 to 73. The general problem where the fixed curve is any given curve whatever, has recently been solved in a very complete and elegant form by Dr. Nöther, in the two notes "Zur Reduction algebraischer Differentialausdrücke auf der Normalförmen" and "Ueber die algebraischen Differentialausdrücke, 2e Note," Sitzungsb. der phys-med. Soc. zu Erlangen, 10 Dec. 1883 and 14 Jan. 1884. I consider here the case of the quartic curve, $n = 4$, and connect his result with my former investigations.

We have the differential

$$\frac{(x, y, z)_{12}^{n-2}\, d\omega}{012}, = \frac{\Omega_{12}\, d\omega}{012}$$

where Ω_{12}, or as I also write it $\Omega\,(0; 1, 2; 3, 4, 5)$, is a rational and integral function of the degree $(n - 2 =)\ 2$ in the current coordinates (x, y, z): it depends also on the parametric points 1, 2, which are points on the quartic, coordinates (x_1, y_1, z_1), (x_2, y_2, z_2) respectively; and on $(p =)\ 3$ other points 3, 4, 5 on the quartic, coordinates (x_3, y_3, z_3), (x_4, y_4, z_4), (x_5, y_5, z_5) respectively. The curve $\Omega = 0$ is a conic which is taken to pass through the dps (none in the present case) and through the $(n - 2 =)\ 2$ residues of the parametric points, and the function Ω is such that on writing therein (x_1, y_1, z_1) for (x, y, z) it becomes $= (n\,.\,1^{n-1}2^* =)$ $4\,.\,1^3 2$ (viz. we have $\Omega\,(1; 1, 2; 3, 4, 5) = 4\,.\,1^3 2$, which implies also $\Omega\,(2; 1, 2; 3, 4, 5)$

---

* $\left(x_2 \dfrac{d}{dx_1} + y_2 \dfrac{d}{dy_1} + z_2 \dfrac{d}{dz_1}\right) f(x_1, y_1, z_1) = n\,.\,1^{n-1}2$. See No. 2.

$=4.12^2$): so defined, the function would contain ($p=$) 3 arbitrary constants, but these are determined so that the curve $\Omega=0$ passes through the 3 points 3, 4, 5 on the quartic: and the function Ω, $=\Omega(0:1,2:3,4,5)$ is thus a completely determinate function, rational and integral of the degree 2 in the coordinates (x, y, z) of the current point, and rational in the coordinates of the other five points respectively. I call to mind that 012 denotes the determinant formed with the coordinates (x, y, z), etc., of the points 0, 1, 2 respectively: the like notation is used throughout.

75. The function $\Omega(0;1,2:3,4,5)$ is in fact the function Ω' of No. 43 with only the further condition in regard to the points 3, 4, 5 of the quartic; viz., Ω is the function determined by the equation

$$\begin{vmatrix} (x, y, z)^2 & , \Omega \\ 1(x_1, y_1, z_1)^2 & , 4.1^22 \\ 2(x_1, y_1, z_1)(x_2, y_2, z_2) & , 6.1^22^2 \\ 1(x_2, y_2, z_2)^2 & , 4.12^2 \\ (x_3, y_3, z_3)^2 & , 0 \\ (x_4, y_4, z_4)^2 & , 0 \\ (x_5, y_5, z_5)^2 & , 0 \end{vmatrix} = 0:$$

this is of the form

$$M\Omega+\square=0,$$

and as appears in No. 46, M is $=123.124.125.345$. Hence writing $\square(0;1,2;3,4,5)$ for $\square$, we have

$$\Omega, =\Omega(0;1,2;3,4,5), =\frac{-\square(0;1,2;3,4,5)}{123.124.125.345}.$$

Hence further writing Q, $=Q(0;1,2;3,4,5)$, $=\dfrac{\Omega(0;1,2;3,4,5)}{012}$, so that the differential is $Qd\omega$, $=Q(0;1,2;3,4,5)\,d\omega$, we have

$$Q=Q(0;1,2;3,4,5), =\frac{-\square(0;1,2;3,4,5)}{012.123.124.125.345},$$

which is of the form

$$Q=\frac{0^2\,\overline{12}^4\,\overline{345}^2}{0^1\,\overline{12}^4\,\overline{345}^2}, =0^1\,\overline{12345}^0,$$

viz., Q is a rational fraction where the numerator is of the degree 2, and the denominator of the degree 1 as regards the coordinates (x, y, z) of the current point: but the numerator and denominator are each of the degree 4 as regards the coordinates of the points 1, 2 separately, and of the degree 2 as regards the coordinates of the points 3, 4, 5 separately: that is Q is of the degree 1 as regards the coordinates (x, y, z), but of the degree 0 as regards the coordinates of the points 1, 2, 3, 4, 5 separately.

76. The signification of the symbol of quasi-differentiation ∂ (applicable only to a function of the degree 0 in the coordinates to which the differentiations have reference) is explained *ante* No. 60, the function Q just mentioned is of the degree 0 in regard to the coordinates of each of the points 1, 2, 3, 4, 5; and it can thus be operated upon by the symbols $\partial_1, \partial_2, \partial_3, \partial_4, \partial_5$ respectively. Observe in particular that we have $\partial_1 Q(0;\,1,\,2;\,3,\,4,\,5) = \overline{01^1}\,\overline{2345^{0}}$, viz., it is of the degree 1 in the coordinates of the points 0 and 1 respectively, but of the degree 0 in regard to the coordinates of the points 2, 3, 4, 5 respectively.

77. This being so we may consider the function

$$\begin{aligned} H(0;\,1,\,2;\,3,\,4,\,5;\,6,\,7,\,8) = {} & \partial_1 Q(0;\,1,\,2;\,3,\,4,\,5) \\ & + \partial_3 Q(1;\,3,\,2;\,6,\,4,\,5).\frac{045}{345} \\ & + \partial_4 Q(1;\,4,\,2;\,3,\,7,\,5).\frac{053}{453} \\ & + \partial_5 Q(1;\,5,\,2;\,3,\,4,\,8).\frac{034}{534}, \end{aligned}$$

where 6, 7, 8 are arbitrary points on the quartic; the functions

$$\partial_3 Q(1;\,3,\,2;\,6,\,4,\,5),\quad \partial_4 Q(1;\,4,\,2;\,3,\,7,\,5),\quad \partial_5 Q(1;\,5,\,2;\,3,\,4,\,8)$$

are functions of the same form as $\partial_1 Q(0;\,1,\,2;\,3,\,4,\,5)$, and derived from it by changing in each case the current point 0 into the parametric point 1, and by further changing in the three cases this parametric point into the points 3, 4, 5 respectively, and replacing the corresponding point 3, 4 or 5 by the new arbitrary point 6, 7 or 8. Further 045, etc., denote determinants as above; so that in H each of the last three terms is in fact as regards the point 0 a mere linear function of the coordinates $(x,\,y,\,z)$ of this point.

We have $\partial_3 Q(1;\,3,\,2;\,6,\,4,\,5) = \overline{13^1}\,\overline{2645^0}$, and hence this function into $\frac{045}{345}$ is $= \overline{01^1}\,\overline{23456^0}$; and so for the third and fourth terms of H: thus each of the four terms of H is $= \overline{01^1}\,\overline{2345678^0}$, of the degree 1 in the coordinates of the points 0 and 1 respectively, but of the degree 0 in the coordinates of the other points 2, 3, 4, 5, 6, 7, 8 respectively.

Nöther's conversion-theorem consists herein, that the function

$$H(0;\,1,\,2;\,3,\,4,\,5;\,6,\,7,\,8)$$

is unaltered by the interchange of the two points 0, 1; or putting for shortness

$$H(0;\,1,\,2;\,3,\,4,\,5;\,6,\,7,\,8) = H_1(0),$$

the theorem is

$$H_1(0) = H_0(1).$$

78. We have No. 59,

$$\frac{\tfrac{1}{2}\Omega_{12}}{012} = \frac{-01^3.02^3 + 01^2 2.012^2 + 0^2 12.1^2 2^2}{012.1^2 2^2}, = \{0^2 12\},$$

or as for greater simplicity I write it $= 0^2 12$,

viz. $0^2 12$ is now written instead of $\{0^2 12\}$ to denote the function just given as the value of $\frac{\tfrac{1}{2}\Omega_{12}}{012}$: Ω_{12} is thus $= 3.012.0^2 12$, viz. this is a particular form of Ω_{12} satisfying the conditions that $\Omega_{12} = 0$ is a conic passing through the residues of the points 1, 2, and such that Ω_{12} on writing therein (x_1, y_1, z_1) for (x, y, z) becomes $= 4.1^3 2$: hence the general form of the function satisfying these conditions is $= 3.012\{0^2 12 + \text{arbitrary linear function of } (x, y, z)\}$. The beforementioned function $\Omega(0; 1, 2; 3, 4, 5)$ is a function satisfying these conditions and the further conditions that the conic $\Omega = 0$ shall pass through the three points 3, 4, 5 on the quartic: these further conditions serve to determine the linear function: and we at once obtain

$$\frac{\tfrac{1}{2}\Omega(0; 1, 2; 3, 4, 5)}{012} = 0^2 12 - 3^2 12\,\frac{045}{345} - 4^2 12\,\frac{053}{453} - 5^2 12\cdot\frac{034}{534},$$

viz. the value of Ω given by this equation, on writing therein $0 = 3$, 4 or 5, becomes as it should do $= 0$.

79. We thus have

$$Q(0; 1, 2; 3, 4, 5) = 0^2 12 - 3^2 12\,\frac{045}{345} - 4^2 12\,\frac{053}{453} - 5^2 12\,\frac{034}{534},$$

and Nöther's conversion-equation becomes

$$\begin{aligned}
&\partial_1\left\{0^2 12 - 3^2 12\,\frac{045}{345} - 4^2 12\,\frac{053}{453} - 5^2 12\,\frac{034}{534}\right\} \\
&+ \partial_3\left\{1^2 32 - 6^2 32\,\frac{145}{645} - 4^2 32\,\frac{156}{456} - 5^2 32\,\frac{164}{564}\right\}\cdot\frac{045}{345} \\
&+ \partial_4\left\{1^2 42 - 3^2 42\,\frac{175}{375} - 7^2 42\,\frac{153}{753} - 5^2 42\,\frac{137}{537}\right\}\cdot\frac{053}{453} \\
&+ \partial_5\left\{1^2 52 - 3^2 52\,\frac{148}{348} - 4^2 52\,\frac{183}{483} - 5^2 48\,\frac{134}{834}\right\}\cdot\frac{034}{534} \\
= {}&\partial_0\left\{1^2 02 - 3^2 02\,\frac{145}{345} - 4^2 02\,\frac{153}{453} - 5^2 02\,\frac{134}{534}\right\} \\
&+ \partial_3\left\{0^2 32 - 6^2 32\,\frac{045}{645} - 4^2 32\,\frac{056}{456} - 5^2 32\,\frac{064}{564}\right\}\cdot\frac{145}{345} \\
&+ \partial_4\left\{0^2 42 - 3^2 42\,\frac{075}{375} - 7^2 42\,\frac{053}{753} - 5^2 42\,\frac{037}{537}\right\}\cdot\frac{153}{453} \\
&+ \partial_5\left\{0^2 52 - 3^2 52\,\frac{048}{348} - 4^2 52\,\frac{083}{483} - 5^2 48\,\frac{034}{834}\right\}\cdot\frac{134}{534}
\end{aligned}$$

an equation where the functions operated on with the ∂'s are only functions such as $0^2 12$; for there is not any determinant operated upon which contains the number which is the suffix of the ∂ which operates upon it.

80. Taking all the terms over to the left-hand side there are in all 32 terms: but of these 3 + 3 destroy each other, and 6 + 6 unite in pairs into 6 terms: there are thus in all 7 + 7 + 6, = 20 terms: viz. multiplying the whole equation by 345, it is found that the equation becomes

	or as this may be written	where
$345(\ \ \partial_1 0^2 12 - \partial_0 1^2 02)$	$345\ \boxed{012}$	$\boxed{012} = \partial_1 0^2 12 - \partial_0 1^2 02$, etc.,
$-045(-\partial_3 1^2 32 + \partial_1 3^2 12)$	$-\ 045\ \boxed{312}$	
$-053(-\partial_4 1^2 42 + \partial_1 4^2 12)$	$-\ 305\ \boxed{412}$	
$-034(-\partial_5 1^2 52 + \partial_1 5^2 12)$	$-\ 340\ \boxed{512}$	
$-145(\ \ \partial_3 0^2 32 - \partial_0 3^2 02)$	$-\ 145\ \boxed{032}$	
$-153(\ \ \partial_4 0^2 42 - \partial_0 4^2 02)$	$-\ 315\ \boxed{042}$	
$-134(\ \ \partial_5 0^2 52 - \partial_0 5^2 02)$	$-\ 341\ \boxed{052}$	
$+013(-\partial_4 5^2 42 + \partial_5 4^2 52)$	$+\ 301\ \boxed{452}$	
$+014(-\partial_5 3^2 52 + \partial_3 5^2 32)$	$+\ 140\ \boxed{532}$	
$+015(-\partial_3 4^2 32 + \partial_4 3^2 42)$	$+\ 015\ \boxed{342}$	
$= 0,$	$= 0,$	

viz. the equation is

$$\Sigma \pm 345\ \boxed{012} = 0,$$

the nine terms which follow the first term $345\ \boxed{012}$ of the sum being obtained by the interchanges of 0, 1 (one or each) with the 3, 4, 5, each interchange giving rise to a sign $-$.

81. In obtaining the foregoing result, we have, for instance, a pair of terms

$$\partial_4\, 5^2 42 \frac{-137.053+037.153}{537},\ =\partial_4\, 5^2 42\frac{-013.537}{537},\ =\partial_4\, 5^2 42(-013),$$

viz. this depends on the equation

$$137.053-037.153-537.013=0,$$

or say

$$-137.035+037.135+013.357=0,$$

an identity which, in a form which will be readily understood, may be written

$$\text{det.}\begin{vmatrix} 0137 \\ 013735 \end{vmatrix}=0.$$

Similarly the two terms which contain $\partial_5\, 4^2 52$ combine into the single term $\partial_5\, 4^2 52(013)$: and the two new terms taken together are

$$013(-\partial_4\, 5^2 42+\partial_5\, 4^2 52).\ =301\ \boxed{452}\,.$$

82. The proof of the identity,

$$\Sigma\pm 345\ \boxed{012}=0,$$

depends on the property of the function

$$\boxed{012},\ =\partial_1\, 0^2 12-\partial_0\, 1^2 02,$$

enuntiated No. 67, and proved *à posteriori* by the tedious calculation Nos. 70 to 73, viz. in No. 67, writing 2 in place of 3, this is:—$\partial_1\, 0^2 12-\partial_0\, 1^2 02$ is equal to the difference of two functions, the first of them linear in the coordinates (x, y, z) of the point 0, but depending also on the coordinates of the points 1 and 2; the second of them linear in the coordinates (x_1, y_1, z_1) of the point 1 but depending also on the coordinates of the points 0 and 2. Or what is the same thing the property is

$$\boxed{012}=A_{12}x+B_{12}y+C_{12}z-(A_{02}x_1+B_{02}y_1+C_{02}z_1),$$

where A_{12}, B_{12}, C_{12} are functions of (x_1, y_1, z_1), (x_2, y_2, z_2), and A_{02}, B_{02}, C_{02} are the like functions of (x, y, z), (x_2, y_2, z_2).

Substituting such values in the sum $\Sigma\pm 345\ \boxed{012}$, but writing down only the terms which contain x, these are

$$\begin{array}{lll} & 345\,(A_{12}x-A_{02}x_1) & \\ -045\,(A_{12}x_3-A_{32}x_1) & -145\,(A_{32}x-A_{02}x_3) & +301\,(A_{52}x_4-A_{42}x_5) \\ -305\,(A_{12}x_4-A_{42}x_1) & -315\,(A_{42}x-A_{02}x_4) & +140\,(A_{32}x_5-A_{52}x_3) \\ -340\,(A_{12}x_5-A_{52}x_1) & -341\,(A_{52}x-A_{02}x_5) & +015\,(A_{42}x_3-A_{32}x_4). \end{array}$$

This is
$$
\begin{aligned}
= \quad & A_{02}(-x_1 345 + x_3 145 + x_4 315 + x_5 341) \\
+ \; & A_{12}(\;\; x\,345 - x_3 045 - x_4 305 - x_5 340) \\
+ \; & A_{32}(\;\; x_1 045 - x\,145 + x_5 140 - x_4 015) \\
+ \; & A_{42}(-x\,315 + x_3 015 + x_1 305 - x_5 301) \\
+ \; & A_{52}(-x\,341 + x_4 301 + x_1 340 - x_3 140)
\end{aligned}
$$
where the coefficient of each of the A's is identically $=0$: and similarly the terms in y and the terms in z are each $=0$. We have thus the proof of the identity
$$\Sigma \pm 345 \boxed{012} = 0,$$
that is, of the conversion-equation $H_1(0) = H_0(1)$.

The Syzygy—Fixed Curve a Quartic. Art. No. 83.

I revert to the theory of the Syzygy, *ante* No. 59.

83. We have
$$Q(0;\,1,\,2;\,3,\,4,\,5) = 0^2 12 - 3^2 12 \,\frac{045}{345} - 4^2 12\,\frac{053}{453} - 5^2 12\,\frac{034}{534},$$
or if for convenience we take instead of 1, 2, the parametric points to be α, β coordinates $(x_\alpha,\, y_\alpha,\, z_\alpha)$ and $(x_\beta,\, y_\beta,\, z_\beta)$ respectively, then this equation is
$$Q_{\alpha\beta} = Q(0;\,\alpha,\,\beta;\,3,\,4,\,5) = 0^2\alpha\beta - 3^2\alpha\beta\,.\,\frac{045}{345} - 4^2\alpha\beta\,\frac{053}{453} - 5^2\alpha\beta\,\frac{034}{534}.$$
Considering a new parametric point γ, and forming the like functions $Q_{\beta\gamma}$ and $Q_{\gamma\alpha}$, it is to be shown that we have identically
$$Q_{\alpha\beta} + Q_{\beta\gamma} + Q_{\gamma\alpha} = 0.$$
To prove this observe that in the equation at the end of No. 59, Δ, ρ, σ, τ denote 123, 023, 031, 012 respectively. Hence writing therein α, β, γ in place of 1, 2, 3 respectively, and putting A, B, C for the coefficients (including therein the factor $\frac{1}{\Delta^2}$) of ρ, σ, τ respectively, the equation is
$$0^2\beta\gamma + 0^2\gamma\alpha + 0^2\alpha\beta = A\,.\,0\beta\gamma + B\,.\,0\gamma\alpha + C\,.\,0\alpha\beta,$$
where A, B, C are absolute constants (functions, that is, of the coefficients of the quartic) each divided by $(\alpha\beta\gamma)^2$. We hence obtain
$$
\begin{aligned}
(Q_{\beta\gamma} + Q_{\gamma\alpha} + Q_{\alpha\beta})\,.\,345 = \; & 345\,(A\,.\,0\beta\gamma + B\,.\,0\gamma\alpha + C\,.\,0\alpha\beta) \\
& - 045\,(A\,.\,3\beta\gamma + B\,.\,3\gamma\alpha + C\,.\,3\alpha\beta) \\
& - 053\,(A\,.\,4\beta\gamma + B\,.\,4\gamma\alpha + C\,.\,4\alpha\beta) \\
& - 034\,(A\,.\,5\beta\gamma + B\,.\,5\gamma\alpha + C\,.\,5\alpha\beta)
\end{aligned}
$$

and on the left-hand side the whole coefficient of A is $=0$; viz. this coefficient has the value det. $\begin{vmatrix} 0345 \\ 0345\beta\gamma \end{vmatrix}$, which is $=0$. Similarly the whole coefficient of B is $=0$, and the whole coefficient of C is $=0$: and we have thus the required result

$$Q_{\beta\gamma}+Q_{\gamma\alpha}+Q_{\alpha\beta}=0.$$

The syzygy is thus obtained in a more perfect form than in No. 59; viz. by considering (instead of $0^2\alpha\beta$) the new form $Q_{\alpha\beta}$, then instead of a sum which is a linear function of the coordinates (x, y, z) we obtain a sum $=0$.

The Fixed Curve a Cubic—Syzygy and Conversion. Art. Nos. 84, 85.

84. In the case fixed curve a cubic (see Nos. 58 and 64) the analogous formulæ are

$$\tfrac{1}{3}\Omega(0;1,2;3)=\frac{1^2 2.023+12^2.031}{123},\ =\widetilde{012}-\frac{\widetilde{123}}{123}012 \text{ (see No. 49)},$$

that is

$$Q(0;1,2;3),\ =\frac{\frac{1}{3}\Omega}{012},\ =\frac{\widetilde{012}}{012}-\frac{\widetilde{123}}{123},\ =\{012\}-\{123\},$$

where 1, 2 are the parametric points: 3 any other point on the cubic: the brackets $\{\ \}$ are of course here necessary in order to distinguish $\{012\}$ from the determinant 012. It will be remembered that $\{012\}$ is an alternate function

$$\{012\},\ =-\{102\},\ =\{120\}, \text{ etc.}$$

If instead of 1, 2 we take the parametric points to be α, β, coordinates $(x_\alpha, y_\alpha, z_\alpha)$ and $(x_\beta, y_\beta, z_\beta)$ respectively, then the formula is

$$Q_{\alpha\beta}=Q(0;\alpha,\beta;3)=\{0\alpha\beta\}-\{3\alpha\beta\}.$$

Hence taking on the cubic a new point γ, coordinates $(x_\gamma, y_\gamma, z_\gamma)$ and forming the functions $Q_{\beta\gamma}$ and $Q_{\gamma\alpha}$ we have

$$\begin{aligned}Q_{\beta\gamma}+Q_{\gamma\alpha}+Q_{\alpha\beta}&=\{0\beta\gamma\}+\{0\gamma\alpha\}+\{0\alpha\beta\}\\&\quad-\{3\beta\gamma\}-\{3\gamma\alpha\}-\{3\alpha\beta\}.\end{aligned}$$

But by the formula No. 58, $\{0\beta\gamma\}+\{0\gamma\alpha\}+\{0\alpha\beta\}=\{\alpha\beta\gamma\}$; hence also $\{3\beta\gamma\}+\{3\gamma\alpha\}+\{3\alpha\beta\}=\{\alpha\beta\gamma\}$: and we have thus

$$Q_{\beta\gamma}+Q_{\gamma\alpha}+Q_{\alpha\beta}=0,$$

the syzygy for the cubic.

85. For the conversion, the definition of H is

$$\begin{aligned}H(0;1,2;3,6)&=\partial_1\,Q(0;1,2;3)\\&\quad+\partial_3\,Q(1;3,2;6).\end{aligned}$$

viz. this is

$$\begin{aligned}H_0(1)=H(0;1,2;3,6)&=\partial_1(\{012\}-\{123\})\\&\quad+\partial_3(\{132\}-\{326\}),\\&=\partial_1\{012\}-(\partial_1+\partial_3)\{123\}-\partial_3\{326\}.\end{aligned}$$

which in virtue of $(\partial_1+\partial_2+\partial_3)\{123\}=0$ (see No. 63) becomes

$$H_0(1)=\partial_1\{012\}+\partial_2\{123\}-\partial_3\{326\}.$$

Interchanging the 0 and 1 we thence have

$$H_1(0)=\partial_0\{102\}+\partial_2\{023\}-\partial_3\{326\}.$$

Hence the difference $H_0(1)-H_1(0)$ is

$$=\partial_1\{012\}-\partial_0\{102\}+\partial_2\{123\}-\partial_2\{023\},$$

viz. this is $=(\partial_1+\partial_0)\{012\}+\partial_2(\{123\}-\{230\}),$

where the first term is $=-\partial_2\{012\}$ and the whole therefore is

$$=\partial_2(\{123\}-\{230\}-\{012\})$$

$$=-\partial_2\{301\},$$

in virtue of $\{123\}-\{230\}+\{301\}-\{012\}=0$, and is consequently $=0$.

We have thus $H_0(1)-H_1(0)=\sigma,$

the conversion-equation in the case of the cubic.

CHAPTER V. MISCELLANEOUS INVESTIGATIONS.

The Differential Symbol $d\omega$. Art. Nos. 86 and 87.

86. The definition is

$$\frac{ydz-zdy}{\frac{df}{dx}}=\frac{zdx-xdz}{\frac{df}{dy}}=\frac{xdy-ydx}{\frac{df}{dz}}=d\omega,$$

and it hence follows that we have

$$d\omega=\frac{\begin{vmatrix} dx, & dy, & dz \\ x, & y, & z \\ \lambda, & \mu, & \nu \end{vmatrix}}{\lambda\frac{df}{dx}+\mu\frac{df}{dy}+\nu\frac{df}{dz}},$$

where (λ, μ, ν) are arbitrary constants, or if we please arbitrary functions of (x, y, z): viz. the expression just written down is altogether independent of the values of λ, μ, ν: and is consequently equal to the value obtained by writing any two of these symbols $=0$, that is, the expression is equal to any one of the foregoing three equal values of $d\omega$. The expression was first given by Aronhold (1863), in the memoir presently referred to.

It is to be remarked that considering (λ, μ, ν) as the coordinates of a point, the denominator $\lambda\frac{df}{dx}+\mu\frac{df}{dy}+\nu\frac{df}{dz}$ equated to 0 is the polar $(n-1)$thic of the point λ, μ, ν in regard to the fixed curve.

If instead of λ, μ, ν we write $bc'-b'c$, $ca'-c'a$, $ab'-a'b$, where (a, b, c) (a', b', c') are constants, then the numerator is $=(ax+by+cz)(a'dx+b'dy+c'dz)$ $-(a'x+b'y+c'z)(adx+bdy+cdz)$, or introducing ρ, σ to denote the arbitrary linear functions $ax+by+cz$ and $a'x+b'y+c'z$ respectively, the numerator is $=\rho d\sigma-\sigma d\rho$: moreover, observing that a, b, c and a', b', c' are the differential coefficients of ρ, σ in regard to the coordinates (x, y, z), the denominator is $=J(f, \rho, \sigma)$; and the value of $d\omega$ is

$$d\omega=\frac{\rho d\sigma-\sigma d\rho}{J(f, \rho, \sigma)},$$

where in accordance with a previous remark, the denominator equated to 0 is the polar $(n-1)$thic of the intersection of the lines $\rho=0$, $\sigma=0$ in regard to the fixed curve.

Obviously by taking for ρ, σ any two of the three coordinates x, y, z, we reproduce the original three forms of $d\omega$.

87. The last mentioned form of $d\omega$ suggests the expression for this symbol in the case where the fixed curve, instead of being a plane curve, is a curve of double curvature defined by two equations $f=0$, $g=0$ between the four coordinates (x, y, z, w): viz. ρ, σ being now arbitrary linear functions

$$ax+by+cz+dw, \text{ and } a'x+b'y+c'z+d'w$$

of the four coordinates, the expression is

$$d\omega=\frac{\rho d\sigma-\sigma d\rho}{J(f, g, \rho, \sigma)}:$$

and by taking for ρ, σ any two of the four coordinates x, y, z, w, we have for $d\omega$ six values which must of course be equal to each other; it is easy to verify *à posteriori* that this is so.

In the case where the curve of double curvature is not the complete intersection of two surfaces, the denominator (regarded as the Jacobian of the curve and of the arbitrary planes ρ, σ) will have a definite meaning, but what this is I do not at present consider.

The last mentioned expression for $d\omega$ will be applied further on to the case of the quadri-quadric curve $y^2+x^2=1$, $z^2+k^2x^2=1$.

Integral Formulæ. Art. Nos. 88 to 90.

88. In what precedes $d\omega$ has been used as a single symbol to denote any one of the equal differential expressions

$$\frac{ydz-zdy}{\frac{df}{dx}},=\frac{zdx-xdz}{\frac{df}{dy}},=\frac{xdy-ydx}{\frac{df}{dz}};$$

there is no quantity ω. These expressions are of the order $-(n-3)$ in the coordinates (x, y, z), and since (x, y, z) are as to their absolute magnitudes altogether arbitrary (only their ratios being determinate), a symbol such as

$$\omega, = \int d\omega, = \int \frac{ydz - zdy}{\frac{df}{dy}},$$

would, except in the case $n = 3$, be altogether meaningless. In fact the integral would be

$$\int \frac{z^2 d\left(\frac{y}{z}\right)}{z^{n-1}\phi\left(\frac{x}{z}, \frac{y}{z}\right)}, = \int \frac{d\left(\frac{y}{z}\right)}{z^{n-3}\phi\left(\frac{x}{z}, \frac{y}{z}\right)},$$

where $\frac{x}{z}$ is by the equation of the fixed curve given as a function of $\frac{y}{z}$, but the other factor z^{n-3} is an absolutely indeterminate variable value, and the expression is meaningless.

But we have integrals $\int Qd\omega$, where Q is a homogeneous function of the order $n-3$ in the coordinates (x, y, z); and in particular we have such integrals where (corresponding to the forms which present themselves in the differential pure and affected theorems respectively) Q is either a rational and integral function $(x, y, z)^{n-3}$, or a rational and integral function $(x, y, z)^{n-3}$ divided by a linear function $(x, y, z)^1$: for in every such case the form of integral is

$$\int \frac{d\left(\frac{y}{z}\right)}{\phi\left(\frac{x}{z}, \frac{y}{z}\right)},$$

where $\frac{x}{z}$ is a given function of $\frac{y}{z}$, and the factor of $d\left(\frac{y}{z}\right)$ is thus a mere function of $\frac{y}{z}$. More definitely, in the integrals $\int Qd\omega$ which are considered, Q is either a minor function $(x, y, z)^{n-3}$, or it is the quotient of a major function $(x, y, z)^{n-3}_{12}$ by the linear function 012.

In the case $n = 2$ there is no rational and integral function $(x, y, z)^{n-3}$, but the function may be of the form belonging to the affected theorem, unity divided by a linear function $(x, y, z)^1$; or say the integral is $\int \frac{d\omega}{\alpha x + \beta y + \gamma z}$, where the (x, y, z) are connected by a quadric equation $(a, \ldots \between x, y, z)^2 = 0$: it will be shown presently that this integral is obtainable as a logarithmic function.

In the case $n=3$, we have the rational and integral function $(x, y, z)^{n-3}$, $=$ a constant, or say $=1$, so that there is here an integral $\int d\omega$: we do not call this ω, but introducing a new letter, say u, and fixing at pleasure the inferior limit of the integral, we write $u=\int d\omega$.

89. In the foregoing form $\int Q d\omega$, so long as we retain the symbol $d\omega$, there is nothing to show what is the variable in regard to which the integration is to be performed; we may for instance writing

$$d\omega=\frac{y^2 d\left(\frac{z}{y}\right)}{\frac{df}{dx}}$$

make it to be $\frac{y}{z}$, or in like manner to be any other of the six quotients. We thus cannot attribute a *value* to the inferior or superior limit of such an integral, but we may take the limits to be each of them a point on the fixed curve: for instance if 1, 0 be points on the fixed curve then the integral $\int_0^1 Q d\omega$, means the integral taken from the value at the point 0 to the value at the point 1, of the variable in regard to which the integration is performed; or when there is no expressed superior limit, then the integral is to be taken from the value for the expressed or known inferior limit to the value at the current point (x, y, z) of the variable in regard to which the integration is performed. The actual value of the integral will of course depend upon the *path* of the variable; but this is a question which is not here entered upon.

If using Cartesian Coordinates x, y, we write for instance

$$d\omega=\frac{dx}{\frac{df}{dy}}, \text{ then } \int\frac{Q dx}{\frac{df}{dy}}$$

will denote an integral $\int \phi x\, dx$ in regard to the variable x, and the inferior and superior limits will be as usual values of x, or if there is no expressed superior limit then the integral $\int_0 \phi x\, dx$ will be the integral taken from the inferior limit x_0 to the current value x.

We may, if we please, consider the coordinates (x, y, z) as depending upon a parameter s, viz. the ratios $x:y:z$ may be regarded as given functions of s, and the integral $\int Q d\omega$, is then an integral $\int \Omega ds$, which taken from a constant inferior limit up to the value ϑ, which belongs to a given point 1 of the curve, is a given function of ϑ_1, or say of the point 1. But except in the case of the cubic (or generally if $p=1$), we do not have the coordinates actually given as known functions of a parameter ϑ (say they are potentially known functions of ϑ), and it is further to be noticed the functions which present themselves are functions not of a single point, but of p or more points: thus in the case of the quartic, $n=4$, $p=3$; we have $\int^1 x d\omega$, $\int^1 y d\omega$, $\int^1 z d\omega$, each standing for a given function of the parameter ϑ_1, but these integrals do not present themselves singly, but in combinations such as $\left(\int^1+\int^2+\int^3+\int^\xi\right)(x d\omega,\ y d\omega,\ z d\omega)$, say these sums of integrals are u, v, w: each of the functions u, v, w is a potentially known function of the parameters ϑ_1, ϑ_2, ϑ_3, ϑ_ξ which belong to the points 1, 2, 3, ξ respectively, and is consequently regarded as a given function of these four points.

90. Consider as before, in the case of a cubic curve, the integral $u=\int d\omega$: it will presently be seen that for the general curve as given by a cubic equation $f=0$ of any form whatever, we arrive at a form of elliptic function: but the ordinary elliptic functions sn, cn, dn connect themselves most readily with the cubic curve $y^2=x.1-x.1-k^2x$. We have here

$$d\omega=\frac{\tfrac{1}{2}dx}{y},=\frac{\tfrac{1}{2}dx}{\sqrt{x.1-x.1-k^2x}},$$

or, in the equation $u=\int d\omega$, taking the inferior limit to be 0, say

$$u=\int_0\frac{\tfrac{1}{2}dx}{\sqrt{x.1-x.1-k^2x}},$$

an equation which determines u as a function of x, or conversely, x as a function of u. We might thence by means of Abel's theorem as applied to the curve in question investigate the properties of the function $x=\lambda(u)$ thus arising, and so establish the theory of elliptic functions: but it is more convenient, treating the elliptic functions as known functions, to write for λu its value; viz. to take for x as given by this equation, the value $x=\text{sn}^2u$: we thence have $y=\text{sn}\,u\,\text{cn}\,u\,\text{dn}\,u$; viz. these values $x=\text{sn}^2u$, $y=\text{sn}\,u\,\text{cn}\,u\,\text{dn}\,u$, satisfy the equation $y^2=x.1-x.1-k^2x$ of the curve, and give, moreover, $d\omega=du=\frac{\tfrac{1}{2}dx}{y}$: and we can with these values, and the formulæ for elliptic functions, verify any results given by Abel's theorem. This will be done in considerable detail: but at present I wish only to remark that the formulæ give the coordinates x, y of a point on the cubic curve expressed as one-valued functions of a parameter or argument u: but that this argument u is not a one-valued function of the coordinate x, or even of the coordinates x, y of the given point on the curve: say the argument u has not a unique value for a given point (x, y) of the curve. There are in fact an infinity of values $u = u_0 + 2mK + 2m'iK'$, where m, m' are any positive or negative integers: that this is so, depends on the multiplicity of values, according to the different paths of the variable, of the integral $u = \int_0 \frac{\frac{1}{2}dx}{\sqrt{x.1-x.1-k^2x}}$; or, regarding the elliptic functions as known functions, it depends upon the double periodicity of these functions.

Aronhold's Quadric Integral. Art. Nos. 91 to 93.

91. I reproduce the investigation contained in Aronhold's paper "Ueber eine neue algebraische Behandlungsweise u. s. w., Crelle, t. lxi, 1863, pp. 95–145. We take f the general quadric function $(a, b, c, f, g, h\between x, y, z)^2$; $\alpha x + \beta y + \gamma z$ an arbitrary linear function of x, y, z: the theorem is $\frac{d\omega}{\alpha x + \beta y + \gamma z} =$ differential of logarithm of an algebraic function of (x, y, z); viz. taking (ξ, η, ζ) for the coordinates of either of the points of intersection of the line $\alpha x + \beta y + \gamma z = 0$ with the quadric $(a, \ldots\between x, y, z)^2 = 0$, and writing also

$$\Omega^2 = -(bc - f^2, ca - g^2, ab - h^2, gh - af, hf - bg, fg - ch\between \alpha, \beta, \gamma)^2,$$

then the theorem is

$$\frac{d\omega}{\alpha x + \beta y + \gamma z} = \frac{1}{\Omega} d.\log \frac{(a, \ldots\between x, y, z\between \xi, \eta, \zeta)}{\alpha x + \beta y + \gamma z},$$

or, what is the same thing,

$$\int \frac{d\omega}{\alpha x + \beta y + \gamma z} = \frac{1}{\Omega} \log \frac{(a, \ldots\between x, y, z\between \xi, \eta, \zeta)}{\alpha x + \beta y + \gamma z} + \text{const.}$$

It is to be observed in regard to this equation that the two sides respectively are in regard to (α, β, γ) homogeneous functions of the degree -1, and in regard to (ξ, η, ζ) homogeneous of the degree 0; viz. on the right-hand side the effect of a change in the absolute magnitudes of ξ, η, ζ, say the change into $k\xi$, $k\eta$, $k\zeta$, is merely to change by $\log k$ the constant of integration.

It is to be remarked also that the equation $(a, \ldots\between \xi, \eta, \zeta\between x, y, z) = 0$, represents the tangent to the conic at the point (ξ, η, ζ) of intersection with the line $\alpha x + \beta y + \gamma z = 0$; calling the linear function in question T, the value of the integral is $\frac{1}{\Omega}\log\frac{T}{ax+\beta y+\gamma z}$; if (ξ_1, η_1, ζ_1), (ξ_2, η_2, ζ_2) are the coordinates of the two points of intersection respectively, then in passing from one of these to the other we change the sign of the radical Ω, and the two values thus are $\frac{1}{\Omega}\log\frac{T_1}{ax+\beta y+\gamma z}$ and $-\frac{1}{\Omega}\log\frac{T_2}{ax+\beta y+\gamma z}$, these must differ by a constant only; viz. we should have $\log\frac{T_1T_2}{(ax+\beta y+\gamma z)^2}=$ a const. And in fact T_1, T_2 being the tangents to the conic f at its intersections with the line $ax+\beta y+\gamma z=0$, we have it is clear $f=\lambda T_1T_2+\mu(ax+\beta y+\gamma z)^2$, that is, (x, y, z) referring to a point of the conic $f=0$, we have $\frac{T_1T_2}{(ax+\beta y+\gamma z)^2}=$ a constant, which is right.

92. We require the coordinates (ξ, η, ζ) of an intersection: these are determined by the equations $a\xi+\beta\eta+\gamma\zeta=0$, $(a, \ldots \between \xi, \eta, \zeta)^2=0$, or as these may be written

$$\begin{array}{c}
a\xi \qquad\qquad\qquad +\beta\eta \qquad\qquad\qquad +\gamma\zeta=0,\\
(a\xi+h\eta+g\zeta)\,\xi+(h\xi+b\eta+f\zeta)\,\eta+(g\xi+f\eta+c\zeta)\,\zeta=0,
\end{array}$$

we have thence ξ, η, ζ proportional to the determinants

$$\left|\begin{array}{ccc} a\xi+h\eta+g\zeta, & h\xi+b\eta+f\zeta, & g\xi+f\eta+c\zeta \\ a\quad, & \beta\quad, & \gamma \end{array}\right|,$$

say these determinants are $\Omega\xi, \Omega\eta, \Omega\zeta$, where Ω is a value as yet undetermined. The equations are $\gamma(h\xi+b\eta+f\zeta)-\beta(g\xi+f\eta+c\zeta)-\Omega\xi=0$, etc., viz. these are

$$\begin{array}{l}
(\gamma h-\beta g-\Omega)\,\xi+(\gamma b-\beta f\qquad)\,\eta+(\gamma f-\beta c\qquad)\,\zeta=0\\
(ag-\gamma a\qquad)\,\xi+(af-\gamma h-\Omega)\,\eta+(ac-\gamma g\qquad)\,\zeta=0\\
(\beta a-ah\qquad)\,\xi+(\beta h-ab\qquad)\,\eta+(\beta g-af-\Omega)\,\xi=0,
\end{array}$$

eliminating (ξ, η, ζ) we have an equation which may be written

$$\left|\begin{array}{ccc} A-\Omega, & B\quad, & C \\ A'\quad, & B'-\Omega, & C' \\ A''\quad, & B''\quad, & C''-\Omega \end{array}\right|=0.$$

that is
$$\left|\begin{array}{ccc} A, & B, & C \\ A', & B', & C' \\ A'', & B'', & C'' \end{array}\right| -\Omega(B'C''-B''C'+C''A-CA''+AB'-A'B) +\Omega^2(A+B'+C'')-\Omega^3=0.$$

We find very easily that the determinant, and $A+B'+C''$ are each $=0$; and the equation thus reduces itself to

$$\Omega^2=B'C''-B''C'+C''A-CA''+AB'-A'B,$$

or substituting for A, B, etc. their values

$$\Omega^2=-(bc-f^2, \ldots \between a, \beta, \gamma)^2,$$

and this being so, the ratios of ξ, η, ζ are determined by means of any two of the foregoing linear equations.

93. We may now verify the theorem; in the general expression for $d\omega$ writing for λ, μ, ν the values ξ, η, ζ, the equation to be verified becomes

$$\frac{\begin{vmatrix} dx, & dy, & dz \\ x, & y, & z \\ \xi, & \eta, & \zeta \end{vmatrix}}{(\alpha x+\beta y+\gamma z)(a,\ldots\between x, y, z\between\xi,\eta,\zeta)} = \frac{1}{\Omega}\left\{\frac{(a,\ldots\between\xi,\eta,\zeta\between dx, dy, dz)}{(a,\ldots\between\xi,\eta,\zeta\between x, y, z)} - \frac{\alpha dx+\beta dy+\gamma dz}{\alpha x+\beta y+\gamma z}\right\},$$

viz. this is

$$\Omega\begin{vmatrix} dx, & dy, & dz \\ x, & y, & z \\ \xi, & \eta, & \zeta \end{vmatrix} \begin{matrix} =(a,\ldots\between\xi,\eta,\zeta\between dx, dy, dz).(\alpha x+\beta y+\gamma z) \\ \\ -(a,\ldots\between\xi,\eta,\zeta\between x, y, z).(\alpha dx+\beta dy+\gamma dz). \end{matrix}$$

Here on the right-hand side the coefficient of dx is

$$(a\xi+h\eta+g\zeta)(\alpha x+\beta y+\gamma z)$$
$$-\alpha\{(a\xi+h\eta+g\zeta)x+(h\xi+b\eta+f\zeta)y+(g\xi+f\eta+c\zeta)z\}$$

which is

$$=y\{\beta(a\xi+h\eta+g\zeta)-\alpha(h\xi+b\eta+f\zeta)\}$$
$$-z\{\alpha(g\xi+f\eta+c\zeta)-\gamma(a\xi+h\eta+g\zeta)\},$$
$$=y.\Omega\zeta-z.\Omega\eta,$$
$$=\Omega(y\zeta-z\eta),$$

which is right; and similarly the coefficients of dy and dz have the same values on the two sides of the equation respectively.

Aronhold's Quadric Integral deduced from the Affected Theorem.
Art. Nos. 94 to 98.

94. Let the fixed curve be a conic, say $f=\frac{1}{2}(a, b, c, f, g, h\between x, y, z)^2, =0$: and let the variable curve be a line meeting the conic in the points 3 and 4. The affected theorem is

$$\Sigma\frac{12d\omega}{012} = -\frac{\delta 134}{134}\frac{\delta 234}{234},$$

where (x_1, y_1, z_1) and (x_2, y_2, z_2) being the coordinates of the points 1 and 2 respectively, 12 denotes the constant $(a,\ldots\between x_1, y_1, z_1\between x_2, y_2, z_2)$: and 012, etc. denote determinants as usual.

The left-hand side is here

$$12\left\{\frac{d\omega_3}{312}+\frac{d\omega_4}{412}\right\}:$$

on the right-hand side δ refers to the variation of the constants of ϕ, that is to the variations of the points 3 and 4; or we may write $\delta = d_3 + d_4$; the points 3, 4 are independent, and the equation, being satisfied at all, must be satisfied separately in regard to the variations of 3, and in regard to the variations of 4: we must therefore have

$$12\frac{d\omega_3}{312} = -\frac{d_3 134}{134} + \frac{d_3 234}{234};$$

and the like equation obtained herefrom by the interchange of the numbers 3 and 4.

95. The equation just written down relates to any four points 1, 2, 3, 4 of the conic; and if for 3, 4 we write 0, 3 respectively, it becomes

$$12\frac{d\omega}{012} = -\frac{d.031}{031} + \frac{d.023}{023},$$

which relates to the points 0, 1, 2, 3 of the conic: writing as before 023, 031, $012 = \rho, \sigma, \tau$, this equation is

$$12\frac{d\omega}{\tau} = -\frac{d\sigma}{\sigma} + \frac{d\rho}{\rho},$$

which may be verified as follows: the equation of the conic is $f = 23.\sigma\tau + 31.\tau\rho + 12.\rho\sigma, = 0$: we have $d\omega = \dfrac{\rho d\sigma - \sigma d\rho}{\dfrac{df}{d\tau}}$, where $\dfrac{df}{d\tau} = 23.\sigma + 31\rho, = -\dfrac{12.\rho\sigma}{\tau}$, that is, $d\omega = \dfrac{12}{\tau}\left(-\dfrac{d\sigma}{\sigma} + \dfrac{d\rho}{\rho}\right)$, the equation in question.

96. We have as a property of any four points 0, 1, 2, 3 of a conic

$$\frac{23}{123.023} = \frac{-01}{012.031}, \text{ or say } \frac{23}{\Delta.\rho} = \frac{-01}{\sigma\tau}, \text{ that is } \frac{23}{\Delta}\frac{\sigma}{\rho} = -\frac{01}{\tau};$$

hence considering 0 as a variable point, and differentiating the logarithms,

$$-d\log\frac{01}{\tau} = -\frac{d\sigma}{\sigma} + \frac{d\rho}{\rho},$$

and the foregoing equation $12\dfrac{d\omega}{\tau} = -\dfrac{d\sigma}{\sigma} + \dfrac{d\rho}{\rho}$, thus becomes $12\dfrac{d\omega}{\tau} = -d\log\dfrac{01}{\tau}$, or restoring for τ its value 012,

$$12\frac{d\omega}{012} = -d\log\frac{01}{012}.$$

Taking now $ax + \beta y + \gamma z = 0$ for the equation of the line 012; this meets the conic in the points 1, 2, coordinates (x_1, y_1, z_1) and (x_2, y_2, z_2) respectively: and we have

$$a, \beta, \gamma = y_1 z_2 - y_2 z_1,\ z_1 x_2 - z_2 x_1,\ x_2 y_1 - x_1 y_2,$$

$$12 = (a, \ldots \between x_1, y_1, z_1 \between x_2, y_2, z_2),$$

and from this last value

$$\overline{12}^2 = \{(a,\ldots\between x_1, y_1, z_1\between x_2, y_2, z_2)\}^2 - (a,\ldots\between x_1, y_1, z_1)^2.(a,\ldots\between x_2, y_2, z_2)^2$$

(the second term being of course $=0$), viz. this is

$$\overline{12}^2 = -(bc-f^2,\ldots\between y_1z_2-y_2z_1,\ z_1x_2-z_2x_1,\ x_1y_2-x_2y_1)^2$$
$$= -(bc-f^2,\ldots\between \alpha, \beta, \gamma)^2,$$

or say $12 = -\Omega$, if $\Omega^2 = -(bc-f^2,\ldots\between \alpha,\beta,\gamma)^2$ as before: and the equation thus is

$$\frac{\Omega d\omega}{\alpha x+\beta y+\gamma z} = d\log\frac{(a\ldots\between x_1, y_1, z_1\between x, y, z)}{\alpha x+\beta y+\gamma z},$$

or finally, writing (ξ, η, ζ) instead of (x_1, y_1, z_1) to denote the coordinates of one or other of the intersections of the line $\alpha x+\beta y+\gamma z=0$ with the conic, the equation becomes

$$\frac{\Omega d\omega}{\alpha x+\beta y+\gamma z} = d\log\frac{(a,\ldots\between \xi, \eta, \zeta\between x, y, z)}{\alpha x+\beta y+\gamma z},$$

which is Aronhold's quadric integral. .

97. (The foregoing property, which may also be written

$$\frac{23}{\overline{023}.\overline{123}} = \frac{01}{\overline{201}.\overline{301}},$$

is verified very simply in the case of four points 0, 1, 2, 3 of a circle: in fact

$$23 = x_2x_3+y_2y_3-1,\ =\cos 23 - 1,\ = -2\sin^2\tfrac{1}{2}23,$$
$$023 = 2\sin\tfrac{1}{2}23\ \sin\tfrac{1}{2}30\ \sin\tfrac{1}{2}02;$$

and so for the other like expressions; each side of the equation is thus reduced to $1\div\sin\frac{1}{2}02\ \sin\frac{1}{2}03\ \sin\frac{1}{2}12\ \sin\frac{1}{2}13$).

98. In particular if the conic is taken to be the circle $x^2+y^2-1=0$, then for the coordinates $\left(\frac{\xi}{\zeta}, \frac{\eta}{\zeta}\right)$ of the intersections with the line $\alpha x+\beta y+\gamma=0$, we have

$$\begin{aligned} \Omega\xi + \gamma\eta + \beta\zeta &= 0, \quad \text{giving } \Omega^2 = \alpha^2+\beta^2+\gamma^2 \\ \gamma\xi - \Omega\eta + \alpha\zeta &= 0, \\ \beta\xi - \alpha\eta + \Omega\zeta &= 0, \end{aligned}$$

and then

$$\begin{aligned} \xi:\eta:\zeta &= -\beta^2+\gamma^2 : \alpha\beta-\gamma\Omega : \alpha\gamma+\beta\Omega \\ &= -\alpha\beta-\gamma\Omega : \alpha^2-\gamma^2 : \beta\gamma+\alpha\Omega \\ &= \alpha\gamma+\beta\Omega : \beta\gamma-\alpha\Omega : -\alpha^2-\beta^2, \end{aligned}$$

and the formula then becomes

$$\int\frac{dx}{(\alpha x+\beta\sqrt{1-x^2}+\gamma)\sqrt{1-x^2}} = \frac{1}{\sqrt{\alpha^2+\beta^2-\gamma^2}}\log\frac{\xi x+\eta\sqrt{1-x^2}-\zeta}{\alpha x+\beta\sqrt{1-x^2}+\gamma},$$

or, retaining Ω, y for the values $\sqrt{\alpha^2+\beta^2-\gamma^2}$, and $\sqrt{1-x^2}$, as this may also be written

$$= \frac{1}{\Omega}\log\frac{\gamma(\alpha x+\beta y+\gamma)+\Omega(\beta x-\alpha y+\Omega)}{\alpha x+\beta y+\gamma}.$$

The form of the integral is still such that the value is not very readily obtainable by ordinary methods: the value just written down can of course be verified, but the verification is scarcely easier than for the original more general form.

In the very particular case $\alpha=0,\ \beta=0,\ \gamma=1$, we have $\Omega=i$; $\xi:\eta:\zeta=1:-i:0$ and the formula becomes

$$\int\frac{dx}{\sqrt{1-x^2}}=\frac{1}{i}\log(x-iy):\ \text{viz. this is}$$

$$\sin^{-1}x=\tfrac{1}{2}\pi+\frac{1}{i}\log(x-i\sqrt{1-x^2}),$$

which is right; for putting $\sin^{-1}x=u$, and therefore $x=\sin u$, the equation becomes $i(u-\frac{1}{2}\pi)=\log.(\sin u-i\cos u)$: that is $\cos(u-\frac{1}{2}\pi)+i\sin(u-\frac{1}{2}\pi)=\sin u-i\cos u$.

Fixed Curve a Cubic: the Parametric Points 1, 2 *consecutive points on the Curve.* Art. Nos. 99 to 106.

99. The major function $(x, y, z)^1_{12}$ is taken to be $=\dfrac{1^2 2.023-12^2.013}{123}$, so that calling the differential $Qd\omega$ we have

$$Q=\frac{1^2 2.023-12^2.013}{123.012},$$

and it is required to find what this becomes when 1, 2 are consecutive points on the curve, or what is the same thing when the line 012 is a tangent at the point 1.

I take for convenience the cubic to be $f,=\frac{1}{3}(x^3+y^3+z^3),=0$. The coordinates of 1 are (x_1, y_1, z_1), those of 2 are $(x_1+\delta x_1,\ y_1+\delta y_1,\ z_1+\delta z_1)$, or as for shortness I write them $(x_1+\alpha,\ y_1+\beta,\ z_1+\gamma)$, where α,β,γ are considered as infinitesimals of the first order: this being so, the denominator of Q is at once seen to be of the second order; it will appear that the numerator is of the third order; whence, Q is of the first order.

100. We have $d\omega=\dfrac{ydz-zdy}{x^2},\ =\dfrac{zdx-xdz}{y^2}=\dfrac{xdy-ydx}{z^2}$, and in analogy herewith we may write $\delta\omega_1=\dfrac{y_1\gamma-z_1\beta}{x_1^2},\ =\dfrac{z_1\alpha-x_1\gamma}{y_1^2},\ =\dfrac{x_1\beta-y_1\alpha}{z_1^2}$; this being so we have

$$012=\begin{vmatrix} x, & y, & z \\ x_1 & y_1 & z_1 \\ \alpha & \beta & \gamma \end{vmatrix}=(xx_1^2+yy_1^2+zz_1^2)\,\delta\omega_1=01^2.\delta\omega_1,$$

and similarly $312=31^2.\delta\omega_1$.

Moreover

$$023=\begin{vmatrix} x, & y, & z \\ x_1 & y_1 & z_1 \\ x_3 & y_3 & z_3 \end{vmatrix}+\begin{vmatrix} x, & y, & z \\ \alpha, & \beta, & \gamma \\ x_3, & y_3, & z_3 \end{vmatrix}=013+0\delta13,$$

as the second term may be written; moreover

$$1^22 = x_1^2(x_1+a) + y_1^2(y_1+\beta) + z_1^2(z_1+\gamma), = ax_1^2+\beta y_1^2+\gamma z_1^2,$$

$$12^2 = x_1(x_1+a)^2 + y_1(y_1+\beta)^2 + z_1(z_1+\gamma)^2, = 2(ax_1^2+\beta y_1^2+\gamma z_1^2)+a^2x_1+\beta^2y_1+\gamma^2z_1,$$

and hence

$$\begin{aligned}1^22.023 - 12^2.013 = &\quad (ax_1^2+\beta y_1^2+\gamma z_1^2) \quad (013 + 0\delta 13)\\ &- [2(ax_1^2+\beta y_1^2+\gamma z_1^2) + (a^2x_1+\beta^2y_1+\gamma^2z_1)]\,013\\ = &- \quad [(ax_1^2+\beta y_1^2+\gamma z_1^2) + (a^2x_1+\beta^2y_1+\gamma^2z_1)]\,013\\ &+ \quad (ax_1^2+\beta y_1^2+\gamma z_1^2).0\delta 13,\end{aligned}$$

or reducing by $3(ax_1^2+\beta y_1^2+\gamma z_1^2) + 3(a^2x+\beta^2y+\gamma^2z) + (a^3+\beta^3+\gamma^3) = 0$,
this is

$$= +\tfrac{1}{3}(a^3+\beta^3+\gamma^3)\,013 - (a^2x_1+\beta^2y_1+\gamma^2z_1)\,0\delta 13,$$

which is of the third order.

101. We may show that each of the terms contains the factor $(\delta\omega_1)^2$: we have in fact

$$y_1z_1(\delta\omega_1)^2 = a\beta\frac{x_1}{y_1} - a^2 - \beta\gamma\frac{x_1^2}{y_1z_1} + \gamma a\frac{x_1}{z_1},$$

$$z_1x_1(\delta\omega_1)^2 = \beta\gamma\frac{y_1}{z_1} - \beta^2 - \gamma a\frac{y_1^2}{z_1x_1} + a\beta\frac{y_1}{x_1},$$

$$x_1y_1(\delta\omega_1)^2 = \gamma a\frac{z_1}{x_1} - \gamma^2 - a\beta\frac{z_1^2}{x_1y_1} + \beta\gamma\frac{z_1}{y_1},$$

and hence first multiplying by a, β, γ and adding we have

$$\begin{aligned}(ay_1z_1+\beta z_1x_1+\gamma x_1y_1)(\delta\omega_1)^2 = &\frac{\beta\gamma}{y_1z_1}(\beta y_1^2+\gamma z_1^2) + \frac{\gamma a}{z_1x_1}(\gamma z_1^2+ax_1^2) + \frac{a\beta}{x_1y_1}(ax_1^2+\beta y_1^2)\\ &- a\beta\gamma\left(\frac{x_1^2}{y_1z_1}+\frac{y_1^2}{z_1x_1}+\frac{z_1^2}{x_1y_1}\right)\\ &- (a^3+\beta^3+\gamma^3).\end{aligned}$$

But in virtue of $ax_1^2+\beta y_1^2+\gamma y_1^2=0$, the first line becomes = the second line, or the two together are $= -2a\beta\gamma\left(\frac{x_1^2}{y_1z_1}+\frac{y_1^2}{z_1x_1}+\frac{z_1^2}{x_1y_1}\right)$ which is = 0 in virtue of $x_1^3+y_1^3+z_1^3=0$; hence the equation is

$$(ay_1z_1+\beta z_1x_1+\gamma x_1y_1)(\delta\omega_1)^2 = -(a^3+\beta^3+\gamma^3),$$

the required expression for the first term.

102. Again, multiplying by x_1, y_1, z_1, and adding we have

$$\begin{aligned}3x_1y_1z_1(\delta\omega_1)^2 = &\frac{\beta\gamma}{y_1z_1}(y_1^3+z_1^3-x_1^3) + \frac{\gamma a}{z_1x_1}(z_1^3+x_1^3-y_1^3) + \frac{a\beta}{x_1y_1}(x_1^3+y_1^3-z_1^3)\\ &-(a^2x_1+\beta^2y_1+\gamma^2z_1),\end{aligned}$$

where in virtue of $x_1^3+y_1^3+z_1^3=0$, the first line is

$$= -\frac{2}{x_1y_1z_1}(\beta\gamma x_1^4+\gamma a y_1^4+a\beta z_1^4),$$

and this again is $= -2(a^2x_1+\beta^2y_1+\gamma^2z_1)$: (in fact we have identically

$$x_1y_1z_1(a^2x_1+\beta^2y_1+\gamma^2z_1) = (ax_1^3+\beta y_1^3+\gamma z_1^3)(ay_1z_1+\beta z_1x_1+\gamma x_1y_1) \\ -(\beta\gamma x_1+\gamma a y_1+a\beta z_1)(x_1^3+y_1^3+z_1^3) \\ +(\beta\gamma x_1^4+\gamma a y_1^4+a\beta z_1^4)$$

which in virtue of $ax_1^3+\beta y_1^3+\gamma z_1^3=0$, and $x_1^3+y_1^3+z_1^3=0$ becomes

$$x_1y_1z_1(a^2x_1+\beta^2y_1+\gamma^2z_1) = (\beta\gamma x_1^4+\gamma a y_1^4+a\beta z_1)).$$

Hence the equation is $3x_1y_1z_1(\delta\omega_1)^2 = -3(a^2x_1+\beta^2y_1+\gamma^2z_1)$, or finally

$$x_1y_1z_1.(\delta\omega_1)^2 = -\quad(a^2x_1+\beta^2y_1+\gamma^2z_1),$$

the required expression for the second term.

103. Writing for shortness $ay_1z_1+\beta z_1x_1+\gamma x_1y_1 = \delta(x_1y_1z_1)$ we have

$$1^22.023 - 12^2.013 = \{-\tfrac{1}{3}\delta(x_1y_1z_1)\,013 + x_1y_1z_1.0\delta13\}(\delta\omega_1)^2,$$

and hence dividing by $\quad 012.123, \quad = 01^2, 31^2.(\delta\omega_1)^2$,

we have
$$Q = \frac{1^22.023-12^2.013}{012.123} = \frac{-\frac{1}{3}\delta(x_1, y_1, z_1).013 + x_1y_1z_1.0\delta13}{01^2.31^2}.$$

But this can be further reduced: the numerator multiplied by 3, is

$$= -(ay_1z_1+\beta z_1x_1+\gamma x_1y_1)\begin{vmatrix} x, & y, & z \\ x_1, & y_1, & z_1 \\ x_3, & y_3, & z_3\end{vmatrix} + 3x_1y_1z_1\begin{vmatrix} x, & y, & z \\ a, & \beta, & \gamma \\ x_3, & y_3, & z_3\end{vmatrix},$$

which is
$$= \begin{vmatrix} x & , & y & , & z \\ x_1(y_1^3-z_1^3), & & y_1(z_1^3-x_1^3), & & z_1(x_1^3-y_1^3) \\ x_3 & , & y_3 & , & z_3 \end{vmatrix} d\omega_1,$$

where $x_1(y_1^3-z_1^3)$, $y_1(z_1^3-x_1^3)$, $z_1(x_1^3-y_1^3)$ are the coordinates of the tangential of the point 1 in regard to the cubic, viz. the point of intersection of the tangent at 1 with the cubic. The determinant may for shortness be called $0t13$; and we thus have

$$Q = \frac{1^22.023-12^2.013}{012.123} = \frac{\frac{1}{3}\delta\omega_1}{31^2}\frac{0t13}{01^2},$$

where observe that $01^2=0$ is the equation of the tangent at the point 0: and $0t13=0$, is the equation of the line joining the tangential of 1 with the arbitrary point 3.

104. The identity just referred to is proved very easily: comparing on each side the coefficient of yz_3-y_3z, the factor x_1 divides out and we ought to have

$$-(ay_1z_1+\beta z_1x_1+\gamma x_1y_1)+3y_1z_1a = (y_1^3-z_1^3)\,\delta\omega_1,$$

that is
$$(y_1^3-z_1^3)\,\delta\omega_1 = 2ay_1z_1-\beta z_1x_1-\gamma x_1y_1,$$

and in fact from $y_1^3\delta\omega_1 = z_1a-x_1\gamma$, $z_1^3\delta\omega_1 = x_1\beta-y_1a$, we have

$$(y_1^3-z_1^3)\,\delta\omega_1 = y_1(z_1a-x_1\gamma)-z_1(x_1\beta-y_1a),$$

which is the value in question: similarly the coefficients of zx_3-z_3x, xy_3-x_3y are equal on the two sides; and the equation is thus verified.

105. The proof has been given in regard to the particular cubic $x^3+y^3+z^3=0$; but it might have been given for the canonical form $x_3+y_3+z_3+6lxyz=0$: and from the invariantive form it is clear that the result in fact applies to any cubic whatever. The result is an important one: we see by it that when the points 1 and 2 are consecutive points on the curve we must in place of the differential $Qd\omega$, which is evanescent, consider a new form $\frac{0t13}{01^2}\,d\omega_1$, where, as already remarked, the denominator represents the tangent at the point 1, and the numerator the line joining the tangential of this point with the point 3.

106. We have $\qquad \{023\}+\{031\}+\{012\}=\{123\},$

or writing this in the form

$$\{012\}-\{312\}+\{023\}-\{013\}=0,$$

suppose 2 is here the consecutive point $1+\delta 1$, $\{012\}-\{312\}$, $=\dfrac{1^2 2.023-12^2.013}{012.312}$ becomes $=\{0t13\}\delta\omega_1$, we have also $\{023\}=\{013\}+\partial_1\{013\}\delta\omega_1$,

and the result is $\qquad -\{0t13\}+\partial_1 013=0,$

that is $\partial_1\{013\}=\{0t13\}$, | (in case of cubic $x^3+y^3+z^3+6lxyz=0$) is $=$

$$-\frac{\begin{vmatrix} x & , & y & , & z \\ x_1(y_1^3-z_1^3), & & y_1(z_1^3-x_1^3), & & z_1(x_1^3-y_1^3) \\ x_3 & , & y_3 & , & z_3 \end{vmatrix}}{3(x_1^2x+y_1^2y+z_1^2z)(x_1^2x_3+y_1^2y_3+z_1^2z_3)},$$

i. e. the differential coefficient of $\{013\}$ in regard to the parametric point 1, is $=\{0t13\}$ the symbol for the case where the parametric line is the tangent at 1.

Fixed Curve a Cubic: the Parametric Points corresponding points.

Art. Nos. 107 to 110.

107. The parametric points 1, 2 are taken to be corresponding points, that is such that the tangents at these points meet at a point, say 3, on the cubic. We may from 3 draw two other tangents, touching the cubic, say at the points 1′ and 2′. The four points 1, 2, 1′, 2′ are then such that the lines 12, 1′2′ meet in a point, say 4, of the cubic; and moreover 3, 4 are corresponding points.

We may take (x, y, z), $=(1, 0, 0)$, $(0, 1, 0)$, $(0, 0, 1)$ for the coordinates of the points 1, 2, 3 respectively: $x=0$, $y=0$ are thus the equations of the lines 32, 31 respectively, and $z=0$ is the equation of the line 12, viz. we have $z=012$. Taking $x-M_1y=0$, $x-M_2y=0$, for the equations of the tangents 31′, 32′ respectively, and $\zeta=0$ for the equation of the line 1′2′ joining their points of contact (ζ a properly determined linear function of (x, y, z)), it is to be shown that the differential $Qd\omega$ may be taken to be $\frac{\zeta d\omega}{z}$, and that this is $= \frac{1}{2}\left(\frac{dx}{x} - \frac{dy}{y}\right)$: the affected theorem thus assumes a special form, which will be noticed.

108. The cubic passes through the points $(x = 0, z = 0)$ and $(y = 0, z = 0)$, the tangents at these points being $x = 0$, and $y = 0$ respectively: also through the point $x = 0$, $y = 0$: its equation thus is

$$f, = gz^2x + 2lzxy + iz^2y + hx^2y + kxy^2, = 0,$$

and writing

$$d\omega = \frac{xdy - ydx}{\frac{df}{dz}}.$$

we have

$$\frac{df}{dz} = 2(gzx + lxy + izy),$$

which from the equation of the curve written in the form

$$z(gzx + lxy + izy) + xy(hx + ky + lz) = 0,$$

or say

$$z(gzx + lxy + izy) + xy\zeta = 0,$$

becomes

$$= \frac{-2xy\zeta}{z},$$

and we thus have

$$d\omega = \frac{-z}{2xy\zeta}(xdy - ydx), = \frac{1}{2}\frac{z}{\zeta}\left(\frac{dx}{x} - \frac{dy}{y}\right),$$

where $\zeta = hx + ky + lz$. To find the meaning of ζ, observe that the line $x - My = 0$, meets the curve in the point $(x = 0, y = 0)$, and in two other points determined by the equation

$$z^2(gm + i) + 2zy \, . \, lM + y^2(hM^2 + kM) = 0;$$

this line will be a tangent if

$$(gM + i)(hM + k) - l^2M = 0,$$

and we then have at the point of contact $(hM + k)y + lz = 0$; and writing this in the form $hx + ky + lz = 0$, we see that the equation $\zeta = 0$ is satisfied at the point of contact of each of the two tangents $x - M_1y = 0$, $x - M_2y = 0$; viz. $\zeta = 0$ is the equation of the line joining the two points of contact. Moreover, from the equation of the curve written in the foregoing form

$$z(gzx + lxy + izy) + xy\zeta = 0,$$

it appears that the lines $z = 0$, $\zeta = 0$, meet on the curve; or what is the same thing, that the line $\zeta = 0$ passes through the residue of the parametric points 1, 2.

109. The function ζ at 1 becomes $=h$, and this is the value of 3.1^22; in fact

$$\begin{aligned}3.1^22 &= \left(x_2\frac{d}{dx_1}+y_2\frac{d}{dy_1}+z_2\frac{d}{dz_1}\right)f_1, \quad (x_2, y_2, z_2)=(0, 1, 0),\\ &=\frac{df_1}{dy_1},\\ &=2lz_1x_1+iz_1^2+hx_1^2+2kx_1y_1, \quad (x_1, y_1, z_1)=(1, 0, 0),\\ &=h.\end{aligned}$$

We have thus ζ, satisfying the required conditions for the major function: and the differential $Qd\omega$ may therefore be taken to be $=\frac{\zeta}{z}d\omega$, that is we have

$$Qd\omega=\tfrac{1}{2}\left(\frac{dx}{x}-\frac{dy}{y}\right).$$

The affected theorem thus becomes

$$\Sigma\,\tfrac{1}{2}\left(\frac{dx}{x}-\frac{dy}{y}\right)=-\frac{\delta\varphi_1}{\varphi_1}+\frac{\delta\varphi_2}{\varphi_2}.$$

110. The meaning of this will be better understood from the integral form. Integrating each side, and assuming that the superior limits are given by a line ϕ which cuts the cubic in the points 4, 5, 6, and the inferior limits by a line ψ which cuts the cubic in the points 7, 8, 9, we find

$$\log\frac{x_4x_5x_6}{x_7x_8x_9}-\log\frac{y_4y_5y_6}{y_7y_8y_9}=2\log\frac{\varphi_2}{\psi_2}\,\frac{\psi_1}{\varphi_1},$$

that is

$$\frac{x_4x_5x_6\;y_4y_5y_6}{x_7x_8x_9\;y_7y_8y_9}=\left(\frac{\psi_2\,\varphi_1}{\psi_2\,\varphi_1}\right)^2,$$

where ϕ_1, ψ_1, ϕ_2, ψ_2 denote the values of the linear functions ϕ, ψ at the points 1 and 2 respectively. We have a cubic cut by the lines ϕ, ψ, x, y in the points 4, 5, 6; 7, 8, 9; 2, 2′, 3 and 1, 1′, 3 respectively: where for the moment 1′, 2′ are written to denote the points on the curve consecutive to 1 and 2 respectively. Hence by a known theorem in transversals

$$\left(\frac{x}{y}\right)_{456}\div\left(\frac{x}{y}\right)_{789}=\left(\frac{\varphi}{\psi}\right)_{22'3}\div\left(\frac{\varphi}{\psi}\right)_{11'3}, \text{ that is}$$

$$\frac{x_4x_5x_6\;y_7y_8y_9}{x_7x_8x_9\;y_4y_5y_6}=\frac{\psi_1\psi_{1'}\psi_3\cdot\varphi_2\varphi_{2'}\varphi_3}{\psi_2\psi_{2'}\psi_3\cdot\varphi_1\varphi_{1'}\varphi_3},$$

which dividing out the $\phi_3\psi_3$, and writing 1, 2 in place of 1′, 2′, becomes

$$=\left(\frac{\varphi_2\psi_1}{\varphi_1\psi_2}\right)^2,$$

agreeing with the result just obtained.

Aronhold's Cubic Transformation. Art. Nos. 111 to 119.

111. This was obtained in the paper "Algebraische Reduction des Integrals $\int F(x, y)\,dx$, u. s. w.," Berl. Monatsb., April, 1861, pp. 462–468. I give in the first place the analytical results, independently of the general theory, with the values for the canonical form $f, = \frac{1}{6}(x^3 + y^3 + z^3 + 6lxyz), = 0$, of the cubic.

T sextic invariant, $= 1 - 20l^3 - 8l^6$,

S quartic „ (Aronhold's) $= -4(l - l^4)$,

R discriminant $= (1 + 8l^3)^3$,

$P = 3h\alpha^2 0 \qquad = \{-3l^2\alpha^2 + (1 + 2l^3)\beta\gamma\}x +$ etc.,

$Q = f\alpha^2 0 \qquad = (\alpha^2 + 2l\beta\gamma)x +$ etc.,

$B = f\mathrm{a}^2 0 \qquad = (\mathrm{a}^2 + 2l\mathrm{bc})x +$ etc.,

$[\mathrm{a}, \mathrm{b}, \mathrm{c} = \alpha(\beta^3 - \gamma^3),\ \beta(\gamma^3 - \alpha^3),\ \gamma(\alpha^3 - \beta^3),]$

$C = f\mathrm{a}0^2 \qquad = \mathrm{a}(x^2 + 2lyz) +$ etc.,

$D = f0^3 \qquad = x^3 + y^3 + z^3 + 6lxyz.$

Then we have $\quad 2TQ^4 + 6SPQ^2 + 8P^3Q = -R^{\frac{1}{2}}(6C^2 - 8BD)$,

viz. this equation, where each side is a quartic function $(x, y, z)^4$, is an identity when (α, β, γ) are connected by the equation, $f\alpha^3, = \alpha^3 + \beta^3 + \gamma^3 + 6l\alpha\beta\gamma, = 0$; and further

$$QdP - PdQ = -R^{\frac{1}{2}}\{\mathrm{a}(ydz - zdy) + \mathrm{b}(zdx - xdz) + \mathrm{c}(xdy - ydx)\}.$$

Hence writing $\quad \lambda = \dfrac{6h\alpha^2 0}{f\alpha^2 0}, = \dfrac{2P}{Q}, = \dfrac{2(\{-3l^2\alpha^2 + (1 + 2l^3)\beta\gamma\}x + \text{etc.})}{(\alpha^2 + 2l\beta\gamma)x + \text{etc.}}$,

we have $\quad Q^4(\lambda^3 - 3S\lambda - 2T) = R^{\frac{1}{2}}(6C^2 - 8BD)$,

and $\quad Q^2 d\lambda, = 2(QdP - PdQ) = -2R^{\frac{1}{2}}\{\mathrm{a}(ydz - zdy) + \text{etc.}\}$.

112. Supposing now that (x, y, z) are the coordinates of a point on the cubic, then $D = 0$; and taking the square root of each side of the first equation, we may write

$$Q^2\sqrt{\lambda^3 - 3S\lambda - 2T} = -R^{\frac{1}{4}}\sqrt{6}.C,$$

$$Q^2 d\lambda = -2R^{\frac{1}{2}}\{\mathrm{a}(ydz - zdy) + \text{etc.}\}.$$

We have $\quad d\omega = \dfrac{ydz - zdy}{x^2 + 2lyz} = \dfrac{zdx - xdz}{y^2 + 2lzx} = \dfrac{xdy - ydx}{z^2 + 2lxy}$,

whence $\quad d\omega = \dfrac{\mathrm{a}(ydz - zdy) + \text{etc.}}{C}$;

and we consequently have $\quad \dfrac{d\lambda}{\sqrt{\lambda^3 - 3S\lambda - 2T}} = \dfrac{2}{\sqrt{6}}$,

or as this may also be written

$$\frac{d\lambda}{\sqrt{4\lambda^3 - 12S\lambda - 8T}} = \frac{1}{\sqrt{6}}\,d\omega,$$

which, if $12S$, $8T$ are put $= g_2, g_3$ respectively, takes the Weierstrassian form

$$\frac{d\lambda}{\sqrt{4\lambda^3 - g_2\lambda - g_3}} = \frac{1}{\sqrt{6}}\,d\omega.$$

The conclusion is that for the cubic curve, taking λ a quotient of two linear functions of (x, y, z), the differential $d\omega$ is transformed into $d\lambda \div$ square root of a cubic function of λ: viz. we have thus a form of differential, not the same, but such as that which belongs to the ordinary theory of elliptic functions, and which has been adopted by Weierstrass as a canonical form.

113. The transformation depends on the arbitrary point (α, β, γ) of the cubic: the point (a, b, c) is the tangential of this point, viz. the point of intersection of the tangent at (α, β, γ) with the cubic: we can from (a, b, c) draw four tangents to the cubic, viz. the tangent at (α, β, γ) and three other tangents: the equations of the four tangents being $\frac{2P}{Q}, = \frac{6ha^{2}0}{fa^{2}0}, = \infty, \lambda_1, \lambda_2, \lambda_3$ respectively; where $\lambda_1, \lambda_2, \lambda_3$ are the roots of the equation $\lambda^3 - 3S\lambda - 2T = 0$.

Suppose for a moment that (α, β, γ) is a point not on the cubic curve, and write $A = \alpha^3 + \beta^3 + \gamma^3 + 6l\alpha\beta\gamma$, we have

$$A^2D^2 + 4AC^3 + 4B^3D - 3B^2C^2 - 6ABCD = 0,$$

for the equation of the six tangents which can be drawn from the point (α, β, γ) to the cubic: when (α, β, γ) is on the cubic $A = 0$, and the equation becomes $B^2(4BD - 3C^2) = 0$, where $B = 0$ is the equation of the tangent at the point (α, β, γ): throwing out the factor B^2 we have $4BD - 3C^2 = 0$, for the equation of the four tangents from (α, β, γ) to the curve; viz. the equation of the four tangents is $2TQ^4 + 6SPQ^3 + 8P^3Q = 0$,
or as this may be written

$$Q(2P - \lambda_1 Q)(2P - \lambda_2 Q)(2P - \lambda_3 Q) = 0,$$

viz. the equations of the four tangents are as is mentioned above; it was in fact by these geometrical considerations that Aronhold obtained his results.

114. The foregoing expression for $QdP - PdQ$, say

$$QdP - PdQ = (1 + 8l^3)\{\mathrm{a}(ydz - zdy) + \mathrm{b}(zdx - xdz) + \mathrm{c}(xdy - ydx)\},$$

may be verified without difficulty. Writing for a moment

$$\begin{aligned} QdP - PdQ = {} & (Ax + By + Cz)(Ldx + Mdy + Ndz) \\ & - (Adx + Bdy + Cdz)(Lx + My + Nz), \\ = {} & (BN - CM)(ydz - zdy) - \text{etc.}; \end{aligned}$$

we have

$$\begin{aligned} BN - CM = {} & \{-3l^2\beta^2 + (1 + 2l^3)\gamma\alpha\}(\gamma^2 + 2l\alpha\beta) \\ & - \{-3l^2\gamma^2 + (1 + 2l^3)\alpha\beta\}(\beta^2 + 2l\gamma\alpha) \\ = {} & -6l^3\alpha\beta^3 + (1 + 2l^3)\alpha\gamma^3 + 6l^3\alpha\gamma^3 - (1 + 2l^3)\alpha\beta^3, \\ = {} & -(1 + 8l^3)\alpha(\beta^3 - \gamma^3) \\ = {} & -(1 + 8l^3)\mathrm{a}, \end{aligned}$$

which proves the theorem.

115. I content myself with a partial verification of the identity

$$2TQ^4 + 6SPQ^3 - 8P^3Q = -(1 + 8l^3)^2(6C^2 - 8BD);$$

writing herein $x, y, z = 1, -1, 0$, we have $D = 0$, and the equation becomes

$$2TQ^4 + 6SPQ^3 - 8P^3Q + 6(1 + 8l^3)^2C^2 = 0,$$

where now

$$Q = (\alpha - \beta)(\alpha + \beta - 2l\gamma),\quad P = (\alpha - \beta)\{-3l^2(\alpha + \beta) - (1 + 2l^3)\gamma\},$$

$$C = \mathrm{a} + \mathrm{b} - 2l\mathrm{c},\ = (\alpha - \beta)\{-\alpha\beta^2 - \alpha^2\beta - \gamma^3 - 2l\gamma(\alpha^2 + \alpha\beta + \beta^2)\},$$

which putting therein

$$-\gamma^3 = \alpha^3 + \beta^3 + 6l\alpha\beta\gamma \text{ becomes } = (\alpha - \beta)^2(\alpha + \beta - 2l\gamma).$$

Hence writing

$$X = \alpha + \beta - 2l\gamma,\quad Y = -3l^2(\alpha + \beta) - (1 + 2l^3)\gamma,$$

we have $\quad Q,\ P,\ C = (\alpha - \beta)X,\ (\alpha - \beta)Y,\ (\alpha - \beta)^2X$:

substituting these values the factor $(\alpha - \beta)^4X$ divides out, and the equation becomes $\quad 2TX^3 + 6SX^2Y - 8Y^3 + 6(1 + 8l^3)^2(\alpha - \beta)^2X = 0$.

To complete the verification observe that we have $Y + 3l^2X = -(1 + 8l^3)\gamma$, whence

$$-Y^3 = (1 + 8l^3)^3\gamma^3 + 9(1 + 8l^3)^2l^2\gamma^2X + 27(1 + 8l^3)l^4\gamma X^2 + 27l^6X^3,$$

and herein $-\gamma^3 = \alpha^3 + \beta^3 + 6l\alpha\beta\gamma$, whence

$$-\gamma^3 + \alpha\beta X = (\alpha + \beta)^3 = (X + 2l\gamma)^3 = X^3 + 6l\gamma X^2 + 12l^2\gamma^2X + 8l^3\gamma^3,$$

that is $\quad -(1 + 8l^3)\gamma^3 = X^3 + 6l\gamma X^2 + (12l^2\gamma^2 - 3\alpha\beta)X.$

Hence the equation to be verified becomes

$$2TX^3 + 6SX^2Y - 8\left\{\begin{array}{l}(1+8l^3)^2([X^3 + 6l\gamma X^2 + (12l^2\gamma^2 - 3\alpha\beta)X]\\ -(1 + 8l^3)^2 9l^2\gamma^2X^2\\ -(1 + 8l^3)27l^4\gamma X\\ -27l^6X^3\end{array}\right\}$$

$$+ 6(1 + 8l^3)^2(\alpha - \beta)^2X = 0;$$

viz. throwing out the factor X, this is

$$\{2T - 8(1 + 8l^3)^3 + 216l^6\}X^2 + 6SXY - 48(1 + 8l^3)^2l\gamma X + 216(1 + 8l^3)l^4\gamma X$$

$$-(1 + 8l^3)^2\{96l^2\gamma^2 - 24\alpha\beta - 72l^2\gamma^2 - 6(\alpha - \beta)^2\} = 0,$$

where the last term is $\quad = +6(1 + 8l^3)^2\{(\alpha + \beta)^2 - 4l^2\gamma^2\}$,

viz. this is $\quad = 6(1 + 8l^3)^2(\alpha + \beta + 2l\gamma)X$,

and there is again the factor X which can be thrown out: the equation thus becomes

$$[2T - 8(1 + 8l^3)^3 + 216l^6]X + 6SY - 48(1 + 8l^3)^2l\gamma + 216(1 + 8l^3)l^4\gamma$$

$$+ 6(1 + 8l^3)^2(\alpha + \beta + 2l\gamma) = 0.$$

This may be written

$$[2T - 8(1+8l^3)^2 + 216l^6]X + 6S[-3l^2X - (1+8l^3)\gamma] - 48(1+8l^3)^2 l\gamma$$
$$+ 216(1+8l^3)l^4\gamma + 6(1+8l^3)^2(X + 4l\gamma) = 0.$$

or finally it is

$$[2T - 8(1+8l^3)^2 + 216l^6 - 18l^2S + 6(1+8l^3)^2]X$$
$$+ [-6l^{-1}S - 48(1+8l^3) + 216l^3 + 24(1+8l^3)](1+8l^3)l\gamma = 0;$$

and substituting for T, S their values $1 - 20l^3 - 8l^6$ and $-4l + 4l^4$ respectively, the coefficients of X and $(1+8l^3)l\gamma$ are separately $= 0$, and the equation is thus verified.

116. The foregoing equation $\lambda = \frac{6ha^20}{fa^20}$ regarding therein λ as an arbitrary parameter and (x, y, z) as current coordinates is the equation of an arbitrary line through the point (a, b, c) of the cubic: it meets the cubic in two other points depending, of course, on the value of λ; and the coordinates of either of these is thus expressible, irrationally, in terms of λ, the expressions involving the radical $\sqrt{\lambda^3 - 3S\lambda - 2T}$: from the values of x, y, z in terms of λ we should be able to deduce the foregoing equation $\frac{2}{\sqrt{6}}\,d\omega = \frac{d\lambda}{\sqrt{\lambda^3 - 3S\lambda - 2T}}$: the expressions assume a peculiarly simple form when (α, β, γ) instead of being an arbitrary point of the cubic is a point of inflexion of the cubic; and it is easy to see *à priori* why this is so: in fact if we assume $x:y:z = u + \alpha\sqrt{\Lambda} : v + \beta\sqrt{\Lambda} : w + \gamma\sqrt{\Lambda}$, where u, v, w are linear functions and Λ a cubic function of λ; then the locus is a cubic curve, and corresponding to the value $\lambda = \infty$, we have $x:y:z = \alpha:\beta:\gamma$, viz. the curve passes through the point (α, β, γ): moreover, it can be shown that this point is an inflexion of the curve; expressions of the foregoing simple form thus only exist in the case where the point (α, β, γ) is an inflexion, and the formulæ referring to an arbitrary point (α, β, γ) of the curve are necessarily of a more complex form.

117. To work this out we start from the foregoing equation

$$\lambda = \frac{6ha^20}{fa^20} = \frac{2(\{-3l^2\alpha^2 + (1+2l^3)\beta\gamma\}x + \text{etc.})}{(\alpha^2 + 2l\beta\gamma)x + \text{etc.}}$$

which putting therein $L = \lambda + 6l^2$, $M = l\lambda - (1 + 2l^3)$, and

$$A,\ B,\ C = L\alpha^2 + 2M\beta\gamma,\ L\beta^2 + 2M\gamma\alpha,\ L\gamma^2 + 2M\alpha\beta,$$

becomes $Ax + By + Cz = 0$, the equation of a line through the point (a, b, c), $= (\alpha(\beta^3 - \gamma^3),\ \beta(\gamma^3 - \alpha^3),\ \gamma(\alpha^3 - \beta^3))$ as before: and we have to find the intersections of this line with the cubic $x^3 + y^3 + z^3 + 6lxyz = 0$. We have

$$C^3(x^3 + y^3) - (Ax + By)^3 - 6lC^2(Ax + By)xy = 0:$$

the cubic function contains as we know the factor $\mathrm{b}x - \mathrm{a}y$, and in the remaining quadric factor it is easy to calculate the coefficients of x^2 and y^2: we thus obtain the identity $\quad C^3(x^3+y^3)-(Ax+By)^3-6lC^2(Ax+By)xy$

$$=(\mathrm{b}x-\mathrm{a}y)\{[(-\beta^2-6l\gamma\alpha)L^3+6\alpha\gamma L^2M-8\beta^2M^3]x^2 \\ +2H\quad xy \\ +[(-\alpha^2-6l\alpha\beta)L^3+6\beta\gamma L^2M-8\alpha^2M^3]y^2\},$$

from which the as yet unknown coefficient $2H$ is to be obtained. This is most easily effected by assuming $x,\ y=\alpha,\ -\beta$; values which give

$$x^3+y^3=\alpha^3-\beta^3,\ Ax+By=L(\alpha^3-\beta^3),\ \mathrm{b}x-\mathrm{a}y=-\alpha\beta(\alpha^3-\beta^3):$$

the whole equation becomes divisible by $\alpha^3-\beta^3$ and omitting this factor we have

$$C^3-L^3(\alpha^3-\beta^3)^2+6lC^2L\alpha\beta \\ =\alpha\beta\{[2\alpha^3\beta^3+6l\gamma(\alpha^3+\beta^3)]L^3-6\gamma(\alpha^3+\beta^3)L^2M+16\alpha^3\beta^3M^3\}+2H\alpha^2\beta^2,$$

where for C is to be substituted its value $L\gamma^2+2M\alpha\beta$, and we may also reduce by $\alpha^3+\beta^3+\gamma^3+6l\alpha\beta\gamma=0$. The left-hand side is

$$C^3-L^3(\alpha^3+\beta^3)^2-4\alpha^3\beta^3L^3+6lC^2L\alpha\beta,$$

which after reduction is found to contain the factor $\alpha\beta$; and omitting this factor and reducing also the right-hand side the whole equation becomes

$$\begin{aligned} &L^3(-6l\gamma^4-36l^2\gamma^2\alpha\beta+4\alpha^2\beta^2) &&=L^3(-6l\gamma^4-36l^2\gamma^2\alpha\beta+2\alpha^2\beta^2)\\ &+L^2M(6\gamma^4+24l\gamma^2\alpha\beta) &&+L^2M(6\gamma^4+36l\gamma^2\alpha\beta)\\ &+LM^2(12\gamma^2\alpha\beta+24l\alpha^2\beta^2) &&\\ &+M^3(8\alpha^2\beta^2) &&+M^3(16\alpha^2\beta^2)\\ & &&+2H\alpha\beta; \end{aligned}$$

omitting here the terms which destroy each other, the equation again divides by $\alpha\beta$, and we thus obtain the value of H; and the required identity is

$$C^3(x^3+y^3)-(Ax+By)^3-6lC^2(Ax+By)xy \\ =(\mathrm{b}x-\mathrm{a}y)\{[(-\beta^2-6l\gamma\alpha)L^3+6\alpha\gamma L^2M-8\beta^2M^3]x^2 \\ +[\alpha\beta L^3-6l\gamma^2L^2M+(6\gamma^2+12l\alpha\beta)LM^2-4\alpha\beta M^3]\,2xy \\ +[(-\alpha^2-6l\beta\gamma)L^3+6\beta\gamma L^2M-8\alpha^2M^3]y^2\}.$$

Hence putting for shortness

$$\begin{aligned} \mathfrak{A}&=(\alpha^2+6l\beta\gamma)L^3-6\beta\gamma L^2M &&+8\alpha^2M^3,\\ \mathfrak{B}&=(\beta^2+6l\gamma\alpha)L^3-6\alpha\gamma L^2M &&+8\beta^2M^3,\\ \mathfrak{H}&=\alpha\beta L^3-6l\gamma^2L^2M+(6\gamma^2+12l\alpha\beta)LM^2-4\alpha\beta M^3, \end{aligned}$$

the equation in (x, y) is

$$\mathfrak{B}x^2-2\mathfrak{H}xy+\mathfrak{A}y^2=0,$$

giving $\qquad \mathfrak{B}x-\mathfrak{H}y=\sqrt{\mathfrak{H}^2-\mathfrak{A}\mathfrak{B}}\,y,$

or say $\qquad x:y=\mathfrak{H}+\sqrt{\mathfrak{H}^2-\mathfrak{A}\mathfrak{B}}:\mathfrak{B}.$

We find without difficulty, reducing always by $\alpha^3+\beta^3+\gamma^3+6l\alpha\beta\gamma=0$,

$$\begin{aligned}\tfrac{1}{6}(\mathfrak{H}^3-\mathfrak{A}\mathfrak{B}) = {} & l\gamma^4 && L^6\\ &+(-\gamma^4+4l\alpha\beta\gamma^2) && L^5M^3\\ &+(6l^2\gamma^4-4\alpha\beta\gamma^2+4l\alpha^2\beta^2) && L^4M^2\\ &+(-4l\gamma^4+24l^2\alpha\beta\gamma^2-4\alpha^2\beta^2) && L^3M^3\\ &+(-2\gamma^4-16l\alpha\beta\gamma^2+24l\alpha^2\beta^2) && L^2M^4\\ &+(\quad -8\alpha\beta\gamma^2-16l\alpha^2\beta^2) && LM^5\\ &+(\quad -8\alpha^2\beta^2) && M^6,\end{aligned}$$

which is $\quad = (lL-M)(L^3+6lLM^2+2M^3)(\gamma^2L+2\alpha\beta M)^2.$

But we have $\quad lL-M=1+8l^3,$

$$L^3+6lLM^2+2M^3=(1+8l^3)(\lambda^3-3S\lambda-2T),$$

and the equation thus is

$$\tfrac{1}{6}(\mathfrak{H}^3-\mathfrak{A}\mathfrak{B})=(1+8l^3)^2(\lambda^3-3S\lambda-2T)\left((\gamma^2+2l\alpha\beta)\lambda+\{6l^2\gamma^2-2(1+2l^3)\alpha\beta\}\right)^2,$$

showing that the solution involves the radical $\sqrt{\lambda^3-3S\lambda-2T}$.

118. If (α,β,γ) is the inflexion $(1,-1,0)$, the expression for λ is here

$$\lambda=\frac{-6l^2x-6l^2y-2(1+2l^3)z}{x+y-2lz};$$

the equation in (x, y) is

$$(L^3+8M^3)x^2+(2L^3+24lLM^2-8M^3)xy+(L^3+8M^3)y^2=0,$$

or as this may be written

$$(L^3+6lLM^2+2M^3)(x+y)^2+(-6lLM^2+6M^3)(x-y)^2=0,$$

say $\quad (L^3+6lLM^2+2M^3)(x+y)^2=6M^2(lL-M)(x-y)^2,$

viz. we thus have $\quad \sqrt{\lambda^3-3S\lambda-2T}\,(x+y)=M\sqrt{6}(x-y),$

or substituting for M its value

$$x=\sqrt{6}\{l\lambda-(1+2l^3)\}+\sqrt{\lambda^3-3S\lambda-2T},$$
$$y=\sqrt{6}\{l\lambda-(1+2l^3)\}-\sqrt{\lambda^3-3S\lambda+2T},$$

whence also $\quad z=\sqrt{6}(\lambda+6l^2),$

these values satisfying identically $(\lambda+6l^2)(x+y)-2[l\lambda-(1+2l^3)]z=0$, and $x^3+y^3+z^3+6lxyz=0$:

119. Starting from these values we in fact easily obtain

$$xdy-ydx=\frac{-\sqrt{6}\,d\lambda}{\sqrt{\lambda^3-3S\lambda-2T}}\times$$
$$\{l\lambda^3+(-3-6l^3)\lambda^2+(-12l^2+12l^5)\lambda+(-8l-92l^4-8l^7)\},$$
$$z^2+2lxy=-2\{\text{Do.}\}.$$

and hence $$d\omega=\frac{xdy-ydx}{z^2+2lxy}=\frac{\tfrac{1}{2}\sqrt{6}\,d\lambda}{\sqrt{\lambda^3-3S\lambda-2T}}.$$

The same result might of course have been obtained from the values of x, z or y, z, the factor which divides out being in each of these cases irrational.

The Cubic $y^2 = x(1-x)(1-k^2x)$. Art. Nos. 120 to 130 (*several sub-headings*).

120. The curve is a semi-cubical parabola, symmetrical in regard to the axis of x; and if as usual k^2 is taken to be real, positive and less than 1, then the curve consists of an oval, and an infinite portion which may be called the anguis. See Figure.

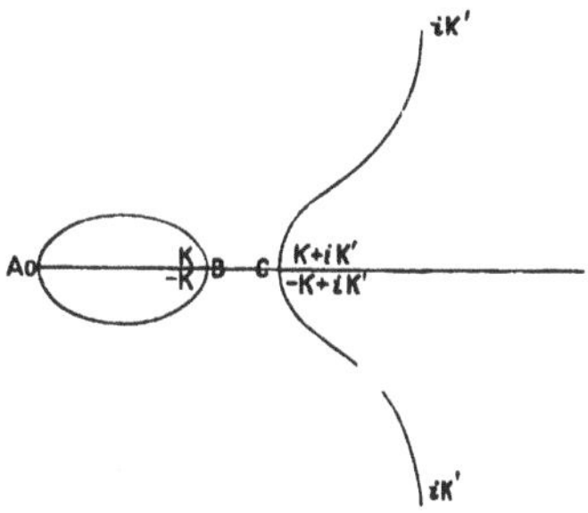

The equation is satisfied by

$$x = \operatorname{sn}^2 u,$$

$$y = \operatorname{sn} u \operatorname{cn} u \operatorname{dn} u.$$

Observe that the periods for these combinations of the elliptic functions are $2K$, $2iK'$; in fact

$$\begin{array}{llll} \operatorname{sn}(u+2K) = & -\operatorname{sn} u, & \operatorname{sn}(u+2iK') = & \operatorname{sn} u, \\ \operatorname{cn} \quad \text{"} \quad = & -\operatorname{cn} u, & \operatorname{cn} \quad \text{"} \quad = & -\operatorname{cn} u, \\ \operatorname{dn} \quad \text{"} \quad = & \operatorname{dn} u, & \operatorname{dn} \quad \text{"} \quad = & -\operatorname{dn} u, \end{array}$$

whence the sn^2, and sn.cn.dn are each unaltered by the change of u into $u+2K$ or $u+2iK'$. Hence to a given point (x, y) on the curve, the argument u is not $=$ a determinate value u_0, for it may be equally well taken to be $= u_0 + 2mK + 2m'iK'$, where m, m' are any positive or negative integers whatever: we express this by $u \equiv u_0$, or say u is congruent to u_0. But when u is thus given by a congruence $u \equiv u_0$ where u_0 has a determinate value, the point on the curve is uniquely determined. It is, however, to be noticed that a congruence $2u \equiv u_0$ does not uniquely determine the point on the curve: there are in fact four incongruent values of u, viz. $\frac{1}{2}u_0$, $\frac{1}{2}u_0 + K$, $\frac{1}{2}u_0 + iK'$, $\frac{1}{2}u_0 + K + iK'$, and the point on the curve is thus one of the four points belonging to these values of u respectively.

121. If to fix the ideas we select for each point of the curve one of the congruent values of the argument, we may assume for the oval, u real: at A, $u=0$; from A to B above the axis u positive and at B, $=K$; below the axis u negative and at B, $=-K$; there is thus a discontinuity, K, $-K$ at B, but the two values are congruent. For the anguis, $u = iK' +$ real value v: above the axis v positive, viz. at infinity $v=0$, and at C, $=K$; below the axis v negative,

viz. at infinity $v=0$ and at C, $=-K$: there thus is a discontinuity $iK'+K$, $iK'-K$ at C, but the two values are congruent. Observe that for points opposite to each other in regard to the axis, the arguments are, for points on the oval $u, -u$; for points on the anguis $iK'+v$, $iK'-v$: but that we have $iK'-v \equiv -(iK'-v)$.

122. The pure theorem gives for three points u_1, u_2, u_3 in a line

$$du_1 + du_2 + du_3 = 0\,;$$

and thence $u_1+u_2+u_3=C$. The constant C cannot have a determinate value (for if it had, then assuming the values of u_1 and u_2 at pleasure u_3 would have the determinate value $=C-u_1-u_2$), but must be given by a congruence: or what is the same thing, assigning to C any admissible value, we must instead of $u_1+u_2+u_3=C$, write $u_1+u_2+u_3\equiv C$. Taking any particular line, for instance the tangent at A, we have $u_1, u_2, u_3 = 0, 0, iK'$; whence $C=iK'$; and we have $u_0+u_2+u_3\equiv iK'$, viz. this is the relation between the arguments u_1, u_2, u_3 belonging to the points of intersection of the cubic with a line: in particular for a line at right angles to the axis we have $u_1, u_2, u_3 = \alpha, -\alpha, iK'$ or $= iK'+\beta, iK'-\beta, iK'$ (according as the line cuts the oval or the anguis) and the congruence is in each case satisfied. But I shall in general instead of $\equiv$ use the sign $=$, understanding it as in general meaning $\equiv$, and only replacing it by this sign when for clearness it seems necessary to do so.

Writing $\operatorname{sn} u_1, \operatorname{cn} u_1, \operatorname{dn} u_1 = s_1, c_1, d_1$, and so in other cases, the condition in order that the three points may be in a line is

$$\begin{vmatrix} s_1^2, & s_1c_1d_1, & 1 \\ s_2^2, & s_2c_2d_2, & 1 \\ s_3^2, & s_3c_3d_3, & 1 \end{vmatrix} = 0,$$

a relation which must be satisfied when the arguments are connected by the foregoing relation $u_1+u_2+u_3\equiv iK'$.

We can show from this equation alone that s_3^2 and $s_3c_3d_3$ are expressible rationally in terms of $s_1^2, s_1c_1d_1, s_2^2, s_2c_2d_2$; in fact, writing therein x_1, y_1 in place of $s_1^2, s_1c_1d_1$, etc., we thence have $x_3, y_3, 1 = \lambda x_1+\mu x_2, \lambda y_1+\mu y_2, \lambda+\mu$, and substituting in $y_3^2=x_3(1-x_3)(1-k^2x_3)$, we obtain an equation $\lambda\mu(F\lambda+G\mu)=0$, that is $F\lambda+G\mu=0$, or say $\lambda, \mu = G, -F$, and thence

$$x_3 = s_3^2 = \frac{Gx_1-Fx_2}{G-F},\quad y_3 = s_3c_3d_3 = \frac{Gy_1-Gy_2}{G-F}\,;$$

the values of F, G are easily found to be

$$F = y_1^2 + 2y_1y_2 - x_1(1-x_1)(1-k^2x_2) - x_1(1-k^2x_1)(1-x_2) - (1-x_1)(1-k^2x_1)x_2,$$

$$G = 2y_1y_2 + y_2^2 - x_2(1-x_2)(1-k^2x_1) - x_2(1-k^2x_2)(1-x_1) - (1-x_2)(1-k^2x_2)x_1,$$

or as these may also be written

$$F = -(y_1-y_2)^2 + (x_1-x_2)^2(1+k^2+k^2(x_1+x_2)) + k^2(x_1-x_2)^2x_1,$$

$$G = -(y_1-y_2)^2 + (x_1-x_2)^2(1+k^2+k^2(x_1+x_2)) + k^2(x_1-x_2)^2x_2,$$

where of course x_1, y_1, x_2, y_2 should be replaced by their values s_1^2, $s_1c_1d_1$, s_2^2, $s_2c_2d_2$. This is in fact the ordinary process of finding the third point of intersection of a cubic by a line which meets it in two given points.

Writing $iK'-u_3=u$, and s, c, d for the sn, cn, dn of u, we have $s_3, c_3, d_3 = -\frac{1}{ks}, \frac{id}{ks}, \frac{ic}{s}$, whence $s_3^2 = \frac{1}{k^2s^2}$, $s_3c_3d_3 = \frac{cd}{k^2s^3}$, and the determinant equation becomes

$$\begin{vmatrix} s_1^2, & s_1c_1d_1, & 1 \\ s_2^2, & s_2c_2d_2, & 1 \\ 1, & \dfrac{cd}{s}, & k^2s^2 \end{vmatrix}$$

that is

$$(1-k^2s^2s_2^2)s_1c_1d_1 - (1-k^2s^2s_1^2)s_2c_2d_2 - (s_1^2-s_2^2)\frac{cd}{s} = 0,$$

corresponding to the relation $u=u_1+u_2$ of the arguments. This is easily verified: we have

$$s_1 = \frac{sc_2d_2 - s_2cd}{1-k^2s^2s_2^2}, \quad s_2 = \frac{sc_1d_1 - s_1cd}{1-k^2s^2s_1^2}, \quad s = \frac{s_1^2-s_2^2}{s_1c_2d_2 - s_2c_1d_1};$$

the equation thus becomes

$$(sc_2d_2 - s_2cd)c_1d_1 - (sc_1d_1 - s_1cd)c_2d_2 - (s_1^2-s_2^2)\frac{cd}{s} = 0.$$

that is

$$cd\left\{-s_2c_1d_1 + s_1c_2d_2 - \frac{s_1^2-s_2^2}{s}\right\} = 0,$$

which is right.

The Four Tangents from a Point of the Cubic.

123. Suppose that the line is a tangent to the cubic, say the line touches the cubic at the point u, and again meets it at the point w: then instead of u_1, u_2, u_3 we have u, u, w: and the relation becomes $2u+w\equiv iK'$.

Here u being given, w is uniquely determined: viz. given the argument u of the point of contact, we have a unique value for the argument w of the tangential. But given w, we have $2u=iK'-w$, and we have thus for u the four values $\frac{1}{2}(iK'-w)$, Do. $+K$, Do. $+iK'$, Do. $+K+iK'$, corresponding to the four tangents which can be drawn from the point w to the cubic.

The tangents are real for a point of the anguis, and for such a point we may write $w = iK' + v$, where v is real and included between the values $\pm K$; the corresponding values of u are

$$u_1 = -\tfrac{1}{2}v,\ u_2 = -\tfrac{1}{2}v + K,\ u_3 = -\tfrac{1}{2}v + iK',\ u_4 = -\tfrac{1}{2}v + K + iK':$$

the first and second of these belong to tangents to the oval, the third and fourth to tangents to the anguis. We may further distinguish a tangent according as it passes between or does not pass between the vertices B and C: say in the former case it is intermediate, and in the latter case extramediate: and we then see that for the tangents from point $iK' + v$ of anguis,

$u = -\tfrac{1}{2}v$ for intermediate to oval,
$u = -\tfrac{1}{2}v + K$ " extramediate to oval,
$u = -\tfrac{1}{2}v + iK'$ " extramediate to anguis,
$u = -\tfrac{1}{2}v + K + iK'$ " intermediate to anguis.

124. We may make a corresponding division of the real lines which meet the curve in three real points: any such line meets the oval twice (and then of course the anguis once), or else it meets the anguis three times: and taking the arguments to be u_1, u_2, u_3 we have

$\tfrac{1}{2}(u_1 + u_2 + u_3) = \tfrac{1}{2}iK'$ for intermediate line meeting oval twice,
" $= \tfrac{1}{2}iK' + K$ " extramediate line, Do.,
" $= -\tfrac{1}{2}iK'$ " extramediate line meeting anguis three times,
" $= -\tfrac{1}{2}iK' + K$ " intermediate line, Do.

125. Returning to the tangents, the point $iK' + v$ may be an inflexion: we have then the point of contact of the intermediate tangent to the anguis coinciding with the point $iK' + v$; viz. $iK' + v \equiv -\tfrac{1}{2}v + K + iK'$, or say $= -\tfrac{1}{2}v \pm K + iK'$: that is $v = \pm\tfrac{2}{3}K$; or $iK' + \tfrac{2}{3}K$ and $iK' - \tfrac{2}{3}K$ are the arguments for the real points of inflexion, above and below the axis respectively.

126. Write for a moment the equation in the form $y^2 = Bx + Cx^2 + Dx^3$, then if (α, β) be a point on the curve $(\beta^2 = B\alpha + C\alpha^2 + D\alpha^3)$, and we consider the intersections of the curve with the line $y - \beta = m(x - \alpha)$ we find for the remaining two intersections

$$B + C(x + \alpha) + D(x^2 + \alpha x + \alpha^2) = 2m\beta + m^2(x - \alpha).$$

If the line be a tangent, this will be satisfied by $x = \alpha$; the condition for this is $2m\beta = B + 2C\alpha + 3D\alpha^2$, and supposing this satisfied, then throwing out the factor $x - \alpha$ we obtain $C + D(x + 2\alpha) = m^2$, giving $Dx = m^2 - C - 2D\alpha$ for the coordinate x of the tangential of the point (α, β).

In the case of an inflexion $x=\alpha$, and we have

$$m^2=C+3D\alpha,=\frac{(B+2C\alpha+3D\alpha^2)^2}{4(B\alpha+C\alpha^2+D\alpha^3)},$$

giving for α the equation

$$3D^2\alpha^4+4CD\alpha^3+6BD\alpha-B^2=0,$$

or for B, C, D writing their values 1, $-(1+k^2)$, k^2 this is

$$3k^4\alpha^4-4k^2(1+k^2)\alpha^3+6k^2\alpha^2-1=0$$

for the x-coordinate of the inflexion. There is one negative root, and one or three positive roots; but only one positive root giving a real value of β, and corresponding hereto we have the two real inflexions on the anguis.

It should be possible, from the formulæ of No. 122, writing therein $(x_2, y_2)=(x_1, y_1)$ to arrive at the foregoing result, say $Dx_3=m^2-C-2Dx_1$, but the functions F and G present themselves in vanishing forms, and the reduction is not immediate.

127. The general condition for an inflexion is $3u\equiv iK'$: the nine inflexions thus are $u=iK'$, inflexion at infinity, $u=iK'\pm\frac{2}{3}K$, the two real inflexions, and besides $u=\pm\frac{1}{3}iK'$, $u=\pm\frac{1}{3}iK'\pm\frac{2}{3}K$, six imaginary inflexions.

The Sextactic Points.

128. The vertices A, B, C are each of them a sextactic point: in fact writing the equation in the form $y^2=x-(1+k^2)x^2+k^2x^3$, we see at once that the conic $y^2=x-(1+k^2)x^2$ meets the curve in the point A counting six times: and there is obviously a like proof for the vertices B and C respectively. Hence for the six intersections with any conic whatever we have the condition

$$u_1+u_2+u_3+u_4+u_5+u_6\equiv 0:$$

and for the sextactic points we have the condition $6u\equiv 0$. This gives the 36 points $u=\frac{1}{6}(2mK+2m'iK')$ or say $=\frac{1}{3}(mK+m'iK')$, $m=0, \pm 1, \pm 2, 3$, $m'=0, \pm 1, \pm 2, 3$: but among these are included the 9 inflexions (each of these being an improper sextactic point, the conic becoming the tangent taken twice) and there remain 27 points: among these are included the three vertices ($u=0$, K, $iK'+K$) points of contact with the tangents from the inflexion at infinity; and of the remaining 24 points 6 are real, viz. these are the points $u=\pm\frac{1}{3}K$, $\pm\frac{2}{3}K$ on the oval, and the points $iK'\pm\frac{1}{3}K$ on the anguis: these are in fact the points of contact of the tangents from the real inflexions, viz. the three tangents from the inflexion $iK'+\frac{2}{3}K$ touch the oval in the points $\frac{2}{3}K$, $-\frac{1}{3}K$, and the anguis in the point $iK'-\frac{1}{3}K$; the three tangents from the inflexion $iK'-\frac{2}{3}K$ touch the oval in the points $-\frac{2}{3}K$, $\frac{1}{3}K$, and the anguis in the point $iK'+\frac{1}{3}K$.

Formulæ Relating to the Tangents from the Vertices.

129. I annex some formulæ relating to the tangents to the curve from the vertices A, B, C respectively. We have from each vertex four tangents say $\rho=0$, $\sigma=0$, symmetrically situate in regard to the axis, and $\rho'=0$, $\sigma'=0$, symmetrically situate in regard to the axis: the line joining the points of contact of ρ, σ is a line $\tau=0$ at right angles to the axis, and that joining the points of contact of ρ', σ' is a line $\tau'=0$ at right angles to the axis.

VERTEX A, TANGENTS IMAGINARY. COORDINATES OF POINT OF CONTACT.

$$\rho,\ \sigma,\ \tau = y-i(1+k)x,\ y+i(1+k)x,\quad kx+1; \qquad x=-\frac{1}{k},\ y=\mp\frac{i(1+k)}{k}$$

$$\rho',\ \sigma',\ \tau' = y-i(1-k)x,\ y+i(1-k)x,\ -kx+1; \qquad x=\frac{1}{k},\ y=\pm\frac{i(1-k)}{k},$$

equation of curve is, $f, =y^2-x(1-x)(1-k^2x), =\rho\sigma-x\tau^2, =\rho'\sigma'-x\tau'^2, =0$.

VERTEX B, TANGENTS IMAGINARY. CONTACTS.

$$\rho',\sigma',\tau'=y-(k-ik')(1-x),\ y+(k-ik')(1-x),\ kx-(k-ik');\ x=1-\frac{ik'}{k},\ y=\mp\frac{ik'}{k}(k+ik')$$

$$\rho',\sigma',\tau'=y-(k-ik')(1-x),\ y+(k-ik')(1-x),\ kx-(k+ik');\ x=1+\frac{ik'}{k},\ y=\pm\frac{ik'}{k}(k-ik'),$$

equation is $f=\rho\sigma+(1-x)\tau^2, =\rho'\sigma'+(1-x)\tau'^2, =0$,

VERTEX C, TANGENTS REAL. CONTACTS.

$$\rho,\ \sigma,\ \tau=y-\frac{1}{1+k'}(1-k^2x),\ y+\frac{1}{1+k'}(1-k^2x),\ x-\frac{1}{1+k'};\quad x=\frac{1}{1+k'},\ y=\frac{\pm k'}{1+k'}$$

$$\rho',\ \sigma',\ \tau'=y-\frac{1}{1-k'}(1-k^2x),\ y+\frac{1}{1-k'}(1-k^2x),\ x-\frac{1}{1-k'};\quad x=\frac{1}{1-k'},\ y=\frac{\pm k'}{1-k'},$$

equation is $f, =\rho\sigma-(1-k^2x)\tau, =\rho'\sigma'-(1-k^2x)\tau', =0$.

130. These linear functions ρ, σ, etc. considering therein x, y as denoting $\text{sn}^2 u$, $\text{sn}\,u\,\text{cn}\,u\,\text{dn}\,u$ respectively present themselves as the numerators and denominators of some formulæ given No. 105 of my Elliptic Functions (1876), see p. 76: viz. we have

$$\text{sn}^2(u+\tfrac{1}{2}K)=\frac{1}{1+k'}\,\frac{1-k^2x+(1+k')y}{1-k^2x+(1-k')y}$$

which is

$$=\frac{1}{1-k'}\,\frac{y+\frac{1}{1+k'}(1-k^2x)}{y+\frac{1}{1+k'}(1-k^2x)},\ =\frac{1}{1-k'}\,\frac{\sigma}{\sigma'},\ \text{(vertex C)}.$$

$$\text{sn}^2(u+\tfrac{1}{2}iK')=\frac{1}{k}\,\frac{(1+k)x+iy}{(1+k)x-iy}$$

which is $$= -\frac{1}{k}\,\frac{y - i(1+k)x}{y + i(1+k)x}, \qquad = -\frac{1}{k}\,\frac{\rho}{\sigma}, \text{ (vertex A).}$$

and

$$\text{sn}^2(u + \tfrac{1}{2}K + \tfrac{1}{2}iK') = \frac{k + ik'}{k}\,\frac{1 - x + (k - ik)y}{1 - x + (k + ik)y}$$

which is $$= \frac{k - ik'}{k}\,\frac{y + (k + ik')(1 - x)}{y + (k - ik')(1 - x)}, \qquad = \frac{k - ik'}{k}\,\frac{\sigma}{\sigma'}, \text{ (vertex B).}$$

Observe here that in the second formula we have a pair of tangents ρ, σ which belong to a chord τ through the inflexion at ∞; but in the first and third formulæ we have tangents σ, σ' not forming such a pair. This is as it should be, for the zero and infinity of $\text{sn}^2(u + \frac{1}{2}iK')$ are $u = -\frac{1}{2}iK'$, $u = \frac{1}{2}iK'$ which belong to points in lineâ with the inflexion at infinity: but for $\text{sn}^2(u + \frac{1}{2}K)$ the zero is $u = -\frac{1}{2}K$, and the infinity is $u = iK' - \frac{1}{2}K$, which do not belong to points in lineâ with the inflexion at infinity: and the like for $\text{sn}^2(u + \frac{1}{2}K + \frac{1}{2}iK')$.

Fixed Curve a Quartic in Space, the Quadri-quadric Curve $y^2 = 1 - x^2$, $z^2 = 1 - k^2x^2$. Art. Nos. 131 to 135.

131. It is assumed that k^2 is real, positive, and less than unity: the curve may be regarded as the intersection of the two cylinders

$$x^2 + y^2 = 1,\ k^2x^2 + z^2 = 1;$$

but there is through it a third cylinder $y^2 - k^2z^2 = k'^2$. The cylinder $k^2x^2 + z^2 = 1$, or say the horizontal cylinder has for its section an ellipse axes $\frac{1}{k}$ and 1 respectively: and it is pierced by the cylinder $x^2 + y^2 = 1$, or say the vertical cylinder, in two detached ovals (of double curvature) lying on opposite sides of the plane of xy: only the upper oval $ABA'B'$ is shown in the figure.

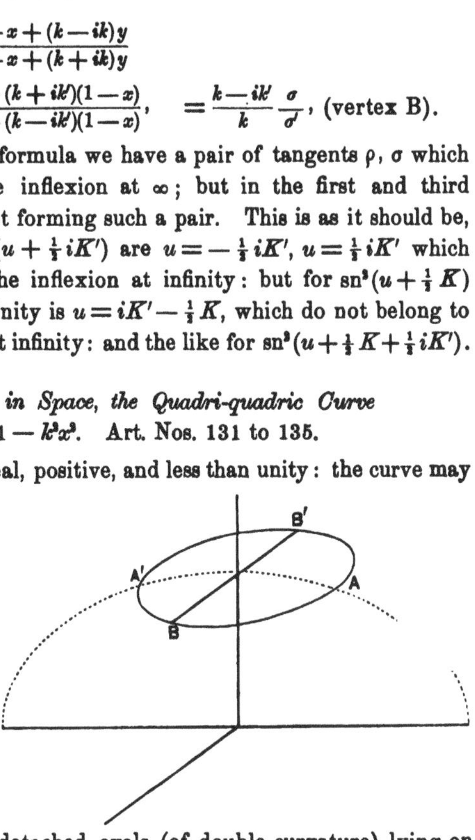

Each of the vertices A, A', B, B' is an inflexion,* viz. a point such that the osculating plane at that point meets the curve in the point counting four times. We may consider two generating lines of the horizontal cylinder, each meeting the oval in two points; the plane through the generating lines meets the curve

* There are in all 16 inflexions, 4 in each of the planes $x = 0$, $y = 0$, $z = 0$, and 4 in the plane infinity.

in these four points, and when the generating lines come each of them to coincide with the tangent at A, we have the osculating plane meeting the curve in the point A counting four times. The like reasoning, with two generating lines of the third cylinder, shows that the vertex B is an inflexion.

132. The equations are satisfied by writing therein $x, y, z = \text{sn}\,u, \text{cn}\,u, \text{dn}\,u$: the periods are here $4K$, $4iK'$: hence to a given point on the curve the argument is not $u =$ a determinate value u_0, but it may equally well be taken to be $= u_0 + 4mK + 4m'iK'$, where m and m' are any positive or negative integers: we express this by $u \equiv u_0$, or say u congruent to u_0. For the upper oval u may be taken to be real, and to be $= 0$ at B, positive for the half oval BAB', and negative for the half oval $BA'B'$; having the values $+K$, $-K$ at A and A' respectively, and the discontinuity $2K$, $-2K$ at B', these two values being congruent. For the lower oval we have $u = 2iK' +$ real value v.

For the intersections of the curve with a plane we have

$$du_1 + du_2 + du_3 + du_4 = 0;\ \text{whence}\ u_1 + u_2 + u_3 + u_4 = C;$$

and by taking the plane to be the osculating plane at B, we find 0 as a value of the constant, and the relation thus is $u_1 + u_2 + u_3 + u_4 \equiv 0$. Writing as before $\text{sn}\,u_1, \text{cn}\,u_1, \text{dn}\,u_1, = s_1, c_1, d_1$ and so in other cases, the relation between the elliptic functions is

$$\begin{vmatrix} s_1, & c_1, & d_1, & 1 \\ s_2, & c_2, & d_2, & 1 \\ s_3, & c_3, & d_3, & 1 \\ s_4, & c_4, & d_4, & 1 \end{vmatrix} = 0;$$

It is important to remark that giving three of the points the fourth point is determined uniquely: that is the equation really gives s_4, c_4, d_4 each as a rational function of the $s_1, c_1, d_1, s_2, c_2, d_2, s_3, c_3, d_3$.

In fact we may write $s_4 = \lambda_1 s_1 + \lambda_2 s_2 + \lambda_3 s_3$, and similarly for c_4 and d_4, and $1 = \lambda_1 + \lambda_2 + \lambda_3$: substituting in $s_4^2 + c_4^2 - 1 = 0$, $k^2 s_4^2 + d_4^2 - 1 = 0$, we have

$$\begin{aligned} X_{23}\lambda_2\lambda_3 + X_{31}\lambda_3\lambda_1 + X_{12}\lambda_1\lambda_2 &= 0, \\ Y_{23}\ \ ,,\ \ + Y_{31}\ \ ,,\ \ + Y_{12}\ \ ,,\ \ &= 0, \end{aligned}$$

where $X_{12} = s_1 s_2 + c_1 c_2 - 1$, $Y_{12} = k^2 s_1 s_2 + d_1 d_2 - 1$, etc.; we thence have

$$\begin{aligned} \lambda_2\lambda_3 : \lambda_3\lambda_1 : \lambda_1\lambda_2 &= X_{31}Y_{12} - X_{12}Y_{31} : X_{12}Y_{23} - X_{23}Y_{12} : X_{23}Y_{31} - X_{31}Y_{23} \\ &= \quad A_1 \quad : \quad A_2 \quad : \quad A_3,\ \text{suppose}; \end{aligned}$$

that is $\quad \lambda_1 : \lambda_2 : \lambda_3 = A_2 A_3 : A_3 A_1 : A_1 A_2.$

and consequently

$$\begin{aligned} s_4, &= -\text{sn}\,(u_1 + u_2 + u_3), = A_2A_3s_1 + A_3A_1s_2 + A_1A_2s_3 \div (A_2A_3 + A_3A_1 + A_1A_2) \\ c_4, &= \quad \text{cn}\,(\quad ,, \quad) = \ ,,\ c_1 + \ ,,\ c_2 + \ ,,\ c_3 \div (\quad ,, \quad) \\ d_4, &= \quad \text{dn}\,(\quad ,, \quad) = \ ,,\ d_1 + \ ,,\ d_2 + \ ,,\ d_3 \div (\quad ,, \quad) \end{aligned}$$

which are the required expressions. If $u_3 = 0$, and consequently $s_3, c_3, d_3 = 0, 1, 1$, the resulting expressions give the sn, cn, and dn of $u_1 + u_2$, but the expressions are in a very complicated form, not easily identifiable with the ordinary ones.

133. The determinant equation may be written

$$(s_1 - s_2)(c_3 d_4 - c_4 d_3) + (s_3 - s_4)(c_1 d_2 - c_2 d_1)$$
$$+ (c_1 - c_2)(d_3 s_4 - d_4 s_3) + (c_3 - c_4)(d_1 s_2 - d_2 s_1)$$
$$+ (d_1 - d_2)(s_3 c_4 - s_4 c_3) + (d_3 - d_4)(s_1 c_2 - s_2 c_1) = 0,$$

and in fact each of the three lines is separately $= 0$. This appears from the following three formulæ

$$\frac{\operatorname{sn}(u_1 + u_2)}{\operatorname{cn}(u_1 + u_2) - \operatorname{dn}(u_1 + u_2)} = \frac{s_1 - s_2}{c_1 d_2 - c_2 d_1},$$

$$\frac{\operatorname{sn}(u_1 + u_2)}{\operatorname{cn}(u_1 + u_2) + 1} = \frac{c_1 - c_2}{d_1 s_2 - d_2 s_1},$$

$$\frac{\operatorname{sn}(u_1 + u_2)}{\operatorname{dn}(u_1 + u_2) + 1} = \frac{-\frac{1}{k^2}(d_1 - d_2)}{s_1 c_2 - s_2 c_1},$$

which are themselves at once deducible from the formulæ,

$$\operatorname{sn}(u_1 + u_2) = s_1^2 - s_2^2, = -(c_1^2 - c_2^2), = -\frac{1}{k^2}(d_1^2 - d_2^2), \div (s_1 c_2 d_2 - s_2 c_1 d_1),$$
$$\operatorname{cn}(u_1 + u_2) = s_1 c_1 d_2 - s_2 c_2 d_1 \qquad\qquad \div \qquad \text{Do.}$$
$$\operatorname{dn}(u_1 + u_2) = s_1 d_1 c_2 - s_2 d_2 c_1 \qquad\qquad \div \qquad \text{Do.}$$

In fact the numerators of $\operatorname{cn}(u_1 + u_2) + \operatorname{dn}(u_1 + u_2)$, $\operatorname{cn}(u_1 + u_2) + 1$, $\operatorname{dn}(u_1 + u_2) + 1$ thus become $= (s_1 + s_2)(c_1 d_2 - c_2 d_1)$, $-(c_1 + c_2)(d_1 s_2 - d_2 s_1)$, $(d_1 + d_2)(s_1 c_2 - s_2 c_1)$ respectively: so that taking the numerator of $\operatorname{sn}(u_1 + u_2)$ under its three forms successively, we have by division the formulæ in question.

134. The above three equations, putting therein $u_1 + u_2 = -2u_3$, and reducing the functions of $2u_3$ become

$$\frac{1}{k^2}\frac{c_3 d_3}{s_3} = \frac{s_1 - s_2}{c_1 d_2 - c_2 d_1}, \text{ giving} - k^2 s_3^2 = \frac{(c_1 - c_2)(d_1 - d_2)}{(d_1 s_2 - d_2 s_1)(s_1 c_2 - s_2 c_1)}$$

$$-\frac{s_3 d_3}{c_3} = \frac{c_1 - c_2}{d_1 s_2 - d_2 s_1}, \quad \text{``} \quad \frac{k^2}{k'^2}c_3^2 = \frac{(d_1 - d_2)(s_1 - s_2)}{(s_1 c_2 - s_2 c_1)(c_1 d_2 - c_2 d_1)}$$

$$\frac{k^2 s_3 c_3}{d_3} = \frac{d_1 - d_2}{s_1 c_2 - s_2 c_1}, \quad \text{``} \quad -\frac{1}{k'^2}d_3^2 = \frac{(s_1 - s_2)(c_1 - c_2)}{(c_1 d_2 - c_2 d_1)(d_1 s_2 - d_2 s_1)},$$

equations which must of course give each of them the same value for s_3^2: the equations belong to the relation $u_1 + u_2 + 2u_3 = 0$, viz. (s_3, c_3, d_3) are the coordinates of a point of contact of the tangent plane drawn through the two given points (s_1, c_1, d_1) and (s_2, c_2, d_2) of the curve.

135. Write

$$a, b, c, f, g, h = s_1 - s_2,\ c_1 - c_2,\ d_1 - d_2,\ c_1 d_2 - c_2 d_1,\ d_1 s_2 - d_2 s_1,\ s_1 c_2 - s_2 c_1,$$
$$a', b', c', f', g', h' = s_3 - s_4,\ c_3 - c_4,\ d_3 - d_4,\ c_3 d_4 - c_4 d_3,\ d_3 s_4 - d_4 s_3,\ s_3 c_4 - s_4 c_3,$$

so that (a, b, c, f, g, h) are the six coordinates of the line 12, and a', b', c', f', g', h' are the six coordinates of the line 34. The determinant equation is nothing else than the condition of the intersection of the two lines, viz. this is

$$af' + a'f + bg' + b'g + ch' + c'h = 0,$$

and by what precedes it appears that not only is this so but that we have separately $af' + a'f = 0$, $bg' + b'g = 0$, $ch' + c'h = 0$, viz. these three equations are satisfied by the coordinates of the lines 12 and 34 which join in pairs the intersections 1, 2 and 3, 4 of the quadri-quadric curve by a plane. But this is a geometrical property depending only on the four points being in a plane: and it is thus a result of Abel's theorem that when the arguments are such that

$$u_1 + u_2 + u_3 + u_4 = 0,$$

then not only the original equation, but each of the three equations, holds good.

The Cubic Curve $xy^2 - 2y + (1 + k^2)x - k^2x^3 = 0$. Art. Nos. 136 and 137.

136. Writing the equation in the forms

$$(xy - 1)^2 = (1 - x^2)(1 - k^2x^2),\ \text{or say}\ xy - 1 = -\sqrt{1 - x^2 . 1 - k^2x^2}.$$
$$x(y^2 - k^2x^2) = 2y - (1 + k^2)x.$$

We see that the general form is as shown in the figure: the real portions of the

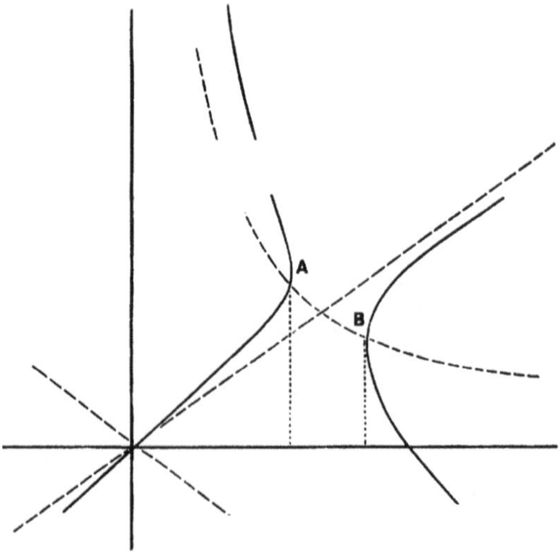

curve lie between the values $x = -\infty, -\frac{1}{k}; -1, +1$; and $\frac{1}{k}, \infty$. The curve may be made to depend on elliptic functions in two different ways: we may write

$$x = \operatorname{sn} u \qquad , = \frac{2i}{1+k}\frac{\operatorname{sn}_1 v}{\operatorname{cn}_1 v \operatorname{dn}_1 v},$$

$$y = \frac{1 - \operatorname{cn} u \operatorname{dn} u}{\operatorname{sn} u}, = \frac{i}{1+k}\frac{\operatorname{sn}_1 v}{\operatorname{cn}_1 v \operatorname{dn}_1 v}\{1 + k^2 - (1-k)^2 \operatorname{sn}_1^2 v\},$$

where the functions sn, cn, dn belong to the modulus k, and the functions sn_1, cn_1, dn_1 to the modulus $\theta, = \frac{1-k}{1+k}$. The first mode is obvious; as to the second, observe that the formulæ give

$$y - kx = \frac{i}{1+k}\frac{\operatorname{sn}_1 v}{\operatorname{cn}_1 v \operatorname{dn}_1 v}(1-k)^2 \operatorname{cn}_1^2 v, = \frac{i(1-k)^2}{1+k}\frac{\operatorname{sn}_1 v \operatorname{cn}_1 v}{\operatorname{dn}_1 v},$$

$$y + kx = \frac{i}{1+k}\frac{\operatorname{sn}_1 v}{\operatorname{cn}_1 v \operatorname{dn}_1 v}(1+k)^2 \operatorname{dn}_1^2 v, = i(1+k)\frac{\operatorname{sn}_1 v \operatorname{dn}_1 v}{\operatorname{cn}_1 v},$$

whence $y^2 - k^2x^2 = -(1-k)^2 \operatorname{sn}_1^2 v$; and therefore

$$x(y^2 - k^2x^2) = \frac{-2i(1-k)^2}{1+k}\frac{\operatorname{sn}_1^3 v}{\operatorname{cn}_1 v \operatorname{dn}_1 v},$$

which is also the value of $2y - (1+k^2)x$.

We find, moreover, $d\omega, = du, = \frac{2i dv}{1+k}$; and thence, u, v each vanishing together, $u = \frac{2iv}{1+k}$. Writing for shortness s_1, c_1, d_1 to denote $\operatorname{sn}_1 v$, $\operatorname{cn}_1 v$, $\operatorname{dn}_1 v$ (that is s_1, c_1, d_1 are the elliptic functions of v to the modulus θ), we have

$$\operatorname{sn}\left(\frac{2iv}{1+k}, k\right) = \frac{2i}{1+k}\frac{s_1}{c_1 d_1},$$

$$\operatorname{cn}\left(\frac{2iv}{1+k}, k\right) = \frac{1}{(1+k)c_1 d_1}\{1 + k + (1-k)s_1^2\},$$

$$\operatorname{dn}\left(\frac{2iv}{1+k}, k\right) = \frac{1}{(1+k)c_1 d_1}\{1 + k - (1-k)s_1^2\}.$$

137. To bring these into a known form, for k write $\frac{1-k}{1+k}$, then θ is changed into k and the formulæ become

$$\operatorname{sn}_1(1+k)iv = (1+k)i\frac{s}{cd},$$

$$\operatorname{cn}_1(1+k)iv = \frac{1}{cd}(1 + ks^2),$$

$$\operatorname{dn}_1(1+k)iv = \frac{1}{cd}(1 - ks^2),$$

where the sn_1, cn_1, dn_1 refer to the modulus θ, and s, c, d denote $\mathrm{sn}\, v$, $\mathrm{cn}\, v$, $\mathrm{dn}\, v$, modulus k.

But from the formulæ, p. 63, of my Elliptic Functions,

$$\mathrm{sn}\,(iu, k),\ \mathrm{cn}\,(iu, k),\ \mathrm{dn}\,(iu, k) = \frac{i\,\mathrm{sn}\,(u, k')}{\mathrm{cn}\,(u, k')},\ \frac{1}{\mathrm{cn}\,(u, k')},\ \frac{\mathrm{dn}\,(u, k')}{\mathrm{cn}\,(u, k')},$$

and herein for u writing $(1+k)v$, and for k writing θ, $=\dfrac{1-k}{1+k}$, whence k' becomes $\dfrac{2\sqrt{k}}{1+k}$, $=\gamma$, and $\gamma' = \dfrac{1-k}{1+k}$, we have

$$\mathrm{sn}_1(1+k)iv,\ \mathrm{cn}_1(1+k)iv,\ \mathrm{dn}_1(1+k)iv = \frac{i\,\mathrm{sn}(\overline{1+k}v, \gamma)}{\mathrm{cn}(\overline{1+k}v, \gamma)},\ \frac{1}{\mathrm{cn}(\overline{1+k}v, \gamma)},\ \frac{\mathrm{dn}(\overline{1+k}v, \gamma)}{\mathrm{cn}(\overline{1+k}v, \gamma)},$$

and the formulæ above obtained are

$$\frac{\mathrm{sn}(\overline{1+k}v, \gamma)}{\mathrm{cn}(\overline{1+k}v, \gamma)} = (1+k)\frac{s}{cd}, \qquad \text{giving } \mathrm{sn}(\overline{1+k}\,v, \gamma) = \frac{(1+k)s}{1+ks^2},$$

$$\frac{1}{\mathrm{cn}(\overline{1+k}v, \gamma)} = \frac{1}{cd}(1+ks^2), \qquad \mathrm{cn}(\overline{1+k}\,v, \gamma) = \frac{cd}{1+ks^2},$$

$$\frac{\mathrm{dn}(\overline{1+k}v, \gamma)}{\mathrm{cn}(\overline{1+k}v, \gamma)} = \frac{1}{cd}(1-ks^2), \qquad \mathrm{dn}(\overline{1+k}\,v, \gamma) = \frac{1-ks^2}{1+ks^2},$$

where as before s, c, d denote $\mathrm{sn}\,(v, k)$, $\mathrm{cn}(v, k)$, $\mathrm{dn}\,(v, k)$: these last are in fact the formulæ of the second line of the table, Elliptic Functions, p. 183.

Fixed Curve the Cubic $y^2 = x(1-x)(1-k^2x)$: *the Function* $\{01'\theta\}$.
Art. Nos. 138 to 142.

138. It was shown, No. 65, how for the affected theorem when the fixed curve is a cubic, the form of differential was $d\omega$ into

$$\{012\} + \left[\int_3^1 d\omega\, \partial_3 \{036\} - \{123\}\right],$$

the last term being the properly determined constant K, attached to the variable term $\{012\}$, in order to obtain a standard form of integral. The object of the present article is to show what these formulæ become for the before-mentioned form of cubic curve $y^2 = x(1-x)(1-k^2x)$ which is most directly connected with elliptic functions, and to exhibit the connection of the formulæ with the ordinary formulæ for elliptic integrals of the second and third kinds respectively.

139. We have in general

$$\{012\}, = \frac{\widetilde{012}}{\overline{012}}, = \left\{\begin{array}{l} (1+k^2)x_1x_2 - (x_1+x_2) + y_1y_1 \\ + x\,[-1 + (1+k^2)(x_1+x_2) \\ + y\,(y_1+y_2) \end{array}\right\} \div \left|\begin{array}{lll} x, & y, & 1 \\ x_1, & y_1, & 1 \\ x_2, & y_2, & 1 \end{array}\right|.$$

Taking here 2, $=\theta$, for the point, coordinates $x_2, y_2 = 0, 0$; we have

$$\{01\theta\} = \frac{-x_1 - x + (1+k^2)xx_1 + yy_1}{xy_1 - x_1 y};$$

and if retaining 1 to denote the point coordinates (x_1, y_1), belonging to the argument u_1, we write $1'$ for the point belonging to the argument $u + iK'$, then the coordinates of $1'$ are $\frac{1}{k^2 x_1}, \frac{-y_1}{k^2 x_1^2}$, and the formula becomes

$$\{01'\theta\} = \frac{x_1 + k^2 xx_1^2 - (1+k^2)xx_1}{xy_1 + x_1 y};$$

the numerator hereof multiplied by x is $= y(xy_1 + x_1 y) - k^2 x^2 x_1 (x - x_1)$, and we thence have

$$\{01'\theta\} = \frac{y}{x} - \frac{k^2 xx_1 (x - x_1)}{xy_1 - x_1 y}$$

which substituting for x, y, x_1, y_1 their values in terms of u, u_1 is

$$= \frac{\operatorname{cn} u \operatorname{dn} u}{\operatorname{sn} u} - k^2 \operatorname{sn} u \operatorname{sn} u_1 \operatorname{sn}(u - u_1).$$

Operating on each side with $\frac{d}{du_1}, = \partial_1$, we obtain

$$\partial_1\{01'\theta\} = k^2 \operatorname{sn}^2 u_1 - k^2 \operatorname{sn}^2(u - u_1),$$

the differentiation being in fact that which occurs in establishing the fundamental property of the elliptic integral of the second kind $Zu = u\left(1 - \frac{E}{K}\right) + k^2 \int_0 \operatorname{sn}^2 u du$, viz. we have $Zu - Zu_1 - Z(u - u_1) = -k^2 \operatorname{sn} u \operatorname{sn} u_1 \operatorname{sn}(u - u_1)$, and thence $\partial_1[-k^2 \operatorname{sn} u \operatorname{sn} u_1 \operatorname{sn}(u - u_1)] = -Z'u_1 + Z'(u - u_1), = k^2 \operatorname{sn}^2 u_1 - k^2 \operatorname{sn}^2(u - u_1)$. Observe that $1'$, referring to $u_1 + iK'$, the subscript 1 might be written $1'$.

The same result should of course be obtainable by the differentiation of the expression of $\{01'\theta\}$ in terms of x, y, x_1, y_1: we have

$$\partial_1 x_1 = 2y_1, \quad \partial_1 y_1 = 1 - 2(1+k^2)x_1 + 3k^2 x_1^2, = \Omega_1$$

for a moment, and we thence obtain

$$\partial_1\{01'\theta\} = \frac{k^2}{(xy_1 + x_1 y)^2}[-2(xy_1 + x_1 y)(x - 2x_1)xy_1 + xx_1(x - x_1)(x\Omega_1 + 2yy_1)]$$

where the term in [ ] is found to be $= x(xy_1 + x_1 y)^2 - xx_1(x - x_1)^2$; whence the equation is

$$\partial_1\{01'\theta\} = k^2 x - \frac{k^2 xx_1 (x - x_1)^2}{(xy_1 + x_1 y)^2},$$

giving the foregoing result.

140. To introduce into the formulæ 1 instead of $1'$ we have only to write $u_1 - iK'$ instead of u_1; putting also for shortness s, c, d, s_1, c_1, d_1 for the functions of u and u_1 respectively, we thus obtain

$$\{01\theta\} = -\frac{cd}{s} + \frac{s}{s_1 \operatorname{sn}(u - u_1)},$$

where observe that interchanging u, u_1 we have

$$\{10\theta\} = -\frac{c_1 d_1}{s_1} - \frac{s_1}{s\,\mathrm{sn}\,(u-u_1)},$$

that is $\{10\theta\} = -\{01\theta\}$ as it should be: the formulæ may be written

$$\{01\theta\}, = -\{10\theta\}, = \frac{s^2 c_1 d_1 + s_1^2 cd}{sc_1 d_1 + s_1 cd}\,\frac{1}{ss_1\,\mathrm{sn}\,(u-u_1)}:$$

and

$$\partial_1\{01\theta\} = \frac{1}{s_1^2} - \frac{1}{\mathrm{sn}^2(u-u_1)},$$

whence

$$\partial_0\{01\theta\} = -\frac{1}{s^2} + \frac{1}{\mathrm{sn}^2(u-u_1)},$$

we have, moreover,

$$\{012\} = \{12\theta\} + \frac{s}{s_1\,\mathrm{sn}\,(u-u_1)} - \frac{s}{s_2\,\mathrm{sn}\,(u-u_1)},$$

and

$$\partial_0\{012\} = \frac{1}{\mathrm{sn}^2(u-u_1)} - \frac{1}{\mathrm{sn}^2(u-u_2)},$$

which last equation gives $(\partial_0 + \partial_1 + \partial_2)\{012\} = 0$ as it should do.

141. Supposing that the differential $d\Pi_{13}$ is defined by the equation

$$d\Pi_{13} = du\,\{012\} + du\left[\int_3^1 du\partial_3\{036\} - \{123\}\right],$$

we have

$$\int_5^4 d\Pi_{13} = \int_5^4 du\{012\} + \int_5^4 du.\left[\int_3^1 du\partial_3\{036\} - \{123\}\right.,$$

and thence

$$\begin{aligned}\partial_1\partial_4\int_5^4 d\Pi_{13} &= \partial_1\{124\} + \partial_3\{136\} - \partial_1\{123\},\\ &= \partial_1\{134\} - \partial_3\{316\},\\ &= \left[\frac{1}{\mathrm{sn}^2(u_1-u_3)} - \frac{1}{\mathrm{sn}^2(u_1-u_4)}\right] - \left[\frac{1}{\mathrm{sn}^2(u_3-u_1)} - \frac{1}{\mathrm{sn}^2(u_3-u_6)}\right],\\ &= \frac{1}{\mathrm{sn}^2(u_3-u_6)} - \frac{1}{\mathrm{sn}^2(u_1-u_4)},\end{aligned}$$

or establishing between the constants u_3, u_6 the relation $\mathrm{sn}^2(u_3-u_6) = \frac{K}{K-E}$,

this becomes

$$= 1 - \frac{E}{K} - k^2\,\mathrm{sn}^2(u_4 - u_1 + iK'),$$

which is

$$= \partial_3\partial_4.\log\frac{\theta(u_4-u_3+iK')\,\theta(u_5-u_1+iK')}{\theta(u_4-u_1+iK')\,\theta(u_5-u_3+iK')},$$

where Θ is Jacobi's theta-function, see Elliptic Functions, p. 144. The expression is in fact $= -\partial_1\partial_4\log\theta(u_4-u_1+iK')$, $= \phi(u_4-u_1+iK')$, if for a moment $\phi v = \partial_v^2\log\Theta v$. But we have

$$\log\Theta v = \log\sqrt{\frac{2k'K}{\pi}} + \tfrac{1}{2}v^2\left(1-\frac{E}{K}\right) - k^2\int_0\int_0 \mathrm{sn}^2 v\,dv^2,$$

that is $\phi v = 1 - \frac{E}{K} - k^2 \operatorname{sn}^2 v$, and consequently we have $\phi(u_4 - u_1 + iK')$ is $=$ $1 - \frac{E}{K} - k^2 \operatorname{sn}^2(u_4 - u_1 + iK')$.

142. In connection with the same curve $y^2 = x(1-x)(1-k^2x)$, we may establish the identity $\frac{d}{du_1}\frac{y_1}{x_1 - x} + \frac{d}{du}\frac{y}{x_1 - x} = k^2(x_1 - x)$,

where as before $x, y = s^2, scd$, and $x_1, y_1 = s_1^2, s_1c_1d_1$. We have

$$(x_1 - x)\frac{dy_1}{du_1} - y_1\frac{dx_1}{du_1} = (s_1^2 - s^2)\{1 - 2(1+k^2)s_1^2 + 3k^2s_1^4\} - 2s_1^2(1-s_1^2)(1-k^2s_1^2)$$
$$= -s^2 - s_1^2 + 2(1+k^2)s^2s_1^2 - 3k^2s^2s_1^4 + k^2s_1^6;$$

and similarly

$$(x - x_1)\frac{dy}{du} - y\frac{dx}{du} = -s^2 - s_1^2 + 2(1+k^2)s^2s_1^2 - 3k^2s^4s_1^2 + k^2s^6.$$

The difference of the two functions on the right-hand side is $= k^2(s_1^2 - s^2)^3$; which is $= k^2(x_1 - x)^3$, and this divided by $(x_1 - x)^2$ is $= k^2(x_1 - x)$; the identity is thus verified.

Fixed Curve the Quartic $y^2 = (1-x^2)(1-k^2x^2)$. Art. Nos. 143 to 145.

143. This is a curve having a tacnode at infinity on the line $x = 0$, as may be seen by writing the equation in the homogeneous form $y^2z^2 = (z^2 - x^2)(z^2 - k^2x^2)$; we have as it were two branches having the line infinity for a common tangent at the point in question. The equation is satisfied by $x = \operatorname{sn} u$, $y = \operatorname{cn} u \operatorname{dn} u$, values which are unaltered by the change of u into $u + 4mK + 2m'iK'$, m and m' any positive or negative integers, and in regard to this curve the sign $\equiv$ is to be understood accordingly. I consider with reference to this curve only the affected theorem, in the particular form in which it most readily connects itself with the ordinary theory of the integral of the third kind.

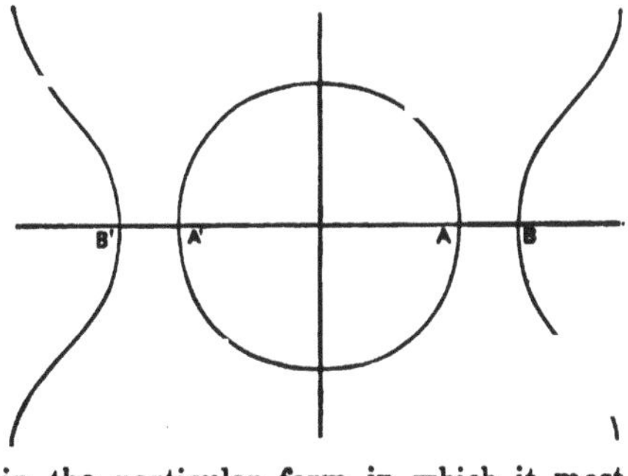

144. I consider the differential $\frac{(x, y, z)_{12}^2\, d\varpi}{012}$ in the particular case where the line 012 is a line parallel to the axis of y: taking its equation to be $x - x_1 = 0$, and putting for shortness $X = \sqrt{1-x^2 . 1-k^2x^2}$, $X_1 = \sqrt{1-x_1^2 . 1-k^2x_1^2}$, the para-

metric points are taken to be $(x_1, \sqrt{X_1})$, $(x_1, -\sqrt{X_1})$, and the residues are the intersections with the two branches at the tacnode. The conic $(x, y, z)^2_{12} = 0$ is to satisfy the conditions of passing through the two nodes of the tacnode, and through the two residues, that is again through the tacnode twice—in all four conditions, and we have thus the form $z\,(x - \theta z) = 0$, containing the arbitrary constant θ: the major function itself is then easily determined, and putting again $z = 1$ we arrive at the form $\dfrac{\sqrt{X_1}}{x_1 - \theta}\,\dfrac{x-\theta}{x - x_1}\,\dfrac{dx}{\sqrt{X}}$.

If the limits are taken to be two points on a line parallel to the axis of x, or what is the same thing, if the limits in regard to x are $x, -x$, we have the integral

$$\int_{-x}^{x} \frac{\sqrt{X_1}}{x_1 - \theta}\,\frac{x-\theta}{x - x_1}\,\frac{dx}{\sqrt{X}},\ = \int_0^{x} \frac{\sqrt{X_1}}{x_1 - \theta}\left(\frac{x-\theta}{x - x_1} + \frac{x+\theta}{x+x_1}\right)\frac{dx}{\sqrt{X}},$$

$$= 2\int_0^{x} \frac{\sqrt{X_1}}{x_1 - \theta}\cdot\frac{x^2 - x_1\theta}{x^2 - x_1^2}\,\frac{dx}{\sqrt{X}}.$$

We have $\dfrac{1}{x_1}\left(\dfrac{x^2}{x_1^2 - x^2} - \dfrac{\theta}{x_1 - \theta}\right) = \dfrac{x^2 - x_1\theta}{x_1^2 - x^2 . x_1 - \theta}$, and the integral thus becomes

$$= -2\int_0^{x} \frac{\sqrt{X_1}\,x^2 dx}{x_1(x_1^2 - x^2)\sqrt{X}} + 2\,\frac{\theta\sqrt{X_1}}{x_1(x_1 - \theta)}\int_0^{x}\frac{dx}{\sqrt{X}}.$$

Taking here $x = \operatorname{sn} u$, $x_1 = \operatorname{sn}(a + iK')$, $= \dfrac{1}{k \operatorname{sn} a}$, we have $dx = \operatorname{cn} u \operatorname{dn} u\, du = \sqrt{X}\, du$, and the result is

$$\int_{-x}^{x} \frac{\sqrt{X_1}}{x_1 - \theta}\,\frac{x-\theta}{x - x_1}\,\frac{dx}{\sqrt{X}} = -2\int_0 \frac{k^2 \operatorname{sn} a \operatorname{cn} a \operatorname{dn} a \operatorname{sn}^2 u\, du}{1 - k^2 \operatorname{sn}^2 a \operatorname{sn}^2 u} + \frac{2k \operatorname{sn} a \operatorname{cn} a}{1 - k\theta \operatorname{sn} a}u,$$

where on the right-hand side the first term is $= -2\Pi(u, a)$, if $\Pi(u, a)$ be Jacobi's form of the integral of the third kind, Elliptic Functions, p. 143.

145. It is to be observed that the proper normal form is not $\Pi(u, a)$, but $\Pi(u, a) - uZa$; say this is $\overline{\Pi}(u, a)$, viz. we then have

$$\overline{\Pi}(u, a) = \Pi(u, a) - u\left[a\left(1 - \frac{E}{K}\right) - k^2\int \operatorname{sn}^2 a\, da\right],$$

and thence

$$\partial_u \overline{\Pi}(u, a) = \frac{k^2 \operatorname{sn} a \operatorname{cn} a \operatorname{dn} a \operatorname{sn}^2 u}{1 - k^2 \operatorname{sn}^2 a \operatorname{sn}^2 u} - a\left(1 - \frac{E}{K}\right) + k^2\int \operatorname{sn}^2 a\, da,$$

$$\partial_a\partial_u \overline{\Pi}(u, a) = k^2 \operatorname{sn}^2 u\, \partial_a \frac{\operatorname{sn} a \operatorname{cn} a \operatorname{dn} a}{1 - k^2 \operatorname{sn}^2 a \operatorname{sn}^2 u} - \left(1 - \frac{E}{K}\right) + k^2 \operatorname{sn}^2 a;$$

or if for shortness we write $\operatorname{sn} u, \operatorname{sn} a = s, \sigma$, this is

$$\partial_a\partial_u \overline{\Pi}(u, a) = \frac{k^2 s^2[1 - 2(1 + k^2)\sigma^2 + 3k^2\sigma^4] + k^4 s^4 \sigma^2}{(1 - k^2 s^2 \sigma^2)^2} - \left(1 - \frac{E}{K}\right) + k^2\sigma^2,$$

which is

$$= \frac{k^2\{(s^2 + \sigma^2)(1 + k^2 s^2 \sigma^2) - 2(1 + k^2) s^2 \sigma^2\}}{(1 - k^2 s^2 \sigma^2)^2} - \left(1 - \frac{E}{K}\right);$$

or this being symmetrical in regard to s, σ, we have

$$\partial_a\partial_u\overline{\Pi}(u, a) = \partial_a\partial_u\overline{\Pi}(a, u),$$

and thence by integration, and a proper determination of the constants,

$$\overline{\Pi}(u, a) = \overline{\Pi}(a, u).$$

CHAPTER VI. THE NODAL QUARTIC.

Nodal Quartic; the General and Fleflecnodal Forms. Art. Nos. 146 to 148.

146. For a cubic, or other curve of deficiency 1, we are concerned with single points on the curve, and corresponding thereto with functions of a single argument (elliptic functions): but for a curve of deficiency 2, we have to consider pairs of points on the curve, and functions of two arguments: there is thus a marked change in the form of the results.

The most simple curve of deficiency 2 is the nodal quartic, $n = 4$, $p = 2$. Using homogeneous coordinates the general form is $Az^2 + 2Bz + C = 0$, where

$$\begin{aligned} A &= (i, j, k\between x, y)^2, \\ B &= (l, m, n, o\between x, y)^3, \\ C &= (p, q, r, s, t\between x, y)^4, \end{aligned}$$

and where we write also

$$B^2 - AC = (l^2 - ip)(x - ay)(x - by)(x - cy)(x - dy)(x - ey)(x - fy).$$

Clearly the equation of the two tangents at the node is $A = 0$; and the equations of the six tangents from the node are $x - ay = 0, \ldots x - fy = 0$: at the points of contact we have $Az + B = 0$, viz. this is the equation of a nodal cubic, the node and the two tangents there being the same with the node and two tangents of the quartic. Hence the node counts as 6 intersections, and there are besides 6 intersections which are the points of contact of the 6 tangents respectively: say these are the points a, b, c, d, e, f: the coordinates of the point a are given by the equations $x : y : z = a : 1 : -\frac{B_a}{A_a}\left(= -\frac{C_a}{B_a}\right)$, where A_a, B_a, C_a are what A, B, C become on writing therein a, 1 for x, y: and similarly for the other points.

147. An important special case is when the equation is $B = 0$; say we have here

$$\begin{aligned} A &= i(x - ey)(x - fy), \\ B &= 0, \\ C &= p(x - ay)(x - by)(x - cy)(x - dy), \end{aligned}$$

or omitting the factors i and p,

$$(x - ey)(x - fy)z^2 = (x - ay)(x - by)(x - cy)(x - dy),$$

the origin is here a fleflecnode; the tangents $x-ey=0$, $x-fy=0$ count as two of the six tangents from the node, and there remain the four tangents

$$x-cy=0,\ x-dy=0,\ x-ay=0,\ x-by=0;$$

the four points of contact are the intersections of the curve with the line $z=0$.

148. The general nodal form depends on 11 constants, but by writing $\alpha x+\beta y$, $\gamma x+\delta y$, ϵz in place of x, y, z, we introduce 5 apoclastic constants, and so reduce the number to $11+1-5, =7$: similarly the fleflecnodal form depends on 7 constants, but we reduce the number in like manner to $7+1-5$, $=3$: the final form might here be taken to be

$$z^2xy=(x-y)(x-by)(x-cy)(x-dy);$$

but it is more convenient to retain the original form

$$z^2(x-ey)(x-fy)=(x-ay)(x-by)(x-cy)(x-dy),$$

bearing in mind that this is reducible to the form just referred to, and thus depends virtually upon only 3 constants.

It is a general property that a curve of deficiency p greater than 1 can be by a rational transformation reduced to a curve of that deficiency depending upon $3p-3$ parameters: in particular if $p=2$, then the form depending upon 3 parameters may be taken to be the fleflecnodal quartic as above: and I proceed to show how the general nodal quartic can in fact be reduced to this fleflecnodal form.

Reduction to the Fleflecnodal Form. Art. Nos. 149 to 152.

149. Consider the general nodal quartic $Az^2+2Bz+C=0$: take $\zeta=0$ for the equation of the line joining the points of contacts of the tangents $x-ey=0$, $x-fy=0$; and then writing $x=\xi$, $y=\eta$, let the curve be transformed in the first instance from x, y, z to the new coordinates ξ, η, ζ.

Writing A_e for the value $(i, j, k\between e, 1)^2$ which A assumes on putting therein $(e, 1)$ for (x, y) respectively, and similarly A_f, B_e, B_f for the other like values, we may take

$$A_eA_f(e-f)\zeta=\begin{vmatrix} x\ , & y\ , & z \\ eA_e, & A_e, & -B_e \\ fA_f, & A_f, & -B_f \end{vmatrix},$$

$$=-x(A_eB_f-A_fB_e)+y(eA_eB_f-fB_eA_f)+z(e-f)A_eA_f,$$

say this equation is $\zeta=-\lambda x-\mu y+z$, the values of λ, μ being

$$\lambda=\frac{A_eB_f-A_fB_e}{(e-f)A_eA_f},\qquad \mu=\frac{-eA_eB_f+fB_eA_f}{(e-f)A_eA_f},$$

and therefore $\qquad \lambda e+\mu=-\dfrac{B_e}{A_e},\qquad \lambda f+\mu=-\dfrac{B_f}{A_f}.$

150. From the values $\zeta, \xi, \eta = -\lambda x - \mu y + z, x, y$, we obtain $z, x, y = \zeta + \lambda\xi + \mu\eta, \xi, \eta$; and the transformed equation is

$$A'(\zeta + \lambda\xi + \mu\eta)^2 + 2B'(\zeta + \lambda\xi + \mu\eta) + C' = 0,$$

where

$$A' = (i, j, k \between \xi, \eta)^2,$$
$$B' = (l, m, n, o \between \xi, \eta)^3,$$
$$C' = (p, q, r, s, t \between \xi, \eta)^4,$$

say this equation is $\mathfrak{A}\zeta^2 + 2\mathfrak{B}\zeta + \mathfrak{C} = 0$, where

$$\mathfrak{A} = A',$$
$$\mathfrak{B} = A'(\lambda\xi + \mu\eta) + B',$$
$$\mathfrak{C} = A'(\lambda\xi + \mu\eta)^2 + 2B'(\lambda\xi + \mu\eta) + C',$$

and thence

$$\mathfrak{B}^2 - \mathfrak{A}\mathfrak{C} = B'^2 - A'C', = (l^2 - ip)(\xi - a\eta)(\xi - b\eta)(\xi - c\eta)(\xi - d\eta)(\xi - e\eta)(\xi - f\eta):$$

We have here $\mathfrak{B}, = A'(\lambda\xi + \mu\eta) + B'$, a cubic function $(\xi, \eta)^3$ containing the factors $\xi - e\eta$ and $\xi - f\eta$: in fact writing $\xi, \eta = e, 1$ it becomes $A_e(\lambda e + \mu) + B_e$ which is $= 0$; and similarly writing $\xi, \eta = f, 1$ it becomes $A_f(\lambda f + \mu) + B_f$, which is $= 0$. Calling the other factor $L\xi + M\eta$, we have thus

$$\mathfrak{B} = (\xi - e\eta)(\xi - f\eta)(L\xi + M\eta),$$

and thence

$$\begin{aligned}\mathfrak{A}\mathfrak{C} &= (\xi - e\eta)^2(\xi - f\eta)^2(L\xi + M\eta)^2 \\ &\qquad - (l^2 - ip)(\xi - a\eta)(\xi - b\eta)(\xi - c\eta)(\xi - d\eta)(\xi - e\eta)(\xi - f\eta), \\ &= (\xi - e\eta)(\xi - f\eta)[(\xi - e\eta)(\xi - f\eta)(L\xi + M\eta)^2 \\ &\qquad - (l^2 - ip)(\xi - c\eta)(\xi - d\eta)(\xi - e\eta)(\xi - f\eta)].\end{aligned}$$

Hence $\mathfrak{C}$ contains the factor $(\xi - e\eta)(\xi - f\eta)$, say we have

$$\mathfrak{C} = \theta(\xi - e\eta)(\xi - f\eta)(\xi - \varepsilon\eta)(\xi - \phi\eta).$$

151. In the equation $\mathfrak{A}\zeta^2 + 2\mathfrak{B}\zeta + \mathfrak{C} = 0$ of the quartic curve, writing $\zeta = 0$, we find $\mathfrak{C} = 0$, that is $(\xi - e\eta)(\xi - f\eta)(\xi - \varepsilon\eta)(\xi - \phi\eta) = 0$: but $\zeta = 0$ is the equation of the line joining the points of contact of the tangents $\xi - e\eta = 0$, $\xi - f\eta = 0$; hence $\xi - \varepsilon\eta = 0$, $\xi - \phi\eta = 0$ are the lines drawn from the node to the two points ε, ϕ which are the residues of these two points of contact. We now have

$$\begin{aligned}\theta\mathfrak{A}(\xi - \varepsilon\eta)(\xi - \phi\eta) &= (\xi - e\eta)(\xi - f\eta)(L\xi + M\eta)^2 \\ &\qquad - (l^2 - ip)(\xi - a\eta)(\xi - b\eta)(\xi - c\eta)(\xi - d\eta),\end{aligned}$$

and thence

$$0 = (\varepsilon - e)(\varepsilon - f)(L\varepsilon + M)^2 - (l^2 - ip)(\varepsilon - a)(\varepsilon - b)(\varepsilon - c)(\varepsilon - d),$$
$$0 = (\phi - e)(\phi - f)(L\phi + M)^2 - (l^2 - ip)(\phi - a)(\phi - b)(\phi - c)(\phi - d),$$

which equations determine L and M; and then with these values of L, M, and for $\mathfrak{A}$ substituting its value $(i, j, k \between \xi, \eta)^2$ the equation must become an identity.

We have in what precedes, by the transformation $z=\zeta+\lambda\xi+\mu\eta$, $x=\xi$, $y=\eta$, passed from the form $Az^2+2Bz+C=0$, to the form

$$\mathfrak{A}\zeta^2+2\mathfrak{B}\zeta+\mathfrak{C}=0,$$

where

$$\begin{aligned}\mathfrak{A}&=(i,\,j,\,k)(\xi,\,\eta)^2,\\ \mathfrak{B}&=(\xi-e\eta)(\xi-f\eta)(L\xi+M\eta),\\ \mathfrak{C}&=\theta(\xi-e\eta)(\xi-f\eta)(\xi-\varepsilon\eta)(\xi-\phi\eta),\end{aligned}$$

viz. $\mathfrak{B}$ and $\mathfrak{C}$ have here the common factor $(\xi-e\eta)(\xi-f\eta)$.

152. Assume now

$$\xi,\,\eta,\,\zeta=X,\,Y,\,\theta\frac{(X-\varepsilon Y)(X-\varphi Y)}{Z-LX-MY},$$

and therefore conversely

$$X,\,Y,\,Z=\xi,\,\eta,\,L\xi+M\eta+\theta\frac{(\xi-\varepsilon\eta)(\xi-\varphi\eta)}{\zeta},$$

then we have in the new coordinates $(X,\,Y,\,Z)$ the equation

$$(i,\,j,\,k)(X,\,Y)^2\theta^2\left\{\frac{(X-\varepsilon Y)(X-\varphi Y)}{Z-LX-MY}\right\}^2$$

$$+\,2(X-eY)(X-fY)(LX+MY)\,\theta\frac{(X-\varepsilon Y)(X-\varphi Y)}{Z-LX-MY},$$

$$+\,\theta(X-eY)(X-fY)(X-\varepsilon Y)(X-\phi Y)=0,$$

that is

$$\begin{aligned}&(i,\,j,\,k)(X,\,Y)^2\theta(X-\varepsilon Y)(X-\phi Y)\\ &+2(X-eY)(X-fY)(LX+MY)(Z-LX-MY)\\ &+\ (X-eY)(X-fY)(Z-LX-MY)^2=0,\end{aligned}$$

where the second and third lines together are

$$=(X-eY)(X-fY)\{Z^2-(LX+MY)^2\},$$

and the equation thus is

$$\begin{aligned}(X-eY)(X-fY)Z^2+\{\theta(i,\,j,\,k)(X,\,Y)^2(X-\varepsilon Y)(X-\phi Y)&\\ -(X-eY)(X-fY)(LX+MY)^2\}&=0.\end{aligned}$$

But the term in { } is identically

$$=-(l^2-ip)(X-aY)(X-bY)(X-cY)(X-dY),$$

and the equation thus becomes

$$(X-eY)(X-fY)Z^2-(l^2-ip)(X-aY)(X-bY)(X-cY)(X-dY)=0;$$

viz. the original equation $Az^2+2Bz+C=0$ of the general nodal quartic is, by the equations

$$x,\,y,\,z=X,\,Y,\,\lambda X+\mu Y+\theta\frac{(X-\varepsilon Y)(X-\varphi Y)}{Z-LX-MY};$$

or conversely

$$X,\,Y,\,Z=x,\,y,\,Lx+My+\frac{\theta(x-\varepsilon y)(x-\varphi y)}{z-\lambda\xi-\mu\eta},$$

transformed into the fleflecnodal form as above.

It originally appeared to me that the fleflecnodal form was more easily dealt with than the general form; and I effected the transformation for this reason: there is, however, the disadvantage that the six points a, b, c, d, e, f enter into the equation unsymmetrically; and I afterwards found that the general form could be dealt with nearly as easily, and in what follows I use therefore the general form. The transformation is given as interesting for its own sake, and as an illustration of the theorem in regard to the number of constants in a curve of deficiency p.

Application of Abel's Theorem. Art. Nos. 153 to 157.

153. Taking the fixed curve to be $f, = \frac{1}{2}(Az^2 + 2Bz + C), = 0$, we have $\frac{df}{dz} = Az + B = \sqrt{(x, y)^6}$, if for shortness we write

$$(x, y)^6 = B^2 - AC, = (l^2 - 4p)(x-ay)(x-by)(x-cy)(x-dy)(x-ey)(x-fy),$$

and we thence have

$$d\omega = \frac{xdy - ydx}{\sqrt{(x, y)^6}}.$$

The minor curve $(x, y, z)^{n-3} = 0$, is an arbitrary line passing through the node, that is the point $x = 0$, $y = 0$; and the pure theorem thus gives the two relations $\Sigma x d\omega = 0$, $\Sigma y d\omega = 0$; where the summation extends to the intersections of the fixed curve $Az^2 + 2Bz + C = 0$, with the variable curve ϕ.

The variable curve is taken to be a cubic $Az + B = (\alpha, \beta, \gamma, \delta \between x, y)^3$, or say $Az + B = \Omega$, where Ω is a given cubic function $(x, y)^3$: viz. this is a nodal cubic, the node and the two tangents there being the same with the node and the two tangents of the quartic: hence it meets the quartic in the node counting 6 times, and in 6 other points, say these are the points 1, 2, 3, 4, 5, 6: hence the differential relations are

$$x_1 d\omega_1 + x_2 d\omega_2 + x_3 d\omega_3 + x_4 d\omega_4 + x_5 d\omega_5 + x_6 d\omega_6 = 0,$$
$$y_1 d\omega_1 + y_2 d\omega_2 + y_3 d\omega_3 + y_4 d\omega_4 + y_5 d\omega_5 + y_6 d\omega_6 = 0.$$

154. Observe that the intersections of the cubic with the fixed curve are given by the equation $\Omega^2 = B^2 - AC$, or say $\Omega^2 = (x, y)^6$; an equation which determines the ratio $x:y$ for the six points respectively, and the ratio $z:x$ is then determined rationally by the original equation $Az + B = \Omega$. Instead of regarding Ω as a given function, we may, if we please, take 1, 2, 3, 4 given points on the quartic: we then have four equations for the determination of the coefficients $(\alpha, \beta, \gamma, \delta)$ of the function Ω; viz. these equations may be taken to be

$$
\begin{aligned}
(\alpha, \beta, \gamma, \delta)(x_1, y_1)^3 &= \sqrt{(x_1, y_1)^6}, \\
(\quad ,, \quad)(x_2, y_2)^3 &= \sqrt{(x_2, y_2)^6}, \\
(\quad ,, \quad)(x_3, y_3)^3 &= \sqrt{(x_3, y_3)^6}, \\
(\quad ,, \quad)(x_4, y_4)^4 &= \sqrt{(x_4, y_4)^6},
\end{aligned}
$$

Ω is hereby completely determined, and this being so the remaining intersections 5 and 6 are also completely determined: there are thus between the six points 2 integral relations, which agrees with the number, $= 2$, of the differential relations obtained above.

155. If we now assume

$$
\begin{aligned}
du &= x_1 d\omega_1 + x_2 d\omega_2,\ du' = x_3 d\omega_3 + x_4 d\omega_4,\ du'' = x_5 d\omega_5 + x_6 d\omega_6, \\
dv &= y_1 d\omega_1 + y_2 d\omega_2,\ dv' = y_3 d\omega_3 + y_4 d\omega_4,\ dv'' = y_5 d\omega_5 + y_6 d\omega_6,
\end{aligned}
$$

or say

$$
u,\ u',\ u'' = \int_{\alpha\beta}^{12},\ \int_{\alpha\beta}^{34},\ \int_{\alpha\beta}^{56} \frac{x(xdy - ydx)}{\sqrt{(x, y)^6}},
$$

$$
v,\ v',\ v'' = \int_{\alpha\beta}^{12},\ \int_{\alpha\beta}^{34},\ \int_{\alpha\beta}^{56} \frac{y(xdy - ydx)}{\sqrt{(x, y)^6}};
$$

that is

$$
u = \left(\int_{\alpha}^{1} + \int_{\beta}^{2}\right) \frac{x(xdy - ydx)}{\sqrt{(x, y)^6}},\quad v = \left(\int_{\alpha}^{1} + \int_{\beta}^{2}\right) \frac{y(xdy - ydx)}{\sqrt{(x, y)^6}},
$$

where α, β are points assumed at pleasure on the quartic: and similarly for u', v', u'', v'': then u, v are hereby determined as functions of the points 1, 2: and we may conversely regard the points 1, 2 as determined in terms of the two arguments u, v. We might, selecting any two symmetrical functions of the degree zero, for instance, $\frac{x_1}{y_1} + \frac{x_2}{y_2}$, $\frac{x_1 x_2}{y_1 y_2}$, represent them as functions $\phi(u, v)$, $\psi(u, v)$; and then $\frac{x_1}{y_1}$ and $\frac{x_2}{y_2}$ will be functions of $\phi(u, v)$, $\psi(u, v)$, but instead of this selection it is proper to consider the ratios of six functions depending on the points a, b, c, d, e, f respectively: viz. we assume

$$
\begin{aligned}
&\sqrt{(x_1 - ay_1)(x_2 - ay_2)} : \sqrt{(x_1 - by_1)(x_2 - by_2)} \ldots : \sqrt{(x_1 - fy_1)(x_2 - fy_2)} \\
&= \quad A(u, v) \quad : \quad B(u, v) \quad \ldots : \quad F(u, v),
\end{aligned}
$$

and of course 3, 4 will be in like manner determined by means of the corresponding functions of u', v', and 5, 6 by means of the corresponding functions of u'', v''. The squared functions A^2, B^2, C^2, D^2, E^2, F^2 are proportional to given linear functions of $x_1 x_2$, $x_1 y_2 + x_2 y_1$, $y_1 y_2$, and are thus connected by three independent linear relations.

156. The differential relations then become

$$
du + du' + du'' = 0,\ dv + dv' + dv'' = 0,
$$

and we have consequently

$$
u + u' + u'' = I,\quad v + v' + v'' = J,
$$

where I, J are constants which are determinable as definite integrals by the consideration that when the cubic is taken to be $Az+B=0$, the six points 1, 2, 3, 4, 5, 6 coincide with the points of contact a, b, c, d, e, f. I do not at present see my way to a proper development of this point of the theory: but in explanation of the nature of the result, I assume for the moment that by a proper determination of the inferior limits α, β, or otherwise, we may take $I=0$, $J=0$. We then have $u''=-u-u'$, $v''=-v-v'$; and the integral equations which determine the points 5, 6 in terms of the points 1, 2 and the points 3, 4, then in effect determine the functions A, B, etc. of $-u-u'$, $-v-v'$, or say those of $u+u'$, $v+v'$ in terms of the like functions of (u, v) and of (u', v'): viz. these equations give the addition theory of the functions $A(u, v)$, etc.

157. We may, in the first instance, disregarding altogether the consideration of the arguments u, v, etc., attend only to the algebraic functions such as $\sqrt{(x_1-ay_1)(x_2-ay_2)}$, etc. of the coordinates of the pairs of points 1, 2; 3, 4, and 5, 6; and we can in regard to these develope a proper theory. This depends only on the equation $\Omega=\sqrt{(x, y)^6}$; it will be convenient to assume herein $y=1$, and slightly modifying the form, to write it

$$(\alpha, \beta, \gamma, \delta)(x, 1)^3=\sqrt{a-x.b-x.c-x.d-x.e-x.f-x};$$

and accordingly to consider the functions $\sqrt{a-x_1.a-x_2}$, etc. These are called the single-letter functions A, etc. but there are certain double-letter functions AB, etc. which have also to be considered; and I will, in the first instance, show how these present themselves in connection with the cubic curve.

Origin of the Double-Letter Functions. Art. Nos. 158, 159.

158. The cubic curve $Az+B=\Omega$ may be taken to be a curve through two of the points of contact, say the points a, b; these will then be two out of the six points, and taking the remaining four points to be the pairs 1, 2 and 3, 4, we have single-letter functions of 3, 4 presenting themselves as double-letter functions of 1, 2. In fact the equation of the curve is

$$Az+B=\lambda(x-ay)(x-by)(x-ky);$$

for the intersections with the quartic we have $\lambda^2(x-ay)^2(x-by)^2(x-ky)^2=\Omega^2$, or throwing out the factor $(x-ay)(x-by)$ and changing the constant λ, this is $(x-ay)(x-by)(x-ky)^2-\lambda(x-cy)(x-dy)(x-ey)(x-fy)=0$; and the quartic function must be a multiple of $(xy_1-x_1y)(xy_2-x_2y)(xy_3-x_3y)(xy_4-x_4y)$.

Putting each of the y's equal 1, we have the identity

$$(a-x)(b-x)(k-x)^2-\lambda(c-x)(d-x)(e-x)(f-x)$$
$$=\mu(x_1-x)(x_2-x)(x_3-x)(x_4-x);$$

and hence, introducing a notation which will be convenient, $a-x=\mathrm{a}$, $a-x_1=\mathrm{a}_1$, and so in other cases, we have by giving different values to x the equations

$$\mathrm{a}_1\mathrm{b}_1\mathrm{k}_1^2=\lambda\mathrm{c}_1\mathrm{d}_1\mathrm{e}_1\mathrm{f}_1,\qquad (a-c)(b-c)(k-c)^2=\mu\mathrm{c}_1\mathrm{c}_2\mathrm{c}_3\mathrm{c}_4,$$
$$\mathrm{a}_2\mathrm{b}_2\mathrm{k}_2^2=\lambda\mathrm{c}_2\mathrm{d}_2\mathrm{e}_2\mathrm{f}_2,\qquad (a-d)(b-d)(k-d)^2=\mu\mathrm{d}_1\mathrm{d}_2\mathrm{d}_3\mathrm{d}_4,$$
$$\mathrm{a}_3\mathrm{b}_3\mathrm{k}_3^2=\lambda\mathrm{c}_3\mathrm{d}_3\mathrm{e}_3\mathrm{f}_3,\qquad (a-e)(b-e)(k-e)^2=\mu\mathrm{e}_1\mathrm{e}_2\mathrm{e}_3\mathrm{e}_4,$$
$$\mathrm{a}_4\mathrm{b}_4\mathrm{k}_4^2=\lambda\mathrm{c}_4\mathrm{d}_4\mathrm{e}_4\mathrm{f}_4,\qquad (a-f)(b-f)(k-f)^2=\mu\mathrm{f}_1\mathrm{f}_2\mathrm{f}_3\mathrm{f}_4,$$
$$-\lambda(c-a)(d-a)(e-a)(f-a)=\mu\mathrm{a}_1\mathrm{a}_2\mathrm{a}_3\mathrm{a}_4,$$
$$-\lambda(c-b)(d-b)(e-b)(f-b)=\mu\mathrm{b}_1\mathrm{b}_2\mathrm{b}_3\mathrm{b}_4.$$

We have thus

$$\frac{(a-c)(b-c)}{(a-d)(b-d)}\left(\frac{c-k}{d-k}\right)^2=\frac{\mathrm{c}_1\mathrm{c}_2\mathrm{c}_3\mathrm{c}_4}{\mathrm{d}_1\mathrm{d}_2\mathrm{d}_3\mathrm{d}_4},$$

and

$$\frac{\mathrm{k}_1^2}{\mathrm{k}_2^2}=\frac{\mathrm{a}_2\mathrm{b}_2\mathrm{c}_1\mathrm{d}_1\mathrm{e}_1\mathrm{f}_1}{\mathrm{a}_1\mathrm{b}_1\mathrm{c}_2\mathrm{d}_2\mathrm{e}_2\mathrm{f}_2},$$

which last equation, writing for a moment $\gamma,\ \delta=\sqrt{\mathrm{a}_2\mathrm{b}_2\mathrm{c}_1\mathrm{d}_1\mathrm{e}_1\mathrm{f}_1},\ \sqrt{\mathrm{a}_1\mathrm{b}_1\mathrm{c}_2\mathrm{d}_2\mathrm{e}_2\mathrm{f}_2}$, gives $\frac{k-x_1}{k-x_2}=\frac{\gamma}{\delta}$, whence $k(\gamma-\delta)=x_2\gamma-x_1\delta$, and thence

$$\frac{c-k}{d-k}=\frac{\gamma\mathrm{c}_2-\delta\mathrm{c}_1}{\gamma\mathrm{d}_2-\delta\mathrm{d}_1}=\frac{\sqrt{\mathrm{c}_1\mathrm{c}_2}\{\sqrt{\mathrm{a}_2\mathrm{b}_2\mathrm{c}_2\mathrm{d}_1\mathrm{e}_1\mathrm{f}_1}-\sqrt{\mathrm{a}_1\mathrm{b}_1\mathrm{c}_1\mathrm{d}_2\mathrm{e}_2\mathrm{f}_2}\}}{\sqrt{\mathrm{d}_1\mathrm{d}_2}\{\sqrt{\mathrm{a}_2\mathrm{b}_2\mathrm{d}_2\mathrm{c}_1\mathrm{e}_1\mathrm{f}_1}-\sqrt{\mathrm{a}_1\mathrm{b}_1\mathrm{d}_1\mathrm{c}_2\mathrm{e}_2\mathrm{f}_2}\}}.$$

or substituting in the first equation

$$\frac{\sqrt{(a-c)(b-c)}}{\sqrt{(a-d)(b-d)}}\cdot\frac{\sqrt{\mathrm{a}_2\mathrm{b}_2\mathrm{c}_2\mathrm{d}_1\mathrm{e}_1\mathrm{f}_1}-\sqrt{\mathrm{a}_1\mathrm{b}_1\mathrm{c}_1\mathrm{d}_2\mathrm{e}_2\mathrm{f}_2}}{\sqrt{\mathrm{a}_2\mathrm{b}_2\mathrm{d}_2\mathrm{c}_1\mathrm{e}_1\mathrm{f}_1}-\sqrt{\mathrm{a}_1\mathrm{b}_1\mathrm{d}_1\mathrm{c}_2\mathrm{e}_2\mathrm{f}_2}}=\frac{\sqrt{\mathrm{c}_3\mathrm{c}_4}}{\sqrt{\mathrm{d}_3\mathrm{d}_4}}.$$

159. Considering the duad DE as an abbreviation for the double triad $ABC.DEF$, the expressed duad being always accompanied by the letter F, we are thus led to the consideration of the double-letter functions

$$AB_{12}=\frac{1}{x_1-x_2}\{\sqrt{\mathrm{a}_1\mathrm{b}_1\mathrm{f}_1\mathrm{c}_2\mathrm{d}_2\mathrm{e}_2}-\sqrt{\mathrm{a}_2\mathrm{b}_2\mathrm{f}_2\mathrm{c}_1\mathrm{d}_1\mathrm{e}_1}\},\text{ etc.}$$

in connection with the already mentioned single-letter functions $A_{12}=\sqrt{\mathrm{a}_1\mathrm{a}_2}$, etc. viz. in this notation the equation just obtained is

$$\frac{C_{34}}{D_{34}}=\sqrt{\frac{(a-c)(b-c)}{(a-d)(b-d)}}\,\frac{DE_{12}}{CE_{12}},$$

and it thus appears that the points 3, 4 being obtained as above from the given points 1, 2, then that the quotient of two of the single-letter functions of 3, 4 is a constant multiple of the quotient of two of the double-letter functions of 1, 2. Observe that the points 3, 4 are derived from 1, 2 by means of the two points a, b: we have DE standing for $ABC.DEF$, CE for $ABD.CEF$, and if the two functions were represented by ABC, ABD respectively, then the form would have been

$$\frac{C_{34}}{D_{34}} = \sqrt{\frac{(a-c)(b-c)}{(a-d)(b-d)}} \cdot \frac{ABC_{12}}{ABD_{12}},$$

which is a clearer expression of the theorem; the apparent want of symmetry of the first form arises only from the arbitrary selection of the letter F to accompany the expressed duad, and is at once removed by substituting for a duad DE the triad $ABC.DEF$ which is thereby signified. The denominator factor $x_1 - x_2$ is introduced in order to make the degree in x_1 or x_2 equal to that of the single-letter functions.

The Addition Theory. Art. Nos. 160 to 163.

160. We have the six single-letter symbols A, B, C, D, E, F; viz. $A_{12} = \sqrt{\mathfrak{a}_1 \mathfrak{a}_2}$, etc.: and the ten double-letter symbols AB, AC, AD, AE, BC, BD, BE, CD, CE, DE, viz.

$$AB_{12} = \frac{1}{x_1 - x_2} \{ \sqrt{\mathfrak{a}_1 \mathfrak{b}_1 \mathfrak{f}_1 \mathfrak{c}_2 \mathfrak{d}_2 \mathfrak{e}_2} - \sqrt{\mathfrak{a}_2 \mathfrak{b}_2 \mathfrak{f}_2 \mathfrak{c}_1 \mathfrak{d}_1 \mathfrak{e}_1} \}, \text{ etc.}$$

these 16 functions being connected by algebraical relations which are immediately deducible from these expressions of the functions in terms of x_1, x_2. The problem is to express the functions of 5, 6 in terms of those of 1, 2 and of those of 3, 4. The relation between the variables x_1, x_2, x_3, x_4, x_5, x_6 consists herein that we have x_1, x_2, x_3, x_4, x_5, x_6 as the roots of the equation

$$[(1, x)^3]^2 \alpha x^3 + \beta x^2 + \gamma x + \delta - \lambda (a-x)(b-x)(c-x)(d-x)(e-x)(f-x) = 0;$$

or what is the same thing, it consists in the identity

$$[\alpha x^3 + \beta x^2 + \gamma x + \delta]^2 - \lambda (a-x)(b-x)(c-x)(d-x)(e-x)(f-x)$$
$$- \mu (x_1 - x)(x_2 - x)(x_3 - x)(x_4 - x)(x_5 - x)(x_6 - x) = 0;$$

or again it may be expressed by the plexus of equations

$$\left\| \begin{array}{cccccc} 1, & 1, & 1, & 1, & 1, & 1 \\ x_1, & x_2, & x_3, & x_4, & x_5, & x_6 \\ x_1^2, & x_2^2, & x_3^2, & x_4^2, & x_5^2, & x_6^2 \\ x_1^3, & x_2^3, & x_3^3, & x_4^3, & x_5^3, & x_6^3 \\ \sqrt{X_1}, & \sqrt{X_2}, & \sqrt{X_3}, & \sqrt{X_4}, & \sqrt{X_5}, & \sqrt{X_6} \end{array} \right\| = 0,$$

(where $X_1 = (a - x_1)(b - x_1) \ldots (f - x_1)$, etc.) equivalent of course to two equations, and serving to determine x_5, x_6 in terms of x_1, x_2, x_3, x_4.

161. The solution is in fact as is given in my paper "On the Addition of the double ϑ-Functions," *Crelle*, t. 88 (1880), pp. 74–81. Writing successively

$x = x_1, x_2, x_3, x_4$, we have

$$\alpha x_1^3 + \beta x_1^2 + \gamma x_1 + \delta = \sqrt{\lambda}\sqrt{X_1},$$
$$\alpha x_2^3 + \beta x_2^2 + \gamma x_2 + \delta = \sqrt{\lambda}\sqrt{X_2},$$
$$\alpha x_3^3 + \beta x_3^2 + \gamma x_3 + \delta = \sqrt{\lambda}\sqrt{X_3},$$
$$\alpha x_4^3 + \beta x_4^2 + \gamma x_4 + \delta = \sqrt{\lambda}\sqrt{X_4},$$

which equations serve to determine in terms of x_1, x_2, x_3, x_4 the ratios $\alpha : \beta : \gamma : \delta$: and we have then the two like equations

$$\alpha x_5^3 + \beta x_5^2 + \gamma x_5 + \delta = \sqrt{\lambda}\sqrt{X_5},$$
$$\alpha x_6^3 + \beta x_6^2 + \gamma x_6 + \delta = \sqrt{\lambda}\sqrt{X_6},$$

which determine the symmetric functions of x_5, x_6.

If reverting to the identity, we write therein for instance $x = a$, we find

$$\alpha a^3 + \beta a^2 + \gamma a + \delta = \sqrt{\mu}\, A_{12} A_{34} A_{56},$$

which equation when properly reduced gives the proportional value of A_{56}.

162. Calling for a moment the function on the left-hand side Ω, we have

$$\begin{vmatrix} x_1^3, & x_1^2, & x_1, & 1, & \sqrt{\lambda}\sqrt{X_1} \\ x_2^3, & x_2^2, & x_2, & 1, & \sqrt{\lambda}\sqrt{X_2} \\ x_3^3, & x_3^2, & x_3, & 1, & \sqrt{\lambda}\sqrt{X_3} \\ x_4^3, & x_4^2, & x_4, & 1, & \sqrt{\lambda}\sqrt{X_4} \\ a^3, & a^2, & a, & 1, & \Omega \end{vmatrix} = 0$$

that is

$$\Omega \begin{vmatrix} x_1^3, & x_1^2, & x_1, & 1 \\ x_2^3, & x_2^2, & x_2, & 1 \\ x_3^3, & x_3^2, & x_3, & 1 \\ x_4^3, & x_4^2, & x_4, & 1 \end{vmatrix} + \sqrt{\lambda} \begin{vmatrix} x_1^3, & x_1^2, & x_1, & 1, & \sqrt{X_1} \\ x_2^3, & x_2^2, & x_2, & 1, & \sqrt{X_2} \\ x_3^3, & x_3^2, & x_3, & 1, & \sqrt{X_3} \\ x_4^3, & x_4^2, & x_4, & 1, & \sqrt{X_4} \end{vmatrix} = 0,$$

viz. this is

$$\begin{aligned} \Omega\,(x_1-x_2)(x_1-x_3)(x_1-x_4)(x_2-x_3)(x_2-x_4)(x_3-x_4) \\ = -\sqrt{\lambda}\{\sqrt{X_1}.x_2-x_3.x_2-x_4.x_2-a.x_3-x_4.x^3-a.x_4-a \\ +\sqrt{X_2}.x_3-x_4.x_3-a.x_3-x_1.x_4-a.x_4-x_1.a-x_1 \\ +\sqrt{X_3}.x_4-a.x_4-x_1.x_4-x_2.a-x_1.a-x_2.x_1-x_2 \\ +\sqrt{X_4}.a-x_1.a-x_2.a-x_3.x_2-x_3.x_2-x_4.x_3-x_4\}, \end{aligned}$$

or as this may be written

$$\begin{aligned} \Omega.x_1-x_3.x_1-x_4.x_2-x_3.x_2-x_4 \\ = \frac{\sqrt{\lambda}.a-x_3.a-x_4}{x_1-x_2}\{x_2-x_3.x_2-x_4.a-x_2.\sqrt{X_1}-.x_1-x_3.x_1-x_4.a-x_1.\sqrt{X_2}\} \\ + \frac{\sqrt{\lambda}.a-x_1.a-x_2}{x_3-x_4}\{x_4-x_1.x_4-x_2.a-x_4.\sqrt{X_3}-.x_3-x_1.x_3-x_2.a-x_3.\sqrt{X}\}, \end{aligned}$$

we have here $\sqrt{\lambda}.a-x_3.a-x_4 = \sqrt{\lambda}\,A_{34}^2$, and the function

$$\frac{1}{x_1-x_2}\{x_2-x_3.x_2-x_4.\mathfrak{a}_2\sqrt{X_1}-.x_1-x_3.x_1-x_4.\mathfrak{a}_1\sqrt{X_2}\},$$

which multiplies this is without difficulty found to be

$$= \frac{-A_{12}}{c-d.d-b.b-c}\Sigma\{c-d.B_{34}^2 BE_{12}.C_{12}.D_{12}\},$$

where the summation extends to the three terms obtained by the cyclical interchange of the letters b, c, d: these being a set of three out of the five letters other than a. Similarly $\sqrt{\lambda}.a-x_1.a-x_2$ is $=\sqrt{\lambda}A_{12}^2$, and the function which multiplies this is

$$= \frac{-A_{34}}{c-d.d-b.b-c}\Sigma\{c-d.B_{12}^2.BE_{34}C_{14}D_{14}\},$$

the expression for Ω thus contains the factor $A_{12}A_{34}$. But we have

$$\Omega, = a\alpha^3 + b\alpha^2 + c\alpha + d, = \sqrt{\mu}A_{12}A_{34}A_{56};$$

this equation contains therefore the factor $A_{12}A_{34}$, and omitting it we find

$$-\frac{\sqrt{\mu}}{\sqrt{\lambda}}(x_1-x_3.x_1-x_4.x_2-x_3.x_2-x_4)(c-d.d-b.b-c)A_{56}$$
$$= A_{34}\Sigma\{c-d.B_{34}^2 BE_{12}C_{12}D_{12}\} + A_{12}\Sigma\{c-d.B_{12}^2 BE_{34}C_{34}D_{14}\},$$

where as before the summations refer each to the three terms obtained by the cyclical interchange of the letters b, c, d; these being any three of the five letters other than a: and the remaining two letters e, f enter into the formulæ symmetrically. The formula thus gives for A_{56} ten values which are of course equal to each other.

By reason of the undetermined factor $\frac{\sqrt{\mu}}{\sqrt{\lambda}}$ the formula gives only the proportional value of A_{56}; viz. combining it with the like formulæ for B_{56}, etc. we have determinate values of the ratios $A_{56} : B_{56} \ldots : F_{56}$. But this being understood, we regard the formula as a formula for each single-letter function of x_5, x_6 in terms of the single and double-letter functions of x_1, x_2 and of x_3, x_4 respectively

163. We require further the expressions for the double-letter functions of x_5, x_6. Consider for example the function DE_{56} which is

$$= \frac{1}{x_5-x_6}\{\sqrt{\mathrm{d}_6\mathrm{e}_5\mathrm{f}_5\mathrm{a}_6\mathrm{b}_6\mathrm{c}_6} - \sqrt{\mathrm{d}_6\mathrm{e}_6\mathrm{f}_6\mathrm{a}_5\mathrm{b}_5\mathrm{c}_5}\}$$

then multiplying by $A_{56}B_{56}C_{56}, = \sqrt{\mathrm{a}_5\mathrm{b}_5\mathrm{c}_5\mathrm{a}_6\mathrm{b}_6\mathrm{c}_6}$, we have

$$DE_{56}A_{56}B_{56}C_{56} = \frac{1}{x_5-x_6}\{\mathrm{a}_6\mathrm{b}_6\mathrm{c}_6\sqrt{X_5} - \mathrm{a}_5\mathrm{b}_5\mathrm{c}_5\sqrt{X_6}\},$$

or recollecting that $\sqrt{\lambda}\sqrt{X_5}$ and $\sqrt{\lambda}\sqrt{X_6}$ are $=\alpha x_5^3+\beta x_5^2+\gamma x_5+\delta$, and $\alpha x_6^3+\beta x_6^2+\gamma x_6+\delta$ respectively, this may be written

$$\sqrt{\lambda}\,DE_{56}A_{56}B_{56}C_{56} = \frac{1}{x_5-x_6}\{a-x_6.b-x_6.c-x_6.(\alpha x_5^3+\beta x_5^2+\gamma x_5+\delta)$$
$$-.a-x_5.b-x_5.c-x_5.(\alpha x_6^3+\beta x_6^2+\gamma x_6+\delta)\}.$$

Using the well-known identity

$$\alpha x_5^3 + \beta x_5^2 + \gamma x_5 + \delta = \Sigma . \alpha a^3 + \beta a^2 + \gamma a + \delta . \frac{b - x_5 . c - x_5 . d - x_5}{b - a . c - a . d - a},$$

where the summation extends to the four terms obtained by the cyclical interchanges of the letters a, b, c, d: and the like identity for $\alpha x_6^3 + \beta x_6^2 + \gamma x_6 + \delta$, there will be terms in $\alpha a^3 + \beta a^2 + \gamma a + \delta$, $\alpha b^3 + \beta b^2 + \gamma b + \delta$, $\alpha c^3 + \beta c^2 + \gamma c + \delta$, but the term in $\alpha d^3 + \beta d^2 + \gamma d + \delta$ will disappear of itself. After some easy reductions the result is

$$\sqrt{\lambda}\, DE_{56} A_{56} B_{56} C_{56} = \Sigma \frac{\alpha a^3 + \beta a^2 + \gamma a + \delta}{b - a . c - a} B_{56}^2 C_{56}^2,$$

where the summation extends to the three terms obtained by the cyclical interchanges of the letters a, b, c. We have $\alpha a^3 + \beta a^2 + \gamma a + \delta = \sqrt{\mu} . A_{12} A_{34} A_{56}$, and similarly for the other two terms; the whole equation thus divides by $A_{56} B_{56} C_{56}$, and we find

$$-\frac{\sqrt{\mu}}{\sqrt{\lambda}} DE_{56} = \frac{1}{b - c . c - a . a - b} \left(\frac{\sqrt{\mu}}{\sqrt{\lambda}}\right)^2 . \Sigma (b - c . A_{12} A_{34} B_{56} C_{56},$$

in which equation, if we imagine $\frac{\sqrt{\mu}}{\sqrt{\lambda}} A_{56}$, $\frac{\sqrt{\mu}}{\sqrt{\lambda}} B_{56}$, $\frac{\sqrt{\mu}}{\sqrt{\lambda}} C_{56}$, each replaced by its value in terms of the single and double-letter functions of x_1, x_2 and x_3, x_4, we have an equation of the form

$$-\frac{\sqrt{\mu}}{\sqrt{\lambda}} (x_1 - x_3 . x_1 - x_4 . x_2 - x_3 . x_2 - x_4)\, DE_{56} = \frac{1}{x_1 - x_3 . x_1 - x_4 . x_2 - x_3 . x_2 - x_4} M,$$

where M is a given rational and integral function of the single and double-letter functions of x_1, x_2 and x_3, x_4. The factor on the left-hand side has been made the same as in the formula for the single-letter functions A_{56}, etc., and to do this it was necessary to bring in on the right-hand side the factor

$$\frac{1}{x_1 - x_3 . x_1 - x_4 . x_2 - x_3 . x_2 - x_4};$$

this disappears in the expression for the ratio of two double-letter functions; but it enters into the expression for the ratio of a single-letter to a double-letter function, and it then requires to be itself expressed in terms of the functions of x_1, x_2 and x_3, x_4: it is easy to see that we have

$$x_1 - x_3 . x_1 - x_4 . x_2 - x_3 . x_2 - x_4 = \Sigma \frac{(A_{34}^2 B_{12}^2 - A_{12}^2 B_{34}^2)(A_{34}^2 C_{12}^2 - A_{12}^2 C_{34}^2)}{(a - b)^2 (a - c)^2},$$

where the summation extends to the three terms obtained by the cyclical interchanges of the letters a, b, c: these being a set of any three out of the six letters.

We have, in what precedes, obtained the expressions for the ratios of the 16 functions $A_{56}, \ldots F_{56}, AB_{56}, \ldots DE_{56}$ in terms of the ratios of the like functions of x_1, x_2 and x_3, x_4.

CHAPTER VII. THE FUNCTIONS T, U, V, θ.

The present chapter is substantially a reproduction of C. & G.'s seventh section, "Die Function $T_{\xi,\eta}$" (borrowing only from the next section the definition of the theta-function), but for greater simplicity I consider for the most part, the case, fixed curve a quartic; $n=4$, $p=3$.

Integral Form of the Affected Theorem. Art. Nos. 164 to 169.

164. Writing for shortness $\dfrac{(x,y,z)_{12}^{n-2}\,d\omega}{\dot{0}\dot{1}2}=d\Pi_{12}$, we are concerned with the integrals $\int_{a'}^{a} d\Pi_{12}$ which present themselves in connection with the affected theorem: the notation is explained, Chap. V; a, a' are points on the curve f; the variable may be any parameter serving for the determination of the current point, and the integral, taken from the value which belongs to the point a' to the value which belongs to the point a, is represented as above by means of the two points a, a' as limits of the integral. It is assumed that the integral is a canonical integral having the limits and the parametric points interchangeable, $\int_{a'}^{a} d\Pi_{12}=\int_{2}^{1} d\Pi_{aa'}$: see Chapter IV.

165. Writing for shortness

$$\left(\int_{a'}^{a}+\int_{b'}^{b}+\int_{c'}^{c}+\ldots\right)d\Pi_{12}=\int\binom{a,\,b,\,c\,\ldots}{a',\,b',\,c'\,\ldots}d\Pi_{12}$$

then if ϕ, ψ are curves each of the order m, the former of them intersecting the fixed curve f in the points $a, b, c \ldots$, and the latter of them intersecting the same curve in the points $a', b', c' \ldots$, and if ϕ_1, ψ_1, ϕ_2, ψ_2 are what the functions ϕ, ψ become on substituting therein in place of the current coordinates the values which belong to the parametric points 1, 2 respectively; the theorem becomes

$$\int\binom{a,\,b,\,c\,\ldots}{a',\,b',\,c'\,\ldots}d\Pi_{12}=\log\frac{\varphi_2\psi_1}{\varphi_1\psi_2}.$$

The superior limits may be interchanged in any manner, and so also the inferior limits may be interchanged in any manner. If a superior limit coincide with an inferior limit, the two may thus be considered as belonging to an integral which will then have the value 0, and the coincident points may therefore be omitted from the expression on the left-hand side: and so in the case of any number of coincidences.

166. If the intersections of the curves ϕ, ψ and the parametric points are situate on a curve of the order m; then taking the equation of this curve to be $\phi+\lambda\psi=0$, we have simultaneously $\phi_1+\lambda\psi_1=0$, $\phi_2+\lambda\psi_2=0$; whence $\phi_2\psi_1=\psi_1\phi_2$, and the logarithmic term disappears: viz. the theorem becomes

$$\int\binom{a,\ b,\ c\ldots}{a',\ b',\ c'\ldots}d\Pi_{12}=0.$$

167. Suppose that the curves ϕ, ψ are each of them a major curve, that is a curve of the order $n-2$ passing through the δ dps, and consequently besides meeting the curve f in $n(n-2)-2\delta$, $=2p+n-2$ points: the theorem is

$$\int\binom{a,\ b,\ c\ldots}{a',\ b',\ c'\ldots}d\Pi_{12}=\log\frac{\varphi_2\psi_1}{\varphi_1\psi_2},$$

where the numbers of the superior and of the inferior points are each $=2p+n-2$.

168. Suppose further that the curves ϕ, ψ, being major curves as above, pass each of them through the $n-2$ residues of 1, 2; they besides meet in $(n-2)(n-3)$ points (viz. these are the δ dps and $(n-2)(n-3)-\delta$ variable points): these $(n-2)(n-3)$ points lie on a minor curve, that is a curve of the order $n-3$ passing through the dps; and the minor curve together with the parametric line 12 make together a major curve passing through the intersections of ϕ, ψ and also through the parametric points 1, 2: viz. these points and the intersections of ϕ, ψ are situate on a curve of the order $n-2$; the logarithmic term thus vanishes. The intersections of ϕ with the fixed curve are the δ dps, the $n-2$ residues and $2p$ other points, say these are $a, b, c\ldots, a^{\times}, b^{\times}, c^{\times},\ldots$; similarly the curve ψ meets the fixed curve in the δ dps, the $n-2$ residues, and in $2p$ other points, say these are $d, e, f,\ldots, d^{\times}, e^{\times}, f^{\times},\ldots$: the theorem is

$$\int\binom{a,\ b,\ c\ldots a^{\times},\ b^{\times},\ c^{\times}\ldots}{d,\ e,\ f\ldots d^{\times},\ e^{\times},\ f^{\times}\ldots}d\Pi_{12}=0,$$

where there are $2p$ superior and inferior points respectively.

169. I introduce the definitions: a minor curve meets the fixed curve in the dps and in $2p-2$ other points, called "cominors": a major curve passing through the $n-2$ residues of the points 1, 2, meets the fixed curve in the δ dps, the $n-2$ residues and in $2p$ other points, called "comajors in regard to the points 1, 2". Observe that $p-1$ of the cominors determine uniquely the remaining $p-1$ cominors; and similarly p of the comajors determine uniquely the remaining p comajors.

The foregoing theorem thus is that the sum $\int\left(\;\right)d\Pi_{12}$ is $=0$, when the superior points and the inferior points are each of them a system of comajors in regard to the parametric points 1, 2.

Fixed Curve a Quartic. Art. No. 170.

170. It would be easy to go on with the general form, but as already mentioned, I prefer to consider the case, fixed curve a quartic; $n=4$, $p=3$. A minor curve is here a line meeting the quartic in 4 points, which are "cominors"; the major curve is a conic, and if this passes through the residues of 1, 2 it besides meets the quartic in 6 points, which are "comajors in regard to the points 1, 2". Two points and their residues are cominors, but this is only by reason that $n-3=1$.

The Function T. Art. No. 171.

171. In conformity with C. & G. I introduce the functional symbol

$$T_{12}\begin{pmatrix} a, & b, & c\ldots \\ a', & b', & c'\ldots \end{pmatrix} = \int \begin{pmatrix} a, & b, & c\ldots \\ a', & b', & c'\ldots \end{pmatrix} d\Pi_{12},$$

so that T denotes a function of the parametric points, and of the sets of superior and inferior points respectively. The foregoing theorem for the quartic thus is

$$T_{12}\begin{pmatrix} a, & b, & c, & a^{\times}, & b^{\times}, & c^{\times} \\ d, & e, & f, & d^{\times}, & e^{\times}, & f^{\times} \end{pmatrix} = 0.$$

Observing that $\int_d^a + \int_{d^\times}^{a^\times} = 2\int_d^a - \int_{a^\times}^a + \int_{d^\times}^d$, and so in other cases, this may be written

$$2T_{12}\begin{pmatrix} a, & b, & c \\ d, & e, & f \end{pmatrix} = T_{12}\begin{pmatrix} a, & b, & c \\ a^{\times}, & b^{\times}, & c^{\times} \end{pmatrix} - T_{12}\begin{pmatrix} d, & e, & f \\ d^{\times}, & e^{\times}, & f^{\times} \end{pmatrix},$$

and if as a definition of $T_{12}(a, b, c)$, we write

$$T_{12}(a, b, c) = T_{12}\begin{pmatrix} a, & b, & c \\ a^{\times}, & b^{\times}, & c^{\times} \end{pmatrix},$$

where $a^{\times}$, $b^{\times}$, $c^{\times}$ are the comajors of a, b, c in regard to 1, 2, then the equation is

$$2T_{12}\begin{pmatrix} a, & b, & c \\ d, & e, & f \end{pmatrix} = T_{12}(a, b, c) - T_{12}(d, e, f),$$

viz. the function of the $(2p+2=)\,8$ points $1, 2, a, b, c, d, e, f$ is here expressed as a difference of two functions each of $(p+2=)\,5$ points: $T_{12}(a, b, c)$ is regarded as a function of the 5 points $1, 2, a, b, c$, because the remaining points $a^{\times}$, $b^{\times}$, $c^{\times}$ depend only on these 5 points.

The Function U. Art. Nos. 172 to 175.

172. We consider on the quartic the points ξ, μ; 1, 2, 3; and taking f, f' for the cominors of 2, 3; g, g' for the cominors of 3, 1; and h, h' for the

cominors of 1, 2, we write

$$T = T_{\xi\mu}(1, 2, 3),$$

$$T_1 = T_{1\mu}(\xi, f, f'),\ T_2 = T_{2\mu}(\xi, g, g'),\ T_3 = T_{3\mu}(\xi, h, h');$$

it is to be shown that there exists a function $U(1, 2, 3; \xi)$ such that

$$\delta U = \tfrac{1}{2}\{\delta_\xi T + \delta_1 T_1 + \delta_2 T_2 + \delta_3 T_3\},$$

viz. considering ξ, 1, 2, 3 as variable points on the quartic, the whole infinitesimal variation of U is the sum of these parts, where $\delta_\xi T$ is the variation of T when only ξ is varied, $\delta_1 T_1$ the variation of T_1 when only 1 is varied, and similarly for $\delta_2 T_2$ and $\delta_3 T_3$. We consider in the proof three other points 4, 5, 6 on the quartic; and taking l, l' for the cominors of 5, 6; m, m' for those of 6, 4; and n, n' for those of 4, 5, we write further

$$X_1 = T_{1\mu}(4, l, l'),\ X_2 = T_{2\mu}(5, m, m'),\ X_3 = T_{3\mu}(6, n, n'),$$

and it then requires to be shown that

$$\tfrac{1}{2}T \qquad = \int \binom{123}{456} d\Pi_{\xi\mu} + \tfrac{1}{2}T_{\xi\mu}(4, 5, 6),$$

$$\tfrac{1}{2}(T_1 - X_1) = \int \binom{\xi 56}{423} d\Pi_{1\mu} + \log\frac{\mu 23.156}{\mu 56.123},$$

$$\tfrac{1}{2}(T_2 - X_2) = \int \binom{\xi 64}{531} d\Pi_{2\mu} + \log\frac{\mu 31.264}{\mu 64.123},$$

$$\tfrac{1}{2}(T_3 - X_3) = \int \binom{\xi 45}{612} d\Pi_{3\mu} + \log\frac{\mu 12.345}{\mu 45.123},$$

where $\mu 23$ is the determinant formed with the coordinates of the points μ, 2, 3 respectively: and so in other cases.

173. We have $$T_{\xi\mu}\begin{pmatrix}1, 2, 3\\ 4, 5, 6\end{pmatrix} = \tfrac{1}{2}T_{\xi\mu}(1, 2, 3) - \tfrac{1}{2}T_{\xi\mu}(4, 5, 6),$$

that is $$= \tfrac{1}{2}T \qquad - \tfrac{1}{2}T_{\xi\mu}(4, 5, 6),$$

and thence the above value of $^1 T$.

The affected theorem gives

$$\int \begin{pmatrix}f, f', 2, 3\\ l, l', 5, 6\end{pmatrix} d\Pi_{1\mu} = \log\frac{F_\mu L_1}{F_1 L_\mu},$$

where $F = 0$ is the equation of the line through f, f', 2, 3; and F_1, F_μ are what the function F becomes on substituting therein for the current coordinates the coordinates of the points 1, μ respectively. And similarly $L = 0$ is the equation of the line through l, l', 5, 6; and L_1, L_μ are what the function L becomes by the same substitutions respectively. The values of F_1, F_2 are 123, $\mu 23$: those of L_1, L_μ are 156, $\mu 56$, and the logarithmic term is thus

$$= \log\frac{\mu 23.156}{\mu 56.123}.$$

We then have

$$\tfrac{1}{2}(T_1-X_1)=\int\begin{pmatrix}\xi, f, f'\\ 4, l, l'\end{pmatrix}d\Pi_{1\mu}, =\int\begin{pmatrix}\xi, 5, 6\\ 4, 2, 3\end{pmatrix}d\Pi_{1\mu}+\int\begin{pmatrix}f, f', 2, 3\\ l, l', 5, 6\end{pmatrix}d\Pi_{1\mu},$$

and in this last expression for $\tfrac{1}{2}(T_1-X_1)$ substituting for the second term the logarithmic value just obtained we have the required expression for $\tfrac{1}{2}(T_1-X_1)$: and those for $\tfrac{1}{2}(T_2-X_2)$ and $\tfrac{1}{2}(T_3-X_3)$, are deduced by mere cyclical permutations of the letters.

174. Returning to the assumed relation $\delta U=\tfrac{1}{2}\{\delta_\xi T+\delta_1 T_1+\delta_2 T_2+\delta_3 T_3\}$; in order to the existence of the function U, it is only necessary to show that $T-T_1$ contains no term in $1, \xi$ (that is no term depending on both these points), and that T_1-T_2 contains no term in $1, 2$: for then by symmetry the like properties hold in regard to $T-T_2$, $T-T_3$, T_1-T_3, T_2-T_3 respectively, and the assumed expression is a complete differential, from which the function U may be obtained by integration.

175. To show that $T-T_1$ contains no term in $1, \xi$.

For T, the only term in $1, \xi$ is $\int_4^1 d\Pi_{\xi\mu}$,

" T_1 " " $\int_4^\xi d\Pi_{1\mu}$,

and it is to be shown that the difference of the two integrals contains no term in $1, \xi$. Considering on the quartic the two new points δ, ϵ, the first integral is

$$\int\begin{pmatrix}1, \delta\\ \delta, 4\end{pmatrix}(d\Pi_{\xi\epsilon}+d\Pi_{\epsilon\mu}), =\int_\delta^1 d\Pi_{\xi\epsilon}+\int_4^\delta d\Pi_{\xi\epsilon}+\int_4^1 d\Pi_{\epsilon\mu},$$

and the second is

$$\int\begin{pmatrix}\xi, \epsilon\\ \epsilon, 4\end{pmatrix}(d\Pi_{1\delta}+d\Pi_{\delta\mu}), =\int_\epsilon^\xi d\Pi_{1\delta}+\int_4^\epsilon d\Pi_{\delta\mu}+\int_\mu^\xi d\Pi_{\delta\mu}.$$

Hence in the difference the only terms which can contain $1, \xi$ is

$$\int_\delta^1 d\Pi_{\xi\epsilon}-\int_\epsilon d\Pi_{1\delta}$$

and this term is $=0$: wherefore there is not in the difference any term in $1, \xi$. This proves the property for $T-T_1$. The property for T_1-T_2 is proved in a similar manner.

Theorems in regard to the Function U. Art. Nos. 176 to 179.

176. Theorem (A).

$$U(1, 2, 3;\ \xi)-U(1, 2, 3;\ \mu)=\tfrac{1}{2}\,T_{\xi\mu}(1, 2, 3), \qquad \text{(A)}$$

we have

$$U(1, 2, 3;\ \xi)-U(1, 2, 3;\ \mu)=\int_\mu^\xi d_\xi U=\tfrac{1}{2}\int_\mu^\xi d_\xi T$$

$$=\tfrac{1}{2}T_{\xi\mu}(1, 2, 3)-\tfrac{1}{2}T_{\mu\mu}(1, 2, 3),$$

and $T_{\mu\mu}(1, 2, 3) = \int \begin{pmatrix} 1, & 2, & 3 \\ 1^\times & 2^\times & 3^\times \end{pmatrix} d\Pi_{\mu\mu}$, where $d\Pi_{\mu\mu} = 0$, viz. considering this as derived from $d\Pi_{\xi\mu}, = Q_{\xi\mu}\, d\omega$, by making the point ξ coincide with μ, then when ξ is indefinitely near to μ, the numerator and denominator of $Q_{\xi\mu}$ are each of them infinitesimal of the orders 3 and 2 respectively, and thus the function $Q_{\xi\mu}$ ultimately vanishes (see as to this, Chap. V. Art. Nos. 99 to 106). We have therefore $T_{\mu\mu}(1, 2, 3) = 0$, and the required theorem is proved.

177. Theorem (B).

$$U(1, 2, 3;\, \xi) - U(4, 2, 3;\, \xi) = \tfrac{1}{2} T_{14}(\xi, f, f'), \qquad \text{(B)}$$

where as before f, f' are the cominors of 2, 3, that is 2, 3, f, f' lie on a line. We have

$$U(1, 2, 3;\, \xi) - U(4, 2, 3;\, \xi) = \int_4^1 d_1 U = \tfrac{1}{2}\int_4^1 d_1 T_1$$

$$= \tfrac{1}{2} T_{1\mu}(\xi, f, f') - \tfrac{1}{2} T_{4\mu}(\xi, f, f'),$$

the point μ is arbitrary, and it may be taken to coincide with 4; but we then have $T_{44}(\xi, f, f') = 0$, and the theorem is thus proved.

178. Theorem (C). $T_{11^\times}(\xi, f, f') + T_{11^\times}(\eta, k, k') = 0$: (C)

where ξ, η, 1, 2, 3 are arbitrary points on the quartic; $1^\times, 2^\times, 3^\times$ are the comajors of 1, 2, 3 in regard to ξ, η, viz. the points 1, 2, 3, $1^\times, 2^\times, 3^\times$ lie on a conic which passes through ξ', η' the residues (or cominors) of ξ, η: f, f' are the cominors of 2, 3; and k, k' are the cominors of $2^\times, 3^\times$.

Taking θ, θ' for the cominors of 1, $1^\times$, the four lines $\xi\eta\xi'\eta'$, $11^\times\theta\theta'$, $23ff'$ and $2^\times 3^\times kk'$ form a quartic cutting the fixed quartic in the 16 points: but of these ξ', η', 1, 2, 3, $1^\times, 2^\times, 3^\times$ lie in a conic: hence the remaining 8 points $\theta, \theta', \xi, \eta, f, f', k, k'$ lie in a conic; that is, ξ, η, f, f', k, k' lie on a conic through θ, θ', the residues of 1, $1^\times$, or they are comajors in regard to 1, $1^\times$; whence the theorem.

179. We have

From A. From B.

$$\tfrac{1}{2} T_{11^\times}(\xi, f, f') = U(\xi, f, f';\, 1) - U(\xi, f, f';\, 1^\times), = U(1, 2, 3;\, \xi) - U(1^\times, 2, 3;\, \xi),$$

$$\tfrac{1}{2} T_{11^\times}(\eta, k, k') = U(\eta, k, k';\, 1) - U(\eta, k, k';\, 1^\times), = U(1, 2^\times, 3^\times;\, \eta) - U(1^\times, 2, 3;\, \eta),$$

viz. we have thus two expressions for each term of the equation (C),

$$T_{11^\times}(\xi, f, f') + T_{11^\times}(\eta, k, k') = 0.$$

In particular we have Theorem (D)

$$U(1, 2, 3;\, \xi) - U(1^\times, 2, 3;\, \xi) = -U(1, 2^\times, 3^\times;\, \xi) + U(1^\times, 2^\times, 3^\times;\, \eta). \qquad \text{(D)}$$

Again we have $T_{\xi\eta}(1, 2, 3) + T_{\xi\eta}(1^\times, 2^\times, 3^\times) = 0$; where 1, 2, 3, $1^\times, 2^\times, 3^\times$ are comajors in regard to ξ, η: and

$$\tfrac{1}{2} T_{\xi\eta}(1, 2, 3) = U(1, 2, 3;\, \xi) - U(1, 2, 3;\, \eta),$$

$$\tfrac{1}{2} T_{\xi\eta}(1^\times, 2^\times, 3^\times) = U(1^\times, 2^\times, 3^\times, \xi) - U(1^\times, 2^\times, 3^\times, \eta),$$

whence Theorem (E),

$$U(1, 2, 3; \xi) - U(1, 2, 3; \eta) = -U(1^\times, 2^\times, 3^\times; \xi) + U(1^\times, 2^\times, 3^\times, \eta). \quad \text{(E)}$$

The Function V. Art. Nos. 180 to 182.

180. It is convenient to consider U as a logarithm, say $-U(1, 2, 3; \xi) = \log V(1, 2, 3; \xi)$, or $V(1, 2, 3; \xi) = \exp. -U(1, 2, 3; \xi)$. V, like U, is a function of the $(p+1=)4$ points $1, 2, 3; \xi$, on the quartic.

The equation (D) thus becomes

$$\frac{V(1, 2, 3; \xi)}{V(1^\times, 2^\times, 3^\times; \eta)} = \frac{V(1^\times, 2, 3; \xi)}{V(1, 2^\times, 3^\times; \eta)},$$

where $1, 2, 3, 1^\times, 2^\times, 3^\times$ are comajors in regard to ξ, η: the equation shows that in the function $\dfrac{V(1, 2, 3; \xi)}{V(1^\times, 2^\times, 3^\times; \eta)}$ we can without alteration of the value interchange a pair of points $1, 1^\times$ out of the system of comajor points; and it of course follows that we can in any manner whatever interchange these points, so as to have any three of them in the numerator function and the remaining three in the denominator function. In particular we have

$$\frac{V(1, 2, 3; \xi)}{V(1^\times, 2^\times, 3^\times; \eta)} = \frac{V(1^\times, 2^\times, 3^\times; \xi)}{V(1, 2, 3; \eta)}.$$

The equation (E) becomes

$$\frac{V(1, 2, 3; \xi)}{V(1, 2, 3; \eta)} = \frac{V(1^\times, 2^\times, 3^\times; \eta)}{V(1^\times, 2^\times, 3^\times; \xi)},$$

and multiplying we find

$$V^2(1, 2, 3; \xi) = V^2(1^\times, 2^\times, 3^\times; \eta),$$

that is $V(1, 2, 3; \xi) = \pm V(1^\times, 2^\times, 3^\times; \eta)$, the sign being determinately $+$ or determinately $-$, according to the precise definition of the function V.

181. Considering η and also $1^\times, 2^\times, 3^\times$ as fixed points on the curve; but ξ as a variable point (that is, the parametric line $\xi\eta$ as rotating about the fixed point η), the points $1, 2, 3$ are then determined as the remaining intersections with the quartic, of the conic which passes through the points $1^\times, 2^\times, 3^\times$ and the points ξ', η' which are the residues of ξ, η. And by the theorem just obtained it appears that, $\xi, 1, 2, 3$ thus varying, the function $V(1, 2, 3; \xi)$ remains constant. This comes to saying that V considered as a function of the points $1, 2, 3, \xi$ satisfies a certain linear partial differential equation of the first order, having a solution $V = F(u, v, w)$, an arbitrary function of u, v, w, determinate functions of the points $1, 2, 3, \xi$. And if we can find u, v, w functions of these points

such that they each of them remain constant when the points $1, 2, 3, \xi$ vary as above, then the arbitrary function of u, v, w will remain constant for the variation in question and will thus be a value of the function V.

182. It is easily seen that such functions are

$$u, v, w = \left(\int^1 + \int^2 + \int^3 - \int^\xi\right) xd\omega,\ yd\omega,\ zd\omega,$$

the inferior limit being given points which are regarded as absolute constants. For by the pure theorem we have

$$\Sigma\,(x, y, z)^1 d\omega = 0,$$

where $(x, y, z)^1$ is an arbitrary linear function, and where the summation extends to all the intersections of the quartic with any given curve. Writing

$$p = \int xd\omega,\ \int yd\omega \text{ or } \int zd\omega, \text{ that is, } p_1 = \int^1 xd\omega,\ \int^1 yd\omega \text{ or } \int^1 zd\omega,$$

the inferior limits being any absolutely fixed point on the curve, and similarly p_2, etc.; the integral form of the theorem is $\Sigma p =$ constant. And applying the theorem successively to the parametric line, and to the conic which determines the points $1, 2, 3$, we have

$$p_\xi + p_\eta + p_{\xi'} + p_{\eta'} = \text{const.},$$
$$p_{\xi'} + p_{\eta'} + p_1 + p_2 + p_3 + p_{1^\times} + p_{2^\times} + p_{3^\times} = \text{const.}$$

Taking the difference of these equations

$$p_1 + p_2 + p_3 - p_\xi + p_{1^\times} + p_{2^\times} + p_{3^\times} - p_\eta = \text{const.},$$

viz. the points $\eta, 1^\times, 2^\times, 3^\times$ being fixed points, this is

$$p_1 + p_2 + p_3 - p_\xi = \text{const.},$$

that is, the functions u, v, w defined as above are each of them constant under the variation in question.

The Function Θ. Art. Nos. 183 and 184.

183. The function $V(1, 2, 3; \xi)$ of the $(p + 1 =) 4$ points $1, 2, 3, \xi$, is thus a function of the $(p =) 3$ arguments

$$u, v, w, = \left(\int^1 + \int^2 + \int^3 - \int^\xi\right) xd\omega,\ yd\omega,\ zd\omega.$$

Disregarding a constant and exponential factor we say that it is a theta-function of these arguments, and we write the result provisionally in the form

$$V(1, 2, 3; \xi) = \Theta(u, v, w),$$

the more precise definition of the theta-function being reserved for further consideration.

184. It appears by what precedes that a sum of $(p =) \, 3$ integrals $\int \begin{pmatrix} a & b & c \\ d & e & f \end{pmatrix} d\Pi_{13}$, otherwise called $T_{13} \begin{pmatrix} a, & b, & c \\ d, & e, & f \end{pmatrix}$ is in the first place expressed (see No. 171) as a difference, $= \frac{1}{2} T_{13}(a, b, c) - \frac{1}{2} T_{13}(d, e, f)$ of two functions T. Each of these is by theorem (A) (No. 176) expressed as a difference of two functions U, that is as the difference of the logarithms, or logarithm of the quotient, of two functions V: such function V is according to its original definition a function of $(p + 1 =) \, 4$ points, but in such wise that the function is expressable as a function of $(p =) \, 3$ arguments, and so expressed it is a Θ-function of these arguments: and the final result thus is that the sum of $(p =) \, 3$ integrals $\int \begin{pmatrix} a, & b, & c \\ d, & e, & f \end{pmatrix} d\Pi_{13}$ is equal to the logarithm of a fraction, whereof the numerator and denominator are each of them a product of two Θ-functions.

END OF CHAPTER VII.

Extrait d'une Lettre de M. Hermite

adressée à un étudiant de l'Université Johns Hopkins.

"... Permettez moi de vous indiquer une remarque qui peut-être vous intéressera et que je place dans mes leçons, immédiatement après avoir donné la formule de Maclaurin:

$$f(x)=f(0)+\frac{x}{1}f'(0)+\frac{x^2}{1.2}\cdot f^{(2)}(0)+\dots\frac{x^{n-1}}{1.2.3\dots n-1}f^{(n-1)}(0)+J,$$

ou l'on a:

$$J=\frac{1}{2i\pi}\int\frac{x^n f(z)\,dz}{z^n(z-x)},$$

l'integrale étant prise le long d'un contour qui comprend à son interieur le point dont l'affixe est x.

"On a donc en designant par σ le périmètre du contour et par λ le facteur de M. Darboux, et par ζ l'affixe d'un point du contour:

$$J=\frac{\lambda\sigma}{2\pi}\frac{x^n f(\zeta)}{\zeta^n(\zeta-x)}.$$

"Prenez maintenant pour contour une circonference de rayon r, vous avez: $\sigma=2\pi r$, $\zeta=re^{i\theta}$, ce qui permet d'écrire: $\sigma=2\pi\zeta e^{-i\theta}$, et par consequent:

$$J=\frac{\lambda e^{-i\theta}x^n f(\zeta)}{\zeta^{n-1}(\zeta-x)}.$$

"Cela étant supposons que la fonction $f(z)$ soit holomorphe on pourra sans altérer J, faire croître indéfiniment le rayon r, et on voit ainsi que lorsque $\frac{f(\zeta)}{\zeta^{n-1}}$ aura une limite finie pour ζ infinie, J est nul, de sorte que sous cette condition, la fonction holomorphe $f(z)$ est un polynôme entier du degrè $n-1$."

The proposition of M. Hermite can also be proved in the two following ways:

In both demonstrations I shall suppose that $f(z)$ does not vanish for $z=0$; which does not lessen the generality of the demonstration, for if $f(0)=0$, I may consider $\frac{f(z)}{(z-a)^{n-1}}$, $f(a)$ being different from zero, and may apply to $\frac{f(z)}{(z-a)^{n-1}}$ the very same reasoning which I use with regard to $\frac{f(z)}{z^{n-1}}$.

1. Because $\frac{f(z)}{z^{n-1}}$ is a holomorphic function in all the plane except at the point $z=0$, which is a pole of the $(n-1)^{th}$ order of multiplicity, I construct a function

$$G\left(\frac{1}{z}\right)=\frac{A_1}{z}+\frac{A_2}{z^2}+\ldots\frac{A_{n-1}}{z^{n-1}};\text{ such that }\left[\frac{f(z)}{z^{n-1}}-G\left(\frac{1}{z}\right)\right]$$

does not become infinite for $z=0$.

Now the function $\left[\frac{f(z)}{z^{n-1}}-G\left(\frac{1}{z}\right)\right]$ being holomorphic throughout all the plane, and being for $z=\infty$, equal to a finite value C; then because $G\left(\frac{1}{z}\right)=0$, and $\frac{f(z)}{z^{n-1}}=C$, for $z=\infty$, the function must be a constant, therefore equal to C; also $f(z)=Cz^{n-1}+A_1z^{n-2}+\ldots A_{n-1}$.

2. The function $\frac{f(z)}{z^{n-1}}$ being meromorphic over the whole sphere must be a rational fraction, which I shall suppose reduced to its most simple expression $\frac{p(z)}{q(z)}$; $p(z)$ and $q(z)$ being two integral polynomials.

Now it is evident that $p(z)$ and $q(z)$ are of the same degree, then for $z=8$, their ratio is finite, and different from zero; $p(z)$ cannot vanish for $z=0$, because $\frac{p(z)}{q(z)}$ admits $z=0$ as a pole of the order $n-1$, and it would not have been reduced to its most simple expression if $p(0)$ were equal to zero. On considering $f(z)=z^{n-1}\frac{p(z)}{q(z)}$ for $z=0$, it is easy to see that $q(z)=z^{n-1}\phi(z)$, because $f(0)$ and $p(0)$ are different from zero.

But $\phi(z)$ must be a constant, otherwise $f(z)$ would become infinite for the roots of the polynomial $\phi(z)$, hence $f(z)=Cp(z)$; but $p(z)$ is of the same degree as $q(z)$, $=Cz^{n-1}$, therefore $p(z)$ is an integral polynomial of the degree $n-1$.

Carlo Veneziani.

Solution of Solvable Irreducible Quintic Equations, without the aid of a Resolvent Sextic.

By George Paxton Young, *University College, Toronto, Canada.*

The Problem Stated.

§1. Jerrard has proved that a quintic equation can always be brought to the trinomial form $F(x) = x^5 + p_4 x + p_5 = 0$. (1)
Hence the problem of the solution of the general equation of the fifth degree is reduced to that of the solution of (1). Let $F(x)$ be irreducible. Then since the equation $F(x) = 0$ cannot be solved algebraically except in particular cases, it is incumbent on the Algebraist, first, to find a criterion of its solvability. In other words, if

$$p_4 = \phi(A, B, \text{etc.}), \text{ and } p_5 = \psi(A, B, \text{etc.}),$$

where ϕ and ψ denote rational functions of certain quantities, A, B, etc., he has to discover the most general forms of the expressions $\phi(A, B, \text{etc.})$ and $\psi(A, B, \text{etc.})$ that are compatible with the possibility of exhibiting the roots of the equation $F(x) = 0$ as algebraical functions of A, B, etc. Next, assuming that the functions ϕ and ψ are such as to render the equation $F(x) = 0$ solvable, he has to solve the equation, that is, to obtain its roots in terms of A, B, etc. These are the two things proposed to be done in the present paper.

The Conclusion Reached.

§2. We may take for granted that p_4 is distinct from zero. Then it will be found that the coefficients p_4 and p_5, in the most general forms they can receive consistent with the solvability of the equation (1), are rational functions of two quantities A and B. More definitely,

$$\left.\begin{aligned} p_4 &= \frac{5A^4(3 - B)}{16 + B^2} \\ \text{and } p_5 &= \frac{A^5(22 + B)}{16 + B^2}. \end{aligned}\right\} \qquad (2)$$

The relation between p_4 and p_5 indicated in (2) is *the necessary condition of the solvability of the equation* (1). Assuming now that this relation subsists, take λ, a root of the equation

$$x^4 - Bx^3 - 6x^2 + Bx + 1 = 0. \tag{3}$$

$$\left.\begin{aligned} \text{Put } a &= \frac{-(\lambda^2+1)}{A\lambda(\lambda-1)} \\ \text{and } \theta &= \frac{-A^5\lambda(\lambda-1)^2}{(16+B^2)(\lambda+1)(\lambda^2+1)}. \end{aligned}\right\} \tag{4}$$

Then *the root r of the equation* $F(x) = 0$ is

$$r = \theta^{\frac{1}{5}} + a\theta^{\frac{2}{5}} + \lambda a^2\theta^{\frac{3}{5}} - \lambda a^3\theta^{\frac{4}{5}}. \tag{5}$$

CRITERION OF SOLVABILITY: THE EQUATIONS (2) SHOWN TO BE NECESSARY.

§3. I will now prove that, if the equation (1) be solvable algebraically, the coefficients must be of the forms given in (2).

§4. The root r of the equation (1) may be expressed as follows:

$$r = \theta^{\frac{1}{5}} + a\theta^{\frac{2}{5}} + b\theta^{\frac{3}{5}} + c\theta^{\frac{4}{5}}; \tag{6}$$

where a, b, c, involve only surds subordinate to $\theta^{\frac{1}{5}}$. In fact, these coefficients are rational functions of θ. Taking the second, third, fourth and fifth powers of r, and arranging according to the powers of $\theta^{\frac{1}{5}}$ lower than the fifth,

$$\left.\begin{aligned} r^2 &= d + d_1\theta^{\frac{1}{5}} + d_2\theta^{\frac{2}{5}} + d_3\theta^{\frac{3}{5}} + d_4\theta^{\frac{4}{5}} \\ r^3 &= g + g_1\theta^{\frac{1}{5}} + g_2\theta^{\frac{2}{5}} + g_3\theta^{\frac{3}{5}} + g_4\theta^{\frac{4}{5}} \\ r^4 &= h + h_1\theta^{\frac{1}{5}} + h_2\theta^{\frac{2}{5}} + h_3\theta^{\frac{3}{5}} + h_4\theta^{\frac{4}{5}} \\ r^5 &= k + k_1\theta^{\frac{1}{5}} + k_2\theta^{\frac{2}{5}} + k_3\theta^{\frac{3}{5}} + k_4\theta^{\frac{4}{5}}; \end{aligned}\right\} \tag{7}$$

where d, d_1, g, etc. are clear of $\theta^{\frac{1}{5}}$. If S_c be the sum of the cth powers of the roots of the equation (1), $S_2 = 0 \therefore d = 0$; and $S_3 = 0 \therefore g = 0$. Also

$$S_4 = 5h, \text{ and } S_5 = 5k \therefore 5h = -4p_4, \text{ and } k = -p_5.$$

The value of d is $2\theta(c + ab)$. But $d = 0$. Therefore

$$c = -ab. \tag{8}$$

Again, $g = 3\theta\{(a^2 + b) + \theta(ac^2 + b^2c)\}$. But $g = 0$. Therefore

$$(a^2 + b) + \theta(ac^2 + b^2c) = 0. \tag{9}$$

Suppose, if possible, that $a = 0$. Then, by (8), $c = 0$. Therefore, by (9), $b = 0$. Hence (6) becomes $r = \theta^{\frac{1}{5}}$; and the equation $F(x) = 0$ is $x^5 - \theta = 0$. But this makes p_4 zero; which is at variance with the assumption in §2. Therefore we cannot have $a = 0$. We may consequently put $b = \lambda a^2$. Therefore, by (8), $c = -\lambda a^3$. This reduces the expression for r in (6) to the form (5); and at the same time changes equation (9) into

$$a^5\theta\lambda^2(\lambda-1)=\lambda+1. \tag{10}$$

§5. The values of d_1, d_2, d_3, d_4, are, keeping in view (10),

$$d_1=\theta(b^2+2ac)=a^4\theta\lambda(\lambda-2),$$

$$d_2=1+2bc\theta=-\frac{\lambda+3}{\lambda-1},$$

$$d_3=c^2\theta+2a=\frac{a(3\lambda-1)}{\lambda-1},$$

$$d_4=a^2+2b\ =a^2(2\lambda+1).$$

The values of g_1, g_2, g_3, g_4, are

$$g_1=\theta(a^3+6ab+3c)+3bc^2\theta^2\ =\frac{a^2\theta(6\lambda^2+\lambda-1)}{\lambda-1},$$

$$g_2=\theta(3a^2b+6ac+3b^2)+c^3\theta^2=\frac{\lambda a^4\theta(3\lambda^2-7\lambda+2)}{\lambda-1},$$

$$g_3=1+\theta(3a^2c+6bc+3ab^2)\ =-\frac{2\lambda^2+7\lambda+3}{\lambda(\lambda-1)},$$

$$g_4=3a+\theta(b^3+3c^2+6abc)\ =\frac{a(\lambda^2+\lambda-6)}{\lambda-1}.$$

§6. It will be convenient to put

$$B=3-\frac{p_4(\lambda-1)}{5a\theta(\lambda+1)},$$

$$f_1(x)=x^4-3x^3-6x^2+3x+1,$$

$$f_2(x)=x^4-Bx^3-6x^2+Bx+1,$$

$$f_3(x)=x^4+Bx^3-6x^2-Bx+1,$$

$$f_4(x)=(x^2+1)(x^4+22x^3-6x^2-22x+1),$$

$$p_4^5(22+B)^4\{f_5(x)\}=p_4^5(22+B)^4(x^2+1)^4-5^5p_5^4(3-B)^5x^2(x^2-1)^2.$$

§7. By squaring the first of equations (7), and putting the result equal to the value of r^4 in the third, and again by multiplying together the first two of equations (7), and putting the result equal to the value of r^5 in the fourth,

$$h=2\theta(d_1d_4+d_2d_3)$$

and

$$k=\theta(g_1d_4+g_2d_3+g_3d_2+g_4d_1).$$

Therefore, because $5h=-4p_4$, and $k=-p_5$,

$$-4p_4=10\theta(d_1d_4+d_2d_3)$$

$$-p_5=\theta(g_1d_4+g_2d_3+g_3d_2+g_4d_1).$$

In these results substitute the values of d_1, g_1, etc., in §5. Then, keeping in view the forms of $f_1(x)$ and $f_4(x)$ in §6,

$$5a\theta\{f_1(\lambda)\}+p_4\lambda(\lambda-1)^2=0, \tag{11}$$

$$\theta\{f_4(\lambda)\}+p_5\lambda^2(\lambda-1)^3=0. \tag{12}$$

Hence it can be shown that

$$f_3(\lambda)=0, \text{ and } f_5(\lambda)=0. \qquad (13)$$

The first of equations (13) is obtained by eliminating $a\theta$ from (11) by means of the value of B in §6. To find the second, eliminate θ from (12) by means of (11) and (10). The result is

$$5^5 p_5^4 \lambda(\lambda^2-1)\{f_1(\lambda)\}^5 + p_4^5\{f_4(\lambda)\}^4 = 0. \qquad (14)$$

Now, because $f_3(\lambda)=0,\ f_4(\lambda)=f_4(\lambda)-(\lambda^2+1)\{f_3(\lambda)\}$.

But $f_4(\lambda)=(\lambda^2+1)(\lambda_4+22\lambda^3-6\lambda^2-22\lambda+1)$,

and $(\lambda^2+1)\{f_3(\lambda)\}=(\lambda^2+1)(\lambda^4-B\lambda^3-6\lambda^2+B\lambda+1)$.

Therefore $$f_4(\lambda)=(\lambda^2+1)(\lambda^3-\lambda)(22+B). \qquad (15)$$

Also, the equation $f_3(\lambda)=0$, otherwise written, is

$$f_1(\lambda)=-(3-B)\lambda(\lambda^2-1). \qquad (16)$$

By substituting in (14) the values of $f_1(\lambda)$ and $f_4(\lambda)$ in (16) and (15),

$$(22+B)^4 p_4^5(\lambda^2+1)^4 - 5^5 p_5^4(3-B)^5\lambda^2(\lambda^2-1)^2 = 0.$$

By reference to §6, this will be seen to give the second of equations (13).

§8. The expression B is rational; that is to say, it is a rational function of the quantities, whatever they may be, that enter rationally into the coefficients of the equation $F(x)=0$. For, the first and fourth separate members of the value of r in (6), namely $\theta^{\frac{1}{5}}$ and $c\theta^{\frac{4}{5}}$, are respectively what are called u_1 and u_4 in an Article entitled "*Resolution of Solvable Equations of the Fifth Degree,*" that appeared in Vol. VI of this Journal. But, p_3 being the coefficient of x^3 in the equation $F(x)=0$, u_1u_4 or what is here called $c\theta$ is (Vol. VI, p. 104) the sum of $-\frac{p_3}{10}$ and a quantity which is the square root of a rational expression. And $p_3=0$. Therefore $(c\theta)^2$ or $\lambda^2a^6\theta^6$ is rational. Therefore, by (10), $\frac{(\lambda+1)a\theta}{(\lambda-1)}$ is rational. Therefore, by the value of B in §6, B is rational.

§9. Putting

$$\left.\begin{aligned} P &= p_4^5(22+B)^4 \\ \text{and}\quad Q &= 5^5 p_5^4(3-B)^5, \end{aligned}\right\} \qquad (17)$$

we have, from the forms of $f_2(x)$ and $f_3(x)$ in §6,

$$P\{f_2(x)\}\{f_3(x)\} = P\{f_5(x)\} + x^2(x^2-1)^2\{Q-P(16+B^2)\}. \qquad (18)$$

By (13), λ is a root of each of the equations $f_3(x)=0$ and $f_5(x)=0$. Therefore, by (18), it must be a root of the equation $x(x^2-1)=0$, unless $Q-P(16+B^2)=0$. But the only roots of the equation $x(x^2-1)=0$ are zero and $1^{\frac{1}{2}}$, neither of which is a root of the equation $f_3(x)=0$. Therefore

$$Q-P(16+B^2)=0; \qquad (19)$$

or, from (17), $$p_4^5(16+B^2)(22+B)^4 = 5^5 p_5^4(3-B)^5. \qquad (20)$$

Since, by §8, B is rational, (20) implies that $\frac{p_4(16+B^2)}{5(3-B)}$ is the fourth power of a rational quantity, say of A. Put

$$\left.\begin{aligned} \frac{p_4(16+B^2)}{5(3-B)} &= A^4, \\ \text{therefore, from (20),} \qquad p_4 A(22+B) &= 5p_5(3-B). \end{aligned}\right\} \qquad (21)$$

The two equations (21) give us the forms of p_4 and p_5 in (2). Hence the necessity of these forms is established.

SOLUTION OF THE EQUATION $F(x)=0$.

§10. Having shown that p_4 and p_5 must be of the forms (2) in order that equation (1) may be solvable algebraically, I will now, taking p_4 and p_5 to be of these forms, deduce the rule given in (2) for the solution of the equation. This will prove the criterion of solvability afforded by the equations (2) to be sufficient, as it has already been seen to be necessary.

§11. Because $p_4 = \frac{5A^4(3-B)}{16+B^2}$, and $p_5 = \frac{A^5(22+B)}{16+B^2}$, the equation (19) subsists. Also, from the forms of $f_2(x)$, $f_3(x)$ and $f_5(x)$, in §6, equation (18) subsists. Therefore, from (19),

$$\{f_2(x)\}\{f_3(x)\} = f_5(x). \qquad (22)$$

Let λ be a root of equation (3). Then $f_3(\lambda)=0$. It follows exactly as in §7 that

$$\left.\begin{aligned} f_1(\lambda) &= -(3-B)\lambda(\lambda^2-1) \\ \text{and } f_4(\lambda) &= (\lambda^2+1)(\lambda^2-\lambda)(22+B) \end{aligned}\right\} \qquad (23)$$

Now, from (22), because $f_3(\lambda)=0$, $f_5(\lambda)=0$. Therefore

$$(22+B)^4 p_4^5(\lambda^2+1)^4 - 5^5 p_5^4(3-B)^5\lambda^2(\lambda^2-1)^2 = 0.$$

Substitute here the values of $22+B$ and $3-B$ in (23). The result is equation (14). Take a and θ as in (4). Then

$$a\theta = \frac{A^4(\lambda-1)}{(16+B^2)(\lambda+1)}.$$

By the first of equations (2), this is equivalent to

$$a\theta = \frac{p_4(\lambda-1)}{5(3-B)(\lambda+1)}.$$

And this again, when we substitute for $3-B$ its value in (23), becomes equation (11).

From the second of equations (4) eliminate $16+B^2$ by means of the second of equations (2). Then $$\theta = \frac{-p_5\lambda(\lambda-1)^2}{(22+B)(\lambda+1)(\lambda^2+1)};$$

and this, when we substitute for $22 + B$ its value in (23), becomes equation (12). Eliminate p_4 and p_5 from (14) by means of (11) and (12). The result is equation (10).

§12. Give r the value it has in (5). Let

$$r_1,\ r_2,\ r_3,\ r_4,\ r_5, \qquad (24)$$

be the five values of r obtained by writing for $\theta^{\frac{1}{5}}$ successively the five expressions $\theta^{\frac{1}{5}}$, $w\theta^{\frac{1}{5}}$, $w^2\theta^{\frac{1}{5}}$, $w^3\theta^{\frac{1}{5}}$, $w^4\theta^{\frac{1}{5}}$, w being a primitive fifth root of unity. Let S_c' be the sum of the cth powers of the terms in (24), while S_c is the sum of the cth powers of the roots of equation (1). Then it can be shown that

$$S_1' = S_1,\ S_2' = S_2,\ S_3' = S_3,\ S_4' = S_4,\ S_5' = S_5. \qquad (25)$$

For, S_1' and S_2' are both identically zero. This makes $S_1' = S_1$, and $S_2' = S_2$. Also

$$S_3' = 15a^3\theta\{(\lambda + 1) - a^5\theta\lambda^2(\lambda - 1)\}.$$

Therefore, from (10), $S_3' = 0 = S_3$. Again, if d_1, d_2, g_1, etc., be taken as in §5, namely $d_1 = a^4\theta\lambda(\lambda - 2)$, and so on,

$$S_4' = 10\theta(d_1d_4 + d_2d_3)$$

$$\text{and } S_5' = 5\theta(g_1d_4 + g_2d_3 + g_3d_2 + g_4d_1).$$

Therefore, as in §7,

$$\left.\begin{aligned} S_4' &= \frac{20a\theta\{f_1(\lambda)\}}{\lambda(\lambda-1)^3} \\ \text{and } S_5' &= \frac{5\theta\{f_4(\lambda)\}}{\lambda^3(\lambda-1)^5}. \end{aligned}\right] \qquad (26)$$

Now $S_4 = -4p_4$, and $S_5 = -5p_5$. Therefore, from (26) compared with (11) and (12), $S_4' = S_4$, and $S_5' = S_5$. Thus all the equations in (25) are established.

§13. Let $X = 0$ be the equation whose roots are the terms in (24). Then

$$X = x^5 - \tfrac{1}{4}S_4'x - \tfrac{1}{5}S_5'.$$

Hence, because $S_4' = S_4$, and $S_5' = 5$,

$$X = x^5 - \tfrac{1}{4}S_4x - \tfrac{1}{5}S_5 = x^5 + p_4x + p_5 = F(x).$$

This makes r a root of the equation $F(x) = 0$.

Verifying Instances.

§14. As A and B may have any values whatever, let $A = 5$, and $B = 2$. Then, by (2), equation (1) is

$$x^5 + \frac{625}{4}x + 3750 = 0.$$

The equation (3), for determining λ, is

$$x^4 - 2x^3 - 6x^2 + 2x + 1 = 0.$$

A root of this quartic is $\lambda = 3.52$. Hence, by (4),

$$\theta = -57.71, \text{ and } a = -.3019.$$

Therefore
$$\theta^{\frac{1}{5}} = -2.250,$$
$$a\theta^{\frac{2}{5}} = -1.529,$$
$$\lambda a^2\theta^{\frac{3}{5}} = -3.656,$$
$$-\lambda a^3\theta^{\frac{4}{5}} = 2.484.$$

Therefore
$$r = \theta^{\frac{1}{5}} + a\theta^{\frac{2}{5}} + \lambda a^2\theta^{\frac{3}{5}} - \lambda a^3\theta^{\frac{4}{5}} = -4.951.$$

§15. Second example. Let $A = 1$, $B = -1$. Then equation (1) is

$$x^5 + \frac{20x}{17} + \frac{21}{17} = 0.$$

The equation for determining λ is

$$x^4 + x^3 - 6x^2 - x + 1 = 0.$$

A root of this quartic is $\lambda = 2.0496$. Hence

$$\theta = -.00837, \text{ and } a = -2.4176.$$

Therefore
$$\theta^{\frac{1}{5}} = -.384,$$
$$a\theta^{\frac{2}{5}} = -.357,$$
$$\lambda a^2\theta^{\frac{3}{5}} = -.679,$$
$$-\lambda a^3\theta^{\frac{4}{5}} = .631,$$
$$r = -.789.$$

§16. Third example. Let $A = 1$, $B = 0$. Then equation (1) is

$$x^5 + \frac{15x}{16} + \frac{11}{8} = 0.$$

The equation for determining λ is

$$x^4 - 6x^2 + 1 = 0.$$

A root of this quartic is $\lambda = 2.414215$. Hence

$$\theta = -.01294, \quad a = -2.$$

Therefore
$$\theta^{\frac{1}{5}} = -.419194,$$
$$a\theta^{\frac{2}{5}} = -.351447,$$
$$\lambda a^2\theta^{\frac{3}{5}} = -.711345,$$
$$-\lambda a^3\theta^{\frac{4}{5}} = .596384,$$
$$r = -.8856.$$

§17. Fourth example. Let $A = 1$, $B = 7$. Then equation (1) is

$$x^5 - \frac{4x}{13} + \frac{29}{65} = 0.$$

The equation for determining λ is

$$x^4 - 7x^3 - 6x^2 + 7x + 1 = 0.$$

A root of this quartic is $\lambda = .7690975$. Hence

$$\theta = -.0002246, \text{ and } a = 8.9619.$$

Therefore

$$\theta^{\frac{1}{5}} = -.186239,$$
$$a\theta^{\frac{2}{5}} = .310846,$$
$$\lambda a^2\theta^{\frac{3}{5}} = -.399025,$$
$$-\lambda a^3\theta^{\frac{4}{5}} = -.665999,$$
$$r = -.940617.$$

§18. Fifth example. Let $A = 1$, $B = -7$. Then equation (1) is

$$x^5 + \frac{10x}{13} + \frac{3}{13} = 0.$$

The equation for determining λ is

$$x^4 + 7x^3 - 6x^2 - 7x + 1 = 0.$$

A root of this quartic is $\lambda = 1.300226$. Hence

$$\theta = -.00029133, \; a = -6.89253.$$

Therefore

$$\theta^{\frac{1}{5}} = -1.963,$$
$$a\theta^{\frac{2}{5}} = -.2655,$$
$$\lambda a^2\theta^{\frac{3}{5}} = -.4671,$$
$$-\lambda a^3\theta^{\frac{4}{5}} = .6319,$$
$$r = -.297.$$

§19. Sixth example. Let $A = 1$, and $B = \sqrt{2}$. Then equation (1) is

$$18x^5 + 5(3 - \sqrt{2})x + (22 + \sqrt{2}) = 0.$$

The equation for determining λ is

$$\left(x^2 - \frac{\sqrt{2}}{2}x - 1\right)^2 = \frac{9x^2}{2}.$$

A root of this quartic is $\lambda = \sqrt{2} + \sqrt{3}$. Therefore

$$\theta^{\frac{1}{5}} = -.4469,$$
$$a\theta^{\frac{2}{5}} = -.3223,$$
$$\lambda a^2\theta^{\frac{3}{5}} = -.7313,$$
$$-\lambda a^3\theta^{\frac{4}{5}} = .5275,$$
$$r = -.973.$$

Notes on the Quintic.

By J. C. Glashan, *Ottawa, Canada.*

---

1. If the quintic

$$x^5 + 10p_2x^3 + 10p_3x^2 + 5p_4x + p_5 = 0$$

be solvable by radicals, the coefficients must be so related that if

$$p_2 = n\theta k^2, \quad p_3 = \alpha\theta^2 k^3, \text{ and } p_4 = \beta\theta^2 k^4,$$

then must

$$p_5 = 2(1+n)\gamma\theta^2 k^5,$$

wherein

$$\theta = (1+n)\{1-(1+m^2)n^2\}$$

$$\alpha = 4g\{m-(1+n)g\}/[\{1+m^2-(1+n)g^2\}^2-(1+m^2)]$$

$$\beta = \frac{3-4m}{1+m^2} + 4n + 5(1+m^2)n^2 + 4(1+n)\{m-(1+m^2)n-(1-mn)g\}\alpha,$$

$$\begin{aligned}\gamma = 2(1+n)\,[1+m+2m^2+m^3 &- 2(1+m^2)(1+m+m^2)n + m(1+m^2)n^2 \\ &-\{1+m+m^2+(1-2m-2m^2)n+(1+m^2)^2n^2\}g \\ &-(1+n)\{m-2(1+m^2)n+m(1+m^2)n^2\}g^2 \\ &+(1+n)\{1-2mn+(1+m^2)n^2\}g^3]\,\alpha^2 \\ +\Big[\frac{2}{1+m^2} &- 3(5+2m)+2(5+6m)n+(1+m^2)(19+6m)n^2-(1+m^2)(3+4m)n^3 \\ &+4(1+n)\{1-2mn+(1+m^2)n^2\}g \\ &+(1+n)\Big\{\Big(\frac{1}{1+m^2}-n^2\Big)\Big(\frac{11+2m}{1+m^2}-4n\Big)+4\Big(\frac{m}{1+m^2}-n\Big)n^2\Big\}g^2\Big]\alpha \\ +2\Big[\{1-(1+m^2)n^2\}&\Big(\frac{11+2m}{1+m^2}-4n\Big)+4\{m-(1+m^2)n\}n^2\Big].\end{aligned}$$

2. If $g = 0$, the solvable quintic becomes

$$x^5 + 10n\theta k^2x^3 + 5\left\{\frac{3-4m}{1+m^2} + 4n + 5(1+m^2)n^2\right\}\theta^2 k^4 x$$
$$+ 4\left[\left(\frac{11+2m}{1+m^2} - 4n\right)\theta + 4\{m-(1+m^2)n\}(1+n)n^2\right]\theta^2 k^5 = 0.$$

3. If both $g = 0$ and $n = 0$, the solvable quintic assumes the form

$$x^5 + 5\left(\frac{3-4m}{1+m^2}\right)k^4x + 4\left(\frac{11+2m}{1+m^2}\right)k^5 = 0,$$

a form communicated to the present writer by Professor G. P. Young, of Toronto University, in May, 1883. As any quintic can by means of the Jerrard-Tschirnhausen transformation be reduced to the form $x^5+ax+b=0$, Professor Young's form is equivalent algebraically to the form given in §1 above.

4. If in §1 $m=\frac{p^2-1}{2p}$, the resulting quintic includes Cockle's solvable quintic (Lady's and Gentleman's Diary, 1858) which itself includes the Binomial, the DeMoivrian (1706) and the Euler-Bezout (1762) quintics. For example, if we make

$$g=0,\ m=\frac{p^2-1}{2p}\ \text{and}\ n=-\frac{2p(2p+1)}{p^2+1}$$

we get DeMoivre's quintic.

In the *Quarterly Journal of Mathematics*, Vol. XVIII, pp. 154–157, Professor Cayley, apparently unaware that he had been anticipated by Mr. Cockle, reduces the solution of Cockle's quintic to depend on that of a cubic, but in reality the solution can be obtained without solving any equation of higher degree than a quadratic.

5. The Gaussian quintics for the solution of the binomial

$$x^n-1=0,\ n=10m+1,$$

m being prime to 5 and n a prime number, can be written

$$x^5-10nx^3-5dnx^2-5enx-gn=0,$$

wherein

$$d=a^2+ab-b^2$$
$$16n=d^2+25(a^2+b^2+5)$$
$$16e=3d^2-25(a^2+b^2+25)$$
$$16g=\{d^2-25(a^2+b^2+5)\}d-1250ab.$$

If m be a multiple of 5,

and

$$4d^2=(a^2+b^2)^2,$$

then

$$g=(e-n-125)d.$$

The following table exhibits the values of a, b, d, e and g for all prime values of n under 200:

n	a	b	d	e	g
11	1	1	1	−42	−89
31	3	1	11	−32	−409
41	3	−3	−9	−52	981
61	5	−3	1	−92	1111
71	1	5	−19	−12	101
(101)	(1)	(−5)	(−29)	(78)	(271)
131	5	7	11	−132	−4009
181	9	−5	11	−132	7059
191	7	−1	41	198	1331
101			21	−22	−2604
151	7	8	−29	28	1798

On the Algebra of Logic:

A CONTRIBUTION TO THE PHILOSOPHY OF NOTATION.

BY C. S. PEIRCE.

I.—*Three kinds of Signs.*

Any character or proposition either concerns one subject, two subjects, or a plurality of subjects. For example, one particle has mass, two particles attract one another, a particle revolves about the line joining two others. A fact concerning two subjects is a dual character or relation; but a relation which is a mere combination of two independent facts concerning the two subjects may be called *degenerate*, just as two lines are called a degenerate conic. In like manner a plural character or conjoint relation is to be called degenerate if it is a mere compound of dual characters.

A sign is in a conjoint relation to the thing denoted and to the mind. If this triple relation is not of a degenerate species, the sign is related to its object only in consequence of a mental association, and depends upon a habit. Such signs are always abstract and general, because habits are general rules to which the organism has become subjected. They are, for the most part, conventional or arbitrary. They include all general words, the main body of speech, and any mode of conveying a judgment. For the sake of brevity I will call them *tokens*.

But if the triple relation between the sign, its object, and the mind, is degenerate, then of the three pairs

sign	object
sign	mind
object	mind

two at least are in dual relations which constitute the triple relation. One of the connected pairs must consist of the sign and its object, for if the sign were not related to its object except by the mind thinking of them separately, it would not fulfil the function of a sign at all. Supposing, then, the relation of the sign to its object does not lie in a mental association, there must be a direct dual relation of the sign to its object independent of the mind using the sign. In the second of the three cases just spoken of, this dual relation is not degenerate, and the sign signifies its object solely by virtue of being really connected with it. Of this nature are all natural signs and physical symptoms. I call such a sign an *index*, a pointing finger being the type of the class.

The index asserts nothing; it only says "There!" It takes hold of our eyes, as it were, and forcibly directs them to a particular object, and there it stops. Demonstrative and relative pronouns are nearly pure indices, because they denote things without describing them; so are the letters on a geometrical diagram, and the subscript numbers which in algebra distinguish one value from another without saying what those values are.

The third case is where the dual relation between the sign and its object is degenerate and consists in a mere resemblance between them. I call a sign which stands for something merely because it resembles it, an *icon*. Icons are so completely substituted for their objects as hardly to be distinguished from them. Such are the diagrams of geometry. A diagram, indeed, so far as it has a general signification, is not a pure icon; but in the middle part of our reasonings we forget that abstractness in great measure, and the diagram is for us the very thing. So in contemplating a painting, there is a moment when we lose the consciousness that it is not the thing, the distinction of the real and the copy disappears, and it is for the moment a pure dream,—not any particular existence, and yet not general. At that moment we are contemplating an *icon*.

I have taken pains to make my distinction* of icons, indices, and tokens clear, in order to enunciate this proposition: in a perfect system of logical notation signs of these several kinds must all be employed. Without tokens there would be no generality in the statements, for they are the only general signs; and generality is essential to reasoning. Take, for example, the circles by which Euler represents the relations of terms. They well fulfil the function of icons, but their want of generality and their incompetence to express propositions must have been felt by everybody who has used them. Mr. Venn has, therefore, been led to add shading to them; and this shading is a conventional sign of the nature of a token. In algebra, the letters, both quantitative and functional, are of this nature. But tokens alone do not state what is the subject of discourse; and this can, in fact, not be described in general terms; it can only be indicated. The actual world cannot be distinguished from a world of imagination by any

* See *Proceedings American Academy of Arts and Sciences*, Vol. VII, p. 294, May 14, 1867.

description. Hence the need of pronoun and indices, and the more complicated the subject the greater the need of them. The introduction of indices into the algebra of logic is the greatest merit of Mr. Mitchell's system.* He writes F_1 to mean that the proposition F is true of every object in the universe, and F_u to mean that the same is true of some object. This distinction can only be made in some such way as this. Indices are also required to show in what manner other signs are connected together. With these two kinds of signs alone any proposition can be expressed; but it cannot be reasoned upon, for reasoning consists in the observation that where certain relations subsist certain others are found, and it accordingly requires the exhibition of the relations reasoned with in an icon. It has long been a puzzle how it could be that, on the one hand, mathematics is purely deductive in its nature, and draws its conclusions apodictically, while on the other hand, it presents as rich and apparently unending a series of surprising discoveries as any observational science. Various have been the attempts to solve the paradox by breaking down one or other of these assertions, but without success. The truth, however, appears to be that all deductive reasoning, even simple syllogism, involves an element of observation; namely, deduction consists in constructing an icon or diagram the relations of whose parts shall present a complete analogy with those of the parts of the object of reasoning, of experimenting upon this image in the imagination, and of observing the result so as to discover unnoticed and hidden relations among the parts. For instance, take the syllogistic formula,

All M is P
S is M
$\therefore$ S is P.

This is really a diagram of the relations of S, M, and P. The fact that the middle term occurs in the two premises is actually exhibited, and this must be done or the notation will be of no value. As for algebra, the very idea of the art is that it presents formulae which can be manipulated, and that by observing the effects of such manipulation we find properties not to be otherwise discerned. In such manipulation, we are guided by previous discoveries which are embodied in general formulae. These are patterns which we have the right to imitate in our procedure, and are the *icons par excellence* of algebra. The letters of applied algebra are usually tokens, but the x, y, z, etc. of a general formula, such as

$$(x+y)z = xz + yz,$$

* *Studies in Logic*, by members of the Johns Hopkins University. Boston: Little & Brown, 1883.

are blanks to be filled up with tokens, they are indices of tokens. Such a formula might, it is true, be replaced by an abstractly stated rule (say that multiplication is distributive); but no application could be made of such an abstract statement without translating it into a sensible image.

In this paper, I purpose to develope an algebra adequate to the treatment of all problems of deductive logic, showing as I proceed what kinds of signs have necessarily to be employed at each stage of the development. I shall thus attain three objects. The first is the extension of the power of logical algebra over the whole of its proper realm. The second is the illustration of principles which underlie all algebraic notation. The third is the enumeration of the essentially different kinds of necessary inference; for when the notation which suffices for exhibiting one inference is found inadequate for explaining another, it is clear that the latter involves an inferential element not present to the former. Accordingly, the procedure contemplated should result in a list of categories of reasoning, the interest of which is not dependent upon the algebraic way of considering the subject. I shall not be able to perfect the algebra sufficiently to give facile methods of reaching logical conclusions: I can only give a method by which any legitimate conclusion may be reached and any fallacious one avoided. But I cannot doubt that others, if they will take up the subject, will succeed in giving the notation a form in which it will be highly useful in mathematical work. I even hope that what I have done may prove a first step toward the resolution of one of the main problems of logic, that of producing a method for the discovery of methods in mathematics.

II.—*Non-relative Logic.*

According to ordinary logic, a proposition is either true or false, and no further distinction is recognized. This is the descriptive conception, as the geometers say; the metric conception would be that every proposition is more or less false, and that the question is one of amount. At present we adopt the former view.

Let propositions be represented by quantities. Let $\mathbf{v}$ and $\mathbf{f}$ be two constant values, and let the value of the quantity representing a proposition be $\mathbf{v}$ if the proposition is true and be $\mathbf{f}$ if the proposition is false. Thus, x being a proposition, the fact that x is either true or false is written

$$(x-\mathbf{f})(\mathbf{v}-x)=0.$$

So
$$(x-\mathbf{f})(\mathbf{v}-y)=0$$
will mean that either x is false or y is true. This may be said to be the same as 'if x is true, y is true.' A hypothetical proposition, generally, is not confined to stating what actually happens, but states what is invariably true throughout a universe of possibility. The present proposition is, however, limited to that one individual state of things, the Actual.

We are, thus, already in possession of a logical notation, capable of working syllogism. Thus, take the premises, 'if x is true, y is true,' and 'if y is true, z is true.' These are written
$$(x-\mathbf{f})(\mathbf{v}-y)=0$$
$$(y-\mathbf{f})(\mathbf{v}-z)=0.$$
Multiply the first by $(\mathbf{v}-z)$ and the second by $(x-\mathbf{f})$ and add. We get
$$(x-\mathbf{f})(\mathbf{v}-\mathbf{f})(\mathbf{v}-z)=0,$$
or dividing by $\mathbf{v}-\mathbf{f}$, which cannot be 0,
$$(x-\mathbf{f})(\mathbf{v}-z)=0;$$
and this states the syllogistic conclusion, "if x is true, z is true."

But this notation shows a blemish in that it expresses propositions in two distinct ways, in the form of quantities, and in the form of equations; and the quantities are of two kinds, namely those which must be either equal to $\mathbf{f}$ or to $\mathbf{v}$, and those which are equated to *zero*. To remedy this, let us discard the use of equations, and perform no operations which can give rise to any values other than $\mathbf{f}$ and $\mathbf{v}$.

Of operations upon a simple variable, we shall need but one. For there are but two things that can be said about a single proposition, by itself; that it is true and that it is false, $x=\mathbf{v}$ and $x=\mathbf{f}$.
The first equation is expressed by x itself, the second by any function, ϕ, of x, fulfilling the conditions $\phi\mathbf{v}=\mathbf{f}\quad\phi\mathbf{f}=\mathbf{v}$.
The simplest solution of these equations is
$$\phi x=\mathbf{f}+\mathbf{v}-x.$$
A product of n factors of the two forms $(x-\mathbf{f})$ and $(\mathbf{v}-y)$, if not zero equals $(\mathbf{v}-\mathbf{f})^n$. Write P for the product. Then $\mathbf{v}-\dfrac{P}{(\mathbf{v}-\mathbf{f})^{n-1}}$ is the simplest function of the variables which becomes $\mathbf{v}$ when the product vanishes and $\mathbf{f}$ when it does not. By this means any proposition relating to a single individual can be expressed.

If we wish to use algebraical signs with their usual significations, the meanings of the operations will entirely depend upon those of $\mathbf{f}$ and $\mathbf{v}$. Boole chose

$\mathbf{v} = 1$, $\mathbf{f} = 0$. This choice gives the following forms:

$$\mathbf{f} + \mathbf{v} - x = 1 - x$$

which is best written $\bar{x}$.

$$\mathbf{v} - \frac{(x - \mathbf{f})(\mathbf{v} - y)}{\mathbf{v} - \mathbf{f}} = 1 - x + xy = \overline{x\bar{y}}.$$

$$\mathbf{v} - \frac{(\mathbf{v} - x)(\mathbf{v} - y)}{\mathbf{v} - \mathbf{f}} = x + y - xy$$

$$\mathbf{v} - \frac{(\mathbf{v} - x)(\mathbf{v} - y)(\mathbf{v} - z)}{(\mathbf{v} - \mathbf{f})^2} = x + y + z - xy - xz - yz + xyz$$

$$\mathbf{v} - \frac{(x - \mathbf{f})(y - \mathbf{f})}{\mathbf{v} - \mathbf{f}} = 1 - xy = \overline{xy}$$

It appears to me that if the strict Boolian system is used, the sign + ought to be altogether discarded. Boole and his adherent, Mr. Venn (whom I never disagree with without finding his remarks profitable), prefer to write $x + \bar{x}y$ in place of $\overline{\bar{x}\bar{y}}$. I confess I do not see the advantage of this, for the distributive principle holds equally well when written

$$\overline{\bar{x}\bar{y}}z = \overline{\overline{xz}\,\overline{yz}}$$

$$\overline{\overline{xy}\,\bar{z}} = \overline{\bar{x}\bar{z}}.\overline{\bar{y}\bar{z}}.$$

The choice of $\mathbf{v} = 1$, $\mathbf{f} = 0$, is agreeable to the received measurement of probabilities. But there is no need, and many times no advantage, in measuring probabilities in this way. I presume that Boole, in the formation of his algebra, at first considered the letters as denoting propositions or events. As he presents the subject, they are class-names; but it is not necessary so to regard them. Take, for example, the equation $t = n + hf$,
which might mean that the body of taxpayers is composed of all the natives, together with householding foreigners. We might reach the signification by either of the following systems of notation, which indeed differ grammatically rather than logically.

Sign.	Signification. 1st System.	Signification. 2d System.
t	Taxpayer.	He is a Taxpayer.
n	Native.	He is a Native.
h	Householder.	He is a Householder.
f	Foreigner.	He is a Foreigner.

There is no *index* to show who the "He" of the second system is, but that makes no difference. To say that he is a taxpayer is equivalent to saying that he is a native or is a householder and a foreigner. In this point of view, the constants 1 and 0 are simply the probabilities, to one who knows, of what is true and what is false; and thus unity is conferred upon the whole system.

For my part, I prefer for the present not to assign determinate values to $\mathbf{f}$ and $\mathbf{v}$, nor to identify the logical operations with any special arithmetical ones, leaving myself free to do so hereafter in the manner which may be found most convenient. Besides, the whole system of importing arithmetic into the subject is artificial, and modern Boolians do not use it. The algebra of logic should be self-developed, and arithmetic should spring out of logic instead of reverting to it. Going back to the beginning, let the writing of a letter by itself mean that a certain proposition is true. This letter is a *token*. There is a general understanding that the actual state of things or some other is referred to. This understanding must have been established by means of an *index*, and to some extent dispenses with the need of other indices. The denial of a proposition will be made by writing a line over it.

I have elsewhere shown that the fundamental and primary mode of relation between two propositions is that which we have expressed by the form

$$\mathbf{v}-\frac{(x-\mathbf{f})(\mathbf{v}-y)}{\mathbf{v}-\mathbf{f}}.$$

We shall write this $$x \prec y,$$
which is also equivalent to $$(x-\mathbf{f})(\mathbf{v}-y)=0.$$
It is stated above that this means "if x is true, y is true." But this meaning is greatly modified by the circumstance that only the actual state of things is referred to.

To make the matter clear, it will be well to begin by defining the meaning of a hypothetical proposition, in general. What the usages of language may be does not concern us; language has its meaning modified in technical logical formulae as in other special kinds of discourse. The question is what is the sense which is most usefully attached to the hypothetical proposition in logic? Now, the peculiarity of the hypothetical proposition is that it goes out beyond the actual state of things and declares what *would* happen were things other than they are or may be. The utility of this is that it puts us in possession of a rule, say that "if A is true, B is true," such that should we hereafter learn something of which we are now ignorant, namely that A is true, then, by virtue of this rule, we shall find that we know something else, namely, that *B* is true. There can be no doubt that the Possible, in its primary meaning, is that which may be true for aught we know, that whose falsity we do not know. The purpose is subserved, then, if, throughout the whole range of possibility, in every state of things in which *A* is true, *B* is true too. The hypothetical proposition may therefore be falsified by a single state of things, but only by one in which *A* is true while *B* is false. States of things in which *A* is false, as well as those in which *B* is true, cannot falsify it. If, then, *B* is a proposition true in every case throughout the whole range of possibility, the hypothetical proposition, taken in its logical sense, ought to be regarded as true, whatever may be the usage of ordinary speech. If, on the other hand, *A* is in no case true, throughout the range of possibility, it is a matter of indifference whether the hypothetical be understood to be true or not, since it is useless. But it will be more simple to class it among true propositions, because the cases in which the antecedent is false do not, in any other case, falsify a hypothetical. This, at any rate, is the meaning which I shall attach to the hypothetical proposition in general, in this paper.

The range of possibility is in one case taken wider, in another narrower; in the present case it is limited to the actual state of things. Here, therefore, the proposition

$$a -\!\!< b$$

is true if a is false or if b is true, but is false if a is true while b is false. But though we limit ourselves to the actual state of things, yet when we find that a formula of this sort is true by logical necessity, it becomes applicable to any single state of things throughout the range of logical possibility. For example, we shall see that from $x \,\overline{-\!\!<}\, y$ we can infer $z -\!\!< x$. This does not mean that because in the actual state of things x is true and y false, therefore in every state of things either z is false or x true; but it does mean that in whatever state of things we find x true and y false, in that state of things either z is false or x is true. In that sense, it is not limited to the actual state of things, but extends to any single state of things.

The *first icon* of algebra is contained in the formula of identity

$$x -\!\!< x.$$

This formula does not of itself justify any transformation, any inference. It only justifies our continuing to hold what we have held (though we may, for instance, forget how we were originally justified in holding it).

The *second icon* is contained in the rule that the several antecedents of a *consequentia* may be transposed; that is, that from

$$x \multimap (y \multimap z)$$

we can pass to $\quad y \multimap (x \multimap z).$

This is stated in the formula

$$\{x \multimap (y \multimap z)\} \multimap \{y \multimap (x \multimap z)\}.$$

Because this is the case, the brackets may be omitted, and we may write

$$y \multimap x \multimap z.$$

By the formula of identity

$$(x \multimap y) \multimap (x \multimap y);$$

and transposing the antecedents

$$x \multimap \{(x \multimap y) \multimap y\}$$

or, omitting the unnecessary brackets

$$x \multimap (x \multimap y) \multimap y.$$

This is the same as to say that if in any state of things x is true, and if the proposition "if x, then y" is true, then in that state of things y is true. This is the *modus ponens* of hypothetical inference, and is the most rudimentary form of reasoning.

To say that $(x \multimap x)$ is generally true is to say that it is so in every state of things, say in that in which y is true; so that we may write

$$y \multimap (x \multimap x),$$

and then, by transposition of antecedents,

$$x \multimap (y \multimap x),$$

or from x we may infer $y \multimap x$.

The *third icon* is involved in the principle of the transitiveness of the copula, which is stated in the formula

$$(x \multimap y) \multimap (y \multimap z) \multimap x \multimap z.$$

According to this, if in any case y follows from x and z from y, then z follows from x. This is the principle of the syllogism in *Barbara.*

We have already seen that from x follows $y \multimap x$. Hence, by the transitiveness of the copula, if from $y \multimap x$ follows z, then from x follows z, or from

$$(y \multimap x) \multimap z$$

follows $\quad x \multimap z,$

or $\quad \{(y \multimap x) \multimap z\} \multimap x \multimap z.$

The original notation $x \multimap y$ served without modification to express the pure formula of identity. An enlargement of the conception of the notation so as to make the terms themselves complex was required to express the principle of the transposition of antecedents; and this new *icon* brought out new propositions. The third *icon* introduces the image of a chain of consequence. We must now again enlarge the notation so as to introduce negation. We have already seen that if a is true, we can write $x \prec a$, whatever x may be. Let b be such that we can write $b \prec x$ whatever x may be. Then b is false. We have here a *fourth icon*, which gives a new sense to several formulæ. Thus the principle of the interchange of antecedents is that from

$$x \prec (y \prec z)$$

we can infer $$y \prec (x \prec z).$$

Since z is any proposition we please, this is as much as to say that if from the truth of x the falsity of y follows, then from the truth of y the falsity of x follows.

Again the formula $$x \prec \{(x \prec y) \prec y\}$$

is seen to mean that from x we can infer that anything we please follows from that things following from x, and *a fortiori* from everything following from x. This is, therefore, to say that from x follows the falsity of the denial of x; which is the principle of contradiction.

Again the formula of the transitiveness of the copula, or

$$\{x \prec y\} \prec \{(y \prec z) \prec (x \prec z)\}$$

is seen to justify the inference $$x \prec y$$

$$\therefore\ \bar{y} \prec \bar{x}.$$

The same formula justifies the *modus tollens*,

$$x \prec y$$

$$\bar{y}$$

$$\therefore\ \bar{x}$$

So the formula $$\{(y \prec x) \prec z\} \prec (x \prec z)$$

shows that from the falsity of $y \prec x$ the falsity of x may be inferred.

All the traditional moods of syllogism can easily be reduced to *Barbara* by this method.

A *fifth icon* is required for the principle of excluded middle and other propositions connected with it. One of the simplest formulæ of this kind is

$$\{(x \prec y) \prec x\} \prec x.$$

This is hardly axiomatical. That it is true appears as follows. It can only be false by the final consequent x being false while its antecedent $(x \prec y) \prec x$ is true. If this is true, either its consequent, x, is true, when the whole formula would be true, or its antecedent $x -\!\!< y$ is false. But in the last case the antecedent of $x -\!\!< y$, that is x, must be true.*

From the formula just given, we at once get

$$\{(x -\!\!< y) -\!\!< \alpha\} -\!\!< x,$$

where the α is used in such a sense that $(x -\!\!< y) -\!\!< \alpha$ means that from $(x -\!\!< y)$ every proposition follows. With that understanding, the formula states the principle of excluded middle, that from the falsity of the denial of x follows the truth of x.

The logical algebra thus far developed contains signs of the following kinds:

1st, Tokens; signs of simple propositions, as t for 'He is a taxpayer,' etc.

2d, The single operative sign $-\!\!<$; also of the nature of a token.

3d, The juxtaposition of the letters to the right and left of the operative sign. This juxtaposition fulfils the function of an index, in indicating the connections of the tokens.

4th, The parentheses, subserving the same purpose.

5th, The letters α, β, etc. which are indices of no matter what tokens, used for expressing negation.

6th, The indices of tokens, x, y, z, etc. used in the general formulae.

7th, The general formulae themselves, which are *icons*, or exemplars of algebraic proceedings.

8th, The fourth *icon* which affords a second interpretation of the general formulae.

We might dispense with the fifth and eighth species of signs—the devices

* It is interesting to observe that this reasoning is dilemmatic. In fact, the dilemma involves the fifth icon. The dilemma was only introduced into logic from rhetoric by the humanists of the *renaissance;* and at that time logic was studied with so little accuracy that the peculiar nature of this mode of reasoning escaped notice. I was thus led to suppose that the whole non-relative logic was derivable from the principles of the ancient syllogistic, and this error is involved in Chapter II of my paper in the third volume of this Journal. My friend, Professor Schröder, detected the mistake and showed that the distributive formulæ

$$(x+y)z -\!\!< xz + yz$$
$$(x+z)(y+z) -\!\!< xy + z$$

could not be deduced from syllogistic principles. I had myself independently discovered and virtually stated the same thing. (*Studies in Logic*, p. 189.) There is some disagreement as to the definition of the dilemma (see Keynes's excellent *Formal Logic*, p. 241); but the most useful definition would be a syllogism depending on the above distribution formulæ. The distribution formulæ

$$xz + yz -\!\!< (x+y)z$$
$$xy + z -\!\!< (x+z)(y+z)$$

are strictly syllogistic. DeMorgan's added moods are virtually dilemmatic, depending on the principle of excluded middle.

by which we express negation—by adopting a second operational sign $\mathrel{=\!\!<}$, such that

$$x \mathrel{=\!\!<} y$$

should mean that $x = \mathbf{v}$, $y = \mathbf{f}$. With this, we should require new indices of connections, and new general formulae. Possibly this might be the preferable notation. We should thus have two operational signs but no sign of negation. The forms of Boolian algebra hitherto used, have either two operational signs and a special sign of negation, or three operational signs. One of the operational signs is in that case superfluous. Thus, in the usual notation we have

$$\overline{x+y} = \bar{x}\,\bar{y}$$
$$\bar{x}+\bar{y} = \overline{xy}$$

showing two modes of writing the same fact. The apparent balance between the two sets of theorems exhibited so strikingly by Schröder, arises entirely from this double way of writing everything. But while the ordinary system is not so analytically fitted to its purpose as that here set forth, the character of superfluity here, as in many other cases in algebra, brings with it great facility in working.

The general formulae given above are not convenient in practice. We may dispense with them altogether, as well as with one of the indices of tokens used in them, by the use of the following rules. A proposition of the form

$$x \mathrel{-\!\!<} y$$

is true if $x = \mathbf{f}$ or $y = \mathbf{v}$. It is only false if $y = \mathbf{f}$ and $x = \mathbf{v}$. A proposition written in the form

$$x \mathrel{=\!\!<} y$$

is true if $x = \mathbf{v}$ and $y = \mathbf{f}$, and is false if either $x = \mathbf{f}$ or $y = \mathbf{v}$. Accordingly, to find whether a formula is necessarily true substitute $\mathbf{f}$ and $\mathbf{v}$ for the letters and see whether it can be supposed false by any such assignment of values. Take, for example, the formula

$$(x \mathrel{-\!\!<} y) \mathrel{-\!\!<} \{(y \mathrel{-\!\!<} z) \mathrel{-\!\!<} (x \mathrel{-\!\!<} z)\}.$$

To make this false we must take

$$(x \mathrel{-\!\!<} y) = \mathbf{v}$$
$$\{(y \mathrel{-\!\!<} z) \mathrel{-\!\!<} (x \mathrel{-\!\!<} z)\} = \mathbf{f}.$$

The last gives $(y \mathrel{-\!\!<} z) = \mathbf{v}$, $(x \mathrel{-\!\!<} z) = \mathbf{f}$, $x = \mathbf{v}$, $z = \mathbf{f}$.

Substituting these values in

$$(x \mathrel{-\!\!<} y) = \mathbf{v} \quad (y \mathrel{-\!\!<} z) = \mathbf{v}$$

we have

$$(\mathbf{v} \mathrel{-\!\!<} y) = \mathbf{v} \quad (y \mathrel{-\!\!<} \mathbf{f}) = \mathbf{v},$$

which cannot be satisfied together.

As another example, required the conclusion from the following premises. Any one I might marry would be either beautiful or plain; any one whom I might marry would be a woman; any beautiful woman would be an ineligible wife; any plain woman would be an ineligible wife. Let

m be any one whom I might marry,
b, beautiful,
p, plain,
w, woman,
i, ineligible.

Then the premises are

$$m -\!\!< (b -\!\!< \mathbf{f}) -\!\!< p,$$
$$m -\!\!< w,$$
$$w -\!\!< b -\!\!< i,$$
$$w -\!\!< p -\!\!< i.$$

Let x be the conclusion. Then,

$$[m -\!\!< (b -\!\!< \mathbf{f}) -\!\!< p] -\!\!< (m -\!\!< w) -\!\!< (w -\!\!< b -\!\!< i) -\!\!< (w -\!\!< p -\!\!< i) -\!\!< x$$

is necessarily true. Now if we suppose $m = \mathbf{v}$, the proposition can only be made false by putting $w = \mathbf{v}$ and either b or $p = \mathbf{v}$. In this case the proposition can only be made false by putting $i = \mathbf{v}$. If, therefore, x can only be made $\mathbf{f}$ by putting $m = \mathbf{v}$, $i = \mathbf{f}$, that is if $x = (m -\!\!< i)$ the proposition is necessarily true.

In this method, we introduce the two special tokens of second intention $\mathbf{f}$ and $\mathbf{v}$, we retain two indices of tokens x and y, and we have a somewhat complex *icon*, with a special prescription for its use.

A better method may be found as follows. We have seen that

$$x -\!\!< (y -\!\!< z)$$

may be conveniently written $\quad x -\!\!< y -\!\!< z;$
while $\quad (x -\!\!< y) -\!\!< z$

ought to retain the parenthesis. Let us extend this rule, so as to be more general, and hold it necessary *always* to include the antecedent in parenthesis. Thus, let us write $\quad (x) -\!\!< y$

instead of $x -\!\!< y$. If now, we merely change the external appearance of two signs; namely, if we use the vinculum instead of the parenthesis, and the sign $+$ in place of $-\!\!<$, we shall have

$$x -\!\!< y \text{ written } \bar{x} + y$$
$$x -\!\!< y -\!\!< z \quad \text{"} \quad \bar{x} + \bar{y} + z$$
$$(x -\!\!< y) -\!\!< z \quad \text{"} \quad \overline{\bar{x} + \bar{y}} + z, \text{ etc.}$$

We may further write for $x \mathrel{=\!\!<} y$, $\overline{\bar{x} + y}$ implying that $x + y$ is an antecedent for whatever consequent may be taken, and the vinculum becomes identified with the sign of negation. We may also use the sign of multiplication as an abbreviation, putting

$$xy = \overline{\bar{x} + \bar{y}} = \overline{x \mathrel{-\!\!<} \bar{y}}\,.$$

This subjects addition and multiplication to all the rules of ordinary algebra, and also to the following:

$$y + x\bar{x} = y \qquad y(x + \bar{x}) = y$$

$$x + \bar{x} = \mathbf{v} \qquad \bar{x}x = \mathbf{f}$$

$$xy + z = (x + z)(y + z).$$

To any proposition we have a right to add any expression at pleasure; also to strike out any factor of any term. The expressions for different propositions separately known may be multiplied together. These are substantially Mr. Mitchell's rules of procedure. Thus the premises of Barbara are

$$\bar{x} + y \text{ and } \bar{y} + z.$$

Multiplying these, we get $(\bar{x} + y)(\bar{y} + z) = \bar{x}\bar{y} + yz$.
Dropping $\bar{y}$ and y we reach the conclusion $\bar{x} + z$.

III.—*First-intentional Logic of Relatives.*

The algebra of Boole affords a language by which anything may be expressed which can be said without speaking of more than one individual at a time. It is true that it can assert that certain characters belong to a whole class, but only such characters as belong to each individual separately. The logic of relatives considers statements involving two and more individuals at once. Indices are here required. Taking, first, a degenerate form of relation, we may write $x_i y_j$ to signify that x is true of the individual i while y is true of the individual j. If z be a relative character z_{ij} will signify that i is in that relation to j. In this way we can express relations of considerable complexity. Thus, if

$$\begin{matrix} 1, & 2, & 3, \\ 4, & 5, & 6, \\ 7, & 8, & 9, \end{matrix}$$

are points in a plane, and l_{123} signifies that 1, 2, and 3 lie on one line, a well-known proposition of geometry may be written

$$l_{159} \mathrel{-\!\!<} l_{267} \mathrel{-\!\!<} l_{348} \mathrel{-\!\!<} l_{147} \mathrel{-\!\!<} l_{258} \mathrel{-\!\!<} l_{369} \mathrel{-\!\!<} l_{123} \mathrel{-\!\!<} l_{456} \mathrel{-\!\!<} l_{789}.$$

In this notation is involved a *sixth icon.*

We now come to the distinction of *some* and *all*, a distinction which is precisely on a par with that between truth and falsehood; that is, it is descriptive, not metrical.

All attempts to introduce this distinction into the Boolian algebra were more or less complete failures until Mr. Mitchell showed how it was to be effected. His method really consists in making the whole expression of the proposition consist of two parts, a pure Booliau expression referring to an individual and a Quantifying part saying what individual this is. Thus, if k means 'he is a king,' and h, 'he is happy,' the Boolian $(\bar{k}+h)$
means that the individual spoken of is either not a king or is happy. Now, applying the quantification, we may write

$$\text{Any } (\bar{k}+h)$$

to mean that this is true of any individual in the (limited) universe, or

$$\text{Some } (\bar{k}+h)$$

to mean that an individual exists who is either not a king or is happy. So

$$\text{Some } (kh)$$

means some king is happy, and $\text{Any } (kh)$
means every individual is both a king and happy. The rules for the use of this notation are obvious. The two propositions

$$\text{Any } (x) \qquad \text{Any } (y)$$

are equivalent to $\text{Any } (xy)$.
From the two propositions $\text{Any } (x) \qquad \text{Some } (y)$
we may infer $\text{Some } (xy)$.*
Mr. Mitchell has also a very interesting and instructive extension of his notation for *some* and *all*, to a two-dimensional universe, that is, to the logic of relatives. Here, in order to render the notation as iconical as possible we may use Σ for *some*, suggesting a sum, and Π for *all*, suggesting a product. Thus $\Sigma_i x_i$ means that x is true of some one of the individuals denoted by i or

$$\Sigma_i x_i = x_i + x_j + x_k + \text{etc.}$$

---

*I will just remark, quite out of order, that the quantification may be made numerical; thus producing the numerically definite inferences of DeMorgan and Boole. Suppose at least $\frac{2}{3}$ of the company have white neckties and at least $\frac{3}{4}$ have dress coats. Let w mean 'he has a white necktie,' and d 'he has a dress coat.' Then, the two propositions are

$$\tfrac{2}{3}\,(w) \text{ and } \tfrac{3}{4}\,(d).$$

These are to be multiplied together. But we must remember that xy is a mere abbreviation for $\overline{\bar{x}+\bar{y}}$, and must therefore write $\overline{\overline{\tfrac{2}{3}w}+\overline{\tfrac{3}{4}d}}$.
Now $\overline{\tfrac{2}{3}w}$ is the denial of $\tfrac{2}{3}w$, and this denial may be written $(>\tfrac{1}{3})\bar{w}$, or more than $\tfrac{1}{3}$ of the universe (the company) have not white neckties. So $\overline{\tfrac{3}{4}d}=(>\tfrac{1}{4})\bar{d}$. The combined premises thus become

$$\overline{(>\tfrac{1}{3})\bar{w}+(>\tfrac{1}{4})\bar{d}}.$$

Now $(>\tfrac{1}{3})\bar{w}+(>\tfrac{1}{4})\bar{d}$ gives $\text{May be } (\tfrac{1}{3}+\tfrac{1}{4})(\bar{w}+\bar{d})$.
Thus we have $\overline{\text{May be } (\tfrac{7}{12})(\bar{w}+\bar{d})}$,
and this is $(\text{At least } \tfrac{5}{12})(\overline{\bar{w}+\bar{d}})$,
which is the conclusion.

In the same way, $\Pi_i x_i$ means that x is true of all these individuals, or

$$\Pi_i x_i = x_i x_j x_k, \text{ etc.}$$

If x is a simple relation, $\Pi_i \Pi_j x_{ij}$ means that every i is in this relation to every j, $\Sigma_i \Pi_j x_{ij}$ that some one i is in this relation to every j, $\Pi_j \Sigma_i x_{ij}$ that to every j some i or other is in this relation, $\Sigma_i \Sigma_j x_{ij}$ that some i is in this relation to some j. It is to be remarked that $\Sigma_i x_i$ and $\Pi_i x_i$ are only *similar* to a sum and a product; they are not strictly of that nature, because the individuals of the universe may be innumerable.

At this point, the reader would perhaps not otherwise easily get so good a conception of the notation as by a little practice in translating from ordinary language into this system and back again. Let l_{ij} mean that i is a lover of j, and b_{ij} that i is a benefactor of j. Then

$$\Pi_i \Sigma_j l_{ij} b_{ij}$$

means that everything is at once a lover and a benefactor of something; and

$$\Pi_i \Sigma_j l_{ij} b_{ji}$$

that everything is a lover of a benefactor of itself.

$$\Sigma_i \Sigma_k \Pi_j (l_{ij} + \bar{b}_{jk})$$

means that there are two persons, one of whom loves everything except benefactors of the other (whether he loves any of these or not is not stated). Let g_i mean that i is a griffin, and c_i that i is a chimera, then

$$\Sigma_i \Pi_j (g_i l_{ij} + \bar{c}_j)$$

means that if there be any chimeras there is some griffin that loves them all; while

$$\Sigma_i \Pi_j g_i (l_{ij} + \bar{c}_j)$$

means that there is a griffin and he loves every chimera that exists (if any exist). On the other hand,

$$\Pi_j \Sigma_i g_i (l_{ij} + \bar{c}_j)$$

means that griffins exist (one, at least), and that one or other of them loves each chimera that may exist; and

$$\Pi_j \Sigma_i (g_i l_{ij} + \bar{c}_j)$$

means that each chimera (if there is any) is loved by some griffin or other.

Let us express: every part of the world is either sometimes visited with cholera, and at others with small-pox (without cholera), or never with yellow fever and the plague together. Let

c_{ij} mean the place i has cholera at the time j.
s_{ij} " " " small-pox " "
y_{ij} " " " yellow fever " "
p_{ij} " " " plague " "

Then we write

$$\Pi_i \Sigma_j \Sigma_k \Pi_l (c_{ij} \bar{c}_{ik} s_{ik} + \bar{y}_{il} + \bar{p}_{il}).$$

Let us express this: one or other of two theories must be admitted, 1st, that no man is at any time unselfish or free, and some men are always hypocritical, and at every time some men are friendly to men to whom they are at other times inimical, or 2d, at each moment all men are alike either angels or fiends. Let

u_{ij}	mean	the man	i is	unselfish	at the time	j,
f_{ij}	"	"	"	free	"	"
h_{ij}	"	"	"	hypocritical	"	"
a_{ij}	"	"	"	an angel	"	"
d_{ij}	"	"	"	a fiend	"	"
p_{ijk}	"	"	"	friendly	"	"

to the man k,

e_{ijk} the man i is an enemy at the time j to the man k;

1_{jm} the two objects j and m are identical.

Then the proposition is

$$\Pi_i \Sigma_h \Pi_j \Sigma_k \Sigma_l \Sigma_m \Pi_n \Pi_p \Pi_q (\bar{u}_{ij} \bar{f}_{ij} h_{hj} p_{kjl} e_{kml} \bar{1}_{jm} + a_{pn} + d_{qn})$$

We have now to consider the procedure in working with this calculus. It is far from being true that the only problem of deduction is to draw a conclusion from given premises. On the contrary, it is fully as important to have a method for ascertaining what premises will yield a given conclusion. There are besides other problems of transformation, where a certain system of facts is given, and it is required to describe this in other terms of a definite kind. Such, for example, is the problem of the 15 young ladies, and others relating to synthemes. I shall, however, content myself here with showing how, when a set of premises are given, they can be united and certain letters eliminated. Of the various methods which might be pursued, I shall here give the one which seems to me the most useful on the whole.

1st. The different premises having been written with distinct indices (the same index not used in two propositions) are written together, and all the Π's and Σ's are to be brought to the left. This can evidently be done, for

$$\Pi_i x_i . \Pi_j x_j = \Pi_i \Pi_j x_i x_j$$
$$\Sigma_i x_i . \Pi_j x_j = \Sigma_i \Pi_j x_i x_j$$
$$\Sigma_i x_i . \Sigma_j x_j = \Sigma_i \Sigma_j x_i x_j .$$

2d. Without deranging the order of the indices of any one premise, the Π's and Σ's belonging to different premises may be moved relatively to one another, and as far as possible the Σ's should be carried to the left of the Π's. We have

$$\Pi_i \Pi_j x_{ij} = \Pi_j \Pi_i x_{ij}$$

$$\Sigma_i \Sigma_j x_{ij} = \Sigma_j \Sigma_i x_{ij}$$

and also

$$\Sigma_i \Pi_j x_i y_j = \Pi_j \Sigma_i x_i y_j .$$

But this formula does not hold when the i and j are not separated. We do have, however,

$$\Sigma_i \Pi_j x_{ij} \prec \Pi_i \Sigma_j x_{ij} .$$

It will, therefore, be well to begin by putting the Σ's to the left, as far as possible, because at a later stage of the work they can be carried to the right but not to the left. For example, if the operators of the two premises are $\Pi_i \Sigma_j \Pi_k$ and $\Sigma_x \Pi_y \Sigma_z$, we can unite them in either of the two orders

$$\Sigma_x \Pi_y \Sigma_z \Pi_i \Sigma_j \Pi_k$$

$$\Sigma_x \Pi_i \Sigma_j \Pi_y \Sigma_z \Pi_k ,$$

and shall usually obtain different conclusions accordingly. There will often be room for skill in choosing the most suitable arrangement.

3d. It is next sometimes desirable to manipulate the Boolian part of the expression, and the letters to be eliminated can, if desired, be eliminated now. For this purpose they are replaced by relations of second intention, such as "other than," etc. If, for example, we find anywhere in the expression

$$a_{ijk} \bar{a}_{xyz} ,$$

this may evidently be replaceable by

$$(n_{ix} + n_{jy} + n_{kz})$$

where, as usual, n means not or other than. This third step of the process is frequently quite indispensable, and embraces a variety of processes; but in ordinary cases it may be altogether dispensed with.

4th. The next step, which will also not commonly be needed, consists in making the indices refer to the same collections of objects, so far as this is useful. If the quantifying part, or Quantifier, contains Σ_x, and we wish to replace the x by a new index i, not already in the Quantifier, and such that every x is an i, we can do so at once by simply multiplying every letter of the Boolian having x as an index by x_i. Thus, if we have "some woman is an angel" written in the form $\Sigma_w a_w$ we may replace this by $\Sigma_i (a_i w_i)$. It will be more often useful to replace the index of a Π by a wider one; and this will be done by adding $\bar{x}_i$ to every letter having x as an index. Thus, if we have "all dogs are animals, and all animals are vertebrates" written thus

$$\Pi_d a_d \; \Pi_a v_a ,$$

where a and α alike mean animal, it will be found convenient to replace the last index by i, standing for any object, and to write the proposition

$$\Pi_i(\bar{a}_i + v_i).$$

5th. The next step consists in multiplying the whole Boolian part, by the modification of itself produced by substituting for the index of any Π any other index standing to the left of it in the Quantifier. Thus, for

$$\Sigma_i \Pi_j l_{ij},$$

we can write

$$\Sigma_i \Pi_j l_{ij} l_{ii}.$$

6th. The next step consists in the re-manipulation of the Boolian part, consisting, 1st, in adding to any part any term we like; 2d, in dropping from any part any factor we like, and 3d, in observing that

$$x\bar{x} = \mathbf{f}, \qquad x + \bar{x} = \mathbf{v},$$

so that

$$x\bar{x}y + z = z \quad (x + \bar{x} + y)\,z = z.$$

7th. Π's and Σ's in the Quantifier whose indices no longer appear in the Boolian are dropped.

The fifth step will, in practice, be combined with part of the sixth and seventh. Thus, from $\Sigma_i \Pi_j l_{ij}$ we shall at once proceed to $\Sigma_i l_{ii}$ if we like.

The following examples will be sufficient.

From the premises $\Sigma_i a_i b_i$ and $\Pi_j(\bar{b}_j + c_j)$, eliminate b. We first write

$$\Sigma_i \Pi_j a_i b_i (\bar{b}_j + c_j).$$

The distributive process gives

$$\Sigma_i \Pi_j a_i (b_i \bar{b}_j + b_i c_j).$$

But *we always have a right to drop a factor or insert an additive term.* We thus get

$$\Sigma_i \Pi_j a_i (b_i \bar{b}_j + c_j).$$

By the third process, we can, if we like, insert n_{ij} for $b_i \bar{b}_j$. In either case, we identify j with i and get the conclusion

$$\Sigma_i a_i c_i.$$

Given the premises

$$\Sigma_h \Pi_i \Sigma_j \Pi_k (a_{hik} + s_{jk} l_{ji})$$

$$\Sigma_u \Sigma_v \Pi_x \Pi_y (\varepsilon_{uyx} + \bar{s}_{yv} b_{vx}).$$

Required to eliminate s. The combined premise is

$$\Sigma_u \Sigma_v \Sigma_h \Pi_i \Sigma_j \Pi_x \Pi_k \Pi_y (a_{hik} + s_{jk} l_{ji})(\varepsilon_{uyx} + \bar{s}_{yv} b_{vx}).$$

Identify k with v and y with j, and we get

$$\Sigma_u \Sigma_v \Sigma_h \Pi_i \Sigma_j \Pi_x (a_{hiv} + s_{jv} l_{ji})(\varepsilon_{ujx} + \bar{s}_{jv} b_{vx}).$$

The Boolian part then reduces, so that the conclusion is

$$\Sigma_u \Sigma_v \Sigma_h \Pi_i \Sigma_j \Pi_x (a_{hiv} \varepsilon_{ujx} + a_{hiv} b_{vx} + \varepsilon_{ujx} l_{ji}).$$

IV.—*Second-intentional Logic.*

Let us now consider the logic of terms taken in collective senses. Our notation, so far as we have developed it, does not show us even how to express that two indices, i and j, denote one and the same thing. We may adopt a special token of second intention, say 1, to express identity, and may write 1_{ij}. But this relation of identity has peculiar properties. The first is that if i and j are identical, whatever is true of i is true of j. This may be written

$$\Pi_i \Pi_j \{ \bar{1}_{ij} + \bar{x}_i + x_j \}.$$

The use of the general index of a token, x, here, shows that the formula is iconical. The other property is that if everything which is true of i is true of j, then i and j are identical. This is most naturally written as follows: Let the token, q, signify the relation of a quality, character, fact, or predicate to its subject. Then the property we desire to express is

$$\Pi_i \Pi_j \Sigma_k (1_{ij} + \bar{q}_{ki} q_{kj}).$$

And identity is defined thus $1_{ij} = \Pi_k (q_{ki} q_{kj} + \bar{q}_{ki} \bar{q}_{kj})$.

That is, to say that things are identical is to say that every predicate is true of both or false of both. It may seem circuitous to introduce the idea of a quality to express identity; but that impression will be modified by reflecting that $q_{ki} q_{jk}$ merely means that i and j are both within the class or collection k. If we please, we can dispense with the token q, by using the index of a token and by referring to this in the Quantifier just as subjacent indices are referred to. That is to say, we may write $1_{ij} = \Pi_x (x_i x_j + \bar{x}_i \bar{x}_j)$.

The properties of the token q must now be examined. These may all be summed up in this, that taking any individuals i_1, i_2, i_3, etc., and any individuals, j_1, j_2, j_3, etc., there is a collection, class, or predicate embracing all the i's and excluding all the j's except such as are identical with some one of the i's. This might be written

$$(\Pi_\alpha \Pi_{i_\alpha})(\Pi_\beta \Pi_{j_\beta}) \Sigma_k (\Pi_\alpha \Sigma_{i'_\alpha}) \Pi_l \; q_{ki_\alpha} (\bar{q}_{kj_\beta} + q_{li'_\alpha} q_{lj_\beta} + \bar{q}_{li'_\alpha} \bar{q}_{lj_\beta}),$$

where the i's and the i''s are the same lot of objects. This notation presents indices of indices. The $\Pi_\alpha \Pi_{i_\alpha}$ shows that we are to take any collection whatever of i's, and then any i of that collection. We are then to do the same with the j's. We can then find a quality k such that the i taken has it, and also such that the j taken wants it unless we can find an i that is identical with the j taken. The necessity of some kind of notation of this description in treating of classes collectively appears from this consideration: that in such discourse we are neither speaking of a single individual (as in the non-relative logic) nor of a small number of individuals considered each for itself, but of a whole class, perhaps an infinity of individuals. This suggests a relative term with an indefinite series of indices as $x_{ijkl\ldots}$. Such a relative will, however, in most, if not in all cases, be of a degenerate kind and is consequently expressible as above. But it seems preferable to attempt a partial decomposition of this definition. In the first place, any individual may be considered as a class. This is written

$$\Pi_i \Sigma_k \Pi_j \; q_{ki}(\bar{q}_{kj} + 1_{ij}).$$

This is the *ninth icon.* Next, given any class, there is another which includes all the former excludes and excludes all the former includes. That is,

$$\Pi_l \Sigma_k \Pi_i (q_{li}\bar{q}_{ki} + \bar{q}_{li}q_{ki}).$$

This is the *tenth icon.* Next, given any two classes, there is a third which includes all that either includes and excludes all that both exclude. That is

$$\Pi_l \Pi_m \Sigma_k \Pi_i (q_{li}q_{ki} + q_{mi}q_{ki} + \bar{q}_{li}\bar{q}_{mi}\bar{q}_{ki}).$$

This is the *eleventh icon.* Next, given any two classes, there is a class which includes the whole of the first and any one individual of the second which there may be not included in the first and nothing else. That is,

$$\Pi_l \Pi_m \Pi_i \Sigma_k \Pi_j \{q_{li} + \bar{q}_{mi} + q_{ki}(q_{kj} + \bar{q}_{lj})\}.$$

This is the *twelfth icon.*

To show the manner in which these formulæ are applied let us suppose we have given that everything is either true of i or false of j. We write

$$\Pi_k (q_{ki} + \bar{q}_{kj}).$$

The tenth icon gives $\Pi_l \Sigma_k (q_{li}\bar{q}_{ki} + \bar{q}_{li}q_{ki})(q_{lj}\bar{q}_{kj} + \bar{q}_{lj}q_{kj})$

Multiplication of these two formulæ give

$$\Pi_l \Sigma_k (q_{ki}\bar{q}_{li} + q_{lj}\bar{q}_{kj}),$$

or, dropping the terms in k $\qquad \Pi_l (\bar{q}_{li} + q_{lj}).$

Multiplying this with the original datum and identifying l with k, we have

$$\Pi_k (q_{ki}q_{kj} + \bar{q}_{ki}\bar{q}_{kj}).$$

No doubt, a much more direct method of procedure could be found.

Just as q signifies the relation of predicate to subject, so we need another token, which may be written r, to signify the conjoint relation of a simple relation, its relate and its correlate. That is, r_{jai} is to mean that i is in the relation a to j. Of course, there will be a series of properties of r similar to those of q. But it is singular that the uses of the two tokens are quite different. Namely, the chief use of r is to enable us to express that the number of one class is at least as great as that of another. This may be done in a variety of different ways. Thus, we may write that for every a there is a b, in the first place, thus:

$$\Sigma_a \Pi_i \Sigma_j \Pi_h \{\bar{a}_i + b_j r_{jai}(\bar{r}_{jah} + \bar{a}_h + 1_{ih})\}.$$

But, by an icon analogous to the eleventh, we have

$$\Pi_\alpha \Pi_\beta \Sigma_\gamma \Pi_u \Pi_v (r_{u\alpha v} r_{u\gamma v} + r_{u\beta v} r_{u\gamma v} + \bar{r}_{u\alpha v} \bar{r}_{u\beta v} \bar{r}_{u\gamma v}).$$

From this, by means of an icon analogous to the *tenth*, we get the general formula

$$\Pi_\alpha \Pi_\beta \Sigma_\gamma \Pi_u \Pi_v \{r_{u\alpha v} r_{u\beta v} r_{u\gamma v} + \bar{r}_{u\gamma v}(\bar{r}_{u\alpha v} + \bar{r}_{u\beta v})\}.$$

For $r_{u\beta v}$ substitute a_u and multiply by the formula the last but two. Then, identifying u with h and v with j, we have

$$\Sigma_a \Pi_i \Sigma_h \Pi_h \{\bar{a}_i + b_j r_{jai}(\bar{r}_{jah} + 1_{ih})\}$$

a somewhat simpler expression. However, the best way to express such a proposition is to make use of the letter c as a token of a one-to-one correspondence. That is to say, c will be defined by the three formulæ,

$$\Pi_a \Pi_u \Pi_v \Pi_w (\bar{c}_a + \bar{r}_{uav} + \bar{r}_{uaw} + 1_{vw})$$
$$\Pi_a \Pi_u \Pi_v \Pi_w (\bar{c}_a + \bar{r}_{uaw} + \bar{r}_{vaw} + 1_{uv})$$
$$\Pi_a \Sigma_u \Sigma_v \Sigma_w (c_a + r_{uav} r_{uaw} \bar{1}_{vw} + r_{uaw} r_{vaw} \bar{1}_{uv}).$$

Making use of this token, we may write the proposition we have been considering in the form

$$\Sigma_a \Pi_i \Sigma_j \; c_a(\bar{a}_i + b_j r_{jai}).$$

In an appendix to his memoir on the logic of relatives, DeMorgan enriched the science of logic with a new kind of inference, the syllogism of transposed quantity. DeMorgan was one of the best logicians that ever lived and unquestionably the father of the logic of relatives. Owing, however, to the imperfection of his theory of relatives, the new form, as he enunciated it, was a down-right paralogism, one of the premises being omitted. But this being supplied, the form furnishes a good test of the efficacy of a logical notation. The following is one of DeMorgan's examples:

Some X is Y,
For every X there is something neither Y nor Z;
Hence, something is neither X nor Z.

The first premise is simply $\Sigma_a x_a y_a$.
The second may be written

$$\Sigma_a \Pi_i \Sigma_j \; c_a(\bar{x}_i + r_{jai} \bar{y}_j \bar{z}_j).$$

From these two premises, little can be inferred. To get the above conclusion it is necessary to add that the class of X's is a finite collection; were this not necessary the following reasoning would hold good (the limited universe consisting of numbers); for it precisely conforms to DeMorgan's scheme.

> Some odd number is prime;
> Every odd number has its square, which is neither prime nor even;
> Hence, some number is neither odd nor even.*

Now, to say that a lot of objects is finite, is the same as to say that if we pass through the class from one to another we shall necessarily come round to one of those individuals already passed; that is, if every one of the lot is in any one-to-one relation to one of the lot, then to every one of the lot some one is in this same relation. This is written thus:

$$\Pi_\beta \Pi_u \Sigma_v \Sigma_s \Pi_t \{\bar{c}_\beta + \bar{x}_u + x_v r_{u\beta v} + x_s(\bar{x}_t + \bar{r}_{t\beta s})\}$$

Uniting this with the two premises and the second clause of the definition of c, we have

$$\Sigma_a \Sigma_a \Pi_\beta \Pi_u \Sigma_v \Sigma_s \Pi_i \Sigma_j \Pi_t \Pi_\gamma \Pi_e \Pi_f \Pi_g \, x_a y_a c_a (\bar{x}_i + r_{jai} \bar{y}_j \bar{z}_j)$$
$$\{\bar{c}_\beta + \bar{x}_u + x_v r_{u\beta v} + x_s(\bar{x}_t + \bar{r}_{t\beta s})\}(\bar{c}_\gamma + \bar{r}_{e\gamma g} + \bar{r}_{f\gamma w} + 1_{ef}).$$

We now substitute a for β and for γ, a for u and for e, j for t and for f, v for g. The factor in i is to be repeated, putting first s and then v for i. The Boolian part thus reduces to

$$(\bar{x}_s + r_{jas} \bar{y}_j \bar{z}_j) c_a x_a y_a r_{aav} x_v r_{jav} \bar{y}_j \bar{z}_j 1_{aj} + r_{jas} \bar{y}_j \bar{z}_j x_s \bar{x}_j (\bar{x}_v + r_{jav} \bar{y}_j \bar{z}_j)(\bar{r}_{aav} + \bar{r}_{jav} + 1_{aj}),$$

which, by the omission of factors, becomes

$$y_a \bar{y}_j 1_{aj} + \bar{x}_j \bar{z}_j.$$

Thus we have the conclusion $\Sigma_j \bar{x}_j \bar{z}_j$.

It is plain that by a more iconical and less logically analytical notation this procedure might be much abridged. How minutely analytical the present system is, appears when we reflect that every substitution of indices of which nine were used in obtaining the last conclusion is a distinct act of inference. The annulling of $(y_a \bar{y}_j 1_{aj})$ makes ten inferential steps between the premises and conclusion of the syllogism of transposed quantity.

---

* Another of DeMorgan's examples is this: "Suppose a person, on reviewing his purchases for the day, finds, by his counterchecks, that he has certainly drawn as many checks on his banker (and maybe more) as he has made purchases. But he knows that he paid some of his purchases in money, or otherwise than by checks. He infers then that he has drawn checks for something else except that day's purchases. He infers rightly enough." Suppose, however, that what happened was this: He bought something and drew a check for it; but instead of paying with the check, he paid cash. He then made another purchase for the same amount, and drew another check. Instead, however, of paying with that check, he paid with the one previously drawn. And thus he continued without cessation, or *ad infinitum*. Plainly the premises remain true, yet the conclusion is false.

Sur les Equations Linéaires aux Différentielles ordinaires et aux Différences finies.

PAR H. POINCARÉ.

§1. *Etude sommaire des Intégrales Irrégulières.*

Les résultats que je vais chercher à démontrer dans le préseut mémoire et qui se rapportent tant à certaines équations différentielles linéaires qu'à des équations analogues, mais à différences finies, ont déjà été énoncés les uns dans un mémoire que j'ai présenté à l'Académie des Sciences pour le concours du Grand Prix des Sciences Mathématiques le 1er Juin 1880 et qui est resté inédit, les autres dans une communication verbale faite à la Société Mathématique de France en Novembre 1882 et dans une note insérée aux Comptes Rendus de l'Académie des Sciences le 5 Mars 1883.

Soit:

$$P_n \frac{d^n y}{dx^n} + P_{n-1} \frac{d^{n-1} y}{dx^{n-1}} + \ldots + P_1 \frac{dy}{dx} + P_0 y = 0 \tag{1}$$

une équation différentielle linéaire où les coëfficients P seront des polynômes en x que je supposerai tous de même degré, à savoir de degré p. J'appellerai A_i le coëfficient de x^p dans le polynôme P_i.

Nous allons étudier la façon dont se comportent les intégrales de l'équation (1) quand x croît indéfiniment d'une certaine manière, par exemple par valeurs réelles positives. Il reste donc convenu jusqu'à nouvel ordre que x est réel et positif, tandis que les intégrales y et les coëfficients des polynômes P peuvent être imaginaires.

Nous allons avoir à considérer l'équation algébrique:

$$A_n z^n + A_{n-1} z^{n-1} + \ldots + A_1 z + A_0 = 0. \tag{2}$$

Nous supposerons d'abord que cette équation n'a pas de racines multiples, et même qu'elle n'a pas deux racines ayant même partie réelle.

Les méthodes de M. Fuchs ne sont pas applicables au problème qui nous occupe, parce que les intégrales de l'équation (1) sont *irrégulières* dans le voisinage du point $x = \infty$. Il faut donc employer des procédés particuliers.

Nous poserons: $$\frac{P_i}{P_n} = Q_i \quad \frac{A_i}{A_n} = B_i$$

et, supposant d'abord l'équation (1) du 2[d] ordre, nous l'écrirons:

$$\frac{d^2y}{dx^2} + Q_1\frac{dy}{dx} + Q_0 y = 0.$$

Posons: $$y = e^{\int u dx}$$

l'équation différentielle deviendra:

$$\frac{du}{dx} + u^2 + Q_1 u + Q_0 = 0.$$

Je dis que quand x croîtra indéfiniment, u tendra vers une des racines de l'équation (2). Soient en effet α et β les deux racines de cette équation de telle sorte que: $$(z-\alpha)(z-\beta) = z^2 + B_1 z + B_0$$

et que la partie réelle de α soit plus grande que celle de β.

Soit: $$V = v + iv' = \log(u-\alpha) - \log(u-\beta).$$

Il viendra $$\frac{dV}{dx} = (\beta-\alpha)\frac{u^2 + Q_1 u + Q_0}{u^2 + B_1 u + B_0}.$$

Nous allons étudier le signe de la partie réelle de $\frac{dV}{dx}$, c'est à dire de $\frac{dv}{dx}$. Si l'on donne à x une valeur très grande, les différences $Q_1 - B_1$ et $Q_0 - B_0$ sont très petites de l'ordre de $\frac{1}{x}$. Cela posé, on peut démontrer successivement les résultats suivants.

Supposons que $Q_1 - B_1$ et $Q_0 - B_0$ aient des valeurs *données* suffisamment petites, et soit K un nombre donné positif. On peut trouver deux nombres ε et ε_1 tels que toutes les fois que:

$$|u-\alpha| > \varepsilon \qquad |u-\beta| > \varepsilon_1$$

on ait également

$$\left|\frac{u(Q_1-B_1)+(Q_0-B_0)}{u^2+B_1 u+B_0}\right| < K. \tag{3}$$

De plus lorsque $Q_1 - B_1$ et $Q_0 - B_0$ tendront simultanément vers 0, K ne variant pas, ε et ε_1 tendront aussi simultanément vers 0.

En second lieu, on peut toujours trouver un nombre K assez petit pour que $\frac{dv}{dx}$ soit négatif comme la partie réelle de $(\beta-\alpha)$ lorsque l'inégalité (3) a lieu.

Il suffit pour cela que l'on ait :

$$K < \cos [\arg (\alpha - \beta)] .$$

Enfin on peut trouver deux nombres k et k_1 tels que les inégalités

$$|u - \alpha| < \varepsilon \qquad |u - \beta| < \varepsilon_1$$

aient lieu toutes les fois que v est compris entre k et $-k_1$.

On conclut de tout cela que si x est suffisamment grand, il existe deux nombres k et k_1 tels que $\frac{dv}{dx}$ soit négatif toutes les fois que v est compris entre k et $-k_1$; de plus lorsque x croît constamment et indéfiniment, k et k_1 croissent aussi constamment et indéfiniment.

Supposons que pour une valeur donnée de x, v ait une certaine valeur initiale comprise entre k et $-k_1$, on est certain que v va décroître tant qu'il sera supérieur à $-k_1$, et que, si après avoir décrû, il arrive qu'il croît de nouveau, il ne pourra jamais en tous cas redevenir supérieur à $-k_1$.

Soit $M(h)$ la plus grande valeur que puisse prendre v quand x varie de h à $+\infty$. Lorsque h croîtra, $M(h)$ décroîtra ou du moins ne pourra jamais croître. Donc quand h grandira indéfiniment, $M(h)$ tendra vers une limite *finie ou infinie* que j'appellerai M. Si $M = -\infty$, on est certain que v tend vers $-\infty$; tandis que si M était fini, il pourrait arriver ou bien que v tendît vers la limite M, ou que v ne tendît vers aucune limite. Dans le cas qui nous occupe on vient de voir qu'on peut prendre h assez grand pour que l'on ait :

$$M(h) < -k_1,$$

d'où :

$$M < -k_1.$$

Mais nous pouvons prendre x assez grand pour que k_1 soit aussi grand que l'on veut. On a donc :

$$M = -\infty$$

ou

$$\lim v = -\infty \qquad \lim u = \alpha. \qquad\qquad C. Q. F. D.$$

Le raisonnement précédent n'est en défaut que si la valeur initiale de v n'est pas comprise entre k et $-k_1$. Mais nous avons choisi arbitrairement la valeur initiale de x, nous aurions pu prendre tout aussi bien une valeur quelconque de cette variable. Pour que le raisonnement soit en défaut, il faut donc que, quel que soit x, v soit plus grand que k et plus petit que $-k_1$. Or quand x tend vers l'infini, il en est de même de k et de k_1. Donc v tend aussi vers $\pm\infty$. Donc u tend vers β ou vers α. En résumé la limite de u est en général α, mais pour une intégrale particulière, elle peut être égale à β.

Faisons encore le raisonnement pour les équations du 3ème ordre. L'équation :

$$\frac{d^3y}{dx^3} + Q_2 \frac{d^2y}{dx^2} + Q_1 \frac{dy}{dx} + Q_0 y = 0,$$

peut s'écrire

$$(4) \qquad \frac{d^2u}{dx^2} + \frac{du}{dx}(3u + Q_2) + u^3 + Q_2u^2 + Q_1u + Q_0 = 0.$$

Soient α, β, γ les trois racines de l'équation (2) rangées par ordre de parties réelles décroissantes. Nous considérerons à côté de l'équation (4) l'équation

$$(4^{\text{bis}}) \qquad \frac{d^2v}{dx^2} + \frac{dv}{dx}(3v + B_2) + v^3 + B_2v^2 + B_1v + B_0 = 0,$$

dont l'intégrale générale est: $v = \dfrac{\lambda\alpha e^{\alpha x} + \mu\beta e^{\beta x} + \nu\gamma e^{\gamma x}}{\lambda e^{\alpha x} + \mu e^{\beta x} + \nu e^{\gamma x}}$

λ, μ et ν étant les constantes introduites par l'intégration. Nous allons chercher une fonction réelle des parties réelles et imaginaires de v et de $\frac{dv}{dx}$ choisie de telle sorte que sa dérivée soit toujours négative. Nous considérerons ensuite une fonction formée de la même manière avec les parties réelles et imaginaires de u et de $\frac{du}{dx}$, et nous reconnaîtrons que la dérivée de cette nouvelle fonction sera aussi toujours négative pourvu que la fonction elle-même soit comprise entre deux limites données, lesquelles limites tendent respectivement vers $\pm\infty$, quand x tend vers $+\infty$. La méthode que nous suivrons sera donc de tout point semblable à celle que nous avons employée pour le cas du 2^d ordre.

La fonction que nous cherchons à former dépend de u et de $\frac{du}{dx}$ et par conséquent de y, de $\frac{dy}{dx}$ et de $\frac{d^2y}{dx^2}$. Il y a avantage à y introduire directement ces éléments.

Employons pour abréger la notation de Lagrange de façon que y' désigne $\frac{dy}{dx}$ et que y'' designe $\frac{d^2y}{dx^2}$ et posons:

$$y = X + Y + Z,$$
$$y' = \alpha X + \beta Y + \gamma Z,$$
$$y'' = \alpha^2X + \beta^2Y + \gamma^2Z.$$

La différentiation nous donnera:

$$y' = X' + Y' + Z',$$
$$y'' = \alpha X' + \beta Y + \gamma Z,$$
$$-Q_2y'' - Q_1y - Q_0y = \alpha^2X' + \beta^2Y' + \gamma^2Z'.$$

Posons encore:

$$\alpha^3 + Q_2\alpha^2 + Q_1\alpha + Q_0 = A(\alpha - \beta)(\alpha - \gamma)$$
$$\beta^3 + Q_2\beta^2 + Q_1\beta + Q_0 = B(\beta - \alpha)(\beta - \gamma)$$
$$\gamma^3 + Q_2\gamma^2 + Q_1\gamma + Q_0 = C(\gamma - \alpha)(\gamma - \beta)$$

il viendra :

$$X' = \alpha X - (AX + BY + CZ)$$
$$Y' = \beta Y - (AX + BY + CZ)$$
$$Z' = \gamma Z - (AX + BY + CZ).$$

On a alors :

$$\frac{d}{dx}\log\frac{Y}{X} = \beta - \alpha + A - B - A\frac{X}{Y} + B\frac{Y}{X} + C\left(\frac{Z}{X} - \frac{Z}{Y}\right) = \beta - \alpha + \Delta$$

avec des expressions analogues pour les dérivées logarithmiques de $\frac{Z}{X}$ et de $\frac{Z}{Y}$.

Lorsque x croît indéfiniment, A, B et C et par conséquent le terme complimentaire Δ tendent vers 0. La variable x ayant une valeur donnée suffisamment grande, on peut trouver un nombre positif ε tel que l'expression :

$$(|A| + |B| + |C|)\left(1 + \frac{2}{\varepsilon}\right)$$

soit plus petite que la partie réelle de $\alpha - \beta$ et que celle de $\beta - \gamma$. Si alors les valeurs absolues :

$$\left|\frac{X}{Y}\right|;\ \left|\frac{Y}{X}\right|,\ \left|\frac{X}{Z}\right|,\ \left|\frac{Z}{X}\right|,\ \left|\frac{Y}{Z}\right|,\ \left|\frac{Z}{Y}\right|$$

sont simultanément plus grandes que ε, on aura :

$$|\Delta| < (|A| + |B| + |C|)\left(1 + \frac{2}{\varepsilon}\right)$$

et par conséquent, les dérivées logarithmiques de $\frac{Y}{X}$, de $\frac{Z}{Y}$ et de $\frac{Z}{X}$ auront leurs parties réelles négatives, de sorte que :

$$\frac{d}{dx}\log\left|\frac{Y}{X}\right| < 0, \qquad \frac{d}{dx}\log\left|\frac{Z}{X}\right| < 0, \qquad \frac{d}{dx}\log\left|\frac{Z}{Y}\right| < 0.$$

De plus lorsque x croîtra indéfiniment, ε tendra vers 0. Ajoutons d'ailleurs que la dérivée logarithmique de $\left|\frac{Y}{X}\right|$ reste négative quand même $\left|\frac{Z}{X}\right|$ ou $\left|\frac{Z}{Y}\right|$ seraient plus petits que ε. De même on aura :

$$\frac{d}{dx}\log\left|\frac{Z}{X}\right| < 0 \qquad \text{même si } \left|\frac{Y}{Z}\right| \text{ ou } \left|\frac{Y}{X}\right| < \varepsilon,$$
$$\frac{d}{dx}\log\left|\frac{Z}{Y}\right| < 0 \qquad \text{même si } \left|\frac{X}{Z}\right| \text{ ou } \left|\frac{X}{Y}\right| < \varepsilon.$$

Soit maintenant H la plus grande des deux quantités $\left|\frac{Y}{X}\right|$ et $\left|\frac{Z}{X}\right|$. Quelle est la condition pour que H soit une fonction décroissante de x? Je dis qu'il suffit que H soit compris entre ε et $\frac{1}{\varepsilon}$, ε étant bien entendu supposé plus petit que 1.

En effet nous pouvons faire deux hypothèses.

1°. $$H=\left|\frac{Y}{X}\right|>\left|\frac{Z}{X}\right|$$

on a alors $$\left|\frac{Y}{X}\right|>\varepsilon, \quad \left|\frac{X}{Z}\right|>\left|\frac{X}{Y}\right|>\varepsilon, \quad \left|\frac{Y}{Z}\right|>1>\varepsilon,$$

et par conséquent: $$\frac{dH}{dx}=\frac{d}{dx}\left|\frac{Y}{X}\right|<0.$$ *C. Q. F. D.*

2°. $$H=\left|\frac{Z}{X}\right|>\left|\frac{Y}{X}\right|.$$

On a $$\left|\frac{Z}{X}\right|>\varepsilon, \quad \left|\frac{Z}{Y}\right|>1>\varepsilon, \quad \left|\frac{X}{Y}\right|>\left|\frac{X}{Z}\right|>\varepsilon,$$

et par conséquent: $$\frac{dH}{dx}=\frac{d}{dx}\left|\frac{Z}{X}\right|<0.$$ *C. Q. F. D.*

Ainsi H décroît toutes les fois qu'il est compris entre ε et $\frac{1}{\varepsilon}$.

Or pour $x=\infty$ on a: $\lim \varepsilon = 0$.

Donc on a aussi:

$$\lim H=0, \quad \lim\frac{Y}{X}=0, \quad \lim\frac{Z}{X}=0,$$

d'où l'on déduit aisément

$$\lim u=\lim\frac{y'}{y}=\alpha.$$ *C. Q. F. D.*

Il n'y aurait d'exception que si H restait constamment supérieur à $\frac{1}{\varepsilon}$, auquel cas sa limite serait infinie. Dans ce cas, on a toujours:

$$\left|\frac{Z}{X}\right|>\varepsilon, \quad \left|\frac{Y}{X}\right|>\varepsilon,$$

toutes les fois que $\left|\frac{Z}{Y}\right|$ est compris entre ε et $\frac{1}{\varepsilon}$. En effet il vient, ou bien

$$H=\frac{Y}{X}>\frac{1}{\varepsilon}>\varepsilon, \quad \left|\frac{Z}{Y}\right|>\varepsilon\left|\frac{Y}{Z}\right|>\varepsilon$$

d'où $$\left|\frac{Z}{X}\right|=\left|\frac{Z}{Y}\right|\,\left|\frac{Y}{X}\right|>1>\varepsilon,$$

ou bien $$H=\left|\frac{Z}{X}\right|>\frac{1}{\varepsilon}>\varepsilon, \quad \left|\frac{Z}{Y}\right|>\varepsilon \quad \left|\frac{Y}{Z}\right|>\varepsilon,$$

d'où $$\left|\frac{Y}{X}\right|=\left|\frac{Y}{Z}\right|\,\left|\frac{Z}{X}\right|>1>\varepsilon.$$

D'où l'on doit conclure que la fonction $\left|\frac{Z}{Y}\right|$ est décroissante toutes les fois qu'elle est comprise entre ε et $\frac{1}{\varepsilon}$; il en résulte, comme nous l'avons fait voir plusieurs fois, que cette fonction tend *en général* vers 0, et qu'elle peut aussi, *mais exceptionnellement* tendre vers l'∞. Dans le premier cas on a :

$$\lim u = \beta,$$

dans le second :

$$\lim u = \gamma. \qquad C.\ Q.\ F.\ D.$$

Il n'est pas besoin d'insister pour faire comprendre que ce raisonnement est applicable à une équation d'ordre quelconque. Dans tous les cas la limite de la dérivée logarithmique de y est une des racines de l'équation (2).

De ce que la limite de $\frac{y'}{y}$ est égale à un nombre fini et déterminé α, il ne s'en suit pas forcément que $\frac{y}{e^{\alpha x}}$ tende vers une limite finie et determinée ; car si l'on avait par exemple $y = xe^{\alpha x}$ il viendrait :

$$\lim \frac{y'}{y} = \alpha, \qquad \lim \frac{y}{e^{\alpha x}} = \infty.$$

Ce n'est que dans un paragraphe ultérieur que nous démontrerons que la limite $\frac{y}{x^m e^{\alpha x}}$ est en général finie et déterminée.

Pour le moment supposons que x tende vers l'infini de façon que l'on ait :

$$x = \rho\lambda \qquad \lambda = e^{i\omega}$$

ρ croissant indéfiniment par valeurs réelles positives et λ étant une quantité constante d'argument ω, et de module 1. Il est facile de ramener ce cas au précédent.

En effet l'équation (1) devient :

$$(1^{\text{bis}}) \qquad \frac{P_n}{\lambda^n}\frac{d^n y}{d\rho^n} + \frac{P_{n-1}}{\lambda^{n-1}}\frac{d^{n-1}y}{d\rho^{n-1}} + \ldots + \frac{P_1}{\lambda}\frac{dy}{d\rho} + P_0 y = 0,$$

où la nouvelle variable ρ croît indéfiniment par valeurs réelles positives.

L'équation (2) relative à la nouvelle variable et à la nouvelle équation (1bis) s'écrit :

$$(2^{\text{bis}}) \qquad A_n z^n + A_{n-1}\lambda z^{n-1} + \ldots + A_1\lambda^{n-1}z + A_0\lambda^n = 0,$$

et si les racines de l'équation (2) étaient :

$$(5) \qquad \alpha_1, \alpha_2, \ldots, \alpha_n,$$

celles de l'équation (2bis) sont :

$$\alpha_1\lambda, \alpha_2\lambda, \ldots, \alpha_n\lambda.$$

Lorsque x croissait par valeurs positives, nous avions:

(6) $$\frac{dy}{ydx} = \alpha_i,$$

α_i étant l'une des racines de (5). De même ici nous aurons:

$$\frac{dy}{yd\rho} = \alpha_k \lambda,$$

α_k étant encore une des racines (5); d'où:

(6bis) $$\frac{dy}{ydx} = \alpha_k.$$

Mais il y a toutefois une différence entre le cas de l'équation (6) et celui de l'équation (6bis). Lorsque x varie par valeurs positives, la limite α_i de $\frac{dy}{ydx}$ est *en général*, et en laissant de côté les cas exceptionnels dont il a été question plus haut, celle des racines de l'équation (5) dont la partie réelle est la plus grande. Si au contraire $x = \rho\lambda$ la limite α_k de $\frac{dy}{ydx}$ sera, *en général*, celle des racines de l'équation (5) qui est telle que la partie réelle de $\alpha_k \lambda$ soit aussi grande que possible.

Nous avons supposé au début de ce paragraphe que l'équation (2) n'a pas de racines multiples et qu'elle n'a pas non plus deux racines ayant même partie réelle. Voyons cependant ce qui arriverait si cette équation avait deux racines ayant même partie réelle.

En premier lieu supposons que ces deux racines ne soient pas celles dont la partie réelle est la plus grande. En particulier, dans le cas du 3e ordre, où nous avons appelé les trois racines en question α, β et γ, supposons que la partie réelle de α soit plus grande que celle de β et γ, la partie réele de β étant égale à celle de γ. En se reportant au raisonnement qui précède, on verrait que y étant l'intégrale *générale* de l'équation (1), le rapport:

$$u = \frac{dy}{ydx},$$

a encore pour limite α et que le raisonnement ne se trouve en défaut que dans les cas exceptionnels dont il a été question plus haut (quand la valeur initiale de H est plus grande que $\frac{1}{\varepsilon}$) et par conséquent pour certaines intégrales particulières de l'équation (1).

En second lieu, si l'équation (2) n'a pas de racines multiples, l'équation (2bis) n'aura deux racines ayant même partie réelle que pour certaines valeurs particulières de λ et par conséquent la difficulté dont nous parlons ici ne se présentera que pour certaines valeurs *exceptionnelles* de l'argument ω de x.

Reste le cas où l'équation (2) a des racines multiples. Reprenons le cas du 3ᵉ ordre où les racines sont α, β et γ et supposons $\alpha = \beta$. Si l'on voulait répéter le raisonnement que nous avons fait en supposant les trois racines distinctes, on poserait :

$$y = X + Y + Z,$$
$$y' = \alpha X + Y\left(\alpha + \frac{1}{x}\right) + \gamma Z,$$
$$y'' = \alpha^2 X + Y\left(\alpha^2 + \frac{2\alpha}{x}\right) + \gamma^2 Z,$$

et on reconnaitraît que la limite de $\frac{y'}{y}$ est égale à α en général, et, pour une certaine intégrale particulière, à γ.

Nous pouvons d'ailleurs embrasser tous ces cas particuliers dans le résultat suivant qui ne comporte aucune exception et dont nous ferons usage plus tard.

Supposons que x tende vers l'infini par valeurs réelles positives. Soit a un nombre dont la partie réelle soit supérieure à celles de toutes les racines de l'équation (2). On aura : $\lim ye^{-ax} = 0$
y étant une quelconque des intégrales de l'équation (1).

On peut alors trouver deux nombres b et c tels que la partie réelle de b soit plus petite que celle de a et plus grande que celle de c et que la partie réelle de c soit supérieure à celles de toutes les racines de l'équation (2).

Cela posé, considérons l'équation différentielle d'ordre $n + 1$

$$(1^{\text{ter}}) \qquad \Sigma c P_k \frac{d^k y}{dx^k} - \Sigma P_k \frac{d^{k+1} y}{dx^{k+1}} = 0.$$

Cette équation admet toutes les intégrales de l'équation (1) et en outre l'intégrale e^{cx} de sorte que son intégrale générale s'écrit :

$$\lambda e^{cx} + y_1$$

λ étant une constante arbitraire et y_1 l'intégrale générale de l'équation (1).

L'équation (2) relative à l'équation (1ter) s'écrit :

$$(2^{\text{ter}}) \qquad (z - c)\Sigma A_k z^k = 0,$$

et admet les mêmes racines que l'équation (2), plus la racine c dont la partie réelle est plus grande que celle de toutes les autres.

Il en résulte que l'expression $\frac{y'}{y}$ a pour limite c, lorsque y est l'intégrale générale de l'équation (1ter).

Il y a exception toutefois pour certaines intégrales particulières de cette équation. Ces intégrales exceptionnelles ne sont autres d'ailleurs que les intégrales de l'équation (1) elle-même.

De là on peut conclure qu' à partir d'une certaine valeur x_0 de x on a: partie réelle

$$\frac{y'}{y} < b,$$

on déduit de là:

$$|y| < |y_0 e^{b(x-x_0)}|$$

y_0 étant la valeur de y pour $x = x_0$, ou bien

$$|ye^{-ax}| < |y_0 e^{-bx_0}| \, |e^{(b-a)x}|,$$

la partie réelle de $b - a$ étant négative, la limite du second membre est nulle, on a donc:

$$\lim ye^{-ax} = 0.$$

Ce résultat ne paraît d'abord s'appliquer qu'aux intégrales qui sont telles que $\frac{y'}{y}$ tend vers c, et ne pas subsister pour les intégrales exceptionnelles de l'équation (1ter), à savoir les intégrales de l'équation (1). Mais une pareille intégrale peut toujours être regardée comme la différence de deux intégrales non exceptionnelles. Le résultat subsiste donc pour une intégrale quelconque de l'équation (1).

C. Q. F. D.

Si

$$x = \rho\lambda$$

et que ρ tende vers l'infini par valeurs réelles positives; si a est un nombre tel que la partie réelle de $a\lambda$ soit plus grande que la partie réelle d'une racine quelconque de l'équation (2) multipliée par λ, on a encore:

$$\lim ye^{-ax} = 0.$$

Il est à remarquer que dans tout ce qui précède nous ne nous sommes nullement appuyés sur ce que les coëfficients P de l'équation (1) sont des polynômes en x. Les résultats énoncés plus haut subsistent donc pourvu que les rapports

$$\frac{P_{n-1}}{P_n}, \frac{P_{n-2}}{P_n}, \ldots, \frac{P_1}{P_n}, \frac{P_0}{P_n}$$

tendent vers des valeurs finies et déterminées quand x croît indefiniment.

Nous avons supposé d'autre part que les polynômes P étaient tous de même degré. Les résultats subsisteraient encore si un ou plusieurs des polynômes

$$P_{n-1}, P_{n-2}, \ldots, P_1, P_0,$$

étaient de degré inférieur à p, P_n restant de degré p. Mais il n'en serait plus de même si le degré de P_n était inférieur à celui d'un quelconque des autres polynômes. Dans ce cas, l'équation (1) rentrerait dans un autre type d'équations linéaires que nous étudierons plus loin.

§2. *Equations aux Différences Finies.*

Avant de poursuivre les conséquences des résultats précédents, nous allons étendre ces résultats aux équations à différences finies de la forme suivante :

(1) $$P_k u_{n+k} + P_{k-1} u_{n+k-1} + \ldots + P_1 u_{n+1} + P_0 u_n = 0$$

les coëfficients P étant des polynômes d'ordre p par rapport au rang n de la fonction u_n. Il est aisé de voir l'analogie de cette équation avec les équations linéaires que nous avons envisagées dans le paragraphe précédent ; car si l'on pose : $\Delta u_n = u_{n+1} - u_n$, $\Delta^2 u_n = \Delta u_{n+1} - \Delta u_n$, etc.

l'équation (1) s'écrira :

$$R_k \Delta^k u_n + R_{k-1} \Delta^{k-1} u_n + \ldots + R_1 \Delta u_n + R_0 u_n = 0,$$

les coëfficients R étant des polynômes entiers en n.

Nous appellerons A_i le coëfficient de n^p dans le polynôme P_i et nous envisagerons l'équation :

(2) $$A_k z^k + A_{k-1} z^{k-1} + \ldots + A_1 z + A_0 = 0.$$

Posons de plus : $$\frac{P_i}{P_k} = Q_i, \qquad \frac{A_i}{A_k} = B_i.$$

Laissant d'abord de côté le cas exceptionnel où l'équation (2) aurait deux racines égales ou deux racines de même module, je vais démontrer le résultat suivant.

Lorsque n tend vers l'infini, le rapport $\frac{u_{n+1}}{u_n}$ tend vers une des racines de l'équation (2) et en général vers celle dont le module est le plus grand.

Supposons l'équation du 3ème ordre pour fixer les idées, elle s'écrira :

$$u_{n+3} + Q_2 u_{n+2} + Q_1 u_{n+1} + Q_0 u_n = 0.$$

Soient α, β, γ les trois racines de l'équation (2) rangées par ordre de module décroissant. Posons :

$$\begin{aligned} u_n &= X_n + Y_n + Z_n, \\ u_{n+1} &= \alpha X_n + \beta Y_n + \gamma Z_n, \\ u_{n+2} &= \alpha^2 X_n + \beta^2 Y_n + \gamma^2 Z_n, \end{aligned}$$

on en conclut :

$$\begin{aligned} u_{n+1} &= X_{n+1} + Y_{n+1} + Z_{n+1}, \\ u_{n+2} &= \alpha X_{n+1} + \beta Y_{n+1} + \gamma Z_{n+1}, \\ u_{n+3} &= \alpha^2 X_{n+1} + \beta^2 Y_{n+1} + \gamma^2 Z_{n+1}. \end{aligned}$$

Posons encore :

$$\begin{aligned} \alpha^3 + Q_2\alpha^2 + Q_1\alpha + Q_0 &= A(\alpha - \beta)(\alpha - \gamma), \\ \beta^3 + Q_2\beta^2 + Q_1\beta + Q_0 &= B(\beta - \alpha)(\beta - \gamma), \\ \gamma^3 + Q_2\gamma^2 + Q_1\gamma + Q_0 &= C(\gamma - \alpha)(\gamma - \beta). \end{aligned}$$

Il viendra:

$$X_{n+1} = \alpha X_n - (AX_n + BY_n + CZ_n),$$
$$Y_{n+1} = \beta Y_n - (AX_n + BY_n + CZ_n),$$
$$Z_{n+1} = \gamma Z_n - (AX_n + BY_n + CZ_n).$$

On tire de là:

$$\frac{Y_{n+1}}{X_{n+1}}\frac{X_n}{Y_n} = \frac{\beta - \left(A\frac{X_n}{Y_n} + B + C\frac{Z_n}{Y_n}\right)}{\alpha - \left(A + B\frac{Y_n}{X_n} + C\frac{Z_n}{X_n}\right)}$$

avec des formules analogues pour:

$$\frac{Z_{n+1}}{X_{n+1}}\frac{X_n}{Z_n} \text{ et } \frac{Z_{n+1}}{Y_{n+1}}\frac{Y_n}{Z_n}.$$

Il résulte de là que l'on peut trouver un nombre ε tel que

$$\text{si } \left|\frac{X_n}{Y_n}\right|, \left|\frac{Y_n}{X_n}\right|, \left|\frac{X_n}{Z_n}\right| \text{ et } \left|\frac{Y_n}{Z_n}\right| > \varepsilon \quad \text{on ait} \left|\frac{Y_{n+1}}{X_{n+1}}\frac{X_n}{Y_n}\right| < 1$$

$$\text{si } \left|\frac{X_n}{Z_n}\right|, \left|\frac{Z_n}{X_n}\right|, \left|\frac{X_n}{Y_n}\right| \text{ et } \left|\frac{Z_n}{Y_n}\right| > \varepsilon \quad \text{on ait} \left|\frac{Z_{n+1}}{X_{n+1}}\frac{X_n}{Z_n}\right| < 1$$

$$\text{si } \left|\frac{Y_n}{Z_n}\right|, \left|\frac{Z_n}{Y_n}\right|, \left|\frac{Y_n}{X_n}\right| \text{ et } \left|\frac{Z_n}{X_n}\right| > \varepsilon \quad \text{on ait} \left|\frac{Z_{n+1}}{Y_{n+1}}\frac{Y_n}{Z_n}\right| < 1$$

D'ailleurs quand n croît indéfiniment, A, B et C tendent vers 0, il en est de même de ε.

Soit H la plus grande des deux quantités $\left|\frac{Y_n}{X_n}\right|$ et $\left|\frac{Z_n}{X_n}\right|$. Je dis que H sera une fonction de n qui sera décroissante toutes les fois qu'elle sera comprise entre ε et $\frac{1}{\varepsilon}$.

En effet deux cas peuvent se présenter:

1°.
$$H = \left|\frac{Y_n}{X_n}\right| > \left|\frac{Z_n}{X_n}\right|$$

$$\left|\frac{X_n}{Y_n}\right| = \frac{1}{H} > \varepsilon \left|\frac{Y_n}{X_n}\right| > \varepsilon \left|\frac{X_n}{Z_n}\right| > \left|\frac{X_n}{Y_n}\right| > \varepsilon \left|\frac{Y_n}{Z_n}\right| > 1 > \varepsilon.$$

Donc $\left|\frac{Y_n}{X_n}\right|$ et par conséquent H est décroissant.

2°.
$$H = \left|\frac{Z_n}{X_n}\right| > \left|\frac{Y_n}{X_n}\right|$$

$$\left|\frac{X_n}{Y_n}\right| > \left|\frac{X_n}{Z_n}\right| > \varepsilon \left|\frac{Z_n}{X_n}\right| > \varepsilon \left|\frac{Z_n}{Y_n}\right| > 1 > \varepsilon.$$

Donc $\frac{Z_n}{X_n}$ et par conséquent H est décroissant.

Il résulte de là que H tend vers 0 comme nous l'avons fait voir dans le paragraphe précédent, à moins qu'il ne reste constamment supérieur à $\frac{1}{\varepsilon}$.

Si H est constamment supérieur à $\frac{1}{\varepsilon}$, je dis que $\left|\frac{Z_n}{Y_n}\right|$ est une fonction constamment décroissante si elle est comprise entre ε et $\frac{1}{\varepsilon}$. On a alors en effet:

1⁰ ou bien:
$$H = \left|\frac{Y_n}{X_n}\right| > \frac{1}{\varepsilon} > \varepsilon \quad \left|\frac{Z_n}{Y_n}\right| > \varepsilon \quad \left|\frac{Y_n}{Z_n}\right| > \varepsilon$$
et par conséquent:
$$\left|\frac{Z_n}{X_n}\right| = \left|\frac{Z_n}{Y_n}\right| \left|\frac{Y_n}{X_n}\right| > 1 > \varepsilon.$$
2⁰ ou bien:
$$H = \left|\frac{Z_n}{X_n}\right| > \frac{1}{\varepsilon} > \varepsilon \quad \left|\frac{Z_n}{Y_n}\right| > \varepsilon \quad \left|\frac{Y_n}{Z_n}\right| > \varepsilon$$
et par conséquent:
$$\left|\frac{Y_n}{X_n}\right| = \left|\frac{Y_n}{Z_n}\right| \left|\frac{Z_n}{X_n}\right| > 1 > \varepsilon.$$

Dans l'un et l'autre cas la fonction $\left|\frac{Z_n}{Y_n}\right|$ est décroissante. On en conclut en répétant le raisonnement que nous avons déja fait bien des fois que $\left|\frac{Z_n}{Y_n}\right|$ tend vers 0 en général et exceptionnellement vers l'infini.

Il y a donc trois cas possibles:

1er cas *général* $\lim H = 0 \quad \lim \left|\frac{Z_n}{X_n}\right| = \lim \left|\frac{Y_n}{X_n}\right| = 0 \qquad \lim \frac{u_{n+1}}{u_n} = \alpha.$

2ème cas *exceptionnel* $\lim H = \infty \quad \lim \left|\frac{X_n}{Y_n}\right| = \lim \left|\frac{Z_n}{Y_n}\right| = 0 \qquad \lim \frac{u_{n+1}}{u_n} = \beta.$

3ème cas plus exceptionnel encore:
$$\lim H = \lim \left|\frac{Z_n}{Y_n}\right| = \infty \quad \lim \left|\frac{X_n}{Z_n}\right| = \lim \left|\frac{Y_n}{Z_n}\right| = 0 \quad \lim \frac{u_{n+1}}{u_n} = \gamma.$$

Le même raisonnement s'applique sans difficulté au cas des équations d'ordre supérieur au 3ème. Je me bornerai à indiquer ici la marche du raisonnement dans le cas du 4ème ordre.

Soient α, β, γ, δ les racines de l'équation (2) rangées par ordre de module décroissant. Nous poserons:
$$u_{n+i} = \alpha^i X_n + \beta^i Y_n + \gamma^i Z_n + \delta^i T_n. \quad (i = 0, 1, 2, 3)$$
Nous démontrerons ensuite qu'il existe un nombre ε tendant vers 0 avec $\frac{1}{x}$ et jouissant des propriétés suivantes:

1° la fonction $\left|\frac{Y_n}{X_n}\right|$ ou $\left|\frac{Y}{X}\right|$, en supprimant l'indice n pour abréger, est décroissante toutes les fois que $\left|\frac{Y}{X}\right|, \left|\frac{X}{Y}\right|, \left|\frac{X}{Z}\right|, \left|\frac{X}{T}\right|, \left|\frac{Y}{Z}\right|, \left|\frac{Y}{T}\right|$ sont plus grands que ε.

2°. Il en est de même de $\left|\frac{Z}{X}\right|$ toutes les fois que $\left|\frac{Z}{X}\right|, \left|\frac{X}{Z}\right|, \left|\frac{X}{Y}\right|, \left|\frac{X}{T}\right|, \left|\frac{Z}{Y}\right|, \left|\frac{Z}{T}\right|$ sont plus grands que ε, et ainsi de suite en considérant successivement les fonctions $\left|\frac{T}{X}\right|, \left|\frac{Z}{Y}\right|, \left|\frac{T}{Y}\right|, \left|\frac{T}{Z}\right|$ qui sont décroissantes à des conditions analogues, faciles à former par des permutations des lettres.

Cela posé, soit H la plus grande des quantités

$$\left|\frac{Y}{X}\right|, \left|\frac{Z}{X}\right|, \left|\frac{T}{X}\right|$$

et H_1 la plus grande des quantités :

$$\left|\frac{Z}{Y}\right|, \left|\frac{T}{Y}\right|.$$

On démontre que H est décroissant quand il est compris entre ε et $\frac{1}{\varepsilon}$. On en conclut qu'en général H tend vers 0 et par conséquent $\frac{u_{n+1}}{u_n}$ vers α.

Il y a exception quand H est toujours plus grand que $\frac{1}{\varepsilon}$, mais alors on démontre que H_1 est toujours décroissant s'il est compris entre ε et $\frac{1}{\varepsilon}$. On en conclut qu'en général H_1 tend vers 0 et $\frac{u_{n+1}}{u_n}$ vers β.

Il y a encore exception quand H_1 est toujours plus grand que $\frac{1}{\varepsilon}$, mais alors on démontre que $\left|\frac{T}{Z}\right|$ est toujours décroissant s'il est compris entre ε et $\frac{1}{\varepsilon}$. On en conclut qu'en général $\left|\frac{T}{Z}\right|$ tend vers 0 et $\frac{u_{n+1}}{u_n}$ vers γ.

Enfin il reste un dernier cas plus exceptionnel encore que les deux précédents, et où $\left|\frac{T}{Z}\right|$ reste toujours plus grand que $\frac{1}{\varepsilon}$. Alors $\frac{u_{n+1}}{u_n}$ tend vers δ.

Il me reste à examiner les cas où l'équation (2) a deux racines égales, ou deux racines de même module.

Supposons d'abord trois racines α, β, γ dont deux égales, par exemple

$$\alpha = \beta \qquad |\alpha| = |\beta| > |\gamma|.$$

Nous poserons: $u = X_n + Y_n + Z_n,$

$$u_{n+1} = \alpha\left(1+\frac{1}{n}\right)X_n + \alpha Y_n + \gamma Z_n,$$

$$u_{n+2} = \alpha^2\left(1+\frac{2}{n}\right)X_n + \alpha^2 Y_n + \gamma^2 Z_n,$$

d'où:

$$u_{n+1} = X_{n+1} + Y_{n+1} + Z_{n+1},$$

$$u_{n+2} = \alpha\left(1+\frac{1}{n+1}\right)X_{n+1} + \alpha Y_{n+1} + \gamma Z_{n+1},$$

$$u_{n+3} = \alpha^2\left(1+\frac{2}{n+1}\right)X_{n+1} + \alpha^2 Y_{n+1} + \gamma^2 Z_{n+1}.$$

Posons maintenant:

$$\alpha^3 + Q_2\alpha^2 + Q_1\alpha + Q_0 = A,$$
$$3\alpha^2 + 2Q_2\alpha + Q_1 = B,$$
$$\gamma^3 + Q_2\gamma^2 + Q_1\gamma + Q_0 = C,$$

il vient:

$$X_{n+1} = \left(1+\frac{1}{n}\right)\alpha X_n + A',$$
$$Y_{n+1} = \alpha Y_n + B',$$
$$Z_{n+1} = \gamma Z_n + C',$$

A', B' et C' étant des fonctions linéaires en A, B et C, ayant des coëfficients dépendant de X_n, Y_n, Z_n. A, B et C tendent vers 0 quand n croît indéfiniment, et on verrait comme précédemment qu'il en est de même, en général, de A', B' et C'. Il en résulte que *en général*

$$\lim \frac{Y_n}{X_n} = \lim \frac{Z_n}{Y_n} = 0, \qquad \lim \frac{u_{n+1}}{u_n} = \alpha.$$

Voici maintenant une propriété qui subsiste alors même que l'équation (2) admet des racines de même module et qui par conséquent ne souffre aucune exception.

Soit a une quantité de module plus grand que toutes les racines de l'équation (2); l'éxpression $\frac{u_n}{a_n}$ tend vers 0, quand n croît indéfiniment.

La démonstration serait la même que pour la propriété correspondante des équations différentielles démontrée à la fin du paragraphe précédent.

§3. *Transformation de Bessel.*

Revenons maintenant aux équations différentielles. Nous avons vu dans le §1 que si l'on envisage l'intégrale générale y de l'équation

$$\Sigma P_k \frac{d^k y}{dx_k} = 0, \tag{1}$$

étudiée dans ce paragraphe, la dérivée logarithmique

$$\frac{1}{y}\,\frac{dy}{dx}$$

tend vers une certaine limite α, mais qu'on n'en pouvait pas conclure *immédiatement* que $ye^{-\alpha x}$ tend vers une limite finie et déterminée. C'est pourtant ce qui a lieu *en général;* mais pour le démontrer, nous serons forcés d'employer la transformation de Bessel.

Voici en quoi consiste cette transformation. On pose

$$y = \int v e^{zx} dz,$$

v étant une fonction de z qu'il reste à déterminer et l'intégrale étant prise le long d'un chemin imaginaire convenablement choisi. L'intégration par parties donne :

$$xy = \int v x e^{zx}\, dz = [v e^{zx}] - \int \frac{dv}{dz} e^{zx}\, dz.$$

Le chemin d'intégration devra être choisi de telle façon que le terme tout connu de cette intégration par parties soit nul, sans cependant que l'intégrale y le soit elle-même.

On aura ensuite :

$$x^2 y = -\int \frac{dv}{dz} x e^{zx}\, dz = -\left[\frac{dv}{dz} e^{zx}\right] + \int \frac{d^2 v}{dz^2} e_{zx}\, dz,$$

ou si le terme tout connu est supposé nul :

$$x^2 y = \int \frac{d^2 v}{dz^2} e_{zx}\, dz.$$

Et ainsi de suite ; on aura :

$$x^i y = (-1)^i \int \frac{d^i v}{dz^i} e^{zx} dz \qquad (i = 0,\,1,\,2\,\ldots\,p)$$

pourvu que le chemin d'intégration ait été choisi de telle sorte que les termes tout connus des intégrations successives par parties :

$$\left[\frac{d^i v}{dz^i} e^{zx}\right] = 0, \quad (i = 0,\,1,\,2,\,\ldots\,p-1).$$

De même, on aura :

$$x^i \frac{d^k y}{dx^k} = (-1)^i \int \frac{d^i (vz^k)}{dz^i} e^{zx} dz,$$

pourvu que les termes tout connus :

$$\left[\frac{d^i v}{dz^i} z^k e^{zx}\right]$$

soient nuls aux deux limites d'intégration.

Si nous écrivons l'équation (1) sous la forme :

$$\Sigma C_{ik} x^i \frac{d^k y}{dx^k} = 0, \quad \begin{pmatrix} i = 0,\,1,\,2,\,\ldots\,p \\ k = 0,\,1,\,2,\,\ldots\,n \end{pmatrix}$$

l'équation transformée s'écrira :

$$\Sigma C_{ik}(-1)^i \frac{d^i(vz^k)}{dz^i} = 0, \tag{3}$$

Pour trouver les points singuliers de cette équation (3), il suffit d'égaler à 0, le coëfficient de $\frac{d^p v}{dz^p}$. On trouve ainsi l'équation :

$$\Sigma C_{pk} z^k = 0,$$

on en reprenant les notations du §1 :

$$\Sigma A_k z^k = 0, \tag{2}$$

ce qui est l'équation (2) du dit paragraphe.

Il faudrait ajouter à ces points singuliers le point ∞ où les intégrales sont irrégulières pour l'équation (3) comme pour l'équation (1). Si l'équation (2) n'a pas de racine multiple, ce que nous supposerons d'abord, le coëfficient de $\frac{d^p v}{dz^p}$ ne s'annule que du premier ordre en chacun des points singuliers, d'où il résulte que pour chacun de ces points l'équation déterminante $p-1$ racines respectivement égales à $0, 1, 2, \ldots p-2$, la $p^{\text{ème}}$ étant quelconque. Ce sont donc des points singuliers *réguliers*.

Il faut maintenant choisir le chemin d'intégration de façon à satisfaire aux conditions que nous nous sommes imposées. Nous devons choisir les deux limites de ce chemin de façon qu'en chacune d'elles on ait :

$$\frac{d^i v}{dz^i} z^k e^{sz} = 0, \quad \begin{pmatrix} i = 0, 1, 2, \ldots p-1 \\ k = 0, 1, 2, \ldots n \end{pmatrix}.$$

Si l'une de ces limites est à distance finie, on devra avoir :

$$\frac{d^i v}{dz^i} = 0, \quad (i = 0, 1, 2, \ldots, p-1),$$

sans que l'intégrale v soit identiquement nulle. *Cette limite devra donc être un point singulier.* Cette condition n'est d'ailleurs pas suffisante. Il faut encore qu'en ce point singulier, où comme nous l'avons vu $p-1$ des racines de l'équation determinante ont pour valeurs $0, 1, 2, \ldots, p-2$, la $p^{\text{ème}}$ racine de cette équation soit plus grande que $p-1$ et de plus que l'intégrale v soit convenablement choisie.

Supposons maintenant une limite à distance infinie. On devra avoir

$$\lim \frac{d^i v}{dz^i} e^{sz} = 0,$$

et d'abord :

$$\lim v e^{sz} = 0.$$

C'est le moment de recourir à la proposition établie à la fin du §1. Formons l'équation qui joue par rapport à l'équation (3) le même rôle que l'équation (2) par rapport à l'équation (1). Elle s'écrira:

(4) $$\Sigma C_{in}(-1)^i x^i = 0,$$

en appelant x l'indéterminée qui entre dans cette équation.

L'équation qui donne les points singuliers de l'équation (1) s'écrit d'autre part:

$$\Sigma C_{in} x^i = 0.$$

Si donc $a_1, a_2, \ldots a_q$ sont les points singuliers distincts de l'équation (1), les q racines distinctes de l'équation (4) sont $-a_1, -a_2, \ldots -a_q$. D'où, en appliquant les principes du §1, on verra que

$$\lim ve^{zx} = 0,$$

si z est réel positif et si en désignant par $R(u)$ la partie réelle d'une quantité imaginaire u, on a:

$$R(x) < R(a_1), \; R(x) < R(a_2), \ldots, \; R(x) < R(a_q).$$

Si maintenant z est imaginaire et si l'on a:

$$\arg z = \lambda.$$

$$R(xe^{-i\lambda}) < R(a_1 e^{-i\lambda}), \quad R(xe^{-i\lambda}) < R(a_2 e^{-i\lambda}), \ldots, \; R(xe^{-i\lambda}) < R(a_q e^{-i\lambda}),$$

le produit ve^{zx} tendra vers 0 quand z croîtra indéfiniment avec l'argument λ. Il est clair d'ailleurs qu'il en sera de même des diverses expressions:

$$\frac{d^i v}{dz^i} z^k e^{zk}.$$

Les hypothèses (5) sont donc suffisantes pour que le chemin d'intégration satisfasse aux conditions que nous nous sommes imposées.

On peut d'ailleurs remarquer que, si le point x est extérieur au polygone convexe qui, ayant pour sommets certains des points a, laisse tous les autres à son intérieur, on pourra toujours trouver une valeur de λ satisfaisant aux inégalités (5).

Supposons donc le point x extérieur à ce polygone que j'appellerai P. Voici quel chemin d'intégration nous ferons suivre du point z. Nous partirons de l'infini avec un argument satisfaisant aux inégalités (5) et après avoir décrit un certain chemin nous reviendrons à l'infini soit avec le même argument, soit avec un autre argument satisfaisant également à ces mêmes inégalités. Il faudra naturellement que le chemin ainsi décrit enveloppe un certain nombre de points singuliers (c'est à dire de racines de l'équation (2)); car sans cela, l'intégrale:

$$y = \int ve^{zx} dz,$$

serait identiquement nulle.

Nous pourrons supposer que ce chemin enveloppe *un seul* point singulier. En effet un contour enveloppant par exemple les points singuliers a_1 et a_2 peut toujours se décomposer en deux autres, enveloppant, le premier seulement le point a_1 et le second seulement le point a_2. Donc les intégrales qu'on obtiendrait par la considération des contours enveloppant plusieurs points singuliers, ne seraient que des combinaisons linéaires de celles que nous allons considérer et qui sont engendrées par des contours enveloppant un seul point singulier.

Soit a le point singulier enveloppé que nous supposerons d'abord être une racine *simple* de l'équation (2). Son équation déterminante qui est de degré p, a comme nous l'avons vu $p-1$ racines égales à $0, 1, 2, \ldots, p-2$, la $p^{\text{ème}}$ étant égale à

$$\mu \gtrless p-1.$$

Il résulte de là, que le point a *n'est pas un point singulier* pour $p-1$ intégrales de l'équation (3) *linéairement indépendantes* et que la $p^{\text{ème}}$ intégrale s'écrit:

$$v_p = (z-a)^\mu \phi(z),$$

$\phi(z)$ étant holomorphe dans le domaine du point a, l'intégrale générale s'écrit donc:

$$v = A(z-a)^\mu \phi(z) + \psi(z) = Av_p + \psi(z)$$

$\psi(z)$ étant holomorphe dans le domaine du point a. On a alors:

$$\int \psi(z) e^{zx} dz = 0,$$

d'où:

$$y = \int v e^{zx} dz = A \int v_p e^{zx} dz.$$

Ainsi, si l'on fait abstraction du facteur constant A, l'intégrale y ne dépend pas du choix de l'intégrale v.

Qu' arrive-t-il maintenant si a est une racine double de l'équation (2)?

Alors l'intégrale générale s'écrira:

$$v = Av_p + Bv_{p-1} + \psi(z),$$

A et B étant deux constantes arbitraires, $\psi(z)$ étant holomorphe dans le domaine du point a, et v_p et v_{p-1} étant deux intégrales particulières. Il vient alors:

$$y = \int v e^{zx} dz = A \int v_p e^{zx} dz + B \int v_{p-1} e^{zx} dz.$$

Ainsi quelle que soit l'intégrale v choisie, on ne pourra jamais obtenir pour y plus de deux intégrales linéairement indépendantes.

De même si a est une racine multiple d'ordre plus élevé.

Ainsi à chaque racine simple de l'équation (2), correspond une intégrale de l'équation (1), à chaque racine multiple d'ordre m, correspondent m intégrales de cette même équation. On obtient donc en tout de la sorte n intégrales de l'équation (1) et comme cette équation est d'ordre n, on en a l'intégrale générale. Il resterait, il est vrai, à démontrer que ces n intégrales sont linéairement indépendantes, mais c'est ce qui ressortira de diverses propositions que nous établirons plus loin.

Nous n'avons examiné jusqu'ici que le cas où les deux limites d'intégration sont infinies; il est aisé de prévoir, d'après ce qui précède, que les intégrales obtenues, en supposant qu'une ou deux des limites soient finies, ne seront que des combinaisons linéaires de celles que nous connaissons déjà.

Considérons de nouveau un point a qui soit une racine simple de (2) et soient:

$$0, 1, 2, \ldots, p-2, \mu$$

les racines de l'équation déterminante correspondante. Soit:

$$v_p = (z-a)^{\mu} \phi(z),$$

une intégrale de (3), où $\phi(z)$ est holomorphe dans le domaine du point a. Soit:

$$y_1 = \int v_p e^{zx} dz,$$

l'intégrale correspondante de (1), la quadrature s'effectuant le long du contour défini plus haut, qui enveloppe le point singulier $z = a$.

Envisageons maintenant l'intégrale:

$$y_2 = \int_a^{\infty} v_p e^{zx} dz.$$

Nous supposerons, ce qui est toujours possible, que le chemin d'intégration reste constamment intérieur au contour le long duquel a été prise l'intégrale y. Si

$$\mu > p-1,$$

l'intégrale y_2 sera finie et sera une des intégrales de l'équation (1) d'après ce qu'on a vu plus haut. Mais on voit aisément qu'on aura:

$$y_1 = y_2 (1 - e^{2i\pi\mu}).$$

Les deux intégrales y_1 et y_2 ne diffèrent donc que par un facteur constant. Il résulte en même temps de là que l'intégrale y_2 est une intégrale de l'équation (1) toutes les fois qu'elle est finie, c'est à dire toutes les fois que:

$$\mu > -1.$$

Soient $a_1, a_2, \ldots, a_n$ les racines de l'équation (2) que nous supposerons toutes simples. Soient $0, 1, 2, \ldots, p-2, \mu_i,$ les racines de l'équation déterminante relative au point singulier a_i. Il existera toujours une intégrale de la forme:

$$V_i = (z-a_i)^{\mu_i} \phi_i(z),$$

$\Phi_i(z)$ étant holomorphe dans le domaine du point a_i, et on en conclura l'existence d'une intégrale de l'équation (1)

$$Y_i = \int_{a_i}^{\infty} V_i e^{zx} dz,$$

pourvu que: $\mu_i > -1$.

Supposons donc d'abord que tous les μ sont plus grands que -1, de façon que les p intégrales Y_i existent, ou mieux encore supposons d'abord que tous les μ sont plus grands que $p-1$.

Joignons un point quelconque b à chacun des points singuliers $a_1, a_2, \ldots a_n$ par des chemins $l_1, l_2, \ldots l_n$. Soit c_{ik} la valeur que prend au point b la dérivée $k^{\text{ème}}$ de V_i quand la variable va du point a_i au point b par le chemin l_i.

Posons maintenant: $$T_i = \int_{a_i}^{b} V_i e^{zx} dz,$$

l'intégrale étant prise bien entendu le long de l_i.

Il viendra, en appliquant à l'intégrale T_i la méthode d'intégration par parties,

$$(6)\quad x^m T_i = x^{m-1} e^{bx} c_{i0} - x^{m-2} e^{bx} c_{i1} + x^{m-3} e^{bx} c_{i2} \ldots \pm e^{bx} c_{i,m-1} \mp \int \frac{d^m V_i}{dz^m} e^{zx} dz.$$

Soit maintenant $d_{i.k.q}$ la valeur que prend au point b la dérivée $k^{\text{ème}}$ de $V_i z^q$; il viendra de même:

$$(7)\quad x^m \frac{d^q T_i}{dx^q} = x^{m-1} e^{bx} d_{i.0.q} - x^{m-2} e^{bx} d_{i.1.q} + \ldots \mp \int \frac{d^m (V_i z^q)}{dz^m} e^{zx} dz.$$

D'ailleurs il est clair que les $d_{i.k.q}$ s'expriment très simplement à l'aide de b et des $c_{i.k}$.

Il résulte de ce qui précède que, si l'on substitue T_i à la place de y dans l'équation (1), le résultat de cette substitution s'écrira:

$$(8)\quad \Delta(T_i) = g_{i.p-1} x^{p-1} e^{bx} + g_{i.p-2} x^{p-2} e^{bx} + \ldots + g_{i.0} e^{bx},$$

les coëfficients $g_{i.k}$ étant faciles à calculer en fonctions des $c_{i.k}$. Si n est plus grand que p on pourra trouver n nombres:

$$h_1, h_2, \ldots, h_n$$

satisfaisant aux conditions suivantes:

$$(9)\quad \sum_1^n{}_i h_i g_{i.k} = 0, \qquad (k = 0, 1, 2, \ldots, p-1)$$

Le nombre des solutions linéairement indépendantes des équations (9) sera alors de $n-p$.

On aura alors: $$\Delta(\Sigma h_i T_i) = 0,$$

ce qui veut dire que $\Sigma h_i T_i$ est une intégrale de l'équation (1). C'est une intégrale prise le long d'un chemin complexe, mais restant toujours à distance finie.

Il existe toujours $n-p$ pareilles intégrales linéairement indépendantes. Ces intégrales diffèrent essentiellement de celles dont une limite est infinie. Ces dernières ne sont valables que si le point x est extérieur au polygone P, d'après ce que nous avons vu plus haut; au contraire les intégrales telles que $\Sigma h_i T_i$, c'est à dire les intégrales prises le long d'un contour à distance finie, sont valables *quel que soit* x. De plus elles sont holomorphes dans toute l'étendue du plan.

Posons:
$$U_i = \int_b^\infty V_i e^{xz}\, dz,$$
d'où:
$$Y_i = T_i + U_i.$$

Les équations (9) peuvent d'ailleurs se remplacer par les équations plus simples qui suivent: $$\Sigma h_i c_{ik} = 0.$$

Il suffit pour s'en convaincre de rechercher quelle est l'expression des coëfficients g_{ik} en fonctions des c_{ik}. Mais des équations ainsi transformées on déduit aisément l'identité suivante: $$\Sigma h_i V_i = 0,$$
qui subsiste quel que soit z. On a par conséquent:
$$\Sigma h_i U_i = 0, \qquad \Sigma h_i T_i = \Sigma h_i Y_i.$$

Ces relations montrent d'abord que la nouvelle intégrale $\Sigma h_i T_i$ n'est pas linéairement indépendante des intégrales déjà connues Y_i; elles font voir ensuite, que lors même que tous les μ ne sont pas plus grands que $p-1$, l'éxpression $\Sigma h_i T_i$ reste une intégrale de l'équation (1) pourvu que les T_i et les Y_i soient finis, c'est à dire pourvu que tous les μ soient plus grands que -1.

Qu' arrive-t-il enfin si tous les μ ne sont pas plus grands que -1? La difficulté est aisée à tourner. Décrivons du point b comme point initial, un contour fermé revenant au point b après avoir enveloppé le point singulier a_i. Opérons de même pour chacun des points singuliers. Nous aurons ainsi n contours fermés $l_1, l_2, \ldots l_n$. Appelons T'_i l'intégrale
$$\int V_i e^{xz}\, dz,$$
prise le long du contour l_i; ou ce qui revient au même, l'intégrale
$$\int v e^{xz}\, dz,$$
le long du même contour, v désignant une intégrale *quelconque* de l'équation (3). Appelons c_{ik} la valeur dont s'accroît la dérivée $k^{\text{ème}}$ de V_i quand on décrit le contour l_i, en partant du point b comme valeur initiale et revenant au point b comme valeur finale; appelons de même $d_{i.k.q}$ la valeur dont s'accroît la dérivée $k^{\text{ème}}$ de $V_i z^q$ dans les mêmes circonstances.

Si l'on emploie ces notations, les équations (6), (7) et (8) subsisteront. Par conséquent si on a n nombres h_i satisfaisant aux équations:

$$\Sigma h_i c_{i.k} = 0, \tag{10}$$

l'expression $\Sigma h_i T_i$ sera une intégrale de l'équation (1). Cette expression jouit d'ailleurs de la propriété remarquable d'être holomorphe dans tout le plan.

Or les équations (10) admettent $n - p$ solutions linéairement indépendantes.

Donc *si* $n > p$, *l'équation* (1) *aura* $n - p$ *intégrales holomorphes dans tout le plan.*

Ce théorème peut d'ailleurs se démontrer directement.

Je n'insisterai pas davantage sur cette transformation de Bessel qui permet, comme on le sait, d'intégrer l'équation (1) lorsque $p = 1$.

§4. *Etude approfondie des Intégrales irrégulières.*

Nous allons maintenant nous servir des expressions précédentes des intégrales de l'équation (1) pour étudier la façon dont elles se comportent quand x croît indéfiniment, d'une manière plus précise et plus approfondie que nous n'avons pu le faire dans le §1.

Démontrons d'abord le résultat suivant. L'intégrale:

$$J = \int v e^{zx} dz,$$

(si x est positif et très grand, si $|v|$ reste constamment inférieur à une certaine quantité M et si le chemin d'intégration reste constamment à gauche de l'axe des parties imaginaires) tendra vers 0 quand x croîtra indéfiniment.

Soit en effet L la longueur totale du chemin d'intégration, et $-\xi$ la plus grande valeur de la partie réelle de z, de telle façon que le long du chemin d'intégration on ait: $R(z) \leq -\xi$.

Il vient alors $\int |dz| = L, \qquad |e^{zx}| \leq e^{-\xi x},$

d'où enfin: $|J| \leq MLe^{-\xi x},$

et $\lim J = 0, \qquad$ pour $x = \infty$. *C. Q. F. D.*

Passons maintenant au cas où le chemin d'intégration restant toujours à gauche de la droite: $R(z) = -\xi$,

s'étend à l'infini par l'une de ses extrémités. Nous supposerons de plus que, quand z croît indéfiniment en suivant le chemin d'intégration, on peut trouver un nombre λ tel que $\lim v e^{\lambda z} = 0$.

Cela est toujours possible comme le prouve le §1, avec les fonctions v que nous avons à considérer.

Faisons encore une hypothèse sur le chemin d'intégration. Nous supposerons qu'il se compose d'un certain arc de courbe situé à distance finie, suivi d'une portion de ligne droite s'étendant à l'infini; pour tous les cas que nous avons déjà considérés ou que nous aurons à considérer dans la suite, rien ne s'oppose à cette hypothèse. Dans ces conditions on peut trouver un nombre μ tel que l'intégrale:

$$\int |e^{\mu z} dz|,$$

soit égale à une quantité finie L. On pourra également trouver un nombre M, tel que l'on ait constamment: $|ve^{\lambda z}| < M.$

Nous pourrons toujours supposer λ et μ réels. On aura alors:

$$J = \int ve^{\lambda z} . e^{z(x-\lambda-\mu)} . e^{\mu z} dz,$$

d'où:

$$|J| < LMe^{-\xi(x-\lambda-\mu)},$$

quand x croît indéfiniment, on a donc:

$$\lim J = 0. \qquad\qquad C. Q. F. D.$$

De même on verrait aisément que $x^m J$ tend encore vers 0, quelque grand que soit l'exposant m.

Nous allons maintenant étudier l'intégrale suivante:

$$J = \int_0^a ve^{zx} dz.$$

La limite supérieure d'intégration peut être une quantité finie a, ou bien être infinie, mais le chemin d'intégration restera toujours à gauche de l'axe des parties imaginaires. La fonction v sera assujettie aux mêmes conditions que plus haut; je supposerai de plus que dans le domaine du point 0, la fonction v peut se développer en série de la forme suivante:

(1) $$A_0 z^\alpha + A_1 z^{\alpha+1} + A_2 z^{\alpha+2} + \ldots$$

α étant quelconque.

Je dis que dans ces conditions:

$$x^{\alpha+1} J \quad \text{tend vers} \quad -\Gamma(\alpha+1) A_0,$$

quand x croît indéfiniment.

En effet nous pouvons toujours supposer:

$$|A_n| < \mu\rho^n,$$

μ et ρ étant deux quantités convenablement choisies de telle façon que le rayon du cercle de convergence de la série (1) soit égal à $\frac{1}{\rho}$.

Nous pourrons décomposer le chemin d'intégration en deux parties, la 1ère intérieure à un cercle décrit du point 0 comme centre avec r pour rayon ($r\rho < 1$), la 2de extérieure à ce cercle. Nous aurons alors :

$$J = K + H,$$

K et H étant les deux parties de l'intégrale correspondant à ces deux parties du chemin d'intégration. D'après ce qui précède, il vient :

$$\lim x^{\alpha+1} H = 0.$$

Il reste à chercher la limite de $x^{\alpha+1}K$.

Nous pouvons écrire :

$$v = A_0 z^\alpha + A_1 z^{\alpha+1} + \ldots + A_m z^{\alpha+m} + R_m,$$

R_m étant le reste de la série (1). Il vient alors :

$$(2) \quad K = A_0 \int z^\alpha e^{xz} dz + A_1 \int z^{\alpha+1} e^{xz} dz + \ldots + A_m \int z^{\alpha+m} e^{xz} dz + \int R_m e^{xz} dz,$$

les intégrales étant prises le long de la première partie du chemin d'intégration. On aura :

$$|R_m| < \frac{\mu (r\rho)^{m+1} r^\alpha}{1 - r\rho} \qquad |e^{xz}| < 1.$$

Si donc l est la longueur de la première partie du chemin d'intégration, il viendra :

$$\left| \int R_m e^{xz} dz \right| < \frac{l\mu (r\rho)^{m+1} r^\alpha}{1 - r\rho}.$$

On peut toujours prendre m assez grand pour que le second membre de cette inégalité soit aussi petit qu'on voudra. Mais on peut aller plus loin encore. Supposons, ce qui est toujours possible, que la première partie du chemin d'intégration soit rectiligne, et pour fixer les idées davantage encore, qu'elle se réduit au segment de droite $0, -r$. Il viendra alors :

$$\left| x^{\alpha+1} \int_0^{-r} z^\alpha e^{xz} dz \right| < \left| x^{\alpha+1} \int_0^{-\infty} z^\alpha e^{xz} dz \right| = \Gamma(\alpha + 1),$$

ou

$$x^{\alpha+1} \int_0^{-r} |z^\alpha e^{xz} dz| < \Gamma(\alpha + 1),$$

ou enfin :

$$\left| x^{\alpha+1} \int R_m e^{xz} dz \right| < \frac{\Gamma(\alpha + 1) \mu (r\rho)^{m+1}}{1 - r\rho}.$$

Comme $r\rho$ est plus petit que 1, on pourra prendre m assez grand, quel que soit x, pour que le second membre de cette inégalité soit plus petit que $\frac{\varepsilon}{r}$, r étant indépendant de x.

Le nombre m est désormais déterminé et nous allons faire varier x. Le $q^{\text{ème}}$ terme du second membre de l'expression (2) s'écrit :

$$T_q = A_q \int_0^{-r} z^{\alpha+q} e^{xz} dz,$$

Cherchons la limite de $T_q x^{a+1}$; pour cela posons:

$$U_q = A_q \int_{-r}^{-\infty} z^{a+q} e^{xz} dz;$$

On aura:

$$\lim x^{a+1} U_q = 0,$$

d'où:

$$\lim x^{a+1} T_q = \lim x^{a+1} A_1 \int_0^{-\infty} z^{a+q} e^{xz} dz = \lim \frac{-\Gamma(a+q+1) A_q}{x^q}.$$

Cette limite est égale à 0 si q est positif et à $-\Gamma(\alpha+1) A_0$ si q est nul. Donc si l'on multiplie par x^{a+1} chacun des m premiers termes du second membre de (2), le premier des produits ainsi obtenus aura pour limite $-A_0 \Gamma(\alpha+1)$ et les autres 0. Or nous pourrons toujours prendre x assez grand pour que chacun de ces produits diffère de sa limite d'une quantité moindre que

$$\frac{\varepsilon}{2m}.$$

On aura alors:

$$|x^{a+1} K + A_0 \Gamma(\alpha+1)| < \varepsilon,$$

d'où

$$\lim x^{a+1} J = \lim x^{a+1} K = -A_0 \Gamma(\alpha+1). \qquad C. Q. F. D.$$

Nous allons enfin considérer l'intégrale suivante:

$$J = \int v e^{xz} dz,$$

prise le long d'un contour enveloppant le point 0.

Je supposerai qu' à l'intérieur de ce contour la fonction v soit partout holomorphe excepté au point 0 et que dans le voisinage de ce point cette même fonction puisse se mettre sous la forme (1). On peut remarquer que, dans cette expression (1), il n'est plus nécessaire de supposer $\alpha > -1$, comme nous avions dû le faire dans l'exemple précédent.

Nous allons faire croître x indéfiniment par valeurs réelles positives. Nous pouvons donc supposer que le contour d'intégration est formé comme il suit:

1° une portion de ligne droite AB venant de l'infini et se terminant à un certain point B.

2° un arc de courbe quelconque BC allant du point B au point $C = -r$, r étant une quantité positive très petite. Ces deux premières portions du contour seront tout entières à gauche de l'axe des parties imaginaires.

3° un cercle décrit du point 0 comme centre avec r pour rayon, commençant au point C pour finir au point C.

4° et 5° l'arc CB et la droite BA parcourus en sens inverse.

Nous poserons alors: $J = H + K + H'$,

H se rapportant à la portion ABC du contour, H' à la portion CBA, et K au petit cercle de rayon r.

Il vient alors d'après ce qui précède :

$$\lim x^{\alpha+1}J = \lim x^{\alpha+1}K.$$

On a d'autre part :

$$x^{\alpha+1}K = A_0 x^{\alpha+1}\int z^{\alpha} e^{xz} dz + \ldots + A_m x^{\alpha+1}\int z^{\alpha+m} e^{xz} dz + \int R_m x^{\alpha+1} e^{xz} dz.$$

On peut toujours supposer que m est assez grand pour que $\alpha + m$ soit positif. Dans ce cas l'intégrale :

$$\int R_m x^{\alpha+1} e^{xz} dz,$$

prise le long du cercle de rayon r est égale à :

$$(1 - e^{2i\pi\alpha}) \int_0^{-r} R_m x^{\alpha+1} e^{xz} dz.$$

Elle est donc plus petite en valeur absolue que :

$$|1 - e^{2i\pi\alpha}| \frac{\Gamma(\alpha+1)\mu(r\rho)^{m+1}}{1 - r\rho},$$

et elle tend uniformément vers 0 quand m croît indéfiniment, et cela quel que soit x.

De même on a

$$\lim x^{\alpha+1}\int z^{\alpha+q} e^{xz} dz = 0, \qquad q > 0,$$

$$\lim x^{\alpha+1}\int z^{\alpha} e^{xz} dz = -(1 - e^{2i\pi\alpha})\Gamma(\alpha + 1).$$

On déduit de là par le même raisonnement que plus haut

$$\lim x^{\alpha+1}J = \lim x^{\alpha+1}K = (e^{2i\pi\alpha} - 1) A_0 \Gamma(\alpha + 1). \qquad C.\ Q.\ F.\ D.$$

Le second membre prend la forme illusoire $0 \times \infty$ lorsque α est entier négatif. Mais dans ce cas il est aisé de voir que la fonction sous le signe $\int$ est méromorphe à l'intérieur du contour d'intégration. On a donc :

$$J = 2i\pi\left(A_0 \frac{x^{-\alpha-1}}{(-\alpha-1)!} + A_1 \frac{x^{-\alpha-2}}{(-\alpha-2)!} + \ldots + A_{-\alpha-2}x + A_{-\alpha-1}\right).$$

Pour passer au cas où le point singulier enveloppé par le contour d'intégration n'est pas 0, mais un point quelconque a, il suffit de changer z en $z + a$. Pour passer au cas où x croît indéfiniment, non plus par valeurs réelles positives, mais avec l'argument λ, il suffit de changer x en $xe^{i\lambda}$ en même temps que z en $ze^{-i\lambda}$. Les résultats se déduisent immédiatement de ceux qui ont été énoncés plus haut.

Il est aisé de voir comment ce qui précède peut s'appliquer aux intégrales de l'équation (1). Soient :

$$a_1, a_2, \ldots, a_n,$$

les n racines de l'équation déterminante relative au point singulier a_i et v_i l'intégrale qui peut se mettre sous la forme :

$$(z - a_i)^{\mu_i}\phi_i(z),$$

ϕ_i étant holomorphe dans le voisinage du point a_i.

Nous allons faire tendre x vers ∞ avec l'argument λ. Soit maintenant l_i un chemin d'intégration dont les deux limites sont rejetées à l'infini et enveloppant le point singulier a_i. Nous supposerons que quand z tend vers l'infini le long de ce contour, son argument tend vers une limite λ' telle que:

$$\frac{\pi}{2} < \lambda + \lambda' < \frac{3\pi}{2}$$

par exemple vers $\pi - \lambda$.

Soit enfin:
$$y_i = \int v_i e^{zx} dz,$$
l'intégrale étant prise le long du chemin l_i.

Pour achever de préciser le contour l_i, nous le formerons 1° de la droite $a_i + Re^{i(\pi-\lambda)}$, $a_i + \varepsilon e^{i(\pi-\lambda)}$, R et ε étant des quantités, la première infiniment grande, la seconde infiniment petite: 2° d'un cercle complet décrit du point a_i comme centre avec ε pour rayon; 3° de la droite $a_i + \varepsilon e^{i(\pi-\lambda)}$, $a_i + Re^{i(\pi-a)}$ parcourue en sens contraire. Ce contour pourra d'ailleurs être remplacé par tout autre contour équivalent.

Dans ces conditions, lorsque x croîtra indéfiniment avec l'argument λ, l'intégrale y_i se comportera comme
$$e^{a_i x} x^{-\mu_i - 1},$$
c'est à dire que le rapport
$$y_i e^{-a_i x} x^{\mu_i + 1}$$
tendra vers une limite finie et déterminée.

Tel est le résultat, plus complet que celui que nous avions obtenu au §1, que nous permet d'atteindre la transformation de Bessel.

On remarquera d'abord le rôle important que joue dans ce résultat l'argument λ avec lequel x croît indéfiniment. On en conclura que les intégrales de l'équation (1) ne se comportent pas de la même manière quelle que soit la façon dont x tend vers l'infini.

Une autre conséquence importante, c'est que les n intégrales
$$y_1, y_2, \ldots, y_n$$
sont linéairement indépendantes.

Faisons croître en effet x par valeurs réelles positives, et supposons que les n quantités
$$a_1, a_2, \ldots, a_n$$
soient rangées par ordre de parties réelles croissantes. (On peut toujours supposer qu'il n'y a pas deux de ces quantités qui aient même partie réelle, sans quoi on ferait croître x indéfiniment avec un argument différent de 0.)

Soit
$$A_i = \lim e^{-a_i x} x^{\mu_i + 1} y_i, \qquad A_i \gtrless 0.$$

Supposons qu'il existe une identité linéaire entre nos n intégrales

(3) $$C_1 y_1 + C_2 y_2 + \ldots + C_n y_n = 0.$$

Multiplions l'identité par: $e^{-a_n x} x^{\mu_n + 1}$,

et faisons croître x indéfiniment. L'identité devient à la limite:

$$C_n A_n = 0, \qquad \text{d'où } C_n = 0.$$

Effaçons le dernier terme de l'identité (3), multiplions la par:

$$e^{-a_{n-1} x} x^{\mu_{n-1} + 1},$$

et faisons $x = \infty$. Il vient encore:

$$C_{n-1} A_{n-1} = 0, \qquad \text{d'où } C_{n-1} = 0.$$

En continuant de la sorte, on démontrerait successivement que tous les coëfficients C sont nuls, ce qui montre que nos n intégrales sont linéairement indépendantes. La transformation de Bessel conduit donc à l'intégrale générale de l'équation (1).

Il est aisé d'étendre ce raisonnement au cas où l'équation (2) a des racines multiples.

Dans le paragraphe (1), nous avons vu que si

$$R(a_n) > R(a_{n-1}) > R(a_{n-2}) > \ldots > R(a_1),$$

il *peut* y avoir certaines intégrales particulières dont la dérivée logarithmique tend non pas vers a_n, comme cela a lieu pour l'intégrale générale, mais vers a_{n-1}, vers a_{n-2}, ... ou vers a_1. Toutefois les principes de ce premier paragraphe ne nous permettaient pas d'affirmer que ces intégrales particulières existaient réellement. Ce que nous venons de dire démontre l'existence de ces intégrales particulières.

Comme application de ce qui précède, posons nous le problème suivant:

Reconnaître si l'équation (1) admet comme intégrale un polynôme entier.

Pour cela il faut d'abord que l'une des racines de l'équation (2) soit nulle. Supposons qu'elle soit simple; soit par exemple:

$$a_i = 0.$$

Il faudra ensuite que la quantité que nous avons appelée μ_i soit entière négative. Quand μ_i est entier, il n'existe pas en général d'intégrale de l'équation (3) de la forme: $$v_i = (z - a_i)^{\mu_i} \phi_i(z),$$

car le point singulier a_i est en général un point singulier *logarithmique*. Si l'intégrale v contient des logarithmes, l'intégrale:

$$y_i = \int v e^{zx} dz,$$

prise le long d'un contour l_i enveloppant le point 0 ne peut se réduire à un polynôme entier.

Mais dans certains cas particuliers, le point singulier 0 n'est pas logarithmique, il existe une intégrale de la forme:

$$v_i = z^{\mu_i}\phi_i(z),$$

ϕ_i étant holomorphe dans le voisinage du point 0. La fonction ve^{zx} est alors méromorphe à l'intérieur du contour l_i, d'où il résulte que l'intégrale y_i se réduit à un polynôme entier.

Ainsi pour que l'équation (1) admette pour intégrale un polynôme entier, il faut et il suffit:

1° que l'équation (2) ait une racine nulle.

2° que l'une des racines de l'équation déterminante relative au point singulier correspondant de l'équation (3) soit entière négative.

3° que ce point singulier ne soit pas logarithmique.

Cela peut d'ailleurs se vérifier directement.

§5. *Etude du groupe de l'équation* (1).

Chacun sait ce qu'on entend par *groupe d'une équation linéaire.* Lorsque la variable indépendante décrit un contour fermé autour d'un point singulier, les intégrales de l'équation subissent une substitution linéaire et c'est la combinaison de ces substitutions qui engendre le groupe de l'équation.

On sait également qu'une substitution linéaire est caractérisée principalement par ses multiplicateurs et que les multiplicateurs de la substitution relative à un point singulier, s'obtiennent immédiatement, lorsque les intégrales sont régulières dans le voisinage de ce point. En effet on les déduit aisément de l'équation déterminante relative à ce point.

Il n'en est plus de même quand le point singulier est irrégulier, c'est à dire quand les intégrales ne sont pas régulières dans le voisinage de ce point. On n'a alors pour le calcul des multiplicateurs que des méthodes d'approximation plus ou moins rapides.

C'est ce qui arrive pour l'un des points singuliers de l'équation (1), à savoir pour le point $x = \infty$. Ce point sera en effet *irrégulier* en général. Pour qu'il fût régulier, il faudrait que, le polynôme P_n étant de degré p, le polynôme P_{n-1} fût de degré $p - 1$ au plus, le polynôme P_{n-2} de degré $p - 2$ au plus, etc. Dans ce cas l'équation (2) aurait toutes ses racines nulles. Si on laisse de côté ce cas très particulier, on n'a pas de méthode rapide pour trouver les multiplicateurs de la substitution S que subissent les intégrales de l'équation (1) quand le point x décrit un cercle de rayon très grand.

Le groupe de l'équation (3) est dérivé de n substitutions fondamentales correspondant aux différents points singuliers de cette équation, c'est à dire aux différentes racines de l'équation (2). Si ces racines sont simples, les points singuliers correspondants sont réguliers. On peut donc trouver aisément les multiplicateurs de ces substitutions fundamentales, mais pour calculer les coëfficients du groupe lui-même, il faut employer des méthodes d'approximation.

Il y a toutefois entre le groupe de l'équation (3) et la substitution S, un lieu que je désirerais faire ressortir. Si nous supposons connu le groupe de l'équation (3), je dis que nous connaîtrons aussi la substitution S.

Voici sous quelle forme nous nous donnerons les coëfficients du groupe de l'équation (3). Considérons un point singulier quelconque a_i de cette équation; soient

$$0, 1, 2, \ldots, p-2, \mu_i$$

les racines de son équation déterminante et

$$v_{i.1}, v_{i.2}, \ldots, v_{i.p}$$

les intégrales correspondantes de telle façon que:

$$v_{ik} = (z-a_i)^{k-1}\phi(z) \qquad (k = 1, 2, \ldots, p-1),$$

$$v_{ip} = (z-a_i)^{\mu_i}\phi(z),$$

$\phi(z)$ étant holomorphe dans le voisinage du point a_i. Soit b_i un point très voisin du point a_i. Opérons de même pour chacun des points singuliers; joignons $b_i b_j$; quand la variable z ira de b_i en b_j en suivant la droite $b_i b_j$, les intégrales v_{ik} prendront certaines valeurs qui pourront s'exprimer linéairement à l'aide des intégrales $v_{j.k}$.

En d'autres termes, il y aura une substitution linéaire S_{ij} qui changera les intégrales v_{ik} dans les intégrales v_{jk}, de telle façon qu'on puisse écrire avec la notation symbolique ordinairement employée:

$$v_{ik} = v_{ik}.S_j.$$

La connaissance des substitutions S_j suffit pour déterminer le groupe de l'équation (3). Ce sont en effet les substitutions que j'ai appelées *auxiliaires* dans mon mémoire sur les groupes des équations linéaires (Acta Mathematica 4:3, p. 207). Il est à remarquer que ces substitutions ne sont pas indépendantes les unes des autres.

On voit aisément que si l'on connait $n-1$ des substitutions S_{ij} (convenablement choisies) on connaîtra toutes les autres (loc. cit. p. 207). Nous conserverons néanmoins, pour plus de symétrie dans les notations, les $n(n-1)$ substitutions S_{ij} et S_{ji}.

Nous achèverons de définir les intégrales v_{ik} grâce à la convention suivante :

1° si $k=p$, $\phi(z)$ se réduit à 1 pour $z=a_i$.

2° si $k<p$, $\phi(z)$ se réduit à 1 et ses $p-1-k$, premières dérivées s'annulent pour $z=a_i$.

Cela posé, supposons d'abord x réel positif et très grand. Supposons que les droites $a_i b_i$ qui sont très petites soient parallèles à l'axe des parties réelles et de telle façon que : $$R(b_i)<R(a_i).$$

Soit D_i une demi-droite parallèle à cet axe, partant du point b_i et s'étendant à l'infini du côté des parties réelles négatives. Soit C_i un cercle décrit du point a_i comme centre, avec $a_i b_i$ pour rayon. Soit l_i un contour formé de la droite D_i, du cercle C_i et de la droite D_i prise en sens contraire. Soit :

$$y_i=\int v_{ip}e^{zx}dz.$$

prise le long du contour l_i.

Supposons maintenant un chemin quelconque E_i partant du point b_i et s'étendant à l'infini de telle façon que l'argument de z tende vers la limite π. Soit L_i un contour formé du chemin E_i, du cercle C_i et du chemin E_i pris en sens contraire. Cherchons à évaluer l'intégrale :

$$J=\int v_{ip}e^{zx}dz,$$

le long du contour L_i.

Je puis toujours supposer que le chemin E_i ait été remplacé par un contour E_i' équivalent; c'est à dire tel que l'on puisse transformer, par une déformation continue, E_i en E_i' sans franchir aucun point singulier. La valeur de l'intégrale J n'en sera pas changée.

Or on pourra toujours trouver un chemin E_i' équivalent à E_i et formé de la façon suivante ; ce chemin se réduira à une ligne brisée dont les sommets seront des points c_j infiniment voisins de divers points singuliers a_j. Le premier de ces sommets sera $b_i=c_i$. Le sommet suivant sera c_j, infiniment voisin d'un point singulier a_j, mais pouvant être différent de b_j. Puis viendra c_k infiniment voisin d'un point singulier a_k, et ainsi de suite. Enfin la ligne brisée E_i' se terminera par une demi-droite partant du dernier sommet, parallèle à l'axe des parties réelles et dirigée du côté des parties réelles négatives.

Nous pourrons supposer que le contour formé de la ligne brisée E_i' et de la demi-droite D_i ne contient pas à son intérieur d'autre point singulier que ceux qui sont infiniment voisins d'un des sommets de E_i'. Il est évidemment possible de déformer d'une manière continue ce contour, jusqu' à ce qu'il aille passer infiniment près de chacun des points singuliers qu'il contient à son intérieur (et cela sans lui faire franchir aucun point singulier).

Supposons maintenant que l'on étudie ce que devient l'intégrale v_{ip} lorsque la variable z partant du point b_i décrit la ligne brisée E_i'. Au moment où nous arriverons en un sommet c_j de cette ligne brisée, infiniment voisin d'un point singulier a_j, et que nous serons par conséquent dans le domaine de ce point singulier, l'intégrale v_{ip} pourra s'exprimer linéairement à l'aide des p intégrales:

$$v_{j.1}, v_{j.2}, \ldots, v_{j.p}$$

de telle façon qu'on aura:

$$v_{i.p} = A^i_{j.1} v_{j.1} + A^i_{j.2} v_{j.2} + \ldots + A^i_{j.p} v_{j.p}. \tag{4}$$

Les coëfficients $A^i_{j.k}$ peuvent être regardés comme connus, car leur valeur découle immédiatement de la connaissance des substitutions S_{ij}, c'est à dire de la connaissance du groupe de l'équation (3).

Cela posé, nous pouvons décomposer le contour L_i de la manière suivante: soit λ_i le contour formé de la ligne brisée E_i' et de la demi-droite D_i. Nous remplacerons L_i par le contour λ_i, par le contour l_i et par le contour C_i pris en sens contraire. Le contour total ainsi obtenu est évidemment équivalent à L_i.

L'intégrale:

$$\int v_{ip} e^{sz} dz,$$

prise le long de l_i n'est autre que y_i.

Si l'on appelle K la même intégrale prise le long de λ_i, on aura:

$$J = K(1 - e^{2i\pi\mu_i}) + y_i.$$

Maintenant si le contour λ_i contient un certain nombre de points singuliers

$$a_j, a_{j'}, a_{j''}, \ldots$$

on pourra le remplacer par les contours correspondants:

$$l_j, l_{j'}, l_{j''}, \ldots$$

L'intégrale

$$\int v_{ip} e^{sz} dz,$$

prise le long de l_j se réduit, en vertu de la relation (4) à:

$$y_i A^i_{jp},$$

il vient donc enfin:

$$J = (1 - e^{2i\pi\mu_i}) \Sigma_j A^i_{jp} y_j + y_i. \tag{5}$$

On voit que si μ_i est entier négatif et si le point a_i n'est pas logarithmique, il reste:

$$J = y_i,$$

quel que soit le chemin L_i.

Cela posé, voyons ce que deviendra l'intégrale y_i lorsque x, partant d'une valeur réelle positive très grande, reviendra à cette valeur après avoir décrit un cercle de rayon très grand. Pendant que x variera de la sorte, nous serons obligés de déformer le contour l_i le long duquel est prise l'intégrale y_i; car si l'on ne changeait pas ce contour, quand l'argument de x serait devenu plus grand que $\frac{\pi}{2}$, l'intégrale aurait cessé d'être finie car la valeur absolue de e^{zx} aurait pu devenir plus grande que toute quantité donnée.

Voici maintenant comment il faut déformer le contour l_i; nous conserverons le cercle C_i mais nous remplacerons la demi-droite D_i parcourue deux fois en sens inverse, par une ligne quelconque E_i qui partira du point l_i et s'étendra à l'infini et qui devra être également parcourue deux fois en sens contraire.

Nous nous arrangerons toujours pour que l'argument de x soit à chaque instant égal à π, moins l'argument que prend z en s'éloignant indéfiniment sur la ligne E_i. De plus il faudra que la ligne E_i dérive de la demi-droite D_i par déformation continue et cela sans jamais franchir aucun point singulier.

Quand l'argument de x sera revenu à la valeur 0, après un cercle complet, la ligne E_i (que d'ailleurs on peut toujours, comme nous l'avons vu, supposer réduite à une ligne brisée E_i') prendra une forme définitive F_i et l'argument de z à l'infini sur F_i sera égal à π.

Ainsi dans la figure (1), on a supposé 5 points singuliers a, b, c, d, e et on a figuré le cercle C_i, la droite D_i et la ligne F_i.

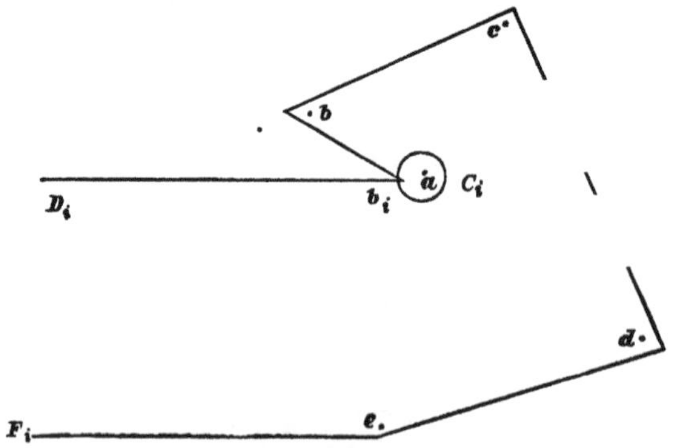

FIGURE 1.

L'intégrale prise le long du contour formé de la ligne F_i, du cercle C_i et de la ligne F_i prise en sens inverse, peut se calculer par le procédé que nous avons exposé un peu plus haut ; elle aura pour valeur :

$$(1 - e^{2i\pi\mu_i}) \Sigma A^i_{j.p} y_j + y_i,$$

en conservant les mêmes notations qu'au commencement de ce paragraphe. Mais cette intégrale n'est autre chose que ce que devient y_i quand x a décrit un cercle très grand.

Nous avons donc la valeur finale de y_i exprimée linéairement à l'aide des valeurs initiales des n intégrales $y_1, y_2, \ldots, y_n$. En d'autres termes, quand nous connaissons le groupe de l'équation (3), nous connaissons aussi la substitution linéaire que subissent les intégrales de l'équation (1), lorsque la variable x décrit un cercle de rayon très grand. *C. Q. F. D.*

On peut d'ailleurs faire la remarque suivante. Si μ_i est entier négatif et que le point a_i ne soit pas logarithmique, la valeur finale de y_i ne diffère pas de la valeur initiale ; cette intégrale n'est pas altérée par la substitution linéaire que nous envisageons. On devait le prévoir puisque nous avons vu que cette intégrale se réduit alors à un polynôme entier.

§6. *Généralisation des* §§1 *et* 2.

Dans le paragraphe (2) nous avons considéré l'équation aux différences finies :

(1) $$P_k u_{n+k} + P_{k-1} u_{n+k-1} + \ldots + P_1 u_{n+1} + P_0 u_n = 0,$$

où les coëfficients sont des polynômes d'ordre p en n. Nous avons vu que la limite du rapport

$$\frac{u_{n+1}}{u_n}$$

était en général celle des racines de l'équation

(2) $$A_k z^k + A_{k-1} z^{k-1} + \ldots + A_1 z + A_0 = 0,$$

dont le module est le plus grand ; A_i désignant le coëfficient de n^p dans P_i.

Nous avons posé ensuite :

$$\frac{P_i}{P_k} = Q_i, \qquad \frac{A_i}{A_k} = B_i,$$

d'où $$B_i = \lim Q_i \qquad (n = \infty)$$

et nous avons vu qu'on peut remplacer les équations (1) et (2) par les suivantes :

(1bis) $$u_{n+k} + Q_{k-1} u_{n+k-1} + \ldots + Q_0 u_n = 0,$$

(2bis) $$z^k + B_{k-1} z_{k-1} + \ldots + B_0 = 0.$$

Nous avons vu également que le résultat subsiste encore, lorsque Q_i au lieu d'être le quotient de deux polynômes entiers de degré p en n, est une fonction *quelconque* de n, tendant vers la limite B_i quand n croît indéfiniment.

Si l'une des quantités B_i est infinie, on en conclut que le rapport $\frac{u_{n+1}}{u_n}$ croît indéfiniment avec n. C'est ce qui arrive en particulier quand le polynôme P_k est de degré inférieur à celui d'un des polynômes suivants P_i.

Il est nécessaire alors d'employer l'artifice suivant:

Posons: $$u_n = (n!)^\mu v_n,$$
μ étant une constante réelle positive qu'il s'agit de déterminer de telle façon que $\frac{v_{n+1}}{v_n}$ tende vers une limite finie.

L'équation (1bis) devient:

$$v_{n+k} + \frac{Q_{k-1}}{(n+k)^\mu} v_{n+k-1} + \frac{Q_{k-2}}{[(n+k)(n+k-1)]^\mu} v_{n+k-2} + \ldots = 0,$$

et il s'agit de déterminer μ de telle façon que, pour $n = \infty$, les expressions:

(3) $$\frac{[(n+i)!]^\mu Q_i}{[(n+k)!]^\mu},$$

soient toutes finies sans être toutes nulles. Pour cela, il suffit d'envisager le degré en n de chacune de ces expressions, c'est à dire l'exposant de la puissance de n par laquelle il faut la diviser pour que le quotient tende vers une limite finie quand n croît indéfiniment. Supposons que le coëfficient P_i de l'équation (1) soit un polynôme entier de degré p_i en n; Q_i sera alors de degré $p_i - p_k$.

Or $$\frac{(n+k)!}{(n+i)!},$$
est un polynôme d'ordre $k - i$ en n. Donc le degré en n de l'expression (3) est:

$$p_i - p_k - \mu(k - i).$$

Il faut donc donner à μ la plus petite valeur qui satisfasse aux inégalités

(4) $$p_k + \mu k \geq p_i + \mu_i.$$

Si l'on choisit justement pour μ cette plus petite valeur, toutes ces inégalites seront satisfaites, de telle sorte que toutes les expressions (3) tendront vers une limite finie et une d'elles au moins se réduira à une égalité, de telle sorte que toutes les expressions (3) ne tendront pas vers 0.

On peut trouver graphiquement cette plus petite valeur de μ de la manière suivante; on marquera tous les points qui ont pour abscisse i et pour ordonnée p_i; on construira le polygone convexe qui enveloppe tous ces points, et celui des côtés de ce polygone qui aboutira au point (k, p_k) nous donnera μ par son coëfficient angulaire. On trouvera dans la figure (2) un exemple de cette détermination de μ en supposant :

$$k=5, \quad p_5=p_4=p_3=2, \qquad p_3=p_1=3, \qquad p_0=4.$$

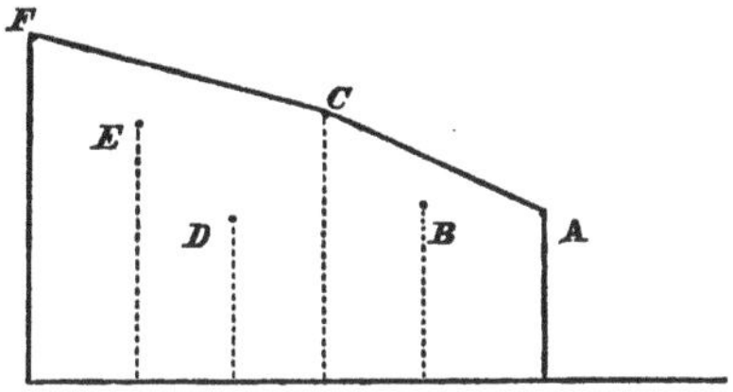

FIGURE 2.

Les points A, B, C, D, E, F correspondent respectivement aux polynômes P_5, P_4, P_3, P_2, P_1, P_0 et c'est le côte AC du polygone $ACFD$ dont le coëfficient angulaire donne la valeur de μ.

Soit alors C_i la limite de l'expression (3) pour $n=\infty$, on formera l'équation

(2ter) $$Cz^k + C_{k-1}z^{k-1} + \ldots + C_1 z + C_0 = 0.$$

Soit α celle des racines de cette équation qui a le module le plus grand, nous aurons :

$$\lim \frac{u_{n+1}}{u_n}(n+1)^{-\mu} = \alpha, \qquad (\text{pour } n=\infty).$$

Supposons maintenant que tous les B_i soient nuls, le rapport $\frac{u_{n+1}}{u_n}$ tendra vers 0. Pour se rendre compte de la façon dont ce rapport tend vers 0, on cherchera encore la plus petite valeur de μ qui satisfasse aux inégalites (4) ; cette valeur sera cette fois négative. On posera :

$$u_n = (n!)^{\mu} v_n,$$

et l'équation (1bis) deviendra :

(1ter) $$\Sigma \left[\frac{(n+i)!}{(n+k)!}\right]^{\mu} Q_i v_{n+i} = 0,$$

on formera l'équation :

(2ter) $$\Sigma C_i z^i = 0,$$

en appellant C_i la limite pour n infini, du coëfficient de v_{n+i} dans l'équation (1ter).

Si l'on appelle ensuite α celle des racines de (2ter) dont le module est le plus grand, on aura :

$$\lim \frac{u_{n+1}}{u_n}(n+1)^{-\mu} = \alpha. \qquad (\text{pour } n=\infty)$$

La même méthode peut s'appliquer aux équations:

(1) $$P_n \frac{d^n y}{dx^n} + P_{n-1} \frac{d^{n-1} y}{dx^{n-1}} + \ldots + P_1 \frac{dy}{dx} + P_0 y = 0,$$

(2) $$A_n z^n + A_{n-1} z^{n-1} + \ldots + A_1 z + A_0 = 0,$$

(1bis) $$\frac{d^n y}{dx^n} + Q_{n-1} \frac{d^{n-1} y}{dx^{n-1}} + \ldots + Q_1 \frac{dy}{dx} + Q_0 y = 0,$$

(2bis) $$z^n + B_{n-1} z^{n-1} + \ldots + B_1 z + B_0 = 0,$$

envisagées dans le paragraphe (1).

Supposons que quelques uns des B deviennent infinis ou que tous les B deviennent nuls. Dans le premier cas la dérivée logarithmique de y tendra vers l'infini, dans le second cas vers 0.

On pourra toujours trouver deux nombres C_i et μ_i tels que

$$\lim \frac{Q_i}{C_i x^{\mu_i}} = 1. \qquad (\text{pour } x = \infty)$$

Si $\mu_i = 0$, $B_i = C_i$; si $\mu_i > 0$, $B_i = \infty$; si $\mu_i < 0$, $B_i = 0$.

On considérera alors l'équation

(2ter) $$z_n + C_{n-1} x^{\mu_{n-1}} z^{n-1} + \ldots + C_1 x^{\mu_1} z + C_0 x^{\mu_0} = 0.$$

Si dans cette équation on pose $z = tx^\lambda$, elle devient;

$$\Sigma C_i x^{\mu_i - \lambda(n-i)} t^i = 0.$$

On donnera à λ une valeur telle que tous les exposants $\mu_i - \lambda(n-i)$ soient tous nuls ou négatifs, sans être tous négatifs; et faisant $x = \infty$ dans l'équation précédente, il viendra:

(2^{quater}) $$\Sigma C'_i t^i = 0,$$

où
$$C'_i = C_i \qquad \text{si } \mu_i = \lambda(n-i),$$
$$C'_i = 0 \qquad \text{si } \mu_i < \lambda(n-i).$$

L'équation (2^{quater}) en t aura alors toutes ses racines finies, sans qu'elles soient toutes nulles. J'appellerai α celle de ces racines dont la partie réelle est la plus grande. Je dis qu'on aura en général:

$$\lim x^{-\lambda} \frac{dy}{y dx} = \alpha.$$

Pour le démontrer, changeons de variable en posant:

$$\xi = x^\rho$$

ρ étant un exposant qu'il reste à déterminer; il viendra:

$$\frac{d^k y}{dx^k} = \Sigma_i D_{ik} x^{i\rho - k} \frac{d^i y}{d\xi^i} \; (i = 1, 2, \ldots k)$$

les D étant des coëfficients numériques.

L'équation (1bis) devient alors :

$$(1^a) \qquad \Sigma Q_k D_{ik} x^{i\rho-k} \frac{d^i y}{d\xi^i} = 0.$$

Dans cette équation le coëfficient de $\frac{d^n y}{d\xi^n}$ s'écrit

$$D_{nn} x^{n\rho-n}.$$

Posons :

$$R_n = 1 \quad R_i = \frac{\Sigma Q_k D_{ik} x^{i\rho-k}}{D_{nn} x^{n\rho-n}}.$$

Remplaçons dans R_i x par sa valeur $\xi^{\frac{1}{\rho}}$, et l'équation (1^a) deviendra

$$(1^b) \qquad \Sigma R_i \frac{d^i y}{d\xi^i} = 0.$$

Quel est le degré du coëfficient R_i en ξ ? Le degré de Q_k est égal à $\frac{\mu_k}{\rho}$; posons :

$$\nu_k = \frac{\mu_k}{\rho} + \frac{n-k}{\rho}.$$

Le degré de R_i en ξ sera la plus grande des $n - i + 1$ quantités

$$\nu_k + i - n \qquad (k = i,\, i+1,\, i+2, \ldots, n).$$

Nous voulons que les degrés de tous les R_i soient tous nuls ou négatifs sans être tous négatifs. Nous choisirons donc ρ de manière à satisfaire aux inégalités

$$\frac{\mu_k + n - k}{\rho} + i - n \leq 0 \qquad (i = 1, 2, \ldots k).$$

Le degré de R_0 est d'ailleurs égal à $\frac{\mu_0 + n}{\rho} - n$. Donc les inégalités précédentes peuvent se ramener aux suivantes :

$$\frac{\mu_k + n - k}{\rho} + k - n \leq 0 \qquad (k, = 0, 1, 2, \ldots, n-1),$$

ou bien

$$\rho \geq \frac{\mu_k + n - k}{n - k}.$$

On prendra pour ρ la plus petite valeur qui satisfasse à ces inégalités. Comparons ρ à la quantité que nous avons appelée plus haut λ. Celle-ci était définie par les inégalités

$$\mu_k - \lambda(n-k) \leq 0,$$

ou

$$\lambda \geq \frac{\mu_k}{n-k}.$$

On a donc :

$$\rho = \lambda + 1.$$

Si l'on donne à ρ cette valeur, les coëfficients R_i tendent vers des limites finies, quand x croît indéfiniment. Les conclusions du §1 sont donc applicables à l'équation (1^b). Formons donc l'équation (2^b) qui joue par rapport à (1^b) le même rôle que (2) par rapport à (1). Si nous posons :

$$\lim R_i = E_i \qquad (x = \infty)$$

cette équation s'écrira

(2^b) $$\Sigma E_i z^i = 0,$$

et si β est celle des racines de cette équation dont la partie réelle est la plus grande, on aura en général:

$$\lim \frac{dy}{yd\xi} = \lim \frac{x^{-\lambda}}{\rho}\frac{dy}{ydx} = \beta.$$

Il reste à déterminer E_i.

Pour cela reprenons l'expression:

$$R_i = \Sigma Q_k \frac{D_{ik}}{D_{nn}} x^{(i-n)\rho-k+n}.$$

Parmi les termes du second membre, il pourra y en avoir dont le degré en x est négatif et d'autres dont le degré en x est nul. Nous n'avons pas à tenir compte des premiers dont la limite est évidemment nulle pour x infini.

Or si $k > i$, les inégalités (5) montrent que le degré en x de

$$Q_k x^{(i-n)\rho-k+n}$$

est toujours négatif. Il reste donc:

$$E_i = \lim Q_i \frac{D_{ii}}{D_{nn}} x^{(i-n)(\rho-1)}.$$

Or il est aisé de voir que: $$D_{ii} = \rho^i$$

il reste donc: $E_i = C_i \rho^{i-n}$ si $\mu_i = \lambda(n-i)$

et $E_i = 0$ si $\mu_i < \lambda(n-i)$

Donc pour passer de l'équation (2^b) à l'équation (2^quater), il suffit de poser:

$$z = \frac{t}{\rho}$$

il vient donc

$$\beta = \frac{\alpha}{\rho}$$

et

$$\lim x^{-\lambda}\frac{dy}{ydx} = \alpha.$$

C. Q. F. D.

On peut tirer de là une conclusion importante. Soit γ un nombre réel positif plus grand que la partie réelle de $\frac{\alpha}{\rho}$. Je dis que

$$v = ye^{-\gamma x^\rho}$$

tendra vers 0 quand x croîtra indéfiniment par valeurs réelles positives. Il viendra en effet

$$\frac{dv}{vdx} = \frac{dy}{ydx} - \gamma\rho x^\lambda$$

d'où

$$\lim x^{-\lambda}\frac{dv}{vdx} = \alpha - \gamma\rho,$$

ou

$$\lim R\left(x^{-\lambda}\frac{dv}{vdx}\right) = R(\alpha - \gamma\rho) < 0$$

$R(u)$ désignant toujours la partie réelle de u. Soit maintenant δ un nombre positif tel que: $R(\alpha - \gamma\rho) < -\delta < 0$.

Donc, à partir d'une certaine valeur x_0 de x, on aura:

$$R\left(\frac{dv}{vdx}\right) < -\delta x^{\lambda}$$

d'où:

$$|v| < |v_0| e^{-\frac{\delta x^{\rho}}{\rho}}$$

v_0 désignant la valeur de v pour $x = x_0$

$$\lim v = 0. \qquad C. Q. F. D.$$

Cette proposition, comme le résultat analogue démontré à la fin du §1 ne souffre aucune exception.

Une dernière remarque: comme ρ est essentiellement réel positif, la méthode précédente se trouve en défaut quand on a pour tous les μ_k

$$\frac{\mu_k}{n-k} + 1 \leq 0$$

ou

$$\mu_k \leq -(k-n).$$

Mais on n'a pas à s'inquiéter de ce cas d'exception, car les intégrales de l'équation (1) sont alors *régulières* pour x très grand.

§7. *Des Séries de Polynômes.*

Les conclusions des paragraphes 2 et 6 sont susceptibles d'une application importante. Elles permettent en effet de résoudre le problème suivant.

Soient: $P_0(x), P_1(x), P_2(x), \ldots, P_n(x), \ldots$

une infinité de polynômes entiers en x, liés entre eux par une relation de récurrence de la forme suivante:

(1) $$Q_k P_{n+k} + Q_{k-1} P_{n+k-1} + \ldots + Q_1 P_{n+1} + Q_0 P_n = 0$$

où les Q sont des polynômes entiers en n et en x. Formons maintenant la série:

(2) $$a_0 P_0 + a_1 P_1 + \ldots + a_n P_n + \ldots$$

où les a sont des coëfficients constants quelconques. Cette série sera convergente tant que le point x restera intérieur à une certaine région du plan, et divergera quand le point x sortira de cette région. On demande quelle est la courbe qui limite cette région de convergence.

J'avais donné une solution de ce problème dans une communication que j'ai faite à la Société Mathématique de France en Novembre 1882 et dans une note que j'ai présentée à l'Académie des Sciences de Paris le 5 Mars 1883.

Voici quelle est la méthode que j'employais. J'envisageais la série suivante:

(3) $$y = P_0 + zP_1 + \ldots + z^n P_n \ldots$$

qui représente une fonction de z et de x. On voit aisément que cette fonction satisfait à une équation différentielle de la forme suivante :

$$(4) \qquad R_p \frac{d^p y}{dz^p} + R_{p-1} \frac{d^{p-1} y}{dz^{p-1}} + \ldots + R_1 \frac{dy}{dz} + R_0 y = S$$

où les coëfficients R et le terme tout connu S sont des polynômes entiers en x et en z. On obtiendra les points singuliers des intégrales de cette équation (en regardant un instant x comme une constante et z comme la seule variable) en égalant à 0 le coëfficient R_p. Soit $z = \alpha$ celle des racines de l'équation $R_p = 0$ (qui est une équation algébrique en z) dont le module est le plus petit. La condition nécessaire et suffisante pour que la série (3) converge (en laissant de côté certains cas exceptionnels) c'est que :

$$\text{mod } z < \text{mod } \alpha.$$

Or α est évidemment une fonction de x. Donc pour une valeur quelconque supposée donnée de z, la courbe qui limitera la région de convergence de la série (3) dans le plan des x aura pour équation :

$$\text{mod } \alpha = \text{mod } z.$$

On en conclut aisément que, si les coëfficients de la série (2) sont tels que $\sqrt[n]{a_n}$ tende vers une limite finie et déterminée quand n croît indéfiniment, la courbe qui limitera la région de convergence de la série (2) aura pour équation :

$$\text{mod } \alpha = \text{const.}$$

Cette méthode a, on le voit, l'inconvénient d'être sujette à objection lorsque $\sqrt[n]{a_n}$ ne tend pas vers une limite déterminée.

Depuis, M. Pincherle a publié dans les Annali di Matematica un mémoire où il traite par la même méthode des questions analogues. (Sui sistemi di functioni analitiche . . . Série II, tome XII.)

M. Pincherle envisage une fonction quelconque $\phi(x, z)$, la développe en série selon les puissances croissantes de x et de z ; il ordonne ensuite cette série suivant les puissances de z de telle façon qui l'on ait :

$$(5) \qquad \phi(x, z) = \phi_0 + \phi_1 z + \phi_2 z^2 + \ldots + \phi_n z^n + \ldots$$

$\phi_0, \phi_1, \phi_2, \ldots$ étant des fonctions de x. Si l'on connaît les points singuliers de $\phi(x, z)$ considérée comme fonction de z, on trouvera aisément, comme nous venons de le voir, les conditions de convergence de la série (5). Considérant ensuite la série plus générale

$$(5^{\text{bis}}) \qquad \alpha_0 \phi_0 + \alpha_1 \phi_1 + \ldots + \alpha_n \phi_n + \ldots$$

où les α sont des coëfficients quelconques, M. Pincherle en détermine les conditions de convergence en recherchant un nombre tel que

$$\lim \alpha_n (R + \varepsilon)^{-n} = 0 \qquad \lim \alpha_n (R - \varepsilon)^{-n} = \infty \qquad (\text{pour } n = \infty)$$

quelque petite que soit la quantité positive ε. Il est clair alors que la série (5bis) sera convergente ou divergente en même temps que

$$\Phi_0 + \Phi_1 R + \Phi_2 R^2 + \ldots + \Phi_n R^n + \ldots$$

Cette méthode, employée presque simultanément par M. Pincherle et par moi, est sujette à l'inconvénient que j'ai signalé plus haut. C'est ce qui m'a décidé à l'abandonner et à employer de préférence les résultats des paragraphes 2 et 6 du présent mémoire.

La relation de récurrence (1) est tout à fait analogue à l'équation (1) du paragraphe (2). Les polynômes P_n y jouent le même rôle que jouaient dans ce paragraphe les quantités inconnues u_n et les coëfficients Q sont des polynômes entiers en n, pourvu que nous regardions un instant x comme une constante.

La règle du paragraphe cité nous permettra donc de déterminer la limite du rapport:

$$\frac{P_{n+1}}{P_n} \quad (\text{pour } n = \infty).$$

Supposons en effet que les polynômes Q soient d'ordre p en n, et soit A_i le coëfficient de n^p dans Q_i. Formons l'équation:

$$A_k z^k + A_{k-1} z^{k-1} + \ldots + A_1 z + A_0 = 0. \tag{6}$$

Elle sera analogue à l'équation (2) du §2.

Il est à remarquer que les coëfficients A sont des fonctions de x.

Imaginons que α soit celle des racines de l'équation (6) dont le module est le plus grand, α sera aussi une fonction de x et on aura en général:

$$\lim \frac{P_{n+1}}{P_n} = \alpha \tag{7}$$

et par conséquent:

$$\lim \left| \frac{P_{n+1}}{P_n} \right| = |\alpha|. \tag{8}$$

Cela posé, quelles sont les conditions de convergence de la série (2)?

Formons la série de puissances:

$$a_0 + a_1 t + a_2 t^2 + \ldots + a_n t^n + \ldots \tag{9}$$

Elle aura un certain rayon de convergence ρ, c'est à dire qu'elle convergera pourvu que

$$|t| < \rho.$$

Je dis que, si nous laissons de côté certains cas exceptionnels, sur lesquels nous reviendrons plus loin, la condition nécessaire et suffisante pour que la série (2) converge s'écrira:

$$|\alpha| < \rho.$$

En effet considérons une valeur de x pour laquelle cette condition soit remplie. On trouvera toujours un nombre t tel que:

$$|\alpha| < |t| < \rho.$$

Pour cette valeur de t la série (9) convergera; de plus on aura à partir d'un certain rang:

$$|P_n| < t^n$$

$$|\alpha_n P_n| < |\alpha_n t^n|.$$

Donc tous les termes de la série (2) seront plus petits en valeur absolue que les termes correspondants d'une série convergente. Donc la série (2) convergera.

C. Q. F. D.

Supposons au contraire $|\alpha| > \rho$.

Nous choisirons t de telle façon que:

$$|\alpha| > |t| > \rho.$$

Il en résultera que la série (9) sera divergente et que la série (2) dont chaque terme est plus grand en valeur absolue que le terme correspondant de la série (9) divergera également. *C. Q. F. D.*

Il résulte de là que les *courbes de convergence* des séries de la forme (2), c'est à dire les courbes qui limitent les régions du plan où les séries de cette forme convergent, ont pour équation générale:

$$|\alpha| = \text{const.}$$

Voici quelques exemples: soit d'abord

$$(n^2 + 1) P_{n+2} - 2n^2 x P_{n+1} + (n^2 + x^2) P_n = 0,$$

la relation de récurrence qui lie trois polynômes P consécutifs.

Formons l'équation (6), elle s'écrira:

$$z^2 - 2zx + 1 = 0$$

et aura pour solution: $z = 2(x \pm \sqrt{x^2 - 1})$

on en conclura que les courbes de convergence ont pour équations

$$|x \pm \sqrt{x^2 - 1}| = \text{const.}$$

si l'on a soin de choisir le signe + ou le signe — de telle façon que le premier membre soit aussi grand que possible.

Posons:

$$x = \tfrac{1}{2}\left(\xi + \frac{1}{\xi}\right)$$

il viendra:

$$x^2 - 1 = \tfrac{1}{4}\left(\xi - \frac{1}{\xi}\right)^2$$

d'où:

$$x \pm \sqrt{x^2 - 1} = \xi \text{ ou } \frac{1}{\xi}.$$

A chaque valeur de x correspondent deux valeurs de ξ dont le produit est égal à 1. L'une d'elles aura donc son module plus grand que 1, l'autre son module plus petit que 1. Nous choisirons celle dont le module est plus grand que 1. On aura alors:

$$|\xi| > \left|\frac{1}{\xi}\right|$$

et par conséquent les courbes de convergence auront pour équation

$$|\xi| = \text{const.}$$

Soit: $|\xi| = t \qquad \xi = te^{i\phi}$

il viendra: $x = \frac{1}{2}\left(t + \frac{1}{t}\right)\cos\phi + \frac{i}{2}\left(t - \frac{1}{t}\right)\sin\phi.$

Les coordonnées du point x auront pour expression:

$$\tfrac{1}{2}\left(t + \frac{1}{t}\right)\cos\phi \text{ et } \tfrac{1}{2}\left(t - \frac{1}{t}\right)\sin\phi.$$

Pour avoir les courbes de convergence, il faudra donner à t une valeur constante et faire varier ϕ de 0 à 2π. On obtiendra ainsi une ellipse ayant pour foyers les points ± 1. Ce sont donc ces ellipses qui sont les courbes de convergence des séries de la forme:

$$a_0 P_0 + a_1 P_1 + a_2 P_2 + \ldots + a_n P_n + \ldots$$

Comme deuxième exemple supposons que la relation de récurrence (1) s'écrive:

$$(n^2 + 1) P_{n+2} - 2n^2 x P_{n+1} + (n^2 x^2 - n^2) P_n = 0.$$

L'équation (6) s'écrira: $z^2 - 2zx + x^2 - 1 = 0,$

et aura pour racines: $z = x \pm 1.$

Si donc ρ est le rayon de convergence de la série $\Sigma a_n t^n$, les conditions de convergence de la série $\Sigma a_n P_n$ s'écriront:

$$|x + 1| < \rho \qquad |x - 1| < \rho.$$

La région de convergence se composera donc de la partie commune à deux cercles décrits avec ρ pour rayon, des points $+1$ et -1 comme centres. Les courbes de convergence seront donc formées de deux arcs de cercle de même rayon, ayant pour centres ces deux points et limités en leurs points d'intersection sur l'axe des parties imaginaires.

Il est à remarquer que ces deux cercles ne se coupent que si leur rayon est plus grand que 1. La série $\Sigma a_n P_n$ ne converge donc pour aucune valeur de x si la série Σa_n n'est pas convergente.

Passons donc maintenant aux cas d'exception dont j'ai parlé plus haut et que nous avions provisoirement laissés de côté. Le premier de ces cas se présente quand l'équation (6) a deux racines qui sans être égales, ont même module et de module plus grand que celui de toutes les autres. Ce cas ne se présentera en général que pour des valeurs particulières de x, à moins que l'équation (6) ne soit de la forme: $[z - \phi(x)][z - e^{i\alpha}\phi(x)]\,\Phi(z, x) = 0$

α désignant une constante, $\phi(x)$ une fonction de x et Φ un polynôme entier en z. Il arrive alors que le cas exceptionnel dont nous parlons se présentera pour toutes les valeurs de x ou pour toute une région du plan. Mais on pourrait voir que les résultats qui ont été exposés dans ce paragraphe n'en subsistent pas moins. Nous n'avons donc pas à nous inquiéter de ce premier cas exceptionnel.

Le second cas est plus important. Nous avons vu dans le §2 que si u_n est l'intégrale générale de l'équation (1) de ce paragraphe, et si $\alpha, \beta, \ldots \lambda$, sont les racines de l'équation (2) rangées par ordre de module décroissant, on a *en général*:

$$\lim \frac{u_{n+1}}{u_n} = \alpha.$$

mais que pour *certaines intégrales particulières* ou peut avoir:

$$\lim \frac{u_{n+1}}{u_n} = \beta, \gamma, \ldots \text{ ou } \lambda.$$

Appliquons cela au cas qui nous occupe. Nous pouvons choisir arbitrairement nos k premiers polynômes $P_0, P_1, \ldots, P_{k-1}$, les polynômes suivants $P_k, P_{k+1}, \ldots$ étant déterminés successivement par la relation de récurrence (1).

Soient $\alpha, \beta, \ldots \lambda$, les racines de l'équation (6) rangées par ordre de module décroissant. On aura *en général* $\lim \frac{P_{n+1}}{P_n} = \alpha$.

c'est à dire que si l'on choisit d'une manière quelconque les k premiers polynômes P_i ce n'est que pour certains choix particuliers que cette relation pourra cesser d'être vraie et qu'on pourra avoir:

$$\lim \frac{P_{n+1}}{P_n} = \beta, \gamma, \ldots \text{ ou } \lambda.$$

Ainsi pour certains choix particuliers des premiers polynômes, il pourra y avoir exception. A quelle condition un pareil cas exceptionnel pourra-t-il se présenter?

Pour nous en rendre compte cherchons à former l'équation (4). Ecrivons la coëfficient Q_i de la relation (1) sous la forme suivante:

$$Q_i = A_{i.p}(n+i)_p + A_{i.p-1}(n+i)_{p-1} + \ldots + A_{i.2}(n+i)_2 + A_{i.1}(n+i)_1 + A_{i.0}(n+i)_0$$

où les A sont des polynômes entiers en x et où:

$$n_q = n(n-1)\ldots(n-q+1), \qquad n_1 = n, \qquad n_0 = 1.$$

Ecrivons de même: $A_{iq} = \Sigma B_{iqh} x^h$,

de telle façon que la relation (1) s'écrive:

$$\Sigma B_{iqh}(n+i)_q x^h P_{n+i} = 0.$$

Il est aisé maintenant d'écrire l'équation (4). Soit en effet:

$$\Sigma C_{mqh} z^m x^h \frac{d^q y}{dz^q}$$

le premier membre d'une équation de la forme (4). Substituons à la place de

y la série $\Sigma P_\nu z^\nu$; ce premier membre deviendra :

$$\Sigma C_{mqh} z^{m+\nu-q} x^h \nu_q P_\nu .$$

Nous devons nous arranger de telle sorte que tous les termes où l'exposant de z dépasse une certaine limite disparaissent. Posons :

$$i = q - m \qquad \nu = n + i \qquad m + \nu - q = m$$

et donnons à n une valeur déterminée suffisamment grande. Il devra venir :

$$\Sigma C_{mqh} x^h (n+i)_q P_{n+i}$$

en comparant avec la relation (1), il vient :

$$C_{mqh} = B_{iqh} .$$

Le premier membre de l'équation (4) s'écrit donc :

$$\Sigma B_{iqh} z^{q-i} x^h \frac{d^q y}{dz^q}$$

quant au second membre, on le trouvera aussi aisément. Le polynôme P_n n'est défini que pour les valeurs positives de n ; convenons, par définition, d'écrire :

$$P_{-1} = P_{-2} = \ldots = P_{-n} = \ldots = 0 .$$

Quand dans le premier membre de la relation (4), on fera $n = -1, -2, \ldots, -k$, le résultat de cette substitution ne sera pas nul ; appelons $\Pi_1, \Pi_2, \ldots, \Pi_k$ le résultat de la substitution dans ce premier membre de ces diverses valeurs négatives de n.

On verra alors que le résultat de la substitution de $y = \Sigma P_n z^n$ dans le premier membre de l'équation (4) s'écrira :

$$\Pi_1 z^{-1} + \Pi_2 z^{-2} + \ldots + \Pi_k z^{-k} .$$

L'équation (4) s'écrira donc :

$$\Sigma B_{iqh} z^{q-i} x^h \frac{d^q y}{dz^q} = \Sigma \Pi_m z^{-m}$$

ou en la mettant sous forme entière :

$$4) \qquad \Sigma B_{iqh} z^{k+q-i} x^h \frac{d^q y}{dz^q} = \Sigma \Pi_m z^{k-m} .$$

Ainsi dans le premier des exemples cités plus haut, le premier membre de (4) s'écrira :

$$\frac{d^2 y}{dz^2}(z^4 - 2xz^3 + z^2) + \frac{dy}{dx}(z^3 - 2xz^2 - 3z) + y\,(x^2 z^2 - 2xz + S) .$$

En général une équation de la forme :

$$\Sigma R_i \frac{d^i y}{dz^i} S$$

(les R et S étant des polynômes entiers en z) présentera dans le voisinage du point $z = 0$ (et par conséquent dans le voisinage d'un point z quelconque) au moins une intégrale particulière holomorphe.

Il n'y aurait d'exception que si tous les polynômes R s'annulaient à la fois pour $z = 0$, ou si le point $z = 0$ était un point singulier logarithmique, ou plus généralement un point singulier dont l'équation déterminante admet des racines entières.

Il résulte de là que si l'équation privée de second membre

$$\Sigma R_i \frac{d^i y}{dz^i} = 0$$

admet p intégrales holomorphes linéairement indépendantes, l'équation à second membre en admettra $p + 1$ (ou n'en admettra aucune, dans les cas exceptionnels dont il vient d'être question).

Ainsi, si nous revenons à l'équation (4) qui nous occupe ici, le point $z = 0$ est pour l'équation sans second membre un point singulier ordinaire dont l'équation déterminante n'a pas en général de racines entières. Donc, en général, l'équation à second membre admettra une intégrale holomorphe et une seule, c'est l'intégrale :

$$\Sigma P_n z^n.$$

Egalons maintenant à 0 le coëfficient de $\frac{d^p y}{dz^p}$ ce qui donne :

$$\Sigma B_{iph} z^{k+p-i} x^h = 0 \tag{9}$$

et, considérant dans cette équation x comme une constante, envisageons les diverses valeurs de z qui annulent le premier membre. Soit α celle de ces valeurs dont le module est le plus petit (à part $z = 0$, bien entendu). Dans le voisinage du point $z = \alpha$, (si α est une racine simple de l'équation (9)) l'équation à second membre (4) admettra en général p intégrales holomorphes linéairement indépendantes $j_1, j_2, \ldots, j_p$ et une $p + 1^{\text{ème}}$ non holomorphe j_{p+1} dont il sera aisé de trouver le développement.

Ces développements seront valables à l'intérieur d'un certain cercle ayant le point α pour centre, et c'est ce cercle que l'on peut appeler le domaine du point α, de même que le cercle qui a le point 0 comme centre et $|\alpha|$ comme rayon. et à l'intérieur duquel la série $\Sigma P_n z^n$ est certainement convergente, s'appellera le domaine du point 0.

Ces deux domaines ont une partie commune. Si dans cette partie commune, $\Sigma P_n z^n$ s'exprime linéairement à l'aide de $j_1, j_2, \ldots, j_p$, la série $\Sigma P_n z^n$ sera convergente pour des modules de z supérieurs à $|\alpha|$ et on aura :

$$\lim \left| \frac{P_{n+1}}{P_n} \right| > \left| \frac{1}{\alpha} \right|.$$

Mais cela n'arrivera pas en général.

Je n'en dirai pas plus long sur ce second cas exceptionnel, qui demanderait une étude plus approfondie, et je passerai à un troisième cas exceptionnel non moins important que les deux premiers.

Reprenons la relation de récurrence (1) et supposons que dans cette relation les coëfficients Q_i regardés comme des polynômes entiers en n, soient tous de degré inférieur au premier d'entre eux Q_k, ou bien que l'un des coëfficients Q_i soit de degré supérieur à Q_k. Il arrivera alors que l'équation (6) aura toutes ses racines nulles, ou bien aura une racine infinie. Nous avons appelé α celle des racines de cette équation (6) dont le module est le plus grand. Nous aurons ici:

$$|\alpha| = 0 \text{ ou bien } \infty$$

et par consequent:

$$\lim \left| \frac{P_{n+1}}{P_n} \right| = 0 \text{ ou bien } \infty.$$

La méthode exposée plus haut pour trouver les courbes de convergence des séries $\Sigma \alpha_n P_n$ se trouve donc en défaut, et c'est le cas d'appliquer les principes du §6. Posons:

$$P_n = P'_n (n!)^\mu.$$

Les séries $\Sigma \alpha_n P_n$ deviennent:

$$\Sigma \alpha_n (n!)^\mu P'_n$$

et sont ordonnées suivant les polynômes P'_n au lieu de l'être suivant les polynômes P_n. Les courbes de convergence des séries de la forme $\Sigma \alpha_n P'_n$ seront donc les mêmes que celles des séries de la forme $\Sigma \alpha_n P_n$.

La relation de récurrence:

(1)
$$\Sigma Q_i P_{n+i} = 0$$

devient:

$$\Sigma Q'_i P'_{n+i} = 0$$

où

$$Q'_i = Q_i \left[\frac{(n+i)!}{n!} \right]^\mu.$$

Nous avons vu dans le §6 qu'on peut toujours trouver une valeur de μ positive ou négative, telle qu'aucune des fonctions Q'_i (considérées comme fonctions de n) ne soit d'ordre supérieur à Q'_k et qu'une d'elles au moins ne soit pas d'ordre inférieur.

Soit alors q le degré de Q'_k de telle sorte que:

$$\lim \frac{Q'_k}{n^q} = A'_k \quad (n = \infty)$$

et soit:

$$\lim \frac{Q'_i}{n^q} = A'_i \quad (n = \infty).$$

Nous formerons l'équation:

(6bis)
$$\Sigma A'_i z^i = 0$$

dont les racines seront toutes finies sans être toutes nulles. Nous appellerons α'

celle d'entre elles dont le module est le plus grand; $|\alpha'|$ sera en général une fonction de x, et les courbes de convergence cherchées auront pour équation générale :

$$|\alpha'| = \text{const.}$$

Je prendrai pour exemple les polynômes de Legendre qui sont liés entre eux par la relation de récurrence bien connue :

$$(1) \qquad P_{n+2} - 2x(2n+3)P_{n+1} + 4(n+1)^2 P_n = 0.$$

Dans ce cas l'équation (6) s'écrit :

$$4 = 0$$

et l'on voit aisément alors qu'elle a deux racines infinies et que par conséquent la méthode générale est en défaut. Posons alors :

$$P_n = P'_n (n!)^\mu.$$

La relation (1) deviendra :

$$P'_{n+2}(n+2)^\mu (n+1)^\mu - 2x(2n+3)(n+1)^\mu P'_{n+1} + 4(n+1)^2 P'_n = 0$$

et si on prend $\mu = 1$, elle s'écrira :

$$(1^{\text{bis}}) \qquad (n^2+3n+2)P'_{n+2} - 2x(2n+3)(n+1)P'_{n+1} + 4(n+1)^2 P'_n = 0$$

d'où l'on déduit l'équation :

$$(6^{\text{bis}}) \qquad z^2 - 4xz + 4 = 0.$$

Cette équation ayant pour racines

$$z = x \pm \sqrt{x^2 - 1}$$

on en déduit comme plus haut que les courbes de convergence sont des ellipses ayant les points ± 1 pour foyers, ce qui est un résultat bien connu.

Un autre cas exceptionnel, que M. Gourier a bien voulu me signaler, est celui où les racines de l'équation (6) ou de l'équation (6bis) sont indépendantes de x.

Prenons pour exemple les polynômes P_n définis par la relation

$$\frac{d^n}{dx^n}(e^{-x^2}) = P_n e^{-x^2}$$

et liés entre eux par la loi de récurrence :

$$(1) \qquad P_{n+2} + 2xP_{n+1} + 2(n+1)P_n = 0.$$

L'équation

$$(6) \qquad 2 = 0$$

ayant ses racines infinies, nous poserons :

$$P_n = P'_n (n!)^{\frac{1}{2}}$$

d'où résulteront les équations :

$$(1^{\text{bis}}) \qquad \sqrt{(n+1)(n+2)}\,P'_{n+2} + 2x\sqrt{n+1}\,P'_{n+1} + 2(n+1)P'_n = 0$$

$$(6^{\text{bis}}) \qquad z^2 + 2 = 0.$$

Les racines de l'équation (6bis) sont $\pm\sqrt{-2}$ et sont par conséquent indépendantes de x, de sorte que les règles précédentes se trouvent encore en défaut.

Pour traiter ce cas exceptionnel, imaginons d'abord une relation de récurrence:

(1) $$\Sigma Q_i P_{n+i} = 0$$

où les coëfficients Q_i sont des polynômes entiers en n et en x (ce qui, comme on le voit, n'est pas le cas de la relation (1bis)) et formons les équations (4) et (6) correspondantes:

(4) $$\Sigma R_q \frac{d^q y}{dz^q} = S$$

(6) $$\Sigma A_i z^i = 0.$$

Soit α celle des racines de (6) dont le module est le plus grand; supposons que cette racine soit indépendante de x.

Que dire alors de la série $\Sigma a_n P_n$? Si le rayon de convergence de la série $\Sigma a_n t^n$ est plus grand que $|\alpha|$, la série $\Sigma a_n P_n$ est *toujours* convergente; si ce rayon est plus petit que $|\alpha|$, la série $\Sigma a_n P_n$ n'est jamais convergente; enfin si ce rayon est égal à $|\alpha|$, nous ne pouvons rien dire, ou plutôt le critérium fondé sur la limite du rapport $\frac{u_{n+1}}{u_n}$ se trouve en défaut. On doit donc recourir à d'autres critères de convergence des séries; par exemple à celui-ci.

On pose: $$\frac{u_{n+1}}{u_n} = 1 - \frac{\beta_n}{n}$$

et on cherche la limite de β_n pour $n = \infty$. Si cette limite a sa partie réelle plus grande que 1, la série est convergente; si elle a sa partie réelle plus petite que 1, la série est divergente.

Appliquons ce principe au problème qui nous occupe.

Ecrivons la relation (1) sous la forme suivante, en ordonnant selon les puissances décroissantes de n:

$$n^p \Sigma A_i P_{n+i} + n^{p-1} \Sigma B_i P_{n+i} + \ldots = 0$$

les A_i et les B_i étant des polynômes en x indépendantes de n. Nous savons que l'équation

(6) $$\Sigma A_i z^i = 0$$

a une racine indépendante de x. Nous pouvons supposer que cette racine est égale à 1, car si elle était égale à α, nous poserions:

$$P_n = \alpha^n P'_n$$

et nous remplacerions les polynômes P_n par les polynômes P'_n ce qui ne changerait pas les courbes de convergence.

On aura donc: $\Sigma A_i = 0$.

J'appelle $F(z)$ le premier membre de l'équation (6), on aura:

$$F(1) = 0.$$

Soit donc:

$$P_{n+1} = P_n\left(1 - \frac{\beta_n}{n}\right) \qquad P_{n+2} = P_n\left(1 - \frac{\beta_n}{n}\right)\left(1 - \frac{\beta_{n+1}}{n+1}\right)\cdots$$

Ecrivons maintenant la relation de récurrence (1) en remplaçant les P par ces valeurs et en ordonnant suivant les puissances décroissantes de n. Nous aurons en divisant par P_n:

$$n^p \Sigma A_i - n^{p-1}\Sigma A_i \gamma_{n.i} + n^{p-1}\Sigma B_i + \text{des termes en } n^{p-2}, n^{p-3}, \ldots = 0.$$

Dans cette formule, on a posé:

$$\gamma_{n.i} = \beta_n + \beta_{n+1} + \ldots + \beta_{n+i-1}.$$

Si $\lim \beta_n = \beta$, on aura $\qquad \lim \gamma_{n.i} = i\beta.$

Si l'on remarque que $\Sigma A_i = 0$, on verra que le terme en n^{p-1} qui est le premier terme s'écrit: $\qquad n^{p-1}(\Sigma B_i - \Sigma A_i \gamma_{n.i})$

A la limite ce terme doit s'annuler, ce qui donne:

$$\beta \Sigma i A_i = \Sigma B_i$$

ou

$$\beta = \frac{\Sigma B_i}{\Sigma i A_i} = \frac{\Sigma B_i}{F'(1)}.$$

Considérons alors une série: $\qquad \Sigma \alpha_n P_n$

où

$$\frac{\alpha_{n+1}}{\alpha_n} = 1 - \frac{\gamma_n}{n} \quad \lim \gamma_n = \gamma.$$

La condition de convergence s'écrira:

$$R(\beta + \gamma) > 1.$$

Il résulte de là que les courbes de convergence ont pour équation générale

$$R(\beta) = \text{const.}$$

ou

$$R\left(\frac{\Sigma B_i}{F'(1)}\right) = \text{const.}$$

Ce résultat peut se rattacher à l'étude de l'équation (4) de la manière suivante. Pour cette équation le point $z = \frac{1}{\alpha}$ est un point singulier, mais nous avons montré plus haut comment on peut toujours supposer $\alpha = 1$. Le point singulier que nous avons à considérer est donc $z = 1$. On a alors:

$$\lim \frac{P_{n+1}}{P_n} = 1$$

et la série $\Sigma P_n z^n$ qui est une intégrale de l'équation (4) est convergente dans le cercle de rayon 1. Nous supposerons que le point $z = 1$ est une racine simple de l'équation (6), alors les racines de l'équation déterminante correspondante seront: $\qquad 0, 1, 2, \ldots, p-2, \mu.$

Cherchons la valeur de μ. Le premier membre de l'équation (4) s'écrit, en reprenant des notations employées un peu plus haut:

$$\Sigma B_{iqh} z^{k+\varrho-i} x^h \frac{d^q y}{dz^q}$$

ou:

$$\Sigma A_{iq} \frac{d^q y}{dx^q} z^{k+\varrho-i}.$$

Or si l'on remarque que ces notations donnent:

$$A_i = A_{ip}$$

$$B_i = A_{i.p-1} + A_{ip}\left(pi - \frac{(p-1)(p-2)}{2}\right)$$

on verra que les deux premiers termes du premier membre de l'équation (4) seront:

$$\Sigma A_i z^{k+p-i} \frac{d^p y}{dz^p} - p\Sigma i A_i z^{k+p-i-1} \frac{d^{p-1}y}{dz^{p-1}}$$

$$+ \frac{(p-1)(p-2)}{2} \Sigma A_i z^{k+p-i-1} \frac{d^{p-1}y}{dz^{p-1}} + \Sigma B_i z^{k+p-i-1} \frac{d^{p-1}y}{dz^{p-1}}$$

pour $z = 1$, le coëfficient de $\frac{d^p y}{dz^p}$ s'annule et si on le divise par $z - 1$, le quotient se réduit à $-F'(1)$; quant au coëfficient de $\frac{d^{p-1}y}{dz^{p-1}}$ il se réduit à

$$\Sigma B_i - pF'(1).$$

L'équation déterminante s'écrit alors:

$$-F'(1)\rho(\rho-1)\ldots(\rho-p+1) + [\Sigma B_i - pF'(1)]\,\rho(\rho-1)\ldots(\rho-p+2) = 0.$$

On tire de là

$$\mu = \frac{\Sigma B_i}{F'(1)} - 1$$

ou

$$\mu = \beta - 1.$$

Il est aisé d'apercevoir le défaut de ce raisonnement. Il *suppose* l'existence de la limite β; je crois qu'il n'y aurait pas de difficulté à *démontrer* cette existence mais cela m'entraînerait trop loin.

Parlons maintenant des cas où la méthode précédente ne s'applique pas, et d'abord revenons sur l'exemple dont nous avons parlé plus haut et considerons les polynômes:

$$P_n = e^{x^2} \frac{d^n}{dx^n}(e^{-x^2}).$$

L'équation (1bis) ordonnée suivant les puissances décroissantes de n s'écrira:

$$n(P'_{n+2} + 2P'_n) + \sqrt{n}.\,2xP'_{n+1} + (\tfrac{3}{2}P'_{n+2} + P'_{n+1}) + \ldots = 0.$$

La présence du terme en $\sqrt{n}$, empêche que la méthode précédente puisse s'appliquer. De plus une autre difficulté, spéciale également au cas qui nous occupe, vient encore s'ajouter à la première. En effet, l'équation:

(6bis) $$z^2 + 2 = 0$$

a deux racines de même module. On en conclut que l'on peut poser

$$P'_n = Q_n + R_n$$

Q_n et R_n étant des fonctions de x telles que :

$$\lim \frac{Q_{n+1}}{Q_n} = +i\sqrt{2} \quad \lim \frac{R_{n+1}}{R_n} = -i\sqrt{2}$$

tandis qu'en général $\frac{P'_{n+1}}{P'_n}$ ne tend vers aucune limite.

De plus Q_n et R_n satisfont à la même relation de récurrence que P'_n. Posons alors :

$$Q_n = Q'_n i^n 2^{\frac{n}{2}} \qquad R_n = R'_n (-i)^n 2^{\frac{n}{2}}$$

il viendra :

(1ter) $$n(-Q'_{n+2} + Q'_n) + \sqrt{2n}\, 2ixQ'_{n+1} + \ldots = 0$$

(1quater) $$n(-R'_{n+2} + R'_n) - \sqrt{2n}\,.\, 2ixR'_{n+1} + \ldots = 0.$$

Posons ensuite :

$$Q'_{n+1} = Q'_n\left(1 - \frac{\beta_n}{\sqrt{n}}\right);\ Q_{n+2} = Q'_{n+1}\left(1 - \frac{\beta_{n+1}}{\sqrt{n+1}}\right).$$

La relation (1ter) ordonnée suivant les puissances décroissantes de n s'écrira :

$$\sqrt{n}(\beta_n + \beta_{n+1}) + \sqrt{2n}\, 2ixQ'_{n+1} + \ldots = 0$$

d'où
$$\lim \beta_n (n = \infty) = -2ix\sqrt{2}.$$

Si de même on pose :
$$R'_{n+1} = R'_n\left(1 - \frac{\beta'_n}{\sqrt{n}}\right)$$

on trouvera :
$$\lim \beta'_n = 2ix\sqrt{2}.$$

Soit maintenant la série :
$$\Sigma a_n Q'_n$$

et
$$\frac{a_{n+1}}{a_n} = 1 - \frac{\gamma_n}{\sqrt{n}} \quad \lim \gamma_n = \gamma = \gamma_0 + i\gamma_1.$$

La condition de convergence sera : partie réelle

$$(\gamma + 2ix\sqrt{2}) > 0.$$

En conséquence les conditions de convergence de la série

$$\Sigma a_n P'_n$$

s'écriront

$$(\text{partie imaginaire d'}x)^2 < \tfrac{1}{8}\gamma_0^2.$$

Les régions de convergence sont donc limitées par deux droites parallèles à l'axe des quantités réelles et situées de part et d'autre à égale distance de cet axe. L'ensemble de deux de ces droites forme une courbe de convergence.

De même, en supposant que les coëfficients de la relation (1) soient des polynômes entiers en n, au quel cas la difficulté précédente serait écartée, la méthode exposée plus haut serait encore en défaut, si $F'(1)$ était nul. Voici comment il faudrait opérer dans ce cas :

1°. Supposons que $F'(1)$ soit nul sans que ΣB_i le soit. On posera :

$$P_{n+1} = P_n\left(1 - \frac{\beta}{\sqrt{n}} - \frac{\gamma_n}{n}\right).$$

Supposons pour fixer les idées, $k = 2$; la relation (1) s'écrira :

$$Q_2\left(1 - \frac{\beta}{\sqrt{n+1}} - \frac{\gamma_{n+1}}{n+1}\right)\left(1 - \frac{\beta}{\sqrt{n}} - \frac{\gamma_n}{n}\right) + Q_1\left(1 - \frac{\beta}{\sqrt{n}} - \frac{\gamma_n}{n}\right) + Q_0 = 0.$$

Il vient en ordonnant suivant les puissances décroissantes de n et en posant :

$$Q_i = A_i n^p + B_i n^{p-1} + \ldots$$

$$n^p(A_2 + A_1 + A_0) - \beta n^{p-\frac{1}{2}}(2A_2 + A_1) + n^{p-1}(B_2 + B_1 + B_0)$$
$$+ A_2 n^{p-1}\beta^2 + n^{p-1}(\gamma_{n+1}A_2 + \gamma_n A_2 + \gamma_n A_1) + \ldots = 0.$$

Soit : $\lim \gamma_n = \gamma$ d'où $\lim(\gamma_{n+1}A_2 + \gamma_n A_2 + \gamma_n A_1) = \gamma F'(1)$

il viendra en tenant compte de :

$$F(1) = A_2 + A_1 + A_0 = 0$$
$$F'(1) = 2A_2 + A_1 = 0$$

et en divisant par n^{p-1}

$$A_2\beta^2 + B_2 + B_1 + B_0 + H = 0$$

H représentant des termes qui s'annulent avec $\frac{1}{n}$. On tire donc de là :

$$\beta^2 = \frac{\Sigma B_i}{A_2}$$

Les courbes de convergence ont pour équation générale :

partie réelle de β = const.

2°. Supposons maintenant que $F'(1)$ et ΣB_i soient nuls à la fois; dans ce cas le point $z = 1$ est un point singulier pour l'équation (4) dans le voisinage duquel les intégrales sont régulières. (Elles sont irrégulières lorsque ($F'(1)$ est nul sans que ΣB_i le soit.) Les racines de l'équation déterminante seront :

$$0, 1, 2, 3, \ldots, p-3, \mu' \text{ et } \mu''.$$

Si l'on pose

$$P_{n+1} = P_n\left(1 - \frac{\beta_n}{n}\right)$$

on aura :

$$\lim \beta_n = \mu + 1$$

μ étant celle des racines μ' et μ'' dont la partie réelle est la plus petite.

§8. *Résumé.*

Dans ce travail je me suis proposé plusieurs buts, mais le premier et le plus important d'entre eux était de contribuer à l'etude des intégrales des équations linéaires dans le voisinage d'un point donné. Si en effet nos connaissances sont assez complètes à ce sujet lorsque le point donné est un point singulier à intégrales régulières, nous ne savons presque rien sur les intégrales irrégulières. J'ai cru qu'il ne serait pas inutile de montrer comment on peut trouver une fonction simple dont le rapport à l'intégrale étudiée tende vers l'unité quand on se rapproche du point singulier. C'était un premier pas dans l'étude de ces intégrales irrégulières.

Pour atteindre ce but, j'ai dû employer comme auxiliaire la transformation de Laplace, et j'ai été amené en passant, à compléter la théorie de cette transformation, comme nous le permettent les progrès récents de nos connaissances sur les variables imaginaires. J'ai rencontré ainsi deux théorèmes qui peuvent d'ailleurs se démontrer aisément sans l'aide de la transformation de Laplace.

En premier lieu, si une équation linéaire d'ordre n a pour coëfficients des polynômes de degré p en x, elle admettra $n-p$ intégrales indépendantes holomorphes dans tout le plan.

Le second théorème peut faciliter la recherche des cas où une équation linéaire admet comme intégrale un polynôme entier.

Les équations différentielles linéaires présentent la plus étroite analogie avec les équations aux différences finies de forme linéaire, ou en d'autres termes, avec les relations linéaires de récurrence entre $k+1$ quantités consécutives:

$$u_{n+k},\ u_{n+k-1},\ \ldots,\ u_{n+1},\ u_n.$$

Cette analogie se poursuit dans les résultats, et la même méthode qui permet d'étudier les intégrales irrégulières des équations différentielles, nous donne, dans les cas des relations de récurrence la limite du rapport $\frac{u_{n+1}}{u_n}$ pour n infini.

Ce résultat a une application immédiate dans la recherche des courbes de convergence des séries ordonnées suivant des polynômes, c'est à dire des séries de la forme:

$$a_0P_0 + a_1P_1 + \ldots + a_nP_n + \ldots$$

lorsque les P sont des polynômes entiers en x et qu'il y a une relation de récurrence entre $k+1$ polynômes consécutifs.

Ces considérations font comprendre comment j'ai été conduit à réunir dans un même travail des recherches en apparence très différentes et expliquent un défaut d'unité que je prie le lecteur de vouloir bien excuser.

PARIS, 10 *Novembre* 1884.

NOTE.—Dans le mémoire précédent il faut remplacer partout le nom de Bessel par celui de Laplace.

A Second Paper on Perpetuants.

By Capt. P. A. MacMahon, R. A.

I here continue the investigation of Perpetuants commenced in Vol. VII, No. 1 of the *American Journal of Mathematics.*

The complete system of the simple or binomial syzygies of the sixth degree is there given, the working out of which led up to the discovery that the simplest sextic perpetuant is of weight 31; for that weight there is one exemplar form, viz.:

$$654^2 3^4$$

and five non-exemplar forms, viz.:

$$6^2 5^2 3^3$$
$$6^2 5 3^4 2$$
$$6^2 4 3^5$$
$$6 5^3 4 3^2$$
$$6 5^2 4 3^3 2\,;$$

the way this came about was that representing the quintic perpetuant forms $54^2 3^4$, $5^3 4 3^2$, $5^2 4 3^3 2$, hyper-symbolically by $\widetilde{124}$, $\widetilde{312}$, $\widetilde{213}$ respectively, that although the two combinations

$$\widetilde{213} + 2\,\widetilde{124},$$
$$\widetilde{312} - 2\,\widetilde{124},$$

were both expressible as sextic syzygants, the forms $\widetilde{124}$, $\widetilde{312}$, $\widetilde{213}$ were not each separately so expressible.

Thus far the generating function for sextic perpetuants was shown to be

$$\frac{x^{31} + 0.x^{32} + \ldots}{2.3.4.5.6.},$$

wherein as usual a number μ in the denominator denotes for brevity $(1 - x^{\mu})$. It remains to prove that there are no more terms in the numerator and that the form of the generating function is in reality

$$\frac{x^{31}}{2.3.4.5.6};$$

this amounts to showing that of weights superior to 25 there exist no quintic perpetuant forms, which, not being symbolised by such a symbol as $\overline{1+\alpha,\ 2+\beta,\ 4+\gamma}$ (α, β, γ being any positive, including zero, integers) are not singly connected through such forms with sextic syzygies; in other words we have to show that every quintic perpetuant not of the above form is expressible by means of such forms as a sextic syzygant; for this will prove that all exemplar sextic perpetuants are comprised in the symbol $6^{*}5^{1+\alpha}4^{2+\beta}3^{4+\gamma}$,
and that consequently the numerator of the generating function is in truth monomial. Firstly, consider the syzygy B_7 of Class 1, Group 5, in the paper above referred to; this is

$$B_7 \qquad 43^5 2^{\kappa-7}.2^2 - 3^6 2^{\kappa-7}.32 \equiv (\kappa-6)\,\overbrace{114} + \overbrace{116} + 2\ \overbrace{124} + \overbrace{213},$$

wherein on the dexter side, reducible quintic forms and forms of lower degree are omitted; in Mr. Hammond's notation we have the operator

$$D_\lambda = \frac{1}{\lambda!}\left(\frac{d}{da_1} + a_1\frac{d}{da_2} + a_2\frac{d}{da_3} + \ldots\right)^\lambda$$

and $$D_\lambda(\lambda_1, \mu_1, \nu_1, \pi_1 \ldots)(\lambda_2, \mu_2, \nu_2, \pi_2 \ldots)(\lambda_3, \mu_3, \nu_3, \pi_3 \ldots)\ldots$$
$$= \Sigma(\mu_1, \nu_1, \pi_1 \ldots)(\lambda_2, \nu_2, \pi_2 \ldots)(\lambda_3, \mu_3, \pi_3 \ldots)\ldots$$
where $$\lambda_1 + \mu_2 + \nu_3 + \ldots = \lambda,$$
the summation including all (including zero) solutions of this equation; so that for instance $$D_4(4^2 3^2 2^3.2^2) = 43^2 2^3.2^2 + 4^2 3^2 2^2.2;$$
take then the operator $D_4 D_3^2$ and operate on each side of the syzygy B_7; thus, putting $\kappa + 5$ for κ to keep the weight $= 2\kappa + 9$, we get

$$3^3 2^{\kappa-3}.2^2 + 43^3 2^{\kappa-3}.2 - 3^4 2^{\kappa-3}.3 - 2(3^5 2^{\kappa-3})$$
$$\equiv (\kappa-1)\,\overbrace{102} + \overbrace{104} + 2\ \overbrace{112} + \overbrace{201}.$$

or since $\overbrace{102}$ and $\overbrace{104}$ are reducible forms, this may be written

$$43^3 2^{\kappa-3}.2 - 3^4 2^{\kappa-3}.3 \equiv 2\ \overbrace{112} + \overbrace{201} \qquad (1)$$

but the sinister being a sextic syzygy it must be possible to express it in terms of exemplar quintic forms, quintic compounds and forms of lower degree; in fact reference to the tables before referred to shows the syzygy

$$A_5 \qquad 43^3 2^{\kappa-3}.2 - 3^4 2^{\kappa-3}.3 \equiv \overbrace{112};$$

whence combining A_5 with (1) we have

$$\overbrace{112} + \overbrace{201} \text{ reducible},$$

which is well known from the previous tables which give the reductions of all the non-exemplar quintic forms by aid of the exemplar; from the syzygy B_7, then, has been derived the formula which gives the reduction of the non-exemplar quintic perpetuant $\overbrace{201}$, and this must necessarily have been so since the syzygy B_7 includes only one form which becomes a non-exemplar quintic perpetuant when operated upon by $D_4 D_3^2$.

Secondly consider the syzygy

$$C_7 \quad 43^5 2^{\varkappa-8}.2^3 - 3^6 2^{\varkappa-8}.32^2 \equiv \tfrac{1}{2}(\varkappa-6)(\varkappa-7)\overbrace{114} + (\varkappa-11)\overbrace{116}$$
$$+ 2(\varkappa-8)\overbrace{124} + 2\ \overbrace{126} + 3\ \overbrace{134} + (\varkappa-7)\overbrace{213} + \overbrace{215} + 2\ \overbrace{223} + \overbrace{312}.$$

Operating with $D_4 D_3^2$ and comparing with B_5

$$2\ \overbrace{211} + \overbrace{300} + (\varkappa-2)\overbrace{201} + \overbrace{203} \equiv -2\ \overbrace{122} - \overbrace{114} - (\varkappa-3)\overbrace{112};$$

since $\overbrace{201} \equiv -\overbrace{112}$,

and from taking B_9 and A_7 together

$$\overbrace{203} \equiv -\overbrace{114},$$

this reduces to $\qquad 2\ \overbrace{211} + \overbrace{300} \equiv -2\ \overbrace{122} + 3\ \overbrace{112},$

which does not exhibit the reductions of the forms $\overbrace{211}$, $\overbrace{300}$ by aid of exemplars, but only the reduction of the combination

$$2\ \overbrace{211} + \overbrace{300};$$

and moreover it will be found impossible to so exhibit each separately by consideration of the binomial syzygies; but as a matter of fact we know that each is separately so expressible and it follows that there must exist capitation syzygies which, in conjunction with the binomial syzygies, will enable such reduction to be exhibited; that is to say, there must exist a syzygy which involves the form $\overbrace{223}$ and no other form which is convertible into a non-exemplar quintic perpetuant through the operation of the operator $D_4 D_3^2$.

It appears from this argument, which is a general one, that syzygies must exist containing one and only one form which the operator $D_4 D_3^2$ converts into a quintic non-exemplar perpetuant; each such form therefore must be expressible in terms of sextic compounds, quintic perpetuants of the form $\overbrace{1+\alpha,\ 2+\beta,\ 4+\gamma}$, and quintic perpetuants which the operator $D_4 D_3^2$ converts into directly reducible quintic forms; as these latter perpetuants have all been exhibited as sextic syzygants (vide Table of Syzygies, *American Journal of Mathematics*, Vol. VII, No. 1) we have the theorem as follows:

"Each quintic perpetuant of an exemplar form which is convertible to the non-exemplar form by the operation of the operator $D_4 D_3^2$ can, in combination with quintic perpetuants of the form $\overbrace{1+\alpha,\ 2+\beta,\ 4+\gamma}$, be expressed as a sextic syzygant."

It results therefore by a sextic capitation that every sextic form is reducible by the aid of such forms as $6^{*}5^{1+\alpha}4^{2+\beta}3^{4+\gamma}$, and that the only exemplar sextic forms are of this type.

Hence their generating function is

$$\frac{x^{31}}{2.3.4.5.6.},$$

and the generating function for sextic syzygies is

$$\frac{x^{6}+x^{13}-2x^{16}-x^{18}+x^{31}}{2.3.4.5.6.}.$$

§2. Proceeding to consider the perpetuants of the seventh degree, or say the septic perpetuants, it is obvious that a form $76^{\kappa}5^{\lambda}4^{\mu}3^{\nu}$ will be such, provided only that the sextic form $6^{\kappa}5^{\lambda}4^{\mu}3^{\nu}$ be singly inexpressible as a septic syzygant.

Suppose the whole series of septic syzygies to be written down and the non-exemplars to be expressed in terms of exemplars as they arise; conceive the operation $D_5 D_4^2 D_3^4$ to be performed throughout on each; this will result in a series of identities and syzygies of the sixth and seventh degrees respectively, and the septic syzygies can be reduced by means of the original syzygies to sextic identities, as in the previous case discussed; as before, exemplar and non-exemplar sextic perpetuant forms will occur, and we must be able to exhibit the reduction of each non-exemplar sextic perpetuant form by the aid of the exemplars; not only so but we must be able to obtain the reduction of every reducible sextic form whatever in a similar manner; *ex. gr.* we have the binomial septic syzygy of weight $2\kappa+31$:

$$54^{3}3^{4}2^{\kappa}.2-4^{4}3^{4}2^{\kappa}.3=654^{2}3^{4}2^{\kappa}+2(5^{2}4^{3}3^{3}2^{\kappa})+3(54^{4}3^{4}2^{\kappa-1})$$
$$+(\kappa+1)\,54^{3}3^{4}2^{\kappa+1}+54^{5}3^{2}2^{\kappa}-4^{5}3^{3}2^{\kappa}.2-5(4^{4}3^{5}2^{\kappa})$$
$$+6(4^{6}3^{3}2^{\kappa-1})+(\kappa+1)\,4^{5}3^{3}2^{\kappa+1}.$$

and operating with $D_5 D_4^2 D_3^4$ and transposing

$$62^{\kappa}=42^{\kappa}\cdot 2-2(4^{2}2^{\kappa-1})-(\kappa+1)\,42^{\kappa+1},$$

giving the reduction of the sextic form 62^{κ}.

Just then as in the former case there was a one to one correspondence between the reducible quintic forms and the sextic syzygies, of a weight higher by ten, that involved quintic perpetuants, so in this case we have a correspondence between the reducible sextic forms and the septic syzygies that involve sextic perpetuants of a weight higher by 25; thus the generating function for reducible sextic forms being

$$\frac{x^6 - x^{31}}{2.3.4.5.6},$$

that for septic syzygies involving sextic perpetuants is

$$\frac{x^{31} - x^{56}}{2.3.4.5.6},$$

and therefore the generating function for sextic perpetuants which are not septic syzygants is

$$\frac{x^{31}}{2.3.4.5.6} - \frac{x^{31} - x^{56}}{2.3.4.5.6} = \frac{x^{56}}{2.3.4.5.6};$$

consequently the theory of capitation shows us that the generating function for septic perpetuants is

$$\frac{x^{63}}{2.3.4.5.6.7}.$$

The form $765^2 4^4 3^8$ may be taken as the exemplar septic form of weight 63, and then every exemplar septic form of higher weight includes these numbers in its symbol.

The reasoning above employed is perfectly general and leads easily to the conclusion that the generating function for perpetuants of degree n is, $(n > 2)$,

$$\frac{x^{2^{n-1}-1}}{2.3.4\ldots.n.};$$

because by operating on the n^{ic} syzygies with the D symbol which corresponds to the simplest $(n-2)^{\text{ic}}$ perpetuant which is not an $(n-1)^{\text{ic}}$ syzygant, we can obtain the identities which give the reduction of every $(n-1)^{\text{ic}}$ reducible form.

The simplest exemplar n^{ic} perpetuant, $(n > 2)$, may be taken of the form

$$n.n-1.\overline{n-2}^2.\overline{n-3}^4.\overline{n-4}^8\ldots 3^{2^{n-4}}.$$

The complete system of groundforms to the quantic of unlimited order, the degree being θ and the weight w, may be stated as the coefficient of $a^\theta x^w$ in the development in ascending powers of x of

$$a + a^2\frac{x^2}{2} + a^3\frac{x^3}{2.3} + a^4\frac{x^7}{2.3.4} + a^5\frac{x^{15}}{2.3.4.5}$$
$$+ a^6\frac{x^{31}}{2.3.4.5.6} + a^7\frac{x^{63}}{2.3.4.5.6.7} + \ldots$$
$$+ a^\theta\frac{x^{2^{\theta-1}-1}}{2.3.4\ldots.\theta} + \ldots$$

ROYAL MILITARY ACADEMY, Woolwich, England, Dec. 12, 1884.

Prüfung grösserer Zahlen auf ihre Eigenschaft als Primzahlen.

Von P. Seelhoff.

Die unten stehende Tabelle enthält eine Zusammenstellung von binären quadratischen Formen, deren Determinante negativ und deren mittlerer Coefficient Null ist, während die äusseren Coefficienten relative Primzahlen sind. Da für die Charaktere, welche ihnen entsprechen, keine andere reducirte Form mit derselben Determinante existirt, so muss sich jede Primzahl N mit den entsprechenden Charakteren entweder durch eine einzige dieser Formen darstellen lassen, wenn diese allein steht oder alternativ durch eine von zweien, wenn sie gepaart vorkommen oder durch eine von vieren, wenn sie zu vieren verbunden sind. Da die Determinanten theilweise verhältnismässig gross sind, so bieten sie ein gutes Mittel dar, um selbst sehr grosse Zahlen ohne zu grossen Aufwand von Arbeit daraufhin zu prüfen, ob sie Primzahlen sind und auch, um die Faktoren zu bestimmen, falls sie zusammengesezt sind, letzteres natürlich nur in dem Falle, wenn die Determinante quadratischer Rest der Zahl, mithin auch ihrer sämmtlichen Faktoren ist. Die Tabelle enthält nur die Formen für Zahlen von der Form $8n+1$, und man wird bei genauerer Prüfung finden, dass diese Formen alle möglichen Fälle decken.

Über die Einrichtung der Tabelle ist Folgendes zu bemerken. Da es erforderlich ist, dass die Determinante quadratischer Rest der zu prüfenden Zahl N ist, so handelt es sich zunächst darum, wie sich die einzelnen Primzahlen, welche erstere constituiren, zu N oder umgekehrt, wie sich N zu den Primzahlen in dieser Beziehung verhält. Ist nun N congruent einem quadratischen Reste nach dem Modulus α, so ist α in der Tabelle unter $+$ eingetragen, im anderen Falle unter $-$. So findet man z. *B.* in der Zeile 33 die Primzahl 3 unter $-$ und 5 unter $+$, in der Rubrik Formen für diese die einzelne Form $(5, 9)$ oder vollständig $(5, 0, 9)$ und die gepaarten Formen $\begin{pmatrix} 9, & 20 \\ 5, & 36 \end{pmatrix}$ d. h. ist $N \equiv 1\,(3)$ und $N \equiv 1$ oder $4\,(5)$, so ist N eine Primzahl, wenn es sich nur auf eine einzige Art durch die Form $(5, 0, 9)$ darstellen lässt, oder wenn man die gepaarten Formen zur Prüfung wählt, wenn nur eine Darstellung entweder durch $(9, 0, 20)$ oder $(5, 0, 36)$ möglich ist. In

Zeile 10 findet man, dass 3 und 5 beide unter + stehen und die zugehörigen Formen sind 4 an der Zahl. Ist also $N \equiv 1\,(3)$ und $\equiv 1$ oder $4\,(5)$, so ist es eine Primzahl, wenn es sich durch eine der vier Formen einmal darstellen lässt. Sowohl in dem ersten wie in dem zweiten Falle gilt ferner, dass wenn man keine Darstellung oder mehr als eine fur N findet, dieses nur eine zusammengesetzte Zahl sein kann, und dass, wenn sich mehr als eine Darstellung findet, aus diesen Darstellungen die Faktoren von N abgeleitet werden können.

Um zugleich zu zeigen, wie vortheilhaft selbst die zu vieren verbundenen Zahlen zur Prüfung sehr grosser Zahlen verwandt werden können, wähle ich für ein erläuterndes Beispiel die Zahl $N = 2^{31} - 1 = 2147470249$. Bekanntlich hat Euler diese Zahl zuerst untersucht und zwar vermittelst Division durch die einzig möglichen Primzahlen von der Form $248z + 1$ und $248z + 63$ bis zu $\sqrt{N} = 46339$ und dieselbe als Primzahl bestimmt.

Wählen wir zu demselben Zwecke eine Form, welche der Tabelle fur die Zahlen von der Form $8n + 7$ angehört. Für $N \equiv 1\,(3)$, $\equiv 1\,(7)$, $\equiv 1\,(11)$, $\equiv 7\,(29)$ hat man die verbundenen Formen (1, 0, 13398), (22, 0, 609), (42, 0, 319), (58, 0, 231). Eine und nur eine von diesen muss eine einzige Darstellung von N geben, falls dieses eine Primzahl ist; dann ist jeder Versuch mit den andern noch übrigen Formen zwecklos. Würde sich keine Darstellung für sämmtliche 4 Formen ergeben, so wäre N keine Primzahl, ebenso nicht, wenn sich für dieselbe Form mehr als eine Darstellung herausstellte. Nun giebt die erste Form (1, 0, 13398) keine Darstellung, ich gehe daher gleich zu der zweiten (22, 0, 609) über, um an ihr das ganze Verfahren im Allgemeinen auseinanderzusetzen. Da also $22x^2 + 609y^2 = N$ sein soll, so muss N in solche zwei Theile zerlegt werden, von denen der eine ein Multiplum von 22, der andere ein solches von 609 ist. Setzt man demgemäss $22a + 609b = 2147483647$,

so ist
$$a = 97612810 - 609k$$
$$b = 3 + 22k.$$

Da $a = x^2$ ist, so müssen die Werthe a für x so genommen werden, dass $97612810 - a^2$ durch 609 theilbar ist. $609 = 3.7.29.97612810 \equiv 1\,(3), \equiv 1\,(7), \equiv 28\,(29)$ und da $1^2 \equiv 1\,(3)$, $1^2 \equiv 1\,(7)$, $12^2 \equiv 28\,(29)$ ist, so ist $x = 3t \pm 1 = 7u \pm 1 = 29v \pm 12$. Hieraus folgen 8 Werthe für x, nämlich $609n + 41$, 104, 244, 302, 307, 365, 505, 568 bis zu der Grenze $\sqrt{97612810} = 9879$. Setzt man diese für x ein, bildet k und hieraus b, so ist eine Darstellung gefunden, wenn b eine Quadratzahl und das zu x gehörige $y = \sqrt{b}$. Übrigens kommen alle geraden Werthe für x nicht in Betracht, weil diese in $22x^2 + 609y^2 = N$ nur Zahlen von der Form $8n + 1$ liefern, und von den ungeraden fallen noch diejenigen mit der Endziffer 5 aus, da im Voraus zu ersehen ist, dass sie keine Quadratzahl für b hervorbringen können. Für $x = 7001$ findet sich dann $b = 1755625$ und $y = 1325$, also $22.7001^2 + 609.1325^2 = 2147483647$. Da sich für diese Form keine weitere Darstellung ergiebt und da die Zahl somit eine Primzahl ist, so ist die Untersuchung abgeschlossen. *Als Beispiel für eine zusammengesetzte Zahl diene $N = 165580141$. Da $N \equiv 6\,(7), \equiv 1\,(11), \equiv 12\,(13)$, so kann man die geparten Formen (14, 0, 143) und (26, 0, 77) benutzen und erhält mit der ersten:

$$14.1399^2 + 143.983^2 = 165580141$$
$$14.3089^2 + 143.473^2 = \quad ``$$

Sind aber α, β und γ, δ zwei Darstellungen der Zahl N durch die Form (m, n), so setze man $\frac{p}{q} = \frac{\alpha \pm \gamma}{\beta \pm \delta}$, und reducire die sich hieraus ergebenden Brüche, so dass p gleich dem Zähler und q gleich dem Nenner ist. Dann bilde man weiter den Bruch $\frac{r}{s} = \frac{mp^2}{nq^2}$ und reducire, so dass hier r gleich dem Zähler und s gleich dem Nenner wird. Dann ist $f = r + s$, oder, wenn dies eine gerade Zahl ist, die Hälfte hiervon ein Faktor, von N.

Also in unserem Beispiele $\frac{p}{q} = \frac{3089 - 1399}{983 - 473} = \frac{169}{51}\;\frac{r}{s} = \frac{14.169^2}{143.51^2}\,.\,f = 59369.$

Aus $\frac{p}{q} = \frac{3089 - 1399}{983 + 473}$ findet man den zweiten Faktor 2789, mithin

$$2789.59369 = 165580141:$$

Die gewählte Zahl ist das 41te Glied der Reihe

$$0,\ 1,\ 1,\ 2,\ 3,\ 5,\ 8,\ 13,\ 21\ \ldots$$

Zum Schlusse meiner Mittheilung möchte ich noch darauf hinweisen, dass neben den Tabellen für die Formen, von welchen die hier gegebene zunächst nur als Beispiel dienen soll, eine genügend weit reichende Tafel der Quadrat-Zahlen und nebenbei eine kleine Tabelle nöthig ist, welche für die in den Determinanten vorkommenden Primzahlen α die Wurzeln der Congruenz $z^2 \equiv r\,(\alpha)$ angiebt.

Die Anzahl der benutzten Determinanten ist 170, davon sind 65 die von Euler sogenannten "numeri idonei;" von den übrigen finden sich einzelne in Legendre: Théorie des nombres oder sonstwie in mathematischen Zeitschriften. Die Mehrzahl derselben habe ich selbst fest stellen müssen.

* NOTE BY EDITOR.—The tables used by the author in the following examples do not appear in the present article. They have, however, been prepared, and, with some additional matter, will appear in a future number of the Journal.

BREMEN im August 1884.

Charaktere und binäre quadratische Formen für $N = 8n + 1$.

Charaktere.								Formen.
+	+	+	+	−	−	−	−	
	.	.	.	.	.	.	.	1,1 1,2 1,4
	.	.	.	.	.	.	.	1,8 1,16
3	.	.	.	.	.	.	.	1,3 1,9
3	.	.	.	.	.	.	.	1,6 1,12
3	.	.	.	.	.	.	.	1,18 1,24
3	.	.	.	.	.	.	.	1,48 1,72
3	.	.	.	.	.	.	.	1,36 1,144 4,9 9,16
3	5	.	.	.	.	.	.	1,15 1,180 1,360 4,45 8,45
3	5	.	.	.	.	.	.	1,30 1,120
3	5	.	.	.	.	.	.	1,225 1,150 9,25 6,25
3	5	.	.	.	.	.	.	1,240 1,600 24,25
3	5	.	.	.	.	.	.	1,900 9,100 4,225 25,36
3	7	.	.	.	.	.	.	1,21 1,42
3	7	.	.	.	.	.	.	1,168 1,63 7,9
3	7	.	.	.	.	.	.	1,84 1,252 4,21 9,28
3	11	.	.	.	.	.	.	1,33 1,66 3,22
3	11	.	.	.	.	.	.	1,198 1,528 9,22 16,33
3	13	.	.	.	.	.	.	1,78 1,312
3	13	.	.	.	.	.	.	1,156 1,39 12,13 3,13
3	13	.	.	.	.	.	.	1,117 9,13
3	17	.	.	.	.	.	.	1,102 1,408
3	19	.	.	.	.	.	.	1,57
3	19	.	.	.	.	.	.	1,228 1,912 4,57 4,228
3	23	.	.	.	.	.	.	1,138 6,23
3	29	.	.	.	.	.	.	1,2088 9,232
3	31	.	.	.	.	.	.	1,93 1,372 4,93
3	37	.	.	.	.	.	.	1,333 9,37
3	43	.	.	.	.	.	.	1,258 6,43
3	47	.	.	.	.	.	.	1,282 3,94
3	59	.	.	.	.	.	.	1,177

Charaktere.								Formen.
+	+	+	+	−	−	−	−	
3	73	.	.	.	.	.	.	1,438 6,73
3	83	.	.	.	.	.	.	1,498 6,83
		.	.	3	.	.	.	2,9
	5	.	.	3	.	.	.	5,9 9,20 5,36
				3	5	.	.	3,5 2,15
				3	7	.	.	3,14
				3	11	.	.	2,33 6,11
				3	13	.	.	2,39
				3	17	.	.	6,17 17,24
				3	43	.	.	2,129 3,86
				3	83	.	.	2,249 6,83
3	5	7	.	.	.	.	.	1,105 1,525 21,25
3	5	7	.	.	.	.	.	1,1680 1,420 16,105 4,105
3	5	7	.	.	.	.	.	1,630 1,210 9,70
3	5	7	.	.	.	.	.	1,660 1,840 4,165
3	5	11	.	.	.	.	.	1,330 1,1320
3	5	11	.	.	.	.	.	1,165
3	5	13	.	.	.	.	.	1,390 1,1170 10,39 9,130
3	5	13	.	.	.	.	.	1,4680 9,520
3	5	17	.	.	.	.	.	1,510 1,765 15,34 9,85
3	5	19	.	.	.	.	.	1,1710 1,570 9,190 19,30
3	5	23	.	.	.	.	.	1,345
3	5	23	.	.	.	.	.	1,690 1,1380 6,115 4,345
3	5	29	.	.	.	.	.	1,3480 1,870 24,145 6,145
3	5	31	.	.	.	.	.	1,2790 9,310
3	5	37	.	.	.	.	.	1,1110 10,111
3	5	43	.	.	.	.	.	1,1290 1,5160 10,129 40,129
3	5	53	.	.	.	.	.	1,1590 10,159 6,265 15,106

Charaktere und Binäre quadratische Formen für $N = 8n + 1$.

Charaktere.								Formen.	Charaktere.								Formen.
+	+	+	+	−	−	−	−		+	+	+	+	−	−	−	−	
3	.	.	.	5	7	.	.	7,90 10,63 3,70		11	.	.	3	5	.	.	5,33 3,110 5,133 20,33
3	.	.	.	5	11	.	.	10,33 33,40		.	.	.	3	5	13	.	3,585 2,585 65,72 13,65
3	.	.	.	5	19	.	.	3,190 10,57		17	.	.	3	5	.	.	2,255 17,30
3	.	.	.	5	29	.	.	10,87 15,58		.	.	.	3	5	17	.	5,153 17,45
3	.	.	.	5	43	.	.	3,430 30,43		.	.	.	3	5	19	.	2,855 13,95
3	7	13	.	.	.	.	.	1,1092 4,273 1,273		7	.	.	3	19	.	.	2,399 14,57
3	7	17	.	.	.	.	.	1,357		11	.	.	3	7	.	.	14,33 33,56
3	7	19	.	.	.	.	.	1,1197 1,798 9,133 7,114		17	.	.	3	7	.	.	17,21
3	7	31	.	.	.	.	.	1,1302 7,186		11	.	.	3	13	.	.	11,78 26,33
3	7	37	.	.	.	.	.	1,3108 21,148 4,777 37,84		11	.	.	3	17	.	.	3,374 11,102
3	7	37	.	.	.	.	.	1,777 21,37		17	.	.	3	11	.	.	2,561 3,561 17,66 17,264
3	.	.	.	7	11	.	.	7,66	5	.	.	.	.	.	.	.	1,5 1,25 1,100 4,25
3	.	.	.	7	13	.	.	13,21 13,84 21,52	5	7	.	.	.	.	.	.	1,70
3	11	13	.	.	.	.	.	1,858 3,286	5	11	.	.	.	.	.	.	1,220 5,44
3	11	17	.	.	.	.	.	1,1122 1,4488 33,34 33,136	5	13	.	.	.	.	.	.	1,130 1,520
3	11	97	.	.	.	.	.	1,6402 66,97 3,2134 22,191	5	17	.	.	.	.	.	.	1,340 4,85
3	.	.	.	11	17	.	.	6,187 22,51	5	19	.	.	.	.	.	.	1,190 1,760
3	13	43	.	.	.	.	.	1,1677 13,129	5	29	.	.	.	.	.	.	1,145 5,29
	.	.	.	.	.	.	.	1,6708 13,516 4,1677 52,129	5	29	.	.	.	.	.	.	4,145 5,116 1,580 20,29
3	13	61	.	.	.	.	.	1,7137 13,549 9,793 61,117	5	31	.	.	.	.	.	.	1,310 10,31
	5	.	.	3	7	.	.	5,21 6,35 5,84 20,21	5	41	.	.	.	.	.	.	1,205 5,41
	5	.	.	3	11	.	.	11,30	5	89	.	.	.	.	.	.	1,445 5,89
	5	.	.	3	13	.	.	6,35 15,26	5	101	.	.	.	.	.	.	1,505 5,101
	5	.	.	3	23	.	.	5,69 5,276 20,69	5	101	.	.	.	.	.	.	1,2020 5,404 4,505 20,101
	5	.	.	3	37	.	.	6,185 15,74			.	.	5	7	.	.	7,10
	7	.	.	3	5	.	.	2,105 3,105			.	.	5	13	.	.	2,65 3,65
	7	.	.	3	11	.	.	2,231			.	.	5	17	.	.	5,68 17,20 5,17
											.	.	5	19	.	.	2,95

Charaktere.								Formen.
+	+	+	+	−	−	−	−	
5	7	11	.	.	.	.	.	1,385
5	7	31	.	.	.	.	.	1,2170 14,155
		11	.	5	7	.	.	5,77
		31	.	5	7	.	.	10,217 35,62
5	11	19	.	.	.	.	.	1,1045 5,209
5	11	19	.	.	.	.	.	1,4180 5,836 4,1045 20,209
5	13	29	.	.	.	.	.	1,1885 29,65
5	13	37	.	.	.	.	.	1,4810 26,185 10,481 65,74
5	13	37	.	.	.	.	.	1,19240 104,185 40,481 65,296
		29	.	5	13	.	.	5,377 13,145
7	.	.	.	.	.	.	.	1,7 1,28
7	.	.	.	.	.	.	.	1,112 1,14 2,7
7	11	.	.	.	.	.	.	1,154 11,14
7	19	.	.	.	.	.	.	1,133 1,532 4,133
7	37	.	.	.	.	.	.	1,1813 37,49
7	47	.	.	.	.	.	.	1,658 2,329
7	47	.	.	.	.	.	.	1,2632 8,329
7	113	.	.	.	.	.	.	1,1582 7,226 2,791 14,113
7	137	.	.	.	.	.	.	1,1918 7,274 2,959 14,137
7	11	13	.	.	.	.	.	1,2002 22,91
7	.	.	.	11	13	.	.	2,1001 11,182
11	.	.	.	.	.	.	.	1,22 1,88
11	23	.	.	.	.	.	.	1,253
13	.	.	.	.	.	.	.	1,13
13	17	.	.	.	.	.	.	1,442 1,1768 17,26 17,104
13	23	.	.	.	.	.	.	1,598 23,26
13	61	.	.	.	.	.	.	1,793 13,61
13	61	.	.	.	.	.	.	1,3172 13,244 4,793 52,61

Charaktere.								Formen.
+	+	+	+	−	−	−	−	
13	17	53	.	.	.	.	.	1,11713 17,689 13,901 53,221
17	.	.	.	.	.	.	.	1,34 1,86 2,17 8,17
23	.	.	.	.	.	.	.	1,46 2,23
29	.	.	.	.	.	.	.	1,58 1.232
31	41	.	.	.	.	.	.	1,2542 31,82 2,1271 41,62
37	.	.	.	.	.	.	.	1,37
41	.	.	.	.	.	.	.	1,82 1,328 2,41 8,41
71	.	.	.	.	.	.	.	1,142 2,71
3	5	7	11	.	.	.	.	1,3465 9,385
3	5	7	13	.	.	.	.	1,1365 1,5460 4,1365
3	5	7	17	.	.	.	.	1,1785 1,3570 21,85 51,70
3	5	7	17	.	.	.	.	1,7140 21,340 4,1785 84,85
3	5	7	23	.	.	.	.	1,19320 105,184
3	5	7	41	.	.	.	.	1,4305 21,205
3	5	.	.	7	17	.	.	6,595 34,105
3	7	11	29	.	.	.	.	1,13398 42,319 22,609 58,231
3	.	.	29	7	11	.	.	7.1914 33,406 6,2233 87,154
3	.	.	13	5	7	.	.	13,105
3	.	.	13	5	7	.	.	13,420 52,105
	5	7	.	3	.	.	13	21,65 21,260 65,84
	5	7	.	3	.	.	23	56,345 120,161
	7	17	.	3	5	.	.	2,1785 35,102
				3	5	7	11	5,693 45,77
				3	5	7	13	5,273 5,1092 20,273
				3	5	7	17	5,357 3,1190 17,105 17,210
				3	5	7	17	5,1428 68,105 20,357 17,420

Solvable Irreducible Equations of Prime Degrees.

By George Paxton Young, *Toronto, Canada.*

Object of the Paper.

§1. Let $F(x) = 0$ be an irreducible solvable equation of the mth degree, m prime, with roots r_1, r_2, etc. The equation being understood to have been deprived of its second term, its roots are of the forms

$$\left.\begin{aligned} mr_1 &= \Delta_1^{\frac{1}{m}} + a_1\Delta_1^{\frac{2}{m}} + b_1\Delta_1^{\frac{3}{m}} + \ldots + c_1\Delta_1^{\frac{m-1}{m}} \\ mr_2 &= \omega\Delta_1^{\frac{1}{m}} + \omega^2 a_1\Delta_1^{\frac{2}{m}} + \omega^3 b_1\Delta_1^{\frac{3}{m}} + \ldots + \omega^{m-1}c_1\Delta_1^{\frac{m-1}{m}} \\ mr_3 &= \omega^2\Delta_1^{\frac{1}{m}} + \omega^4 a_1\Delta_1^{\frac{2}{m}} + \omega^6 b_1\Delta_1^{\frac{3}{m}} + \ldots + \omega^{2(m-1)}c_1\Delta_1^{\frac{m-1}{m}}, \end{aligned}\right\} \quad (1)$$

and so on; where ω is a primitive m^{th} root of unity; and a_1, b_1, etc., are rational functions of Δ_1. If we call

$$\Delta_1^{\frac{1}{m}},\ a_1\Delta_1^{\frac{2}{m}},\ b_1\Delta_1^{\frac{3}{m}},\ \ldots,\ c_1\Delta_1^{\frac{m-1}{m}}, \quad (2)$$

the separate members of mr_1, I propose first of all to establish the fundamental theorem, that *the separate members of the root r_1 can be arranged in groups G_1, G_2, etc., such that any symmetrical function of the terms in any one of the groups is a rational function of the root* (§8). The groups G_1, G_2, etc., may be defined more exactly as follows. The mth powers of the terms in (2) are the roots of a rational equation of the $(m-1)^{\text{th}}$ degree auxiliary to $F(x) = 0$. Should the auxiliary not be irreducible, it can be broken, after the rejection of roots equal to zero, into rational irreducible sub-auxiliaries. This being so, the terms constituting any one of the groups G_1, G_2, etc., are those separate members of r_1, which, severally multiplied by m, are m^{th} roots of the roots of the auxiliary, provided the auxiliary be irreducible; but, when the auxiliary is not irreducible, the terms constituting any one of the groups G_1, G_2, etc., are m^{th} roots of the roots of a sub-auxiliary. *From the fundamental theorem above enunciated can be deduced as a corollary the theorem of Galois, that r_1 is a rational function of r_2 and r_3.* In fact, any symmetrical function of those separate members of r_1 which constitute any one of the groups G_1, G_2, etc., is a rational function of r_2 and r_3 (§13). Not only is it proved that r_1 is a rational function of r_2 and r_3, but *the investigation shows how the function is formed.* An instance in verification is given (§15). It incidentally appears that if c be the number of terms in any one of the groups G_1, G_2, etc., *the sum of a cycle of c primitive m^{th} roots of unity is a rational function of r_1 and r_2* (§17).

PRELIMINARY STATEMENTS.

§2. Use will be made of certain general laws of the structure of the roots of equations, that were established in an article published in this Journal (Vol. VI), entitled "Principles of the Solution of Equations of the Higher Degrees." It was there shown that if

$$\Delta_1,\ \Delta_2,\ \ldots,\ \Delta_e, \tag{3}$$

be the unequal particular cognate forms (see "Principles," §9) of the generic expression Δ under which Δ_1 falls, there are m^{th} roots

$$\Delta_1^{\frac{1}{m}},\ \Delta_2^{\frac{1}{m}},\ \ldots,\ \Delta_e^{\frac{1}{m}}, \tag{4}$$

of the expressions in (3), such that the value of r_1 can be exhibited not only as in the first of equations (1), but also in the following ways:

$$\left.\begin{aligned} mr_1 &= \Delta_2^{\frac{1}{m}} + a_2\Delta_2^{\frac{2}{m}} + \ldots + c_2\Delta_2^{\frac{m-1}{m}} \\ mr_2 &= \Delta_3^{\frac{1}{m}} + a_3\Delta_3^{\frac{2}{m}} + \ldots + c_3\Delta_3^{\frac{m-1}{m}}, \end{aligned}\right\} \tag{5}$$

and so on; where a_2, b_2, etc. are what a_1, b_1, etc. become in passing from Δ_1 to Δ_2; and a_3, b_3, etc. what they become in passing to Δ_3; and so on. The separate members of mr_1, as it is expressed in the first line of (5), are

$$\Delta_2^{\frac{1}{m}},\ a_2\Delta_2^{\frac{2}{m}},\ \ldots,\ c_2\Delta_2^{\frac{m-1}{m}}, \tag{6}$$

§3. The sum of the terms in (6) is m times the same root of the equation $F(x)=0$ as the sum of those in (2). This implies, as was proved in the "Principles," that the terms in (6) are severally equal, in some order, to those in (2). Because Δ_2 and Δ_1 are unequal, $\Delta_2^{\frac{1}{m}}$ and $\Delta_1^{\frac{1}{m}}$ are unequal. Therefore they are equal to distinct members of mr_1 as these are expressed in (2). In like manner the terms in (4) are severally equal to distinct separate members of mr_1.

§4. It can be shown that a cycle of c primitive roots of unity

$$\omega,\ \omega^{\lambda},\ \omega^{\lambda^2},\ \ldots,\ \omega^{\lambda^{c-1}}, \tag{7}$$

can be formed; and that the terms in (2) to which those in (4) are equal are those in which the indices of the powers of $\Delta_1^{\frac{1}{m}}$ are the numbers

$$1,\ \lambda,\ \lambda^2,\ \dots,\ \lambda^{e-1}, \tag{8}$$

with multiples of m rejected. When (7) is called a cycle, the meaning is that no term in the series after the first is equal to the first, but $\omega^{\lambda^e}=\omega$. For brevity's sake I may be allowed, where there is no danger of mistake, if $g_1\Delta_1^{\frac{n}{m}}$ be a term in (2), to speak of it as $g_1\Delta_1^{\frac{\lambda^a}{m}}$, n being λ^a with multiples of m left out. In like manner if

$$\omega,\ \omega^b,\ \omega^{b^2},\ \dots,\ \omega^{b^{s-1}}, \tag{9}$$

be a cycle of primitive m^{th} roots of unity, and if there be a term in (2) in which the index of the power of $\Delta_1^{\frac{1}{m}}$ is b^a, the term may be spoken of as $\sigma_1\Delta_1^{\frac{b^a}{m}}$, where multiples of m must be understood to be rejected from b^a. Let then $\Delta_1^{\frac{1}{m}}$ and $\alpha_1\Delta_1^{\frac{b}{m}}$ in (2) be equal to distinct terms in (4). I will first show that there are terms in (2) in which the indices of the powers of $\Delta_1^{\frac{1}{m}}$ are the indices of the powers of ω in (9). Let $\Delta_2^{\frac{1}{m}}$ be the term in (4) to which $\alpha_1\Delta_1^{\frac{b}{m}}$ is by hypothesis equal. The term in (6) to which $\alpha_1\Delta_1^{\frac{b}{m}}$ in (2) corresponds is $\alpha_2\Delta_2^{\frac{b}{m}}$. Because $\Delta_2^{\frac{1}{m}}=\alpha_1\Delta_1^{\frac{b}{m}}$, $\alpha_2\Delta_2^{\frac{b}{m}}=\alpha_2\alpha_1^b\Delta_1^{\frac{b^2}{m}}$. Hence the term in (2) to which $\alpha_2\Delta_2^{\frac{b}{m}}$ in (6) is equal must be $\beta_1\Delta_1^{\frac{b^2}{m}}$; for, if it were any other term than that mentioned, say $\tau_1\Delta_1^{\frac{n}{m}}$ we should have

$$\tau_1\Delta_1^{\frac{n}{m}}=\alpha_2\alpha_1^b\Delta_1^{\frac{b^2}{m}}, \tag{10}$$

where b^2 with multiples of m left out, is not equal to n. But, from the state in which algebraical expressions are supposed in the "Principles" to be presented, since no surds occur in τ_1, α_1 or α_2 except such as are found in Δ_1 or Δ_2, the equation (10) would require τ_1 and $\alpha_2\alpha_1^b$ to be separately zero; and this again would make $\alpha_2\Delta_2^{\frac{b}{m}}$, and therefore $\alpha_1\Delta_1^{\frac{b}{m}}$, and therefore $\Delta_2^{\frac{1}{m}}$, and therefore $\Delta_1^{\frac{1}{m}}$, zero; which is impossible. Therefore $\alpha_2\Delta_2^{\frac{b}{m}}=\beta_1\Delta_1^{\frac{b^2}{m}}$. But, because $\alpha_1\Delta_1^{\frac{b}{m}}=\Delta_2^{\frac{1}{m}}$, $\alpha_1\Delta_1^{\frac{b}{m}}$ is one of the particular cognate forms of $\Delta^{\frac{1}{m}}$. Therefore also $\alpha_2\Delta_2^{\frac{b}{m}}$ is a particular cognate form of $\Delta^{\frac{1}{m}}$, which may be taken to be $\Delta_3^{\frac{1}{m}}$. Therefore $\beta_1\Delta_1^{\frac{b^2}{m}}$ is equal to $\Delta_3^{\frac{1}{m}}$, a term in (4). In like manner it follows that all the terms in (2) in which the indices of the powers of $\Delta_1^{\frac{1}{m}}$ are any of the indices of the powers of ω in (9) are equal to terms in (4). Let

$$\Delta_1^{\frac{1}{m}},\ \alpha_1\Delta_1^{\frac{b}{m}},\ \beta_1\Delta_1^{\frac{b^2}{m}},\ \dots,\ \gamma_1\Delta_1^{\frac{b^{s-1}}{m}}, \tag{11}$$

be terms in (2) severally equal to the terms in (4),

$$\Delta_1^{\frac{1}{m}},\ \Delta_2^{\frac{1}{m}},\ \Delta_3^{\frac{1}{m}},\ \ldots,\ \Delta_z^{\frac{1}{m}}. \tag{12}$$

We may assume $\alpha_1\Delta_1^{\frac{b}{m}}$ to have been so chosen that there is no term in (2), as $\sigma_1\Delta_1^{\frac{h}{m}}$, equal to a term in (4), and such that when the cycle

$$\omega,\ \omega^h,\ \omega^{h^2},\ \ldots,\ \omega^{h^{v-1}}, \tag{13}$$

is formed, v is greater than z. In that case, z must be equal to c. For suppose if possible that z is less than c. Then there is a term in (4) distinct from those in (12), say $\Delta_{z+1}^{\frac{1}{m}}$, equal to a term in (2) in which the index of the power of $\Delta_1^{\frac{1}{m}}$ is not a power of b, which term in (2) may be taken to be $g_1\Delta_1^{\frac{bd}{m}}$, d not being a power of b. Then, just as we proved that, because $\Delta_1^{\frac{1}{m}}$ and $\alpha_1\Delta_1^{\frac{b}{m}}$ are terms in (2) equal to terms in (4), any term in (2) having for the index of the power of $\Delta_1^{\frac{1}{m}}$ any of the indices of the powers of ω in (9) must be equal to a term in (4), we can show that because $\Delta_1^{\frac{1}{m}}$ and $\Delta_1^{\frac{bd}{m}}$ are terms in (2) equal to terms in (4), there must be a term in (4) equal to one in (2) in which the index of the power of $\Delta_1^{\frac{1}{m}}$ is $b^W d^W$, W being any whole number. Hence there is a distinct term in (4) equal to a term in (2) corresponding to each distinct term in the cycle

$$\omega,\ \omega^{bd},\ \omega^{b^2d^2},\ \text{etc.}$$

Putting h for bd, this cycle is identical with (13). And since d is not a power of b, the number of terms in the cycle ω, ω^{bd}, etc. is greater than that in (9). Hence the number of terms in (13) exceeds that in (9). That is, v is greater than z; which, by hypothesis, is impossible. Hence z cannot be less than c. And it is not greater, because all the terms in (12) are contained in (4). Therefore $z = c$. Therefore there is a cycle of c primitive m^{th} roots of unity, which may be taken to be (7); and, comparing this with (9), λ may be taken to be b; and the series (11), which may now be written

$$\Delta_1^{\frac{1}{m}},\ \alpha_1\Delta_1^{\frac{\lambda}{m}},\ \beta_1\Delta_1^{\frac{\lambda^2}{m}},\ \ldots,\ \gamma_1\Delta_1^{\frac{\lambda^{c-1}}{m}}, \tag{14}$$

has the same number of terms as (4). Consequently the terms in (14) are those terms in (2) which are severally equal to terms in (4).

§5. Take E_1 a rational function of Δ_1; let the generic expression (§2) of which it is a particular form be E; and when Δ_1 passes successively into the c terms in (4), let E_1 become successively

$$E_1,\ E_2,\ \ldots,\ E_c. \tag{15}$$

By the "Principles," Prop. III, each of the unequal particular cognate forms of Δ occurs the same number of times in the series of the cognate forms. Therefore the entire series of the particular cognate forms is made up of k groups of c terms each, the terms in any one of the groups being equal to those in each of the others. These k groups may be written

$$\left.\begin{matrix} \Delta_1, & \Delta_2, & \ldots, \Delta_c, \\ \Delta_{c+1}, & \Delta_{c+2}, & \ldots, \Delta_{2c}, \\ \Delta_{2c+1}, & \Delta_{2c+2}, & \ldots, \Delta_{3c}, \end{matrix}\right\} \qquad (16)$$

and so on. The entire series of the particular cognate forms of E must consist of k corresponding groups of c terms each,

$$\left.\begin{matrix} E_1, & E_2, & \ldots, E_c, \\ E_{c+1}, & E_{c+2}, & \ldots, E_{2c}, \\ E_{2c+1}, & E_{2c+2}, & \ldots, E_{3c}, \end{matrix}\right\} \qquad (17)$$

and so on; E_a being what E_1 becomes when Δ_1 becomes Δ_a.

§6. It is plain that if $\Delta_a = \Delta_s$, $E_a = E_s$. For, since Δ_a is a root of an equation of the cth degree, any rational function of Δ_a may be expressed without using powers of Δ_a above the $(c-1)^{\text{th}}$. And E_a is a rational function of Δ_a. Therefore we may put

$$E_a = s + s_1\Delta_a + s_2\Delta_a^2 + \ldots + s_{c-1}\Delta_a^{c-1}$$

$$\text{and} \quad E_s = s + s_1\Delta_s + s_2\Delta_s^2 + \ldots + s_{c-1}\Delta_s^{c-1},$$

where s, s_1, etc. are rational. But, by hypothesis, $\Delta_a = \Delta_s$. Therefore $E_a = E_s$.

§7. This leads to the conclusion that any symmetrical function of the terms in (15) is rational. For, by §5, the terms in any line of (16) under the first are severally equal to those in the first line. Therefore, by §6, the terms in any line of (17) under the first are severally equal to those in the first. Let the unequal terms in the first line of (17) be $E_1, E_2, \ldots, E_n$. Let E_1 and E_2 occur α and β times respectively in the first line of (17); then they occur αk and βk times respectively in the k groups of (17). But, by the "Principles," Prop. III, each of the unequal particular cognate forms of E occurs the same number of times in the entire series. Therefore αk and βk are equal, and $\alpha = \beta$. That is to say, E_1 and E_2 occur the same number of times in the first line of (17). In like manner all the unequal terms in the first line of (17) occur the same number of times in that line. Therefore, if $X_1 = 0$ be the equation whose roots are E_1, $E_2, \ldots, E_n$, and $X = 0$ be the equation whose roots are $E_1, E_2, \ldots, E_c$, $X = X_1^{\alpha}$. But, by the "Principles," Prop. III, X_1 is rational. Therefore X is rational. This implies that any symmetrical function of the roots of the equation $X = 0$, that is, of the terms in (14), is rational.

Symmetrical Functions of the terms in (4).

§8. I will now establish the fundamental theorem that *any symmetrical function of those separate members of* mr_1, *which are* m^{th} *roots of the roots of the equation auxiliary or of an equation sub-auxiliary to the equation* $F(x) = 0$ *is a rational function of* r_1. When $c = m - 1$, the terms in (3) are the roots of the irreducible auxiliary (see §1) to $F(x) = 0$. When c is less than $m - 1$, they are the roots of a sub-auxiliary. What we need then to make out is, that *any symmetrical function of the terms in* (4) *is a rational function of* r_1.

§9. From the first of equations (1), $\Delta_1^{\frac{1}{m}}$ is a root of the equation

$$c_1 x^{m-1} + \ldots + a_1 x^2 + x - mr_1 = 0, \tag{18}$$

being at the same time a root of the equation

$$x^m - \Delta_1 = 0. \tag{19}$$

Now $\omega\Delta_1^{\frac{1}{m}}$ is not a root of (18); for, if it were, we should have

$$c_1(\omega\Delta_1^{\frac{1}{m}})^{m-1} + \ldots + (\omega\Delta_1^{\frac{1}{m}}) - mr_1 = 0;$$

and therefore, by comparison with the second of equations (1), $r_2 = r_1$, which is impossible. In the same way no root of (19) except $\Delta_1^{\frac{1}{m}}$ is a root of (18). Therefore the highest common measure of the expressions on the left of (18) and (19) is $x - Q$, where Q is a rational function of r_1, Δ_1, a_1, etc., and therefore, by §1, a rational function of r_1 and Δ_1. We may express this, since $\Delta_1^{\frac{1}{m}}$ is the value of Q, by putting

$$\left.\begin{aligned} \Delta_1^{\frac{1}{m}} &= f(r_1, \Delta_1). \\ \text{Similarly, from (5),}\quad \Delta_2^{\frac{1}{m}} &= f(r_1, \Delta_2), \\ \Delta_3^{\frac{1}{m}} &= f(r_1, \Delta_3), \end{aligned}\right\} \tag{20}$$

and so on. Since f here denotes a rational function, if the sum of the c expressions $\Delta_1^{\frac{1}{m}}$, $\Delta_2^{\frac{1}{m}}$, etc., be $\frac{N}{D}$, both N and D must, from (20), be composed of terms of the type $E' r_1^q$; where E' is a symmetrical function of the c expressions, Δ_1, Δ_2, etc., and is therefore, by §7, rational. Consequently the sum of the c terms $\Delta_1^{\frac{1}{m}}$, $\Delta_2^{\frac{1}{m}}$, etc. is a rational function of r_1. In the same way any symmetrical function of these terms is a rational function of r_1. Thus the fundamental theorem is established.

§10. Setting out from $\Delta_1^{\frac{1}{m}}$, one of the separate members of mr_1, and taking the c unequal particular cognate forms of the generic expression Δ under which Δ_1 falls, we have found that certain m^{th} roots of these, being separate members of mr_1, satisfy equations (20), and therefore that any symmetrical function of these m^{th} roots is a rational function of r_1. If now we set out from an mth root of Δ_1 distinct from $\Delta_1^{\frac{1}{m}}$, say $\omega\Delta_1^{\frac{1}{m}}$, one of the separate members of mr_2, we can in the same way demonstrate that there is another group of mth roots of the terms in (3), say

$$\omega\Delta_1^{\frac{1}{m}}, \text{ or } D_1^{\frac{1}{m}},\ D_2^{\frac{1}{m}},\ D_3^{\frac{1}{m}}, \ldots, D_c^{\frac{1}{m}}, \tag{21}$$

by means of which equations corresponding to (20) can be formed.

§11. It is readily seen that the series (21) is identical with

$$\omega\Delta_1^{\frac{1}{m}},\ \omega^{\lambda}\Delta_2^{\frac{1}{m}},\ \omega^{\lambda^2}\Delta_3^{\frac{1}{m}}, \ldots, \omega^{\lambda^{c-1}}\Delta_c^{\frac{1}{m}}. \tag{22}$$

For, by §4, the series (4) is identical with (14), which may again be written down:

$$\Delta_1^{\frac{1}{m}},\ \alpha_1\Delta_1^{\frac{\lambda}{m}},\ \beta_1\Delta_1^{\frac{\lambda^2}{m}}, \ldots, \gamma_1\Delta_1^{\frac{\lambda^{c-1}}{m}}. \tag{23}$$

Taking the term $\Delta_2^{\frac{1}{m}}$ in (4), we saw that $\Delta_2^{\frac{1}{m}} = \alpha_1\Delta_1^{\frac{\lambda}{m}}$. Therefore $\alpha_1(\omega\Delta_1^{\frac{1}{m}})^{\lambda} = \omega^{\lambda}\Delta_2^{\frac{1}{m}}$. But, $\alpha_1\Delta_1^{\frac{\lambda}{m}}$ being one of the separate members of mr_1 in (1), $\alpha_1(\omega\Delta_1^{\frac{1}{m}})^{\lambda}$ is the corresponding separate member of mr_2. Therefore $\omega^{\lambda}\Delta_2^{\frac{1}{m}}$ is equal to one of the members of mr_2 in (1). And its mth power is Δ_2, one of the particular cognate forms of Δ. Therefore it must be a term in (21), because (21) is made up of those separate members of mr_2 whose m^{th} powers are particular cognate forms of Δ. We may take $D_2^{\frac{1}{m}}$ to be equal to $\omega^{\lambda}\Delta_2^{\frac{1}{m}}$. In the same way $D_3^{\frac{1}{m}} = \omega^{\lambda^2}\Delta_3^{\frac{1}{m}}$, and so on.

§12. Hence the equations corresponding to (20), which can be formed by means of the terms in (21), are

$$\left.\begin{aligned} \omega\Delta_1^{\frac{1}{m}} &= f(r_2, \Delta_1) \\ \omega^{\lambda}\Delta_2^{\frac{1}{m}} &= f(r_2, \Delta_2) \\ \omega^{\lambda^2}\Delta_3^{\frac{1}{m}} &= f(r_2, \Delta_3), \end{aligned}\right\} \tag{24}$$

and so on. In the functions on the right of (24), Δ_1, Δ_2, etc. remain as in (20), because the passage from Δ_1 to D_1 or $(\omega\Delta_1^{\frac{1}{m}})^m$, and so on, makes no change in Δ_1, Δ_2, etc. In like manner,

$$\left.\begin{aligned} \omega^2\Delta_1^{\frac{1}{m}} &= f(r_3, \Delta_1) \\ \omega^{2\lambda}\Delta_2^{\frac{1}{m}} &= f(r_3, \Delta_2) \\ \omega^{2\lambda^2}\Delta_3^{\frac{1}{m}} &= f(r_3, \Delta_3), \end{aligned}\right\} \tag{25}$$

and so on.

GALOIS' THEOREM.

§13. We can now deduce Galois' Theorem, that r_1 *is a rational function of* r_2 *and* r_3. In fact, the separate members of r_1 can be arranged in groups such that any symmetrical function of the members in each group is a rational function of r_2 and r_3. One of the groups is obtained by dividing the terms in (4) severally by m. What we have to prove therefore is that *any symmetrical function of the terms in* (4) is a rational function of r_2 and r_3.

§14. Square both sides of (24), and divide by $\omega^2 \Delta_1^{\frac{1}{m}}$ in the case of the first line, by $\omega^{2\lambda} \Delta_2^{\frac{1}{m}}$ in the case of the second, and so on. Then, keeping (25) in view,

$$\left.\begin{aligned} \Delta_1^{\frac{1}{m}} &= \frac{\{f(r_2, \Delta_1)\}^2}{\omega^2 \Delta_1^{\frac{1}{m}}} = \frac{\{f(r_2, \Delta_1)\}^2}{f(r_3, \Delta_1)} \\ \Delta_2^{\frac{1}{m}} &= \frac{\{f(r_2, \Delta_2)\}^2}{\omega^{2\lambda} \Delta_2^{\frac{1}{m}}} = \frac{\{f(r_2, \Delta_2)\}^2}{f(r_3, \Delta_2)}, \end{aligned}\right] \tag{26}$$

and so on. By §7, the sum of the c expressions on the extreme right of (26), only two of which are written down, is a rational function of r_2 and r_3. Calling this $\phi(r_2, r_3)$,

$$\Delta_1^{\frac{1}{m}} + \Delta_2^{\frac{1}{m}} + \ldots + \Delta_c^{\frac{1}{m}} = \phi(r_2, r_3). \tag{27}$$

Thus the sum of the c separate members of mr_1 forming the group (4) is a rational function of r_2 and r_3. If $c = m - 1$, this is Galois' theorem. If c be less than $m - 1$, it may be shown as above that the sum of another quite distinct group of separate members of mr_1 is a rational function of r_2 and r_3. And so on till the series (2) is exhausted, so that Galois' theorem still holds. It is obvious that, in the same way in which (27) was obtained, any symmetrical function of the c expressions, $\Delta_1^{\frac{1}{m}}$, $\Delta_2^{\frac{1}{m}}$, etc. can be shown to be a rational function of r_2 and r_3.

LAW OF THE FORMATION OF THE FUNCTION; VERIFYING INSTANCE.

§15. It will be observed that, in the preceding section, *the law of the formation of the function* $\phi(r_2, r_3)$ *comes to light.* The rule is this: Take $x - Q$, the highest common measure of the expressions on the left of (18) and (19). The expression Q is $f(r_1, \Delta_1)$. Then

$$\phi(r_2, r_3) = \Sigma\left[\frac{\{f(r_2, \Delta_1)\}^2}{f(r_3, \Delta_1)}\right], \tag{28}$$

the expression on the right of (28) being the sum of the c expressions on the extreme right of (26).

§16. A simple verification is afforded by the equation

$$x^3 - \frac{x}{3} - \frac{4}{27} = 0.$$

Putting $\Delta_1 = 2 + \sqrt{3}$ and $\Delta_2 = 2 - \sqrt{3}$, the roots of the equation are

$$\begin{aligned} 3r_1 &= \Delta_1^{\frac{1}{3}} + \Delta_2 \Delta_1^{\frac{2}{3}} \\ 3r_2 &= \omega \Delta_1^{\frac{1}{3}} + \omega^2 \Delta_2 \Delta_1^{\frac{2}{3}} \\ 3r_3 &= \omega^2 \Delta_1^{\frac{1}{3}} + \omega \Delta_2 \Delta_1^{\frac{2}{3}}, \end{aligned}$$

ω being a primitive third root of unity. This gives $\dfrac{3r_1 + \Delta_2}{3r_1 \Delta_2 + 1}$ as the value of Q. Then (28) becomes

$$3r_1 = \phi(r_2, r_3) = \frac{\Delta_1 (3r_2 + \Delta_2)^2 (3r_3 + \Delta_1)}{(3r_2 + \Delta_1)^2 (3r_3 + \Delta_2)} + \frac{\Delta_2 (3r_2 + \Delta_1)^2 (3r_2 + \Delta_2)}{(3r_2 + \Delta_2)^2 (3r_3 + \Delta_1)}. \quad (29)$$

This result will perhaps most easily be seen to be accurate, if, by means of equations (30) immediately to be established, (29) be changed into

$$3r_1 = \frac{\omega^2 \Delta_1 (3r_2 + \Delta_2)}{3r_2 + \Delta_1} + \frac{\omega \Delta_2 (3r_2 + \Delta_1)}{3r_2 + \Delta_2}.$$

CYCLE OF c PRIMITIVE ROOTS OF UNITY.

§17. A result incidentally presenting itself is, that *the sum of a cycle of c primitive m^{th} roots of unity is a rational function of two of the roots of the equation* $F(x) = 0$. For, from (24), (25) and (20)

$$\left.\begin{aligned} \omega &= \frac{f(r_2, \Delta_1)}{f(r_1, \Delta_1)} = \frac{f(r_3, \Delta_1)}{f(r_2, \Delta_1)} = \dots \\ \omega^{\lambda} &= \frac{f(r_2, \Delta_2)}{f(r_1, \Delta_2)} = \frac{f(r_3, \Delta_2)}{f(r_2, \Delta_2)} = \dots \\ &\cdot \quad \cdot \quad \cdot \quad \cdot \quad \cdot \quad \cdot \quad \cdot \quad \cdot \quad \cdot \quad \cdot \quad \cdot \quad \cdot \\ \omega^{\lambda^{c-1}} &= \frac{f(r_2, \Delta_c)}{f(r_1, \Delta_c)} = \frac{f(r_3, \Delta_c)}{f(r_2, \Delta_c)} = \dots \end{aligned}\right\} \quad (30)$$

$$\therefore \omega + \omega^{\lambda} + \dots + \omega^{\lambda^{c-1}} = \Sigma \left\{ \frac{f(r_2, \Delta_1)}{f(r_1, \Delta_1)} \right\} = \Sigma \left\{ \frac{f(r_3, \Delta_1)}{f(r_2, \Delta_1)} \right\} = \dots$$

By §7, the sum of the mth roots of unity in the cycle, ω, ω^{λ}, etc., as the sum is here obtained, is a rational function of r_2 and r_3, of r_3 and r_4, and so on.

On a Certain Class of Linear Differential Equations.

By Thomas Craig.

The subject of linear differential equations whose coefficients are singly or doubly periodic functions of the first kind has been studied by Picard* and Floquet,† and it has been shown that such equations always admit of at least one integral which is a periodic function of the second kind.

The case where the coefficients are periodic functions of the second kind does not seem to have been attempted by any one—and indeed in the general case the problem seems almost impossible of solution—though when the multipliers are roots of unity the solution (at least for singly periodic functions) can be easily led back to the case where the coefficients are periodic of the first kind.

In the following paper I have determined the conditions which are necessary in order that a linear differential equation shall admit of an integral which is a periodic function of the third kind—in the beginning I limit myself to singly periodic functions.

I shall start with the case of the linear differential equation of the second order,

1. $$\frac{d^2y}{dx^2} + p_1 \frac{dy}{dx} + p_2 y = 0,$$

where the coefficients p_1 and p_2 are uniform functions of x. What must be the form of these functions in order that (1) may have an integral which is a periodic function of the third kind?

Suppose $y = F(x)$ to be such an integral, and $F(x+\omega) = e^{-\lambda x + \lambda_0}$ where λ and λ_0 are constants; then if we change x into $x+\omega$ we must have that $F(x+\omega)$ is an integral of

2. $$\frac{d^2y}{dx^2} + p_1(x+\omega)\frac{dy}{dx} + p_2(x+\omega)\, y = 0.$$

---

* Crelle, Vol. XC. Sur les équations différentielles linéaires à coëfficients doublement périodiques.

† Annales de l'Ecole Normale, Feb. 1883. Sur les équations différentielles linéaires à coëfficients périodiques. *Ibid.* May, 1884. Sur les équations différentielles linéaires à coëfficients doublement périodiques.

Writing then in (2) $y = F(x+\omega)$, $= e^{-\lambda x + \lambda_0}\dot{F}(x)$ we have, writing for brevity F instead of $\dot{F}(x)$,

$$\frac{d^2F}{dx^2} + [-2\lambda + p_1(x+\omega)]\frac{dF}{dx} + [\lambda^2 - \lambda p_1(x+\omega) + p_2(x+\omega)]F = 0,$$

also

$$\frac{d^2F}{dx^2} + p_1(x)\frac{dF}{dx} + p_2 F = 0.$$

Subtracting we have

$$[-2\lambda + p_1(x+\omega) - p_1(x)]\frac{dF}{dx} + [\lambda^2 - \lambda p_1(x+\omega) + p_2(x+\omega) - p_2(x)]F = 0,$$

which must be satisfied identically; therefore

3. $$p_1(x+\omega) = p_1(x) + 2\lambda$$
$$p_2(x+\omega) = p_2(x) + \lambda p_1(x) + \lambda^2$$

and obviously

4. $$p_1(x+n\omega) = p_1(x) + 2n\lambda$$
$$p_2(x+n\omega) = p_2(x) + n\lambda p_1(x) + n^2\lambda^2.$$

The first of equations (3) is obviously satisfied by $p_1(x) = \frac{2\lambda x}{\omega}$ and the second by $p_2(x) = \frac{\lambda^2 x^2}{\omega^2}$. If we use these values for p_1 and p_2 we have the equation (replacing λ by $-\mu$)

$$\frac{d^2y}{dx^2} - \frac{2\mu x}{\omega}\frac{dy}{dx} + \frac{\mu^2 x^2}{\omega^2}y = 0,$$

of which the general integral is

$$y = e^{\frac{\mu x^2}{2\omega}}\left[C_1\cos\left(x\sqrt{\frac{\mu}{\omega}}\right) + C_2\sin\left(x\sqrt{\frac{\mu}{\omega}}\right)\right]$$

and on giving μ the value $\frac{4\pi^2}{\omega}$, *i. e.*, $\lambda = -\frac{4\pi^2}{\omega}$ this is obviously a periodic function of the third kind. Changing x into $x+\omega$ and it is easy to verify that

$$y(x+\omega), = e^{\frac{\mu(x+\omega)^2}{2\omega}}\left\{C_1\cos\left(x\sqrt{\frac{\mu}{\omega}} + \sqrt{\mu\omega}\right) + C_2\sin\left(x\sqrt{\frac{\mu}{\omega}} + \sqrt{\mu\omega}\right)\right\}$$

satisfies the equation

$$\frac{d^2y}{dx^2} - \frac{2\mu}{\omega}(x+\omega)\frac{dy}{dx} + \frac{\mu^2}{\omega^2}(x+\omega)^2 y = 0.$$

The most general values of p_1 and p_2 are easily found. Suppose $\phi_j^{(i)}(x)$ to be a singly periodic function of the first kind, *i. e.* $\phi_j^{(i)}(x+\omega) = \phi_j^{(i)}(x)$ and give i the values $2, 3 \ldots n$ and j the values $1, 2 \ldots n$. It is obvious now that the most general values of p_1 and p_2 are given by

$$p_1(x) = \phi_1^{(2)}(x) + \frac{2\lambda x}{\omega},$$

$$p_2(x) = \phi_2^{(2)}(x) + \frac{\lambda\varphi_1^{(2)}(x)}{\omega}x + \frac{\lambda^2 x^2}{\omega^2}.$$

Take as another illustration the equation of the third order

$$\frac{d^3y}{dx^3} + p_1 \frac{d^2y}{dx^2} + p_2 \frac{dy}{dx} + p_3 y = 0 .$$

Here we must have

$$p_1(x + \omega) = p_1(x) + 3\lambda$$
$$p_2(x + \omega) = p_2(x) + 2\lambda p_1(x) + 3\lambda^2$$
$$p_3(x + \omega) = p_3(x) + \lambda p_2(x) + \lambda^2 p_1(x) + \lambda^3,$$

these are satisfied by the values

$$p_1(x) = \frac{3\lambda x}{\omega}, \quad p_2(x) = \frac{3\lambda^2 x^2}{\omega^2}, \quad p_3(x) = \frac{\lambda^3 x^3}{\omega^3}.$$

Replacing λ by $-\mu$ we have the equation

$$\frac{d^3y}{dx^3} - \frac{3\mu x}{\omega} \frac{d^2y}{dx^2} + \frac{3\mu^2 x^2}{\omega^2} \frac{dy}{dx} - \frac{\mu^3 x^3}{\omega^3} y = 0,$$

the general integral of which is

$$y = e^{\frac{\mu x^2}{2\omega}} \left\{ C_1 \cos\left(x\sqrt{3\frac{\mu}{\omega}}\right) + C_2 \sin\left(x\sqrt{\frac{3\mu}{\omega}}\right) + C_3 \right\}$$

which for $\mu = \frac{4\pi^2}{3\omega}$ is a function of the third kind. The general integrals of these two particular equations of the second and of the third orders are thus periodic functions of the third kind—this does not hold in the case of an equation of this form of any higher order: for example take the equation

$$\frac{d^4y}{dx^4} - 4\frac{\mu x}{\omega}\frac{d^3y}{dx^3} + \frac{6\mu^2 x^2}{\omega^2}\frac{d^2y}{dx^2} - 4\frac{\mu^3 x^3}{\omega^3}\frac{dy}{dx} + \frac{\mu^4 x^4}{\omega^4} y = 0,$$

the general integral is

$$y = e^{\frac{\mu x^2}{2\omega}} [A \cos(m_1 x + \alpha) + B \cos(m_2 x + \beta)],$$

A, B, α, β being constants and m_1^2, m_2^2 the roots with their signs changed of the equation

$$\xi^2 + \frac{6\mu}{\omega}\xi + \frac{3\mu^2}{\omega^2} = 0.$$

The general integral in this case is not a periodic function of the third kind—but each of the particular integrals

$$e^{\frac{\mu x^2}{2\omega}} \cos(m_1 x + \alpha), \quad e^{\frac{\mu x^2}{2}} \cos(m_2 x + \alpha)$$

is such a function if λ be properly determined. The general form of the coefficients in any case is readily found. Take the equation of the nth order

$$P(x) \equiv \frac{d^n y}{dx^n} + p_1 \frac{d^{n-1}y}{dx^{n-1}} + p_2 \frac{d^{n-2}y}{dx^{n-2}} + \ldots + p_n y = 0.$$

If $F(x)$ be an integral of $P(x)=0$, then $F(x+\omega)$ will be an integral of $P(x+\omega)=0$.

Expressing the necessary conditions for this and writing $F(x+\omega)=e^{-\lambda x}F(x)$ we find

$$p_1(x+l\omega)=p_1(x)+nl\lambda$$

$$p_2(x+l\omega)=p_2(x)+(n-1)\,l\lambda p_1(x)+\frac{n(n-1)}{1.2}\,l^2\lambda^2$$

. . . .

$$p_i(x+l\omega)=p_i(x)+(n-i+1)\,l\lambda p_{i-1}(x)+\frac{(n-i+2)(n-i+1)}{1.2}\,l^2\lambda^2 p_{i-2}(x)$$
$$+\ldots\frac{(n-i+a)(n-i+a-1)\ldots(n-i+1)}{a!}\,l^a\lambda^a p_{i-a}(x)$$
$$+\ldots\frac{n.n-1\;n-2\ldots n-i+1}{i!}\,l^i\lambda^i$$

$$p_n(x+l\omega)=p_n(x)+l\lambda p_{n-1}(x)+l^2\lambda^2 p_{n-2}(x)+\ldots+l^n\lambda^n$$

where l is any integer; the general term may be written briefly

$$p_i(x+l\omega)=\sum_{a=0}^{a=i}\frac{n-i+a|n-i+1}{a!}\,l^a\lambda^a p_{i-a}(x).$$

These conditions are all satisfied if we write

$$p_i(x)=\frac{n.n-1\ldots n-i+1}{i!}\,\frac{\lambda^i x^i}{\omega^i}$$

i. e.

$$p_1(x)=\frac{n\lambda x}{\omega},\quad p_2(x)=\frac{n.n-1}{1.2}\,\frac{\lambda^2 x^2}{\omega^2},\ \text{etc.}$$

the general values for the coefficients are now easily seen to be

$$p_1(x)=\phi_1^{(n)}(x)+\frac{n\lambda x}{\omega}$$

$$p_2(x)=\phi_2^{(n)}(x)+(n-1)\frac{\lambda\varphi_1^{(n)}(x)}{\omega}\,x+\frac{n.n-1}{1.2}\,\frac{\lambda^2 x^2}{\omega^2}$$

. . . .

$$p_i(x)=\phi_i^{(n)}(x)+(n-i+1)\frac{\lambda\varphi_{i-1}^{(n)}(x)}{\omega}\,x+\frac{n-i+1.n-i+2}{1.2}\,\frac{\lambda^2\varphi_{i-2}^{(n)}(x)}{\omega^2}\,x^2$$
$$+\ldots+\frac{n-i+a.n-i+a-1\ldots n-i+1}{a!}\,\frac{\lambda^a\varphi_{i-a}^{(n)}(x)}{\omega^a}\,x^a$$
$$+\ldots+\frac{n.n-1\ldots n-i+1}{i!}\,\frac{\lambda^i x^i}{\omega^i}$$

. . . .

$$p_n(x)=\phi_n^{(n)}(x)+\frac{\lambda\varphi_{n-1}^{(n)}(x)}{\omega}\,x+\frac{\lambda^2\varphi_{n-2}^{(n)}(x)}{\omega^2}\,x^2+\ldots+\frac{\lambda^n x^n}{\omega^n}$$

or briefly

$$p_i(x) = \sum_{a=0}^{a=i} \frac{n-i+a\,|\,n-i+1}{a\,!} \; \frac{\lambda^a \varphi_{i-a}^{(n)}(x)}{\omega^a}\, x^a$$

$$i = 1, 2, 3 \ldots n.$$

It remains now to show that a differential equation whose coefficients are of the kind just found possesses at least one integral which is a periodic function of the third kind. It will be quite sufficient to do this for equations of the second order. The equation

$$\frac{d^2y}{dx^2} + p_1 \frac{dy}{dx} + p_2 y = 0$$

has for coefficients

$$p_1(x) = P_1(x) + \frac{2\lambda x}{\omega}$$

$$p_2(x) = P_2(x) + \frac{\lambda P_1(x)}{\omega} x + \frac{\lambda^2 x^2}{\omega^2}$$

where $P_1(x)$ and $P_2(x)$ are periodic functions of the first kind having ω for their common period. Assume $y = e^{\frac{-\lambda x^2}{2\omega}+\lambda'} \phi(x)$,

where $\phi(x)$ is a simply periodic function of either the first or second kind. Substituting this value of y in the above differential equation gives

$$\frac{d^2\varphi}{dx^2} + \left(p_1 - \frac{2\lambda x}{\omega}\right)\frac{dy}{dx} + \left(p_2 - p_1 \frac{\lambda x}{\omega} + \frac{\lambda^2 x^2}{\omega^2}\right)\phi = 0.$$

Introducing now the above values of p_1 and p_2 this becomes

$$\frac{d^2\varphi}{dx^2} + P_1 \frac{d\varphi}{dx} + P_2 \phi = 0.$$

Now Picard has shown that this equation always possesses at least one integral which is a periodic function of the second kind (becoming of the first kind for the multiplier $= 1$). We conclude therefore that it is always possible to find at least one integral of our given differential equation which is a periodic function of the third kind.

Next we seek to determine the maximum number of integrals of this kind that our equation may possess. In the differential equation

$$\frac{d^ny}{dx^n} + p_1 \frac{d^{n-1}y}{dx^{n-1}} + \ldots + p_n y = 0.$$

Assume $y = e^{\frac{\lambda x^2}{2\omega}} \Phi$ where $\Phi(x)$ is, in general, a periodic function of the second kind, and give p_i the value

$$p_i = \sum_{=0}^{a=i} \frac{n-i+a\,|\,n-i+1}{a\,!} \; \frac{\lambda^a x^a}{\omega^a} \phi_{i-a}(x)$$

where ϕ_{i-a} is a periodic function of the first kind. The equation now becomes

$$\Phi \equiv \frac{d^n \Phi}{dx^n} + \phi_1 \frac{d^{n-1}\Phi}{dx^{n-1}} + \phi_2 \frac{d^{n-2}\Phi}{dx^{n-2}} + \ldots + \phi_n \Phi = 0.$$

The solution of our problem is therefore led to the investigation of a linear differential equation whose coefficients are singly periodic functions of the first kind. The theory of this equation has been completely developed by Floquet and it is only necessary therefore to quote his results.

Denote by $F_1(x), F_2(x) \ldots F_n(x)$, n distinct solutions of $\Phi = 0$. If the variable describes any path from x to $x + \omega$ the uniform functions $F(x)$ become $F(x+\omega)$ while the coefficients ϕ_i resume their original values, consequently the functions $F_1(x+\omega), F_2(x+\omega) \ldots F_n(x+\omega)$ constitute another fundamental system of integrals of $\Phi = 0$.

We have then

$$\begin{aligned} F_1(x+\omega) &= A_{11}F_1(x) + A_{12}F_2(x) + \ldots + A_{1n}F_n(x) \\ F_2(x+\omega) &= A_{21}F_1(x) + A_{22}F_2(x) + \ldots + A_{2n}F_n(x) \\ & \ldots\ldots\ldots\ldots\ldots \\ F_n(x+\omega) &= A_{n1}F_1(x) + A_{n2}F_2(x) + \ldots + A_{nn}F_n(x) \end{aligned}$$

where the determinant

$$\begin{vmatrix} A_{11} & A_{12} & \ldots & A_{1n} \\ A_{21} & A_{22} & \ldots & A_{2n} \\ \cdot & \cdot & \cdot & \cdot \\ A_{n1} & A_{n2} & \ldots & A_{nn} \end{vmatrix}$$

does not vanish.

Suppose now that $\Phi = 0$ admits as integral a periodic function of the second kind, say $\mathbf{F}(x)$, where $\mathbf{F}(x+\omega) = \varepsilon \mathbf{F}(x)$, we have then

$$\mathbf{F}(x) = n_1 F_1(x) + n_2 F_2(x) + \ldots + n_n F_n(x),$$

where the constants n cannot all vanish. · Now since $\mathbf{F}(x+\omega) = \mathbf{F}(x)$ we have immediately

$$F_1(x)\Sigma A_{i1}n_i + F_2(x)\Sigma A_{i2}n_i + \ldots + F_n(x)\Sigma A_{in}n_i = \varepsilon \Sigma n_i F_i(x).$$

This requires that the constants n satisfy the equations

$$\begin{aligned} (A_{11} - \varepsilon)n_1 + A_{21}n_2 \quad &+ \ldots + A_{n1}n_n \quad = 0 \\ A_{12}n_1 \quad + (A_{22} - \varepsilon)n_2 &+ \ldots + A_{n2}n_n \quad = 0 \\ \ldots\ldots\ldots\ldots & \ldots\ldots \\ A_{1n}n_1 \quad + A_{2n}n_2 \quad &+ \ldots + (A_{nn} - \varepsilon)n_n = 0, \end{aligned}$$

consequently

$$\Delta = \begin{vmatrix} A_{11}-\varepsilon & A_{21} & \dots & A_{n1} \\ A_{12} & A_{22}-\varepsilon & \dots & A_{n2} \\ \cdot & \cdot & \cdot & \cdot \\ A_{1n} & A_{2n} & \dots & A_{nn}-\varepsilon \end{vmatrix} = 0.$$

The equation $\Delta = 0$ is called the fundamental equation, and its roots, which are the multipliers of the periodic functions of the second kind, are evidently all different from zero. It is also well known that the roots of the fundamental equation are independent of the choice of the fundamental system of integrals, a fact which was proved by Fuchs.* Floquet has shown that the general integral of this equation is of the form

$$\begin{aligned} &\psi_{11}(x) + x\psi_{12}(x) + \dots + x^{\mu_1-1}\psi_{1\mu_1}(x) \\ &+ \psi_{21}(x) + x\psi_{22}(x) + \dots + x^{\mu_2-1}\psi_{2\mu_2}(x) \\ &+ \dots \\ &+ \psi_{m1}(x) + x\psi_{m2}(x) + \dots + x^{\mu_m-1}\psi_{2m\mu_m}(x), \end{aligned}$$

where ψ_{ik} is a periodic function of the second kind whose multiplier is ε_i. In this $\varepsilon_1 \varepsilon_2 \dots \varepsilon_m$ are the distinct roots of $\Delta = 0$ of the orders $\mu_1, \mu_2 \dots \mu_m$ respectively.

The general conclusions arrived at by Floquet concerning the equation $\Phi = 0$ are as follows:

I. Let $\varepsilon_1\ \varepsilon_2 \dots \varepsilon_m$ be the distinct roots of the fundamental equation $\Delta = 0$; let λ_i denote the order parting from which the minors of Δ cease to be *all* zero for $\varepsilon = \varepsilon_i$:

1° $\Phi = 0$ admits as distinct integrals

$$\lambda_1 + \lambda_2 + \dots \lambda_m$$

periodic functions of the second kind and no more;

2° There exists a fundamental system of solutions consisting of

$$\lambda_1 + \lambda_2 + \dots + \lambda_m$$

periodic functions of the second kind, and also

$$m - (\lambda_1 + \lambda_2 + \dots + \lambda_m)$$

expressions each of which is of the form of a polynomial in x with coefficients which periodic functions of the second kind all having the same multiplier;

3° The multipliers which appear in this fundamental system either as elements or as coefficients in the elements are equal to the different roots $\varepsilon_1, \varepsilon_2 \dots \varepsilon_m$ of the fundamental equation.

---

* Crelle, Vol. LXVI, Zur Theorie der linearen Differentialgleichungen mit veränderlichen Coefficienten.

II. In order that $\Phi = 0$ may admit of n distinct integrals which are periodic functions of the second kind, it is necessary and sufficient that each root of $\Delta = 0$ shall annul all the minors of Δ up to the order which is equal to the degree of multiplicity of this root.

We have therefore in the case of the differential equation

$$\frac{d^n y}{dx^n} + p_1 \frac{d^{n-1} y}{dx^{n-1}} + \ldots + p_n y = 0,$$

$$\left(\text{where } p_i = \sum_{\alpha=0}^{\alpha=i} \frac{n-i+\alpha \,|\, n-i+1}{\alpha\,!} \frac{\lambda^\alpha x}{\omega} \phi_{i-\alpha}(x)\right)$$

$$\lambda_1 + \lambda_2 + \ldots \lambda_m$$

linearly independent integrals which are periodic functions of the third kind. It remains now to find the remaining $n - (\lambda_1 + \lambda_2 + \ldots \lambda_m)$ integrals, which, with those just mentioned, constitute a fundamental system of the given equation.

It is easy, and, for present purposes, quite enough to verify what these integrals are. Assume

$$e^{\frac{-\lambda x^2}{2\omega}} y_i = e^{\frac{-\lambda x^2}{2\omega}} \left[\Theta_{i1}(x) + x\Theta_{i2}(x) + x^2\Theta_{i3}(x) + \ldots + x^{\mu_j - 1}\Theta_{i\mu_j}(x)\right],$$

give μ_j the meaning above assigned to it and let $\Theta_{ik}(x)$ denote a periodic function of the second kind having ϵ_i for a multiplier. Substitute this value in the differential equation

$$\frac{d^n y}{dx^n} + p_1 \frac{d^{n-1} y}{dx^{n-1}} + p_2 \frac{d^{n-2} y}{dx^{n-2}} + \ldots + p_n y = 0,$$

and give $p_1, p_2 \ldots p_n$ the above values, viz.

$$p_i = \sum_{\alpha=0}^{\alpha=i} \frac{n-i+\alpha \,|\, n-i+1}{\alpha\,!} \frac{\lambda^\alpha x^\alpha}{\omega^\alpha} \phi_{i-\alpha}(x);$$

we have then $\quad \dfrac{d^n y_i}{dx^n} + \phi_1 \dfrac{d^{n-1}}{dx^{n-1}} y_i + \phi_2 \dfrac{d^{n-2} y_i}{dx^{n-2}} + \ldots + \phi_n y = 0,$

and this as we know possesses

$$n - (\lambda_1 + \lambda_2 + \ldots + \lambda_m)$$

integrals of the form

$$y_i = \Theta_{i1}(x) + x\Theta_{i2}(x) + x^2\Theta_{i3}(x) + \ldots + x^{\mu_j - 1}\Theta_{i\mu_j}(x).$$

The solution of the given differential equation is thus seen to depend in all cases upon the solution of an equation of the form

$$\frac{d^n y}{dx^n} + \phi_1 \frac{d^{n-1} y}{dx^{n-1}} + \phi_2 \frac{d^{n-2} y}{dx^{n-2}} + \ldots \phi_n y = 0,$$

where $\phi_1, \phi_2 \ldots \phi_n$ are singly periodic functions of the first kind. The results arrived at then are as follows:

I. In order that the linear differential equation

$$\frac{d^n y}{dx^n} + p_1 \frac{d^{n-1} y}{dx^{n-1}} + p_2 \frac{d^{n-2} y}{dx^{n-2}} + \ldots + p_n y = 0,$$

may possess integrals which are periodic functions of the third kind, it is necessary and sufficient that the coefficients p be of the form

$$p_i = \sum_{\alpha=0}^{\alpha=i} \frac{n-i+\alpha \,|\, n-i+1}{\alpha\,!} \, \frac{\lambda^\alpha x^\alpha}{\omega^\alpha} \, \phi_{i-\alpha}(x),$$

i. e. of the form of polynomials in x whose coefficients, apart from certain determinate constants, are periodic functions of the first kind having ω for a period.

II. If a linear differential equation has the above form it always possesses at least one integral which is a periodic function of the third kind. The total number of linearly independent integrals of this kind is always

$$= \lambda_1 + \lambda_2 + \ldots \lambda_m,$$

where λ, s and the fundamental equation $\Delta = 0$ have the meanings assigned to them above.

III. There exists a fundamental system of integrals consisting of

$$\lambda_1 + \lambda_2 + \ldots + \lambda_m$$

periodic functions of the third kind and in addition $n - (\lambda_1 + \lambda_2 \ldots + \lambda_m)$ functions, which are each of the form of [a polynomial in x whose coefficients are periodic functions of the second kind, all having the same multiplier] $\times e^{\frac{-\lambda x^2}{2\omega}}$ *i. e.* a polynomial in x whose coefficients are periodic functions of the third kind.

IV. The different multipliers of the periodic functions of the second kind are roots of the fundamental equation.

V. In order that the differential equation may have n linearly independent integrals which are periodic functions of the third kind, it is necessary and sufficient that each root of $\Delta = 0$ annul all the minors of Δ up to the order equal to the degree of multiplicity of the root.

Note sur les Nombres de Bernoulli.

PAR F. GOMES-TEIXEIRA,

Professeur à l'Université de Coimbra et à l'École Polytechnique du Porto.

Dans un intéressant mémoire intitulé, *Some Notes on the Numbers of Bernoulli and Euler*, publié dans le Volume V du *American Journal of Mathematics*, Mr. G. S. Ely obtient au moyen des séries qui résultent du développement de tang x, de cot x, de $\sec^p x$, etc., quelques relations entre les nombres de Bernoulli et entre les nombres de Euler. Le but de la présente note est de signaler encore quelques résultats relatifs aux mêmes nombres qu'on peut trouver au moyen du développement de sec x, de $(1+e^x)^{-1}$, de $\sec^p x$, etc.

1. Considérons premièrement la fonction

$$y = (1+e^x)^{-1}.$$

On sait que la dérivée d'ordre n de cette fonction est donnée par la formule

$$y^{(n)} = \Sigma(-1)^i \cdot \frac{n!\, i!\, e^{ix}(1+e^x)^{-i-1}}{\alpha!\, \beta! \ldots \lambda!\, (2!)^\beta (3!)^\gamma \ldots (n!)^\lambda},$$

où $\alpha, \beta, \ldots \lambda$ représentent tous les solutions entières positives de l'équation :

$$\alpha + 2\beta + 3\gamma + \ldots + n\lambda = n;$$

et où $$i = \alpha + \beta + \gamma + \ldots + \lambda.$$

Si on fait maintenant $x = 0$, on trouve

$$y_0^{(n)} = \Sigma(-1)^i \cdot \frac{n!\, i!}{2^{i+1}.\alpha!\, \beta! \ldots \lambda!\, (2!)^\beta (3!)^\gamma \ldots (n!)^\lambda}.$$

D'un autre côté, nous avons

$$y_0^{(2n-1)} = (-1)^n \cdot \frac{2^{2n}-1}{2n} B_{2n-1},$$

et par conséquent

$$(1) \qquad B_{2n-1} = \frac{(2n)!}{2^{2n}-1} \cdot \Sigma(-1)^{i+n} \cdot \frac{i!}{2^{i+1}.\alpha!\, \beta! \ldots \lambda!\, (2!)^\beta (3!)^\gamma \ldots (2n-1!)^\lambda},$$

étant $\alpha + 2\beta + 3\gamma + \ldots + (2n-1)\lambda = 2n-1$, $i = \alpha + \beta + \ldots + \lambda$.

Nous avons donc une formule pour le calcul des nombres de Bernoulli. Cette formule fait encore voir *que le dénominateur des nombres de Bernoulli ne peut contenir d'autres facteurs premiers que* 2 *et ceux de* $2^{2n}-1$, *et que le facteur* 2 *ne peut pas y être élevé à une puissance supérieure à* n.

En effet, on sait par la théorie des dérivées d'ordre quelconque que la fonction numérique

$$(2)\qquad \frac{(2n-1)!}{\alpha!\,\beta!\ldots\lambda!\,(2!)(3!)\ldots(2n-1!)^{\lambda}},$$

donne un nombre entier toutes les fois que α, $\beta, \ldots \lambda$ représentent une solution quelconque de l'équation

$$\alpha + 2\beta + 3\gamma + \ldots + (2n-1)\lambda = 2n-1.$$

On peut encore envisager sous un autre point de vue, que nous ne ferons qu'indiquer ici, la formule (1). Elle établit une rélation entre les nombres de Bernoulli et les nombres (2), dont l'étude est importante parce que ils entrent dans l'expression analytique des dérivées d'ordre quelconque.*

2. Il est évident que chaque formule qui donne une expression de la dérivée d'ordre n d'une fonction, donne une expression correspondante des nombres de Bernoulli. Nous allons donc employer à cette fin la formule†

$$y^{(n)} = n!\sum_{i=1}^{i=n}\frac{A_i}{i!}f^{(i)}(u)$$

$$A_i = \frac{1}{n!}\left[(u^i)^{(n)} - \frac{i}{1}(u^{i-1})^{(n)}.u + \frac{i(i-1)}{1.2}(u^{i-2})^{(n)}.u^2\ldots\right]$$

où $y=f(u)$, $u=\phi(x)$.

Étant
$$y = (1+e^x)^{-1} = u^{-1},\ u = 1+e^x$$
nous avons
$$y^{(n)} = n!\sum_{i=1}^{i=n}(-1)^i.\frac{A_i}{(1+e^x)^{i+1}},$$
où
$$A_i = \frac{1}{n!}\sum_{k=0}^{k=i}(-1)^k.\frac{i(i-1)\ldots(i-k+1)}{k!}\left[(1+e^x)^{i-k}\right]^{(n)}.(1+e^x)^k.$$

Il faut donc chercher la dérivée d'ordre n de $(1+e^x)^{i-k}$, ce qu'on peut faire au moyen de la formule de Leibnitz, qui donne

$$\left[(1+e^x)^{i-k}\right]^{(n)} = \left[(1+e^x)+(1+e^x)+\ldots\right]^{(n)}$$
$$= S\,\frac{n!\,e^{(i-k)x}}{\alpha!\,\beta!\ldots\lambda!}$$

où S représente la somme correspondante à toutes les solutions entières positives de l'équation $\alpha+\beta+\gamma+\ldots+\lambda=n$,
le nombre des quantités α, β, $\gamma, \ldots \lambda$ étant $i-k$.

---

* Voyez notre note *sur les dérivées d'ordre quelconque* dans le Giornale de Battaglini, tome XVIII.

† M. C. Hermite, *Cours d'Analyse*, page 59.

En posant $x = 0$, on trouve

$$y_0^{(n)} = n! \sum_{i=1}^{i=n} (-1)^i . \frac{A_i}{2^{i+1}},$$

où

$$A_i = \sum_{k=0}^{k=i} (-1)^k . \frac{i(i-1)\dots(i-k+1)\,2^k}{k!} \, S \, \frac{1}{\alpha!\,\beta!\dots\lambda!}.$$

D'un autre côté, nous avons

$$B_{2n-1} = (-1)^n \frac{2n}{2^{2n}-1} \, y_0^{(2n-1)},$$

par conséquent

$$B_{2n-1} = \frac{(2n)!}{2^{2n}-1} \sum_{k=0}^{k=2n-1} (-1)^{n+i} . \frac{A_i}{2^{i+1}},$$

où

$$A_i = \sum_{k=0}^{k=i} (-1)^k . \frac{i(i-1)\dots(i-k+1)\,2^k}{k!} \, S \, \frac{1}{\alpha!\,\beta!\dots\lambda!},$$

et

$$\alpha + \beta + \gamma + \dots + \lambda = 2n - 1.$$

Mais

$$(i-k)^{2n-1} = (1 + 1 + \dots)^{2n-1} = S \, \frac{(2n-1)!}{\alpha!\,\beta!\dots\lambda!} . \quad \alpha + \beta + \dots + \lambda = 2n - 1,$$

donc

$$A_i = \sum_{k=0}^{k=i} (-1)^k . \frac{i(i-1)\dots(i-k+1)\,2^k\,(i-k)^{2n-1}}{k!}.$$

Nous avons donc l'expression des nombres de Bernoulli :

$$(3) \quad B_{2n-1} = (-1)^n . \frac{2n}{2^{2n}-1} \, \Sigma \, (-1)^{k+i} . \frac{i(i-1)\dots(i-k+1)\,2^{k-i-1}(i-k)^{2n-1}}{k!},$$

où k et i doivent recevoir toutes les valeurs entières et positives depuis 1 jusqu'à $2n - 1$, et k doit aussi recevoir la valeur zéro avec la condition de substituer alors le facteur

$$\frac{i(i-1)\dots(i-k+1)}{k!}$$

par l'unité.

3. Des expressions analogues à (1) et (3) représentent les coëfficients du développement de $(1 + e^x)^{-p}$ en série. Les résultats qu'on obtient alors en faisant usage des nombres de Bernoulli sont bien moins simples.

En représentant par $\frac{C_{2n-1}}{2n-1!}$ les coëfficients du développement de $(1 + e^x)^{-p}$, nous avons donc premièrement la formule

$$C_{2n-1} = \Sigma \, (-1)^i . \frac{p(p+1)\dots(p+i-1)(2n-1)!}{2^{i+p}\,\alpha!\,\beta!\dots\lambda!\,(2!)^\beta\dots(2n-1!)^\lambda}$$

$$\alpha + 2\beta + 3\gamma + \dots + n\lambda = 2n - 1, \quad i = \alpha + \beta + \dots + \lambda$$

analogue à la formule (1); et ensuite la formule

$$C_{2n-1} = (2n-1)! \sum_{i=1}^{i=2n-1} (-1)^i . \frac{p(p+1)\dots(p+i-1)A_i}{2^{i+p}.i!}$$

$$A_i = \sum_{k=0}^{k=i} (-1)^k . \frac{i(i-1)\dots(i-k+1)\,2^k(i-k)^{2n-1}}{k!},$$

analogue à la formule (3).

4. Considérons maintenant les nombres de Euler. On peut calculer ces nombres au moyen d'une formule analogue à la formule (1). En effet, l'expression analytique de la dérivée d'ordre n de $y=(\cos x)^{-1}$ par rapport à x est

$$y^{(n)} = \Sigma (-1)^i . \frac{n!\,i!\cos^\alpha\left(x+\frac{\pi}{2}\right)\cos^\beta\left(x+2\frac{\pi}{2}\right)\dots\cos^\lambda\left(x+n\frac{\pi}{2}\right)}{\alpha!\,\beta!\dots\lambda!\,(2!)^\beta(3!)^\gamma\dots(n!)^\lambda\cos^{i+1}x}.$$

Nous avons donc

$$(4) \qquad E_{2n} = \Sigma (-1)^i . \frac{(2n)!\,i!\cos^\alpha\frac{\pi}{2}\cos^\beta 2\frac{\pi}{\pi}\dots\cos^\lambda n\frac{\pi}{2}}{\alpha!\,\beta!\dots\lambda!\,(2!)^\beta(3!)^\gamma\dots(n!)^\lambda}$$

où $\alpha, \beta, \dots \lambda$ représentent toutes les solutions entières positives de l'équation

$$\alpha + 2\beta + 3\gamma + \dots + 2n\lambda = 2n$$

et

$$i = \alpha + \beta + \dots + \lambda.$$

Cette formule est encore plus simple que la formule (1) parce que

$$\cos^\alpha\frac{\pi}{2}\cdot\cos^\beta 2\frac{\pi}{2}\dots\cos^\lambda 2n\frac{\pi}{2} = 0, \text{ ou} = +1, \text{ ou} = -1$$

suivant les valeurs de $\alpha, \beta, \dots \lambda$.

On peut aussi calculer les nombres de Euler au moyen d'une formule analogue à (3), mais qui est moins simple. En effet, en posant

$$y = (\cos x)^{-1},$$

nous avons

$$E_{2n} = (2n)! \sum_{i=1}^{i=2n} A_i$$

$$A_i = \sum_{k=0}^{k=i} (-)^k . \frac{i(i-1)\dots(i-k+1)}{k!}\, S\, \frac{\cos\alpha\frac{\pi}{2}\cos\beta\frac{\pi}{2}\dots\cos\lambda\frac{\pi}{2}}{\alpha!\,\beta!\dots\lambda!}$$

où

$$\alpha + \beta + \dots + \lambda = 2n,$$

$\alpha, \beta, \dots$ ne devant recevoir que les valeurs pairs.

5. Considérons maintenant la fonction dont M. Ely s'occupe principalement, à savoir

$$y = (\cos x)^{-p},$$

pour la développer en série. Nous avons premièrement

$$y^{(n)} = \Sigma(-1)^i . \frac{n!\, p(p+1)\ldots(p+i-1)\cos^\alpha\left(x+\frac{\pi}{2}\right)\cos^\beta\left(x+2\frac{\pi}{2}\right)\ldots\cos^\lambda\left(x+n\frac{\pi}{2}\right)}{\alpha!\,\beta!\ldots\lambda!(2!)^\beta(3!)^\gamma\ldots(n!)^\lambda . \cos^{p+i}x},$$

et par conséquent, en appelant C_{2n} le coëfficient de $\frac{x^{2n}}{(2n)!}$ dans le développement considéré, nous aurons

$$C_{2n} = \Sigma(-1)^i . \frac{(2n)!\, p(p+1)(p+2)\ldots(p+i-1)\cos^\alpha\frac{\pi}{2}\cos^\beta 2\frac{\pi}{2}\ldots\cos^\lambda n\frac{\pi}{2}}{\alpha!\,\beta!\ldots\lambda!(2!)^\beta(3!)^\gamma\ldots(n!)^\lambda},$$

où $\alpha + 2\beta + 3\gamma + \ldots + 2n\lambda = 2n, \quad i = \alpha + \beta + \ldots + \lambda,$

et $\cos^\alpha\frac{\pi}{2}\cdot\cos^\beta 2\frac{\pi}{2}\ldots\cos^\lambda n\frac{\pi}{2} = 0, \text{ ou } = +1, \text{ ou } = -1,$

suivant les valeurs de $\alpha, \beta, \ldots \lambda$.

Cette formule va nous conduire à un résultat important.

En y posant $p = p' + 1$, et remarquant que, comme nous avons déjà dit,

$$\frac{(2n)!}{\alpha!\,\beta!\ldots\lambda!\,(2!)^\beta\ldots(n!)^\lambda}$$

est un nombre entier, et que

$p(p+1)\ldots(p+i-1) = (p'+1)(p'+2)\ldots(p'+i) = \text{multiple de } p' + 1.2\ldots i,$

nous avons $\quad C_{2n} = \text{multiple de } p' + E_{2n},$

ou le théorème :

(5) $$C_{2n} \equiv E_{2n} \quad (\text{mod} = p-1).$$

De cette congruence il résulte que le théorème de M. Lucas, à savoir

$$/(-1)^{\frac{p-1}{2}} E_{2n} \equiv /E_{2n+p-1} \quad (\text{mod} = p),$$

a aussi lieu pour les coëfficients C_{2n}, c'est-à-dire que

$$C_{2n} \equiv (-1)^{\frac{p-1}{2}} . C_{2n+p-1},$$

C_{2n} et C_{2n+p-1} étant les coëfficients du développement de $(\cos x)^{-(p+1)}$.

De la formule (5) comparée avec la formule (11) de la mémoire de M. Ely on tire la congruence:

$$\frac{1}{p-1!}[S_t E_{2n} + S_{t-1}E_{2n+2} + \ldots + S_1 E_{2n+p-3} + E_{2n+p-1}] \equiv E_{2n}$$

où S_t représente la somme des combinaisons de $1^2, 3^2, \ldots (p-2)^2$ n à n et où $t = \frac{p-1}{2}$.

PORTO, 2 Janvier, 1885.

A Memoir on Biquaternions.

By Arthur Buchheim, M. A.

Clifford's "preliminary sketch of biquaternions" contains an outline of a calculus devised by him for the analytical treatment of the theory of screws. Besides this sketch there are four fragments dealing with the same subject. Clifford's object was, apparently, so to extend Hamilton's quaternion calculus that it might afford the same help in the study of the screw (that is, of the linear complex), as in its original form it affords in the study of the point and straight line. This end he attained by the invention of the *biquaternion*. Hamilton's biquaternion was a quantity of the form $q + \sqrt{-1}\,q'$, where q, q' are quaternions with real coefficients: modern analysis, however, considers all quantities as complex that are not expressly assumed to be real, and accordingly it is unnecessary to give a distinctive name to a quaternion with complex coefficients. Clifford's biquaternion is also a quantity of the form $q + \omega q'$, where q, q' are ordinary quaternions, but ω is no longer a scalar: it is an operator, commutative with all other operators, and such that its square is a scalar. This being so it appears that a *bivector* represents a motor (screw), and that a biquaternion represents the *quotient* of two motors, that is an operator which changes a given motor into another given motor.

In the first part of the "preliminary sketch" Clifford gives these definitions, and also defines the operator ω: the definition gives $\omega^2 = 0$: when we come to consider the biquaternion we are stopped by a difficulty which, we are told, can be explained by considering our geometry as a particular case of a geometry in which that difficulty does not occur. We now come to the second part of the paper: here we have, first of all, an explanation of the fundamental conceptions of the non-euclidean geometry: this is followed by a statement, for the most part without proof, of the fundamental theorems in the geometry and kinematics of elliptic space. The operator ω is next introduced by a definition giving $\omega^2 = 1$, and then by the introduction of two new symbols we are enabled to write the biquaternion in a form which does not present the difficulty that stopped us in our consideration of parabolic geometry. The rest of the paper consists of investigations of some of the fundamental formulæ in the theory of elliptic space.

Of the fragments above referred to, two contain nothing that is not in the "preliminary sketch"; a third contains the beginning of an investigation of the motion of a rigid body in elliptic space: in this the ideas of the "sketch" are employed, and the velocity-system of the body is represented by a bivector. Lastly there is a fragment in which two problems of the theory of screws are considered: the first problem is that of finding the axis of a given screw, and this is completely solved; the second must, I think, have been the investigation of the cylindroid: all that is preserved is an expression for the axis of the sum of two screws whose axes intersect at right angles, and for the angular distance of this axis from the intersection of the other two.

There are, besides, a few notes dealing chiefly with the geometry of elliptic space.

In a paper "On the application of the Ausdehnungslehre and of Quaternions to the different kinds of uniform space" published in the Cambridge Philosophical Society's Transactions, Mr. Homersham Cox has added the value of ω^2 for hyperbolic space to what had been done by Clifford, but though his paper is interesting, it cannot, I think, be considered as containing any new development of Clifford's calculus.

I have considered the case of elliptic space in a paper, "On the Theory of Screws in Elliptic Space," published in the "Proceedings" of the London Mathematical Society. In this paper I solve the fundamental problems of the theory of screws, and prove most of Clifford's theorems: but the methods used are Grassmann's and not Clifford's. In the present paper I give what appears to me to be a tolerably complete development of Clifford's calculus, in the hope that, if it serves no other useful purpose, it may at least have some interest as a commentary on the "preliminary sketch."

Starting with the definition of a biquaternion I investigate the fundamental metric functions: I then consider the analytical problems answering to the elementary problems in the theory of screws, and give some formulæ relating to parallels. In all this I consider the general biquaternion, and there is no attempt at geometrical interpretation. The results of this first part are used in the second part, where they receive a geometrical interpretation.

It will be seen that I have found it necessary to introduce several new symbols: this is to be regretted, but it was quite unavoidable. One of these symbols is of such fundamental importance that it will be worth while to consider it here. I mean the e which occurs in all the formulæ when developed: this is defined by the equation $\omega^2 = e^2$: in elliptic space we have $e = 1$, in parabolic space $e = 0$, in hyperbolic space $e = \sqrt{-1}$: this quantity e is in fact the reciprocal of the radius of curvature of the space: it is the $1/k$ which occurs in Lobatchewsky's formulæ: it is only by the introduction of this scalar that the formulæ can become applicable to the three kinds of space.

Another change that I have made calls for some remark. I differ from Clifford in representing the point not by a vector, but by a biquaternion: the result is that the biquaternion represents all the *forms* that occur in a space of three dimensions as well as their quotients. Lastly, I remark that the operator ω is essentially a matrix.

PART I. BIQUATERNIONS.

1. *The Biquaternion.*

The whole theory of biquaternions depends upon the introduction of a symbol ω of which the geometrical meaning need not at present concern us: all that we require to know is that in all combinations it may be treated as if it were a mere scalar multiplier, and that its square is a scalar. This scalar I denote by e^2.

The biquaternion Q is defined by the equation

$$Q = q + \omega q',$$

where q, q' are ordinary quaternions.

It follows that $\qquad \omega Q = e^2 q' + \omega q.$

It is convenient to denote q, q' by functional symbols involving Q, and accordingly I write

$$q = \mho Q$$
$$q' = \Omega Q,$$

so that

$$Q = \mho Q + \omega \Omega Q,$$

and then

$$\omega Q = e^2 \Omega Q + \omega \mho Q.$$

We can therefore say that

$$\mho(\omega Q) = e^2 \Omega Q$$
$$\Omega(\omega Q) = \mho Q.$$

Now take two biquaternions

$$Q = q + \omega q'$$
$$R = r + \omega r'.$$

Then

$$QR = (qr + e^2 q'r') + \omega(qr' + q'r).$$

Therefore
$$\mho(QR) = \mho Q.\mho R + e^2 \Omega Q.\Omega R$$
$$\Omega(QR) = \mho Q.\Omega R + \Omega Q.\mho R.$$
These equations are of fundamental importance.

Let $A = a + \omega b$ be a *biscalar*: then
$$(a + \omega b)(a - \omega b) = a^2 - e^2 b^2,$$
therefore
$$a + \omega b . \frac{a - \omega b}{a^2 - e^2 b^2} = 1.$$
Therefore
$$(a + \omega b)^{-1} = \frac{a - \omega b}{a^2 - e^2 b^2}.$$

2. *Distances.*

In all that follows I use Hamilton's symbols SVK in their usual sense. I also use his N to denote QKQ. T is not used here to denote the square root of this, and is defined below.

We obviously have $\chi Q = \chi \mho Q + \omega \chi \Omega Q$, or
$$\mho \chi = \chi \mho$$
$$\Omega \chi = \chi \Omega$$
if χ denotes S, V or K.

I now define all the metric functions used in this paper.

Q being any biquaternion I write
$$TQ = +\sqrt{\mho NQ} = +\sqrt{\mho(QKQ)}$$
so that
$$T^2 Q = \mho Q \mho KQ + e^2 \Omega Q \Omega KQ$$
$$= \mho QK(\mho Q) + e^2 \Omega QK(\Omega Q)$$
$$= N(\mho Q) + e^2 N(\Omega Q).$$
Two biquaternions Q, R determine three angles defined as follows:
$$\cos\ (QR) = \mho SQKR\ (\div)$$
$$\sin \epsilon [QR] = \epsilon \Omega SQKR\ (\div)$$
$$\sin\ \{QR\} = TVQKR\ (\div)$$
Divisor $= TQ.TR$ in each case.

I write $[Q]$ for $[Q^2]$: we have by definition
$$NQ = T^2 Q(1 + \omega \epsilon^{-1} \sin \epsilon [Q]).$$
We have $NV(QKR) = NQ.NR - S^2 QKR.$

Therefore $\mho NV(QKR) = \mho(NQ.NR) - \mho S^2(QKR)$
$$= \mho NQ.\mho NR + e^2 \Omega NQ \Omega NR - (\mho SQKR)^2 - e^2(\Omega SQKR)^2,$$
or $T^2 Q T^2 R \sin^2\{QR\} = (1 + \sin \epsilon [Q] \sin \epsilon [R] - \cos^2(QR) - \sin^2 \epsilon [QR]) T^2 Q T^2 R.$

Therefore $\sin^2\{QR\} = 1 + \sin \epsilon [Q] \sin \epsilon [R] - \cos^2(QR) - \sin^2 \epsilon [QR].$

If either $[Q]$ or $[R]$ vanishes, and $[QR]$ also vanishes, we get

$$\sin^2\{QR\} = 1 - \cos^2(QR)$$
$$\{QR\} = (QR).$$

I now define as follows:

(QR) is the *angle* of Q, R.
$[QR]$ is the *moment* of Q, R.
$\{QR\}$ is the *distance* of Q, R.
$[Q]$ is the *pitch* of Q.

3. *Axes.*

A biquaternion of zero pitch is called a *special* biquaternion: that is, R is a special biquaternion if $\mho NR = 0$.
Let Q be any biquaternion: then a special biquaternion R such that $Q = AR$, where A is a biscalar is called an axis of Q: it will be seen that the determination of the axis leads to a quadratic equation, so that in general a biquaternion has two axes, and it appears that they are of the form R, ωR.
We have $$R = A^{-1} Q$$
and A must be determined so that
$$\Omega NR = 0.$$
That is $$\Omega N(A^{-1}Q) = 0$$
or $$\Omega (A^{-2}NQ) = 0.$$
Therefore $$\Omega A^{-2}\mho NQ + \mho A^{-2}\Omega NQ = 0.$$
But if $$A = a + \omega b$$
$$A^{-1} = \frac{a - \omega b}{a^2 - e^2 b^2}$$
$$A^{-2} = \frac{a^2 + e^2 b^2 - 2\omega ab}{(a^2 - e^2 b^2)^2}.$$
Therefore we have $$0 = (a^2 + e^2 b^2)\Omega NQ - 2ab\mho NQ$$
or $$\frac{2ab}{a^2 + e^2 b^2} = \frac{\Omega NQ}{\mho NQ}$$
$$= \frac{1}{e} \sin e[Q]$$
or $$\frac{2eab}{a^2 + e^2 b^2} = \sin e[Q].$$
This is obviously satisfied by
$$a = \cos e \frac{[Q]}{2}$$
$$b = e^{-1} \sin e \frac{[Q]}{2}.$$

And here I stop to make a remark of some importance: *all biquaternions are only determined to a scalar factor près*, and therefore we can take the values just written as the actual values of a, b. I now introduce two symbols which are constantly used in the sequel, viz. I write

$$\text{es}\,\phi = e^{-1}\sin e\phi$$
$$\text{ec}\,\phi = \cos e\phi.$$

I also write ϕ for $[Q]$ the pitch of Q: we therefore have

$$a = \text{ec}\,\frac{\varphi}{2}$$

$$b = \text{es}\,\frac{\varphi}{2}.$$

And then $$a^2 - e^2b^2 = \text{ec}\,\phi.$$

We therefore get $$Q = \left(\text{ec}\,\frac{\varphi}{2} + \omega\,\text{es}\,\frac{\varphi}{2}\right)R,$$

$$\text{ec}\,\phi R = \left(\text{ec}\,\frac{\varphi}{2} - \omega\,\text{es}\,\frac{\varphi}{2}\right)Q.$$

This gives $$\text{ec}\,\phi.\omega R = \left(-e^2\,\text{es}\,\frac{\varphi}{2} + \omega\,\text{ec}\,\frac{\varphi}{2}\right)Q,$$

and it can be immediately verified that

$$-e^2\,\text{es}\,\frac{\varphi}{2},\quad -\text{ec}\,\frac{\varphi}{2}$$

are the other roots of the equation giving $\frac{a}{b}$, and therefore, as was stated above, the two axes of Q are of the form R, ωR.

There is an important case in which the investigation fails, viz. the case in which $\text{es}\,\phi = \pm e^{-1}$ or $e\phi = \frac{\pi}{2}$ or $\frac{3\pi}{2}$: we have $\text{ec}\,\phi = 0$, and the expression for the axis becomes infinite: the fact is that, as will be seen below, the axis is really indeterminate.

Let Q, Q' be two biquaternions; R, R' two of their axes, and ϕ, ϕ' their pitches.

Let $$Q = AR$$
$$Q' = A'R'$$

we have $$A = \text{ec}\,\frac{\varphi}{2} + \omega\,\text{es}\,\frac{\varphi}{2}$$

$$A' = \text{ec}\,\frac{\varphi'}{2} + \omega\,\text{es}\,\frac{\varphi'}{2}.$$

Therefore
$$A^2 = 1 + \omega \text{ es } \phi$$
$$AA' = \text{ec}\frac{\varphi - \varphi'}{2} + \omega \text{ es } \frac{\varphi + \varphi'}{2}$$
$$NQ = A^2 NR.$$
Therefore
$$T^2 Q = \mho A^2 \mho NR$$
$$= \mho NR$$
$$= T^2 R$$
since $\Omega NR = 0$, by definition moreover
$$QKQ' = AA'.RKR'.$$
Therefore
$$\mho SQKQ' = \mho AA'.\mho SRKR' + e^2 \Omega AA'.\Omega SRKR'$$
or
$$\cos(QQ') = \text{ec}\frac{\varphi - \varphi'}{2}.\cos(RR') + e^2.\text{es}\frac{\varphi + \varphi'}{2}.\text{es}[RR'].$$
We have also
$$\Omega SQKQ' = \mho AA'.\Omega SRKR' + \Omega AA' \mho SRKR'$$
or
$$\sin[QQ'] = \text{ec}\frac{\varphi - \varphi'}{2}\text{es}[RR'] + \text{es}\frac{\varphi + \varphi'}{2}\cos(RR')$$

4. *Vectors.*

I now introduce two operators, ξ, η, defined as follows:
$$\xi = \frac{1 + e^{-1}\omega}{2}$$
$$\eta = \frac{1 - e^{-1}\omega}{2}.$$
It is important to notice that, like ω, ξ and η can be treated as if they were mere scalar multipliers.

We have
$$\xi^2 = \frac{1 + 2e^{-1}\omega + e^{-2}\omega^2}{4}$$
$$= \frac{1 + 2e^{-1}\omega + 1}{4}$$
$$= \frac{1 + e^{-1}\omega}{2}$$
$$= \xi.$$
Similarly
$$\eta^2 = \eta.$$
Moreover
$$\xi\eta = \frac{(1 + e^{-1}\omega)(1 - e^{-1}\omega)}{4}$$
$$= \frac{1 - e^{-2}\omega^2}{4}$$
$$= 0.$$

If $\xi Q = 0$, Q is called a ξ-vector; if $\eta Q = 0$, Q is called an η-vector, as the word *vector* when used in this sense is always accompanied by a reference to the two species of vectors there need be no confusion between this kind of vector and the ordinary vector of quaternions.

We have
$$\xi Q = \frac{1 + e^{-1}\omega}{2} Q.$$
Therefore
$$\Omega \xi Q = \frac{\mathit{\Omega} Q + e^{-1}\mho Q}{2}$$
$$\mho \xi Q = \frac{\mho Q + e\mathit{\Omega} Q}{2}.$$

If Q is a ξ-vector we must have $\Omega \xi Q = \mho \xi Q = 0$: these equations agree in giving
$$\Omega Q = - e^{-1} \mho Q.$$
Therefore
$$Q = (1 - e^{-1}\omega) \mho Q$$
$$= 2\eta \mho Q.$$
Therefore any ξ-vector can be written in the form ηQ: and in fact it is obvious that ηQ is a ξ-vector, because $\xi\eta = 0$: in the same way any η-vector can be written in the form ξQ. It is as well to notice that in these expressions, ξQ, ηQ, Q is a simple quaternion, not a biquaternion, and that, if $R = \xi Q$ is an η-vector, we have
$$Q = 2\mho R,$$
and this is of course also true if $R = \eta Q$ is a ξ-vector.

Any biquaternion can be written, and in one way only, as the sum of a ξ-vector and an η-vector. Let Q be a given biquaternion, then if we are to have
$$Q = R + S$$
where R is a ξ-vector, and S an η-vector, we have also
$$e^{-1}\omega Q = e^{-1}\omega R + e^{-1}\omega S$$
$$= - R + S,$$
by definition: therefore
$$S = \frac{Q + e^{-1}\omega Q}{2}$$
$$= \xi Q$$
and
$$R = \eta Q.$$
Therefore
$$Q = \eta Q + \xi Q$$
is the only decomposition of a biquaternion into a ξ-vector and an η-vector: it is obvious *a priori* that this is such decomposition, and what we have proved is that there is no other.

At the end of (4) it was stated that if $[Q] = \frac{\pi}{2e}$ or $\frac{3\pi}{2e}$ we cannot find the axes of Q: I show now that this happens if Q is an η-vector or a ξ-vector.

Let ξq be any η-vector: q being a simple quaternion.

We have
$$N(\xi q) = \xi q . K\xi q$$
$$= \xi q K q$$
$$= \xi (Nq).$$

Now $\Omega NQ = 0$ because Q is a simple quaternion.

Therefore
$$\mho (N\xi q) = \frac{Nq}{2}$$
$$\Omega (N\xi q) = \frac{e^{-1}Nq}{2}.$$

Therefore
$$\text{es } [\xi q] = e^{-1}$$
$$\sin e\, [\xi q] = 1$$
$$[\xi q] = \frac{\pi}{2e}.$$

In the same way we get
$$[\eta q] = \frac{3\pi}{2e}.$$

It should be noticed that if $Q = \xi q$ is an η-vector we have $\xi Q = Q$, and in the same way if $Q = \eta q$ is a ξ-vector we have $\eta Q = Q$; and, conversely, it is obvious that these equations $Q = \xi Q$, $Q = \eta Q$ define Q as an η-vector or a ξ-vector respectively.

Let ξq, ηr be two vectors of different species: we have
$$\xi q . \eta r = \xi\eta . qr$$
$$= 0.$$

This of course breaks up into a scalar and a vector equation: and we can say that if Q, R are two vectors of different species we have
$$VQR = 0$$
$$SQR = 0.$$

Each of these equations breaks up into two, giving two vector equations and two scalar equations; the scalar equations give
$$(QR) = \frac{\pi}{2}$$
$$[QR] = 0.$$

5. *Parallels.*

Let Q, R be two biquaternions: then, if $\xi Q = \xi R$, Q, R are said to be ξ-parallel: if $\eta Q = \eta R$ they are said to be η-parallel. All ξ-vectors are ξ-parallel, and all η-vectors are η-parallel. If two biquaternions are both ξ (η) parallel to the same biquaternion they are ξ (η) parallel to each other. Since all biquaternions are only determined to a scalar factor *près*, so that λR is the same as R if λ is a simple scalar, we can say that QR are parallel if $\xi Q = \lambda \xi R$, or if $\eta Q = \lambda \eta R$.

If Q is both ξ-parallel and η-parallel to R, Q is of the form $\lambda \xi R + \mu \eta R$: if we have $\xi Q = \xi R$, $\eta Q = \eta R$ we have

$$\begin{aligned} Q &= \xi Q + \eta Q \\ &= \xi R + \eta R \\ &= R \end{aligned}$$

and the two biquaternions are identical. Now let

$$\begin{aligned} \xi Q &= \lambda \xi R \\ \eta Q &= \mu \eta R. \end{aligned}$$

Then
$$\begin{aligned} Q &= \xi Q + \eta Q \\ &= \lambda \xi R + \mu \eta R. \end{aligned}$$

This gives
$$Q = \frac{\lambda + \mu}{2} R + \frac{\epsilon^{-1}(\lambda - \mu)}{2} \omega R.$$

Conversely if we have $\quad Q = aR + b\omega R$

where a and b are scalars, Q is ξ-parallel and η-parallel to R.

To prove this I observe that

$$\begin{aligned} \xi\omega &= \frac{1 + \epsilon^{-1}\omega}{2} \omega \\ &= \frac{\omega + \epsilon^{-1}\omega^2}{2} \\ &= \frac{\omega + e}{2} \\ &= e\xi. \end{aligned}$$

Similarly
$$\eta\omega = -e\eta.$$

Therefore if
$$\begin{aligned} Q &= aR + b\omega R \\ \xi Q &= a\xi R + b\xi\omega R \\ &= (a + be)\,\xi R \end{aligned}$$

and
$$\eta Q = (a - be)\,\eta R.$$

Therefore, as was stated above, Q is ξ-parallel and η-parallel to R. In particular, it follows that the axis of a biquaternion is ξ-parallel and η-parallel to the biquaternion.

Now let ξr be an η-vector, and Q any biquaternion: we have

$$\begin{aligned} Q\xi r &= (\xi Q + \eta Q)\,\xi r \\ &= \xi Q . \xi r. \end{aligned}$$

Similarly if Q' is any other biquaternion,

$$Q'\xi r = \xi Q' . \xi r.$$

Therefore if Q is ξ-parallel to Q' so that $\xi Q = \xi Q'$ we have

$$Q.\xi r = Q'.\xi r.$$

In the same way if Q is η-parallel to Q' we have

$$Q.\eta r = Q'.\eta r.$$

If we have

$$\xi Q = \xi R$$

we get

$$N\xi Q = \chi N(\xi R).$$

But

$$\begin{aligned} N\xi Q &= \xi Q.\xi KQ \\ &= \xi QKQ \\ &= \xi NQ. \end{aligned}$$

Therefore

$$\mho(N\xi Q) = \frac{\sigma NQ + e\Omega NQ}{2}$$

$$\Omega(N\xi Q) = \frac{e^{-1}\sigma NQ + \Omega NQ}{2}.$$

Therefore we have

$$T^2Q(1 + \sin e[Q]) = T^2R(1 + \sin e[R]).$$

In the same way if

$$\xi U = \xi W$$

$$T^2U(1 + \sin e[U]) = T^2W(1 + \sin e[W]).$$

Moreover

$$\xi(QKU) = \xi(RKW).$$

Therefore

$$TUTQ(\cos(QU) + \sin e[QU]) = TRTW(\cos(RW) + \sin e[RW]).$$

Therefore, finally,

$$\frac{\cos(QU) + \sin e[QU]}{(1 + \sin e[Q])^{\frac{1}{2}}(1 + \sin e[U])^{\frac{1}{2}}} = \frac{\cos(RW) + \sin e[RW]}{(1 + \sin e[R])^{\frac{1}{2}}.(1 + \sin e[W])^{\frac{1}{2}}}.$$

If

$$aQ + b\omega Q + cR + d\omega R = 0,$$

and Q, R, ωR are not connected by a linear relation, Q, R are either ξ-parallel or η-parallel: this can only be the case if $e^2 = 1$. For operating with ω on the equation

$$aQ + b\omega Q + cR + d\omega R = 0$$

we get

$$a\omega Q + e^2bQ + c\omega R + e^2dR = 0.$$

Substituting for ωQ in the first equation from the second we get

$$Q\left(a - \frac{e^2b^2}{a}\right) + R\left(c - \frac{e^2bd}{a}\right) + \omega R\left(d - \frac{e^2bc}{a}\right) = 0.$$

Therefore, by what was stipulated as to Q, R, ωR,

$$\begin{aligned} a^2 &= e^2b^2 \\ ac &= e^2bd \\ ad &= e^2bc. \end{aligned}$$

The first equation gives $a = \pm eb$: substituting in the second and third we get either $a = 0$, $e = 0$ (this case will be considered below), or

$$c = \pm ed$$
$$d = \pm ec.$$

These equations are inconsistent unless $e^2 = 1$: assuming this to be the case, and substituting, we get

$$eb(1 \pm e^{-1}\omega)\, Q + ed(1 \pm e^{-1}\omega)\, R = 0.$$

That is
$$b \begin{smallmatrix}\xi \\ \eta\end{smallmatrix} Q + d \begin{smallmatrix}\xi \\ \eta\end{smallmatrix} R = 0.$$

That is Q, R are either ξ-parallel or η-parallel.

If we have $a = 0$, $e = 0$, we get

$$b\omega Q + cR + d\omega R = 0,$$

and then operating with ω we get $c = 0$, and the equation reduces to

$$b\omega Q + d\omega R = 0.$$

Now if $e = 0$, $2\xi Q = (1 + e^{-1}\omega)\, Q$ becomes $e^{-1}\omega Q$, and $2\eta Q$ becomes $-2e^{-1}\omega Q$: therefore the last equation is

$$b \begin{smallmatrix}\xi \\ \eta\end{smallmatrix} Q + d \begin{smallmatrix}\xi \\ \eta\end{smallmatrix} R.$$

Therefore the theorem holds in this case also, except that the two species of parallelism coincide.

In the other excluded case in which we have a relation

$$aQ + bR + c\omega R = 0$$

we get the case considered above in which Q, R are both ξ-parallel and η-parallel.

Therefore we can neglect all distinctions of cases and say generally that Q, R are parallel if there is any relation of the form

$$aQ + b\omega Q + cR + d\omega R = 0.$$

I now prove that, as was stated above, the axis of a vector of either species is indeterminate. Let ξq be a vector: R an axis: we are to have

$$\xi q = aR + b\omega R.$$

Therefore $$0 = \xi\eta q = a\eta R - eb\eta R.$$

Therefore $$a = eb.$$

Therefore $$\xi q = a(1 + e^{-1}\omega)\, R.$$

Therefore R may be any special biquaternion satisfying this condition.

Therefore, the axis of an η-vector is any special biquaternion ξ-parallel to it, and the axis of a ξ-vector is any special biquaternion η-parallel to it.

6. *The Cylindroid.*

Let A, B be two biquaternions: they determine a linear singly infinite series of biquaternions $\lambda A+\mu B$ where λ, μ are scalars: this set is called a cylindroid, so that if C is any biquaternion of the cylindroid (A, B) we have

$$C=\lambda A+\mu B.$$

Every cylindroid contains two special biquaternions (v. def. in (3)): for we have

$$C=\lambda A+\mu B.$$

Therefore
$$KC=\lambda KA+\mu KB$$
$$NC=CKC=\lambda^2 NA+2\lambda\mu SAKB+\mu^2 NB.$$

Therefore if C is to be a special biquaternion, so that $\Omega NC=0$, we must have

$$\lambda^2\Omega NA+2\lambda\mu\Omega SAKB+\mu^2\Omega NB=0,$$

or
$$\frac{\lambda^2}{T^2B}\,\mathrm{es}\,[A]+\frac{2\lambda}{TB}\frac{\mu}{TA}\,\mathrm{es}\,[AB]+\frac{\mu^2}{T^2A}\,\mathrm{es}\,[B]=0.$$

This equation determines the two values of $\frac{\lambda}{\mu}$ corresponding to the two special biquaternions: if A, B are special biquaternions and such that $[AB]=0$, the equation is an identity: every biquaternion of the cylindroid is a special biquaternion, and it can be verified at once that we have also $[AC]=[BC]=0$. The roots coincide if $[AB]=e^{-1}\sin^{-1}e$.

Every cylindroid contains in general two biquaternions C, C', satisfying the *two* conditions $SCKC'=0$.

Let
$$C=\lambda A+\mu B$$
$$C'=\lambda' A+\mu' B.$$

Then
$$SCKC'=\lambda\lambda' NA+(\lambda\mu'+\lambda'\mu)\,SAKB+\mu\mu' NB.$$

But we are to have $\Omega SCKC'=\mho SCKC'=0$: therefore

$$\lambda\lambda'\mho NA+(\lambda\mu'+\lambda'\mu)\,\mho SAKB+\mu\mu'\mho NB=0$$
$$\lambda\lambda'\Omega NA+(\lambda\mu'+\lambda'\mu)\,\Omega SAKB+\mu\mu'\Omega NB=0.$$

Therefore $\frac{\lambda}{\mu}$, $\frac{\lambda'}{\mu'}$ are the two roots of

$$\begin{vmatrix} 1 & -x & x^2 \\ \mho NA & \mho SAKB & \mho NB \\ \Omega NA & \Omega SAKB & \Omega NB \end{vmatrix}=0.$$

Every cylindroid contains, in general, two biquaternions C, C' such that

$$TC=TC'=0.$$

If
$$C=\lambda A+\mu B$$

we are to have
$$0=\mho NC$$
$$=\lambda^2\mho NA+2\lambda\mu\mho SAKB+\mu^2\mho NB.$$

This is an identity if $TA = TB = 0$, $(AB) = \frac{\pi}{2}$: the roots coincide if $(AB) = 0$.

If a cylindroid contains an infinite number of special biquaternions, or an infinite number of biquaternions whose tensor vanishes, it contains an infinite number of pairs such that $SCKC' = 0$. This is obvious from what precedes.

Let Q, R be any two biquaternions: we have

$$V(\lambda Q + \mu R) K(\lambda' Q + \mu' R) = (\lambda\mu' - \lambda'\mu) VQKR.$$

Therefore $VQKR$ is the same (to a scalar factor *près*) for every pair of biquaternions of the cylindroid (Q, R): this bivector $VQKR$ is called the axis of the cylindroid.

If $\xi Q = \xi R = U$ we have $\xi VQKR = V(\xi Q . \xi KR) = VUKU = 0$: therefore if two biquaternions are parallel, the axis of the cylindroid they determine is a vector of the same species as the parallelism. Moreover, if $P = \lambda Q + \mu R$ we have $\xi P = \lambda\xi Q + \mu\xi R = (\lambda + \mu) U$: therefore all the biquaternions of the cylindroid are parallel.

If $\xi A = \xi B$, $\xi C = \xi D$, we have

$$\xi(AKC) = \xi(BKD),$$

and therefore

$$\xi V(AKC) = \xi V(BKD).$$

Now if E, F are two biquaternions of the cylindroids (A, C), (B, D) respectively, and such that

$$E = \lambda A + \mu C$$
$$F = \lambda B + \mu D.$$

We have

$$\xi E = \lambda\xi A + \mu\xi C$$
$$= \lambda\xi B + \mu\xi D$$
$$= \xi F.$$

Therefore to every biquaternion of one cylindroid corresponds a parallel biquaternion of the other cylindroid: we may therefore say that two parallel pairs of biquaternions determine two parallel cylindroids, and the axes of parallel cylindroids are parallel. In all that precedes ξ stands for ξ or η.

PART II. GEOMETRY.

In the first part we have considered the analytical theory of the biquaternion, apart from any interpretation. In the second part we make use of the results of the first part, and interpret them geometrically. The greater part of the first two sections of Part II is foreign to the object of this paper, which is the development of metric geometry from the definition of the biquaternion: but they will, I think, make the rest of the paper more easily intelligible.

For convenience the sections of the whole paper are numbered consecutively.

7. *The Absolute.*

Suppose the tetrahedron of reference self-conjugate with respect to the absolute; and suppose, moreover, that the equation of the absolute is

$$e^2(x^2+y^2+z^2)+\omega^2=0,$$

then the plane-equation is $l^2+m^2+n^2+e^2p^2=0$, and the line equation (the condition that a line may touch the surface is $f^2+g^2+h^2+e^2(a^2+b^2+c^2)=0$, we know that in elliptic space the absolute is $x^2+y^2+z^2+\omega^2=0$, and that in hyperbolic space it is $\omega^2-(x^2+y^2+z^2)=0$: in parabolic space the absolute becomes a plane conic (an infinitesimal quadric) in the plane at infinity, so that its point equation is $(\text{const.})^2=0$. if $(\text{const.})=0$ is the plane at infinity, and its plane equation is $l^2+m^2+n^2=0$: we therefore see that we can represent the three geometries by taking for the three equations of the absolute

$$e^2(x^2+y^2+z^2)+\omega^2=0$$
$$l^2+m^2+n^2+e^2p^2=0$$
$$f^2+g^2+h^2+e^2(a^2+b^2+c^2)=0$$

with the following stipulations.

I. In elliptic space $e^2=1$
In parabolic space $e^2=0$
In hyperbolic space $e^2=-1$.

II. In parabolic space $\delta=1$ for all points at a finite distance and $p=1$ for planes not passing through (0001).

III. In parabolic space a line is considered to touch the absolute if it meets the "circle at infinity": for the absolute consists of this curve taken twice over.

The generators of the absolute are the tangents of the "circle at infinity": each tangent represents two generators, one of each system.

If $(\alpha,\beta,\gamma,\delta)$ is any point, its polar plane with respect to the absolute is $(e^2\alpha, e^2\beta, e^2\gamma, \delta)$, and if $(abcfgh)$ is any line its polar is (fgh, e^2a, e^2b, e^2c): and we can say more generally that this is the polar of the screw $(abcfgh)$.

If a screw is its own conjugate we must have

$$a=\lambda f,\ b=\lambda g,\ c=\lambda h,\ f=e^2\lambda a,\ g=e^2\lambda b,\ h=e^2\lambda c.$$

This gives
$$\lambda^2e^2=1$$
$$\lambda=\pm e^{-1}.$$

Therefore if a screw is its own conjugate its coordinates are of the form

$$(\pm e^{-1}f,\ \pm e^{-1}g,\ \pm e^{-1}h,\ f,\ g,\ h).$$

If a line is its own conjugate it is a generator of the absolute and we get as the coordinates of a generator of one system

$$(e^{-1}f,\ e^{-1}g,\ e^{-1}h,\ f,\ g,\ h)$$

and as the coordinates of a generator of the other system

$$(-e^{-1}f,\ -e^{-1}g,\ -e^{-1}h,\ f,\ g,\ h).$$

8. *Interpretations.*

There is in three dimensional space a ∞^3 series of points, a ∞^3 series of planes, a ∞^5 series of screws (motors, linear complexes): the line is a particular case of the screw, and need not, at present, be considered separately. The biquaternion $q + \omega q'$ contains eight scalar constants, but it is determined (not but by their absolute values but) by their ratios: there is therefore a ∞^7 series of biquaternions. Therefore, there is a ∞^3 series satisfying four scalar conditions, and a ∞^5 series satisfying two conditions: therefore we can make a biquaternion satisfying four conditions represent a point or a plane, and a biquaternion satisfying two conditions represent a screw.

These conditions are chosen as follows: The biquaternion $q + \omega q'$ represents

a *point* if $Vq = 0,\ Sq' = 0$

a *plane* if $Sq = 0,\ Vq' = 0$

a *screw* if $Sq = 0,\ Sq' = 0.$

Moreover, if $(\alpha\beta\gamma\delta)$ are scalars, we say that

$$\delta + \omega(\alpha i + \beta j + \gamma k)$$

$$\alpha i + \beta j + \gamma k\, \omega\delta$$

represent the point and plane respectively whose coordinates in the system of (7) are $(\alpha\beta\gamma\delta)$.

Lastly the screw $(abcfgh)$ is represented by

$$(fi + gj + \gamma k) - \omega(ai + bj + \gamma k).$$

I justify this by showing, in a single instance, that these determinations give metric formulæ agreeing with those in (7): the representation of a screw will be justified by the expression for a line obtained below.

The angle and distance of two points are given by

$$\cos(PP') = \frac{\varpi SPKP'}{TP.TP'},$$

$$\sin\{PP'\} = \frac{\Omega SPKP'}{TP.TP'}.$$

Now let

$$P = \delta + \omega\rho = \delta + \omega(\alpha i + \beta j + \gamma k),\ P' = \delta' + \omega\rho' = \delta' + \omega(\alpha' i + \beta' j + \gamma' k),$$

we have
$$PKP' = (\delta + \omega\rho)(\delta' + \omega\rho')$$
$$= \delta\delta' - e^2\rho\rho' + \omega(\rho\delta' - \rho'\delta),$$
$$PKP = \delta^2 - e^2\rho^2.$$

Therefore
$$\cos(PP') = \frac{\delta\delta' - e^2 S\rho\rho'}{(\delta^2 - e^2\rho^2)^{\frac{1}{2}}.(\delta'^2 - e^2\rho'^2)^{\frac{1}{2}}}$$
$$= \frac{\delta\delta' + e^2(\alpha\alpha' + \beta\beta' + \gamma\gamma')}{\{\delta^2 + e^2(\alpha^2 + \beta^2 + \gamma^2)\}^{\frac{1}{2}}.\{\delta'^2 + e^2(\alpha'^2 + \beta'^2 + \gamma'^2)\}^{\frac{1}{2}}}.$$

But this is the known expression for cos PP' with $\delta^2 + e^2(\alpha^2 + \beta^2 + \gamma^2) = 0$ as the equation of the absolute.

Moreover, we get at once $[PP'] = 0$, and therefore by (1) $\{PP'\} = (PP')$.

Now, writing $(\alpha\beta\gamma)$ for $\alpha i + \beta j + \gamma k$, let

$$P = \delta + \omega(\alpha\beta\gamma)$$

be a point: then
$$\omega P = e^2(\alpha\beta\gamma) + \omega\delta.$$

But this is the plane $(e^2\alpha, e^2\beta, e^2\gamma, \delta)$ which is the polar plane of $(\alpha\beta\gamma\delta)$ with respect to the absolute: therefore ωP is the polar of P. In the same way if Q is a plane or a screw, we see that ωQ is the polar point or screw with respect to the absolute.

This is the geometrical interpretation of ω referred to in (1).

9. *Lines.*

If P, P' are two points, the line joining them is the axis of the cylindroid $(\omega P, P')$: for this is the same for all pairs $\lambda P + \mu P'$, and for such pairs only: but $\lambda P + \mu P'$ is a linear ∞^1 series of points containing P, P': therefore it is the points of the line (PP'): therefore $V(\omega P.KP')$ can be taken to represent the line.

We have, if $P = \delta + \omega\rho$, $P' = \delta' + \omega\rho'$

$$V(\omega P.KP') = V(e^2\rho + \omega\delta)(\delta' - \omega\rho')$$
$$= -e^2\omega V\rho\rho' + e^2(\rho\delta' - \rho'\delta),$$

or, dividing by e^2, the line joining PP' is

$$\rho\delta' - \delta\rho' - \omega V\rho\rho'.$$

This is a bivector $(\alpha + \omega\alpha')$, and we can verify at once that $S\alpha\alpha' = 0$: therefore $\alpha + \omega\alpha'$ is a line only if $S\alpha\alpha' = 0$, that is $[\alpha + \omega\alpha'] = 0$; moreover it will appear below that if $S\alpha\alpha'$ vanishes, $\alpha + \omega\alpha'$ is always a line.

The conditions that a point may be on a line can be found as follows: Let $\delta + \omega\rho$, $\delta' + \omega\rho'$ be two points on the line $\alpha + \omega\alpha'$: then we must have

$$\rho\delta' - \delta\rho' = \alpha$$
$$V\rho\rho' = -\alpha'.$$

The required conditions are to be found by eliminating (ρ', δ'), the first equation gives

$$\rho' = \frac{\rho\delta' - \alpha}{\delta}.$$

Substituting in the second we get $\frac{-V\rho\alpha}{\delta} = -\alpha'$,

or
$$V\rho\alpha - \delta\alpha' = 0.$$

This gives three conditions: we know that this is one more than we require, but it is convenient to keep them all, and indeed to add a fourth, obtained by operating with $S.\rho$, we thus get as the complete set

$$V\rho\alpha - \delta\alpha' = 0$$
$$S\rho\alpha' = 0.$$

If we operate on the first with $S\alpha$ we get the known condition $S\alpha\alpha' = 0$.

The last condition gives $\rho = V\alpha'\epsilon$,
where ϵ is some vector: substituting in the other we get

$$V\alpha V\alpha'\epsilon + \delta\alpha' = 0$$

or
$$\epsilon S\alpha\alpha' - \alpha'(S\alpha\epsilon - \delta) = 0.$$

This equation is possible, if, and only if $S\alpha\alpha' = 0$, and then it gives $\delta = S\alpha\epsilon$.

This proves that $\alpha + \omega\alpha'$ is a line if $S\alpha\alpha' = 0$, and that any point on the line can be represented by $S\alpha\epsilon + \omega V\alpha'\epsilon$.
Moreover it can be verified at once that the point $0 + \omega\alpha$ is always on the line.

To find the intersection of two concurrent lines, and the condition that two straight lines may intersect, let $\alpha + \omega\alpha'$, $\beta + \omega\beta'$ be the lines: $\delta + \omega\rho$ their intersection: we must have

$$S\rho\alpha' = 0$$
$$S\rho\beta' = 0$$
$$V\rho\alpha - \delta\alpha' = 0$$
$$V\rho\beta - \delta\beta' = 0.$$

The first two give $\rho = xV\alpha'\beta'$, and then

$$xV.\alpha V\alpha'\beta' + \delta\alpha' = 0$$
$$xV.\beta V\alpha'\beta' + \delta\beta' = 0$$

or
$$x(\beta' S\alpha\alpha' - \alpha' S\alpha\beta') + \delta\alpha' = 0$$
$$x(\beta' S\alpha'\beta - \alpha' S\beta\beta') + \delta\beta' = 0.$$

Now $Sa\alpha' = S\beta\beta' = 0$, and therefore these equations are consistent if, and only if

$$S\alpha\beta' + S\alpha'\beta = 0,$$

and then they agree in giving

$$\delta = xS\alpha\beta' = -xS\alpha'\beta.$$

Therefore we can take

$$\omega V\alpha'\beta' + S\alpha\beta'$$

for the intersection of $\alpha + \omega\alpha'$, $\beta + \omega\beta'$, and the condition of intersection is

$$S\alpha\beta' + S\alpha'\beta = 0.$$

It can be verified at once that this is

$$[(\alpha + \omega\alpha')(\beta + \omega\beta')] = 0.$$

Therefore if Q is a bivector

$$[Q] = 0$$

is the condition that it may represent a line

$$[QQ'] = 0$$

is the condition that Q, Q' may intersect.

In exactly the same way the conditions that a plane $\omega\delta + \rho$ may pass through a line are

$$S\rho\alpha = 0$$

$$V\rho\alpha' - \delta\alpha = 0$$

and the plane of two complanar lines is $V\alpha\beta + \omega S\alpha'\beta$.

If $P = \delta + \omega\rho$ is a point, and $\Pi = \rho' + \omega\delta'$ is a plane, P is in Π if

$$\delta\delta' - S\rho\rho' = 0.$$

Now consider the plane

$$\Pi = V\rho\alpha - \delta\alpha' - \omega S\rho\alpha'.$$

We can verify at once that this plane passes through P and that if $\alpha + \omega\alpha'$ is a line Π also passes through the line. That is, if $A = P + \omega\alpha'$ is a line the plane Π is the plane (PA). If A is not a line Π is said to be conjugate to WP with respect to the screw A.

Let $\rho = (\alpha\beta\gamma)$: let $\Pi = (\alpha'\beta'\gamma') + \omega\delta'$: let $\alpha = (fgh)$, $\alpha' = -(abc)$, we get

$$(\alpha'\beta'\gamma'\delta') = \left|\begin{matrix} 0 & h & -g & a \\ -h & 0 & f & b \\ g & -f & 0 & c \\ -a & -b & -c & 0 \end{matrix}\right|(\alpha\beta\gamma\delta).$$

If $\Pi\Pi'$ are two planes, their line of intersection is the axis of the cylindroid (Π, Π'): that is, if $\Pi = \varpi + \omega\varepsilon$, $\Pi' = \varpi' + \omega\varepsilon'$, their intersection is

$$-V\varpi\varpi' + \omega(\varpi\varepsilon' - \varpi'\varepsilon).$$

To show that this agrees with the former definition of a line I prove that if P, P' are points and both are in both of the planes $\Pi\Pi'$, that the line joining PP' is the same as the line of intersection of Π, Π'.

We have if $P=\delta+\omega\rho$, $P'=\delta'+\omega\rho'$

$$S\rho\varpi-\delta\epsilon=0$$
$$S\rho'\varpi-\delta'\epsilon=0$$
$$S\rho\varpi'-\delta\epsilon'=0$$
$$S\rho'\varpi'-\delta'\epsilon'=0.$$

The first two equations give, if we eliminate ϵ

$$0=\delta' S\rho\varpi-\delta S\rho'\varpi$$
$$=S\varpi(\rho\delta'-\rho'\delta).$$

Similarly the third and fourth give

$$0=S\varpi'(\rho\delta'-\rho'\delta).$$

Therefore $$\rho\delta'-\rho'\delta=xV\varpi\varpi',$$

where x is a scalar.

Operating with $V\rho$ we get

$$-\delta V\rho\rho'=xV.\rho V\varpi\varpi'$$
$$=x(\varpi' S\rho\varpi-\varpi S\rho\varpi')$$
$$=\delta x(\varpi'\epsilon-\epsilon'\varpi)$$

or $$V\rho\rho'=x(\varpi\epsilon'-\varpi'\epsilon).$$

Therefore $$\frac{\rho\delta'-\rho'\delta}{V\varpi\varpi'}=\frac{V\rho\rho'}{\varpi\epsilon'-\varpi'\epsilon},$$

which proves the theorem.

It is obvious that $$\mho(Q.\omega Q')=e^2\Omega QQ'$$
$$\Omega(Q.\omega Q')=\mho QQ'.$$

Therefore $$\mho S(Q.K\omega Q')=e^2\Omega S(QKQ')$$
$$\Omega S(Q.K\omega Q')=\mho S(QKQ')$$
$$T(\omega Q)=eTQ.$$

These equations give $$\cos(Q.\omega Q')=\sin e\,[QQ']$$
$$\sin e\,[Q.\omega Q']=\cos(QQ').$$

Now let Q, Q' represent lines: then since $[QQ']=0$ is the condition that QQ' may intersect we can say that Q is at right angles to Q if it meets ωQ: or calling $\omega Q'$ the conjugate of Q' we can say that if one line cuts another at right angles, it cuts that line and its conjugate.

10. *Screws.*

We have seen that a biscalar represents a screw, and that a bivector represents a line: therefore the axis of a screw Q is a line R such that

$$Q=AR$$

where A is a biscalar.

Let Q' be another screw, R' its axis, and let $Q' = A'R'$. Then

$$SQKQ' = AA'SRKR'.$$

Therefore if $SRKR' = 0$, $SQKQ' = 0$, and conversely, unless AA' vanishes: but it can be easily verified that AA' cannot vanish unless QQ' are lines.

Therefore if the axes of two screws cut at right angles we have $(QQ') = \frac{\pi}{2}$, $[QQ'] = 0$, and conversely.

Therefore if we denote the axis of the cylindroid (QQ') by R we see at once (since $S(Q.VQKQ') = 0$) that the axes of all screws of the cylindroid cut the axis of R at right angles.

It is worth while to show that the expression for the axis of a screw agrees with the known formulæ for parabolic space.

We have, if $Q = \alpha + \omega\alpha'$

$$\text{ec}\,\phi R = \left(\text{ec}\frac{\varphi}{2} - \omega\ \text{es}\,\frac{\varphi}{2}\right)(\alpha + \omega\alpha')$$

$$= \left(\alpha.\text{ec}\frac{\varphi}{2} - e^2\alpha'\ \text{es}\,\frac{\varphi}{2}\right) + \omega\left(\alpha'\ \text{ec}\,\frac{\varphi}{2} - \alpha\ \text{es}\frac{\varphi}{2}\right).$$

Now put $e = 0$: we get $\quad \text{ec}\,\frac{\varphi}{2} = 1 : \text{es}\,\frac{\varphi}{2} = \frac{\varphi}{2}$

$$R = \alpha + \omega\left(\alpha' - \alpha\cdot\frac{\varphi}{2}\right).$$

Now $\quad \dfrac{\varphi}{2} = \dfrac{-2S\alpha\alpha'}{T^2\alpha}.$

Therefore $\quad R = \alpha + \omega\left(\alpha' - \dfrac{S\alpha\alpha'}{T^2\alpha}\,\alpha\right),$

which agrees with the quaternion expression I have given in Vol. XII of the Messenger of Mathematics, p. 130, if we write α for α' and $-\beta$ for α.

11. *The Cylindroid.*

The word *cylindroid* is used in two senses: it means either the set $\lambda Q + \mu Q'$, or taking Q, Q' to be bivectors, so that their axes are lines, it means the surface which is the locus of the axes of the screws of the set $\lambda Q + \mu Q'$.

I proceed to find the equation of the cylindroid.

I premise that, as is easily proved, the six edges of the tetrahedron of reference are i, j, k, ωi, ωj, ωk.

The cylindroid contains two screws whose axes intersect at right angles: take these as the screws defining the cylindroid, and their axes as edges of the tetrahedron of reference.

Let the screws be $$A = fi + a\omega i$$ $$B = gj + b\omega j.$$

Let $X = \lambda A + \mu B$ be any screw of the cylindroid: let $Y = xX + y\omega X$ be its axis. Then

$$X = \lambda fi + \mu gj + \omega(\lambda ai + \mu bj)$$
$$Y = \lambda i(fx + e^2 ay) + \mu j(gx + e^2 by) + \omega\{\lambda i(ax + fy) + \mu j(bx + gy)\}.$$

Therefore the coordinates of Y are

$$-\lambda(ax + fy),\ -\mu(bx + gy),\ 0,\ \lambda(xf + e^2 ay),\ -\mu(xg + e^2 by),\ 0.$$

Therefore if the point $(\alpha\beta\gamma\delta)$ is on the axis we must have

$$\text{(i)}\quad \lambda\alpha(xa + yf) + \mu\beta(xb + yg) = 0$$
$$\text{(ii)}\quad \lambda\delta(xa + yf) - \mu\gamma(xg + e^2 by) = 0$$
$$\text{(iii)}\quad \lambda\gamma(xf + e^2 ay) + \mu\delta(xb + yg) = 0$$
$$\text{(iv)}\quad \lambda\beta(xf + e^2 ay) - \mu\alpha(xg + e^2 by) = 0.$$

We have to eliminate λ, μ, x, y.

Eliminating λ, μ, first between (i) (ii) and then between (i) (iii) we get, after a rearrangement of the terms,

$$x(ga\gamma + b\beta\delta) + y(e^2 ba\gamma + g\beta\delta) = 0$$
$$x(aa\delta - f\beta\gamma) + y(fa\delta - e^2 a\beta\gamma) = 0.$$

Eliminating x, y we get the equation of the cylindroid in the form

$$0 = (ga\gamma + b\beta\delta)(fa\delta - e^2 a\beta\gamma) - (aa\delta - f\beta\gamma)(e^2 ba\gamma + g\beta\delta)$$
$$= (\alpha^2 + \beta^2)\gamma\delta(fg - e^2 ab) - \alpha\beta(e^2\gamma^2 + \delta^2)(ag - bf).$$

Now let ϕ, ϕ' be the pitches of the screws

$$\sin e\phi = \frac{2eaf}{a^2 e^2 + f^2}$$
$$\sin\frac{e\varphi}{2} = \frac{ea}{\sqrt{(e^2 a^2 + f^2)}}$$
$$\cos\frac{e\varphi}{2} = \frac{f}{\sqrt{(e^2 a^2 + f^2)}}.$$

Therefore the equation of the cylindroid is

$$\text{ec}\,\tfrac{1}{2}(\phi + \phi')(\alpha^2 + \beta^2)\gamma\delta - \text{es}\,\tfrac{1}{2}(\phi - \phi')(e^2\gamma^2 + \delta^2)\alpha\beta = 0.$$

For parabolic space this is

$$(\alpha^2 + \beta^2)\gamma\delta - \frac{\varphi - \varphi'}{2}\delta^2\alpha\beta = 0.$$

Thus there is a factor δ, and putting $\delta = 1$ in the other factor we get

$$\gamma(\alpha^2 + \beta^2) - \frac{\varphi - \varphi'}{2}\alpha\beta = 0.$$

But it is obvious from the expression for ϕ in the last section that ϕ is double what is generally called the pitch, therefore this last equation agrees with the known form for parabolic space.

Writing

$$\alpha = \beta \tan \vartheta$$
$$\gamma = \delta \operatorname{et} \psi$$
$$\left(\operatorname{et} \psi = \frac{\tan e\psi}{e}\right)$$

we get from the equation of the cylindroid

$$\operatorname{es} 2\psi = \frac{\operatorname{es} \frac{1}{2}(\varphi - \varphi')}{\operatorname{ec} \frac{1}{2}(\varphi + \varphi')} \sin 2\vartheta.$$

Now we have $$X = \lambda A + \mu B.$$

Therefore $$SAKX = \lambda NA.$$
$$NX = \lambda^2 NA. + \mu^2 NB.$$
$$SBKX = \mu NB.$$

Therefore $$\cos(AX) = \frac{\lambda}{\sqrt{\lambda^2 + \mu^2}}$$
$$\sin(AX) = \frac{\mu}{\sqrt{\lambda^2 + \mu^2}}.$$

Let $(AX) = l$. Then $\operatorname{es}[X] = \cos^2 l \operatorname{es}[A] + \sin^2 l \operatorname{es}[B]$.

It can be shown that we have

$$\operatorname{et} 2\psi = \frac{\sin 2l . \operatorname{es} \frac{1}{2}(\varphi - \varphi')}{\cos^2 l \operatorname{ec} \varphi + \sin^2 l \operatorname{ec} \varphi'}.$$

It is hardly worth while to stop to point out how these formulæ agree with the known formulæ of the theory of screws.

Prof. Ball has* shown that the cylindroid can be represented in a plane by means of a circle: the relation he uses is

$$(p - p_0)^2 + z^2 = m^2,$$

where p is the pitch of any screw of a cylindroid: z is the intercept its axis makes on the axis of the cylindroid, and, P_α, P_β being the pitches of the two screws of reference,

$$P_0 = \tfrac{1}{2}(P_\alpha + P_\beta); \; m = \tfrac{1}{2}(P_\alpha - P_\beta).$$

The general formula is, if χ is the pitch of any screw of the cylindroid, and ϕ, ϕ', ψ have the same meanings as before

$$4(\operatorname{es}\phi - \operatorname{es}\chi)(\operatorname{es}\chi - \operatorname{es}\phi') \operatorname{es}^2 \frac{\varphi - \varphi'}{2}$$
$$= \operatorname{et}^2 2\psi \{\operatorname{cc}\phi(\operatorname{es}\phi - \operatorname{es}\chi) + \operatorname{ec}\phi'(\operatorname{es}\chi - \operatorname{es}\phi)\}^2.$$

* Ball. On a plane dynamical representation of certain dynamical problems in the theory of a rigid body. 4 Proc. R. I. A., 2nd S. 29.

This equation is got by combining the value of et 2ψ with the expression for the pitch of any screw in terms of l: it is not hard to show that this reduces to Prof. Ball's equation if we take $e = 0$.

12. *Vectors.*

Clifford says that the vector of either species answers to Hamilton's vector: it is worth while to see how this comes about.

Let $\alpha + \omega\alpha'$ be a bivector, and suppose it is to be an η-vector: we must have if $\alpha = (fgh)$, $\alpha' = (abc)$

$$fi + gj + hk + \omega(ai + bj + ck) = e(ai + bj + ck) + \omega e^{-1}(fi + gj + ck).$$

Therefore
$$f = ae$$
$$g = be$$
$$h = ce.$$

Therefore in parabolic space, that is for $e = 0$, we get $f = g = h = 0$: that is, in parabolic space the vector is of the form $\omega\alpha$, which is a more precise form of Clifford's statement.

13. *Parallels.*

In this section we shall have to make more explicit use of the absolute than in what precedes, and the methods used will be mixed: partly biquaternions, partly geometricals, and partly algebraic.

We must bear in mind the stipulations in (7), especially the third.

Any straight line meets four generators of the absolute: for it meets the absolute in two points, and there are two generators through each point.

A generator is either a ξ-vector or an η-vector (7); call these two systems ξ-generators and η-generators respectively. Two lines are parallel if they meet the same two generators of the absolute: there are two cases: the generators may be of the same species or they may be of different species. In the former case the lines are said to be α-parallel: in the latter case they are said to be β-parallel. The conception of α-parallel lines is due to Clifford and Lindemann:* β-parallels are what is generally known as parallels. I proceed to find the conditions for parallelism.

The coordinates of any ξ- or η-vector are of the form

$$(abc \pm ae, \pm be, \pm ce),$$

* Clifford. Preliminary Sketch. Lindemann, 7 Math. Ann.

Moreover, if the vector is a line, we must have

$$af + bg + ch = 0,$$

and this gives $$a^2 + b^2 + c^2 = 0.$$

We can therefore take $a : b : c = i : \cos\vartheta : \sin\vartheta$,

and the coordinates of any generator will be

$$(i, \cos\vartheta, \sin\vartheta, \pm ei, \pm e\cos\vartheta, \pm e\sin\vartheta).$$

Now we must stop to consider the distinction into species.

If $\alpha + \omega\alpha'$ is an η-vector we have

$$\begin{aligned}\alpha + \omega\alpha' &= e^{-1}\omega(\alpha + \omega\alpha') \\ &= e\alpha' + e^{-1}\omega\alpha.\end{aligned}$$

That is $\alpha = e\alpha'$: now if $(abcfgh)$ are the coordinates of the vector we have $\alpha = (fgh)$, $\alpha' = -(abc)$: therefore for an η-vector we have $f = -ea$, etc.: therefore the coordinates of an η-vector are

$$(abc - ea - eb, - ec)$$

and the coordinates of a ξ-vector are

$$(abc\ ea\ eb\ ec).$$

Moreover

$$2\,{}^{\xi}_{\eta}(\alpha + \omega\alpha') = (f \mp ea, g \mp eb, h \mp ec, -a \pm e^{-1}f, -b \pm e^{-1}g, -c \pm e^{-1}h).$$

Now let $(abcfgh)$ be any line: if it is to meet the ${}^{\xi}_{\eta}$ generator ϑ, we must have

$$\begin{aligned}0 &= fi + g\cos\vartheta + h\sin\vartheta \pm eai \pm eb\cos\vartheta \pm ec\sin\vartheta \\ &= i(f \pm ea) + (g \pm eb)\cos\vartheta + (h \pm ec)\sin\vartheta.\end{aligned}$$

That is the line $A = (abcfgh)$ meets the ξ-generators, for which ϑ has the values determined by $i(f + ae) + \cos\vartheta(g + eb) + \sin\vartheta(h + ec) = 0$.

Similarly $A' = (a'b'c'f'g'h')$ meets the ξ-generators, for which ϑ has the values determined by $i(f' + a'e) + \cos\vartheta(g' + eb') + \sin\vartheta(h' + ec') = 0$.

Therefore if A, A' are to meet the same two ξ-generators these two equations must coincide, and we get

$$\frac{ae + f}{a'e + f'} = \frac{be + g}{b'e + g'} = \frac{ce + h}{c'e + h'} = \lambda.$$

That is $$\eta A = \lambda\eta A'.$$

That is A, A' are η-parallel.

In the same way if A, A' meet the same two η-generators they are ξ-parallel.

It follows, from what was proved in (5), that parallel lines meet the same generators: we have now proved the converse of this.

What we have proved is that α-parallelism, as now defined, is the same as what we called parallelism in (5).

I now consider the condition for β-parallelism: this is known to be that the lines intersect and that their angle vanishes: I show how this comes out from the definition given above, viz. that the two lines meet the same ξ-generator and also the same η-generator. Take the equation determining the ξ-generator cut by a line, viz. $i(f+ae)+\cos\vartheta\,(g+eb)+\sin\vartheta\,(h+ec)=0$.

Write

$$2\cos\vartheta = x+\frac{1}{x}$$

$$2\sin\vartheta = \frac{1}{i}\left(x-\frac{1}{x}\right)$$

we get $$-2x(f+ae)+i(x^2+1)(g+eb)+(h+ec)(x^2-1)=0,$$

or $$x^2\{i(g+eb)+(h+ec)\}-2x(f+ae)+\{i(g+eb)-(h+ec)\}=0.$$

Say this is $$Ax^2 \qquad + 2Bx \qquad + C \qquad =0.$$

In the same way the line $(a'b'c'f'g'h')$ gives an equation

$$A'x^2 \qquad + 2B'x \qquad + C' \qquad =0.$$

If the lines are β-parallel the resultant of these equations must vanish: that is we must have

$$(AC'+A'C-2BB')^2=4(AC-B^2)(A'C'-B'^2).$$

We must remember that

$$0=af+bg+ch=a'f'+b'g'+c'h'.$$

We have $$\begin{aligned} AC-B^2 &= \{i(g+eb)+(h+ec)\}\{i(g+eb)-(h+ec)\}-(f+ae)^2 \\ &= -(g+eb)^2-(h+ec)^2-(f+ae)^2 \\ &= -(f^2+g^2+h^2+a^2e^2+b^2e^2+c^2e^2) \\ &= -X^2, \text{ say.}\end{aligned}$$

Similarly $$\begin{aligned} A'C'-B'^2 &= -(f'^2+g'^2+h'^2+a'^2e^2+b'^2e^2+c'^2e^2) \\ &= -X'^2.\end{aligned}$$

$$\begin{aligned} AC'+A'C-2BB' &= \{i(g+eb)+(h+ec)\}\{i(g'+eb')-(h'+ec')\} \\ &\quad +\{i(g+eb)-(h+ec)\}\{i(g'+eb')+(h'+ec')\}-2(f+ae)(f'+a'e) \\ &= -2(g+eb)(g'+eb')+2(h+ec)(h'+ec') \\ &\qquad -2(f+ae)(f'+a'e) \\ &= -2(ff'+gg'+hh'+aa'e^2+bb'e^2+cc'e^2) \\ &\quad -2e(ch'+c'h+bg'+b'g+af'+a'f) \\ &= -2Y-2eZ, \text{ say.}\end{aligned}$$

Therefore we must have $$(Y+eZ)^2=X^2X'^2$$

or $$Y+eZ=XX'.$$

But the two lines must also meet the same η-generator: this gives another equation which is obviously got by writing $-e$ for e: therefore we must have

$$Y + eZ = XX'$$
$$Y - eZ = XX'.$$

Therefore we must have

$$Z = 0$$
$$Y = XX'.$$

But $Z = 0$ gives $af' + a'f + bg' + b'g + ch' + c'h = 0$,
that is to say, $Z = 0$ is the condition that the lines may intersect: and $Y = XX'$ gives

$$\frac{1 = ff' + gg' + hh' + e^2(aa' + bb' + cc')}{(f^2 + g^2 + h^2 + e^2a^2 + e^2b^2 + e^2c^2)^{\frac{1}{2}}.(f'^2 + g'^2 + h'^2 + e^2a'^2 + e^2b'^2 + e^2c'^2)^{\frac{1}{2}}} = \cos\phi,$$

if ϕ is the angle between the lines.

Therefore if two lines are β-parallel they intersect, and the angle between them vanishes. It is worth noticing that in parabolic space the condition $Z = 0$ disappears, since both resultants only give

$$Y = XX'.$$

Moreover, the condition for α-parallelism is

$$\frac{f}{f'} = \frac{g}{g'} = \frac{h}{h'},$$

which gives $\cos\phi = 1$: so that the conditions for the two species of parallelism coincide, as of course they should do.

The investigations which follow are mostly developments of (4) on p. 192 of Clifford's papers, and of Note (i) on p. 642.

Through any point we can draw two lines parallel to a given line, viz. they are the two lines drawn through the point, one cutting the two η-generators cut by the line, and the others cutting the two ξ-generators cut by the line. To find the locus of parallels to a given line drawn through the points of a given line. Let the given lines be A, B: suppose the lines are to be drawn η-parallel to A: then they must meet the two ξ-generators cut by A: but they also meet B: therefore they are one system of generators of a quadric and B and the two ξ-generators are three lines of the other system.

Now B meets two η-generators: these also cut the two ξ-generators: therefore the quadric contains two ξ-generators and two η-generators: all its generators of the one system cut B and the two ξ-generators, and all its generators of the other system cut A and the two η-generators.

Therefore one system of generators of the new quadric (call it the quadric Σ) consists of lines through B, η-parallel to A, and the other system consists of lines through A, ξ-parallel to B.

Let ξ_1, ξ_2, η_1, η_2 be the two ξ-generators, and the two η-generators of Σ: let A, A' be two generators of Σ, of the same system as ξ_1, ξ_2: let B, B' be two

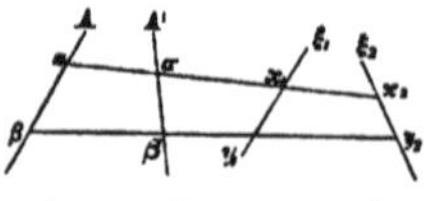

other generators, cutting $AA'\xi_1\xi_2$ in α, α', x_1, x_2, β, β', y_1, y_2, respectively: it need hardly be remarked that the figure is not supposed to represent the actual state of things, but only to show on what lines the points are supposed to lie.

We have by the fundamental properties of a ruled quadric

$$\{\alpha\alpha' x_1 x_2\} = \{\beta\beta' y_1 y_2\}$$

if the $\{\}$ denote anharmonic ratios: but x_1, x_2, y_1, y_2 are the points in which $\alpha\alpha'$, $\beta\beta'$ cut the absolute. Therefore, remembering the anharmonic ratio definition of distance, we get $\qquad \alpha\alpha' = \beta\beta'.$

That is to say, any two parallels of one system cut off equal intercepts on two fixed parallels of the other system.

Now consider the equation proved in (5), viz.

$$\frac{\cos(Qu) + \sin e[Qu]}{(1 + \sin e[Q])^{\frac{1}{2}}.(1 + \sin e[u])^{\frac{1}{2}}} = \frac{\cos(RW) + \sin e[RW]}{(1 + \sin e[R])^{\frac{1}{2}}.(1 + \sin e[W])^{\frac{1}{2}}}.$$

Let $UVWR$ be lines, and let Q, U intersect, and also R, W: then we have

$$0 = [U] = [V] = [W] = [R] = [QU] = [RW]$$

and we get $\qquad \cos(QU) = \cos(RW).$

Then we can take $U = W$, and we get

$$\cos(QU) = \cos(RU).$$

That is to say, if two lines intersect, they make the same angle as any two parallels to them that intersect, and in particular, any line meeting two parallels makes equal angles with them. We see from all this that the geometry of the surface Σ is the same as the geometry of a parallelogram: for in a parallelogram we have two parallelisms, and every line of one parallelism makes equal angles with every line of the other parallelism, and two fixed lines of one parallelism make a constant intercept on all lines of the other parallelism: and all this is true for the generators of the surface Σ.

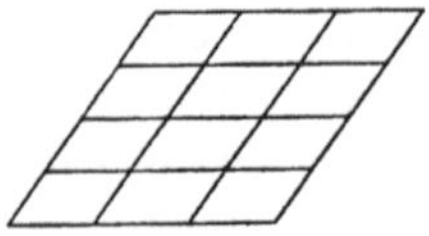

I shall now find the equation to the surface Σ, referred to the rectangular system used in this paper.

Let the line through which the parallels are to be drawn be the intersection of the planes $lx + my + nz + p\omega = 0$, $l'x + m'y + n'z + p'\omega = 0$.

Let $x'y'z'\omega'$ be a point on this line: $(xyz\omega)$ a point on the required locus: then, since the line joining the two points is parallel to a fixed line, and since $(x'y'z'\omega')$ is on the given line, we have a set of equations

$$\begin{aligned} x'\omega + ey'z - ez'y - \omega'x &= \lambda\alpha \\ -ex'z + y'\omega + ez'x - \omega'y &= \lambda\beta \\ ex'y - ey'x + z'\omega \quad - \omega'z &= \lambda\gamma \\ lx' + my' + nz' \quad + p\omega' &= 0 \\ l'x' + m'y' + n'z' \quad + p'\omega' &= 0. \end{aligned}$$

Eliminating λ, x', y', z', ω' and writing $abcfgh$ for the coordinates of the intersection of the two planes we get

$$\begin{vmatrix} \omega & ez & -ey & -x\alpha \\ -ez & \omega & ex & -y\beta \\ ey & -ex & \omega & -z\gamma \\ l & m & n & p0 \\ l' & m' & n' & p'0 \end{vmatrix} = 0.$$

or $\alpha\left[a(e^2x^2 + \omega^2) + ef(y^2 + z^2) + (ec - h)(ezx - y\omega) + (eb - g)(exy + z\omega)\right]$
$+ \beta\left[b(e^2y^2 + \omega^2) + eg(z^2 + x^2) + (ea - f)(exy - z\omega) + (ec - h)(eyz + x\omega)\right]$
$+ \gamma\left[c(e^2z^2 + \omega^2) + eh(x^2 + y^2) + (eb - g)(eyz - x\omega) + (ea - f)(ezx + y\omega)\right]$
$= 0.$

It is worth while to see how this quadric reduces to a plane in parabolic space: we have to put $e = 0$, and then there is a factor ω, and putting $\omega = 1$ in the remaining factor, we get

$$0 = \alpha(a + hy - gz) + \beta(b + fz - hx) + \gamma(c + gx - fy).$$

In this form we see that the plane contains the given line $(abcfgh)$, and writing it in the form

$$0 = \alpha a + \beta b + \gamma c + x(g\gamma - h\beta) + y(h\alpha - f\gamma) + z(f\beta - g\alpha)$$

we see that it is parallel to $(\alpha\beta\gamma)$, as, of course, it should be.

I now change the axes of reference, and take

$$S = x\omega - yz = 0$$

as the equation of the absolute, and

$$\Sigma = x\omega - \lambda yz = 0$$

as the equation of the surface Σ: so that the four common generators are

$$(xy),\ (xz),\ (\omega y),\ (\omega z).$$

The polar of $(x''y''z''\omega'')$ with respect to S is

$$x\omega'' - yz'' - zy'' + \omega x'' = 0.$$

Therefore the pole of $(lmnp)$ is $(p, -n, -m, l)$.
The tangent plane to Σ at $(x'y'z'\omega')$ is

$$x\omega' - \lambda yz' - \lambda zy' + \omega x'.$$

Therefore its pole with respect to S is $(x', \lambda y', \lambda z', \omega')$, and the line joining this pole to $(x'y'z'\omega')$, that is the normal at $(x'y'z'\omega')$ has for its coordinates the minors of

$$\left\| \begin{matrix} x' \; y' \; z' \; \omega' \\ x' \; \lambda y' \; \lambda z' \; \omega' \end{matrix} \right\|,$$

or $$(0, xz(1-\lambda), xy(\lambda-1), 0, y\omega(1-\lambda), z\omega(1-\lambda)),$$
or dividing by $(1-\lambda)$ $$(0, xz, -xy, 0, y\omega, z\omega).$$
Therefore all the normals meet yz and $x\omega$; therefore through any point not on the surface we can only draw one normal (instead of *six*) to the surface: this normal is the line through the point cutting yz, $x\omega$.

The coordinates of the generators of Σ can be easily proved to be of the form $$\rho, -\rho^2, 0, \lambda\rho, \lambda, 0)$$
for the one system: call this the η-system. And

$$(-\rho, 0, \rho^2, \lambda\rho, 0, \lambda)$$

for the other system: call this the ξ-system.

For the generators of S we have, therefore,

$$(\rho, -\rho^2, 0, \rho, 1, 0)$$

and $$(-\rho, 0, \rho^2, \rho, 0, 1).$$
Therefore a line meets the η-generators determined by

$$f\rho - g\rho^2 + a\rho + b = 0,$$

or $$g\rho^2 - (a+f)\rho - b = 0.$$
Therefore two lines are ξ-parallel if

$$\frac{a+f}{a'+f'} = \frac{b}{b'} = \frac{g}{g'}.$$

In the same way two lines are η-parallel if

$$\frac{a-f}{a'-f'} = \frac{c}{c'} = \frac{h}{h'}.$$

The normal at a point $(xyz\omega)$ of Σ was found to be

$$(0, xz, -xy, 0, y\omega, z\omega).$$

Therefore the normals at $(xyz\omega)$, $(x'y'z'\omega')$ are ξ-parallel if

$$\frac{xz}{x'z'} = \frac{y\omega}{y'\omega'}. \tag{A}$$

They are η-parallel if $$\frac{xy}{x'y'} = \frac{z\omega}{z'\omega'}. \tag{B}$$

Now take equation (A): this asserts that the normals at all points of the intersection of (Σ) $x\omega - \lambda yz$ with $y\omega - \mu zx$ are parallel: and in the same way the normals at all points of the intersection of $x\omega - \lambda yz$ with $z\omega - \nu xy$ are parallel: now I say that all these intersections are straight lines, and moreover that the systems ($x\omega - \lambda yz$, $y\omega - \mu zx$) are orthogonal, as also the systems ($x\omega - \lambda yz$, $z\omega - \nu xy$), and that at every point of $x\omega - \lambda yz$ the other two surfaces cut at a constant angle.

Consider the surface $y\omega - \mu zx = 0$: we have

$$\mu = \frac{y\omega}{zx}.$$

But we have also at any point of Σ, $x\omega - \lambda yz$, $= 0$, or

$$\frac{\omega}{z} = \lambda \frac{y}{x}.$$

Therefore at any point of the intersection of the two surfaces we have

$$\mu = \lambda \frac{y^2}{x^2}.$$

Therefore the surface $\mu =$ const. cuts the surface $\lambda =$ const. in two generators determined by

$$\frac{y}{x} = \pm \sqrt{\frac{\mu}{\lambda}}.$$

Moreover the surfaces have the lines (xy), $(z\omega)$ in common: therefore $\mu =$ const. cuts $\lambda =$ const. in four straight lines: in the same way we can show that $\nu =$ const. cuts $\lambda =$ const. in four straight lines, two being (zx), $(y\omega)$, and the other two given by

$$\frac{z}{x} = \pm \sqrt{\frac{\nu}{\lambda}}.$$

Now to show that $\mu =$ const., $\lambda =$ const. are orthogonal. The tangent planes at $x'y'z'\omega'$ are

$$x\omega' - \lambda yz' - \lambda zy' + \omega x' = 0$$

$$- \mu xz' + y\omega' - \mu zx' + \omega y' = 0.$$

Now the plane equation of the absolute is $lp - mn = 0$: therefore two planes $(lmnp)(l'm'n'p')$ are at right angles if

$$lp' + l'p - mn' - m'n = 0.$$

In the present case this is

$$\begin{aligned} 0 &= \omega' y' - \mu x'z' - \lambda\mu x'z' + \lambda y'\omega' \\ &= (\omega' y' - \mu x'z')(1 + \lambda), \end{aligned}$$

which is right, since $\omega' y' - \mu x'z' = 0$.

We can say more generally, that if we take any point on $\mu =$ const. its tangent plane is at right angles to its polar with respect to $\lambda =$ const.

The same is obviously true for the surface $\nu =$ const.

Now consider the two surfaces

$$y\omega - \mu zx = 0$$
$$z\omega - \nu xy = 0.$$

Their tangent planes are

$$-\mu xz' + y\omega' - \mu zx' + \omega y' = 0$$
$$-\nu xy' - \nu yx' + z\omega' + \omega z' = 0.$$

Therefore the angle between them is given by

$$\cos\vartheta = \frac{1}{2}\cdot\frac{-\mu z'^2 - \nu y'^2 - \omega'^2 - \mu\nu x'^2}{\sqrt{(\mu x'\omega' - \mu y'z')(\nu x'\omega' - \nu y'z')}}$$
$$= -\frac{1}{2\sqrt{\mu\nu}}\cdot\frac{\mu\nu x'^2 + \nu y'^2 + \mu z'^2 + \omega'^2}{(x'\omega' - y'z')}.$$

Now, leaving out accents, we have

$$\mu = \frac{y\omega}{xz}$$
$$\nu = \frac{z\omega}{xy},$$

and therefore

$$\mu\nu = \frac{\omega^2}{x^2}.$$

Therefore the numerator of $\cos\vartheta$ becomes

$$\omega^2 + \frac{yz\omega}{x} + \frac{yz\omega}{x} + \omega^2$$
$$= \frac{2\omega}{x}(x\omega + yz)$$
$$= 2\sqrt{\mu\nu}\,(x\omega + yz).$$

Therefore

$$\cos\vartheta = \frac{yz + x\omega}{yz - x\omega}$$
$$= \frac{1+\lambda}{1-\lambda},$$

if $x\omega = \lambda yz$ is the surface of the λ-system passing through the point: therefore all we can say is that for all the intersections of three surfaces, $\lambda =$ const., $\mu =$ const., $\nu =$ const., the tangent planes to the last two cut at the same angle.

There is another way of treating the theory of parallels, which should be noticed.

If we take any two lines, Q, R, there are in general two lines meeting them both at right angles: for it is obvious from what was proved at the end of (9) that these are the four lines meeting Q, R, ωQ, ωR: it may, however, happen that there is an infinite number of such lines: if this is the case we have Q, R, ωQ, ωR connected by a linear relation of the form

$$aQ + b\omega Q + cR + d\omega R = 0.$$

And then it follows by what was proved in (5) that Q, R are parallel.

14. *Biquaternions.*

If Q is any biquaternion its reciprocal is defined by

$$Q^{-1} = \frac{KQ}{NQ},$$

where $\frac{1}{NQ} = (NQ)^{-1}$ as defined at the end of (1). We then get

$$\frac{\alpha|}{|\beta} = \alpha\beta^{-1},$$

so that α/β is a biquaternion.

Now taking α, β as two screws we have to see what the actual operation is which changes β into α: but this can obviously be done as follows: take the shortest distances of the axes of α, β, move an axis of β along them and make it coincide with an axis of α: then alter $N\beta$ until it is equal to $N\alpha$.

And this operation is a biquaternion: and we get an equation

$$(q + \omega q')(\alpha + \omega\alpha') = \beta + \omega\beta':$$

and we have

$$S.V(q + \omega q')(\alpha + \omega\alpha') = 0.$$

We can say that a biquaternion Q can operate upon the bivector α if $S\alpha VQ = 0$: that is, if the axes of α cut the axes of Q at right angles. In this way, I think, we can get an explanation of a difficulty noticed by Clifford (p. 179).

Clifford gets the equation

$$(q + \omega r)(\alpha + \omega\beta) = \gamma + \omega\delta.$$

Then the difficulty is that the expression "$q + \omega r$ does not denote the sum of geometrical operations, which can be applied to the motor as a whole; and the ratio of two motors is only expressed by a symbol as the sum of two parts, each of which separately has a definite meaning in certain other cases, but not in the case in point."

Now I submit that we have no more right to break $npq + \omega r$ into its components q, r and expect each to operate upon $\alpha + \omega\beta$, than we should have to take the equation $q\alpha = \gamma$, to write q in the form

$$a + xi + yj + zk,$$

and to expect each of the three parts a, xi, yj, zk to operate separately upon α.

The cases are entirely analogous: a biquaternion $q+\omega r$, considered as an operator, is one and indivisible in exactly the same way as the quaternion $a+xi+yj+zk$ is one and indivisible when considered as an operator.

Clifford says that the difficulty does not occur in elliptic space: this is only because we can evade it by writing the biquaternion in the form $\xi q+\eta r$, and then we have $(\xi q+\eta r)(\xi\alpha+\eta\beta)=\xi(q\alpha)+\eta(r\beta)$ and both quaternions can operate if α, β are perpendicular to the axes of q, r respectively. If we say that two bivectors A, B are at right angles if $SAB=0$, and if we call VQ (a bivector) for the moment the axis of RQ we can say that a biquaternion can operate upon any bivector at right angles to its axis, and changes it into another bivector at right angles to its axis.

In this sense, the axis of α/β is the axis of the cylindroid $(\alpha\beta)$, and just as in bivectors the cylindroid takes the place of the plane, so the axis of the cylindroid takes the place of the normal to the plane.

THE GRAMMAR SCHOOL, MANCHESTER, *Sept.* 22, 1884.

On the Syzygies of the Binary Sextic and their Relations.

BY J. HAMMOND.

(I). *The Irreducible Syzygies of the Binary Sextic, as far as the eighth degree, with the Linear Relations connecting Compound Syzygies, as far as the ninth degree.*

1. Following Prof. Cayley's notation, "Tables for the Binary Sextic" (*American Journal of Mathematics*, Vol. IV, p. 380), the capital letters A, B, ... Z are used to denote the 26 covariants. These are identical with those given by Prof. Cayley, with two exceptions, viz. J′ and R′, which are connected with them by the relations $J' = BC - J$, $8R' = BK - 6Q + R$.

The syzygies are as follows:

Deg. Order.	
(5.16)	$AL + CH - DG = 0$
(6.8)	$AM - 2ABE + B^2D - BC^2 - 3CJ' - 36DI + 36EF = 0$
(6.10)	$AN - CK + EG = 0$
(6.12)	$3A^2I - 3ABF + 2ACE + BCD - C^3 - 3DJ' + 27F^2 = 0$
(6.14)	$AO - 2CL + 3FG = 0$
(6.14)	$AO - DK + EH = 0$
(6.16)	$A^2J' - A^2BC + 12ACF + 4C^2D + G^2 = 0$
(6.18)	$A^2K - ABH + ACG - 12DL + 18FH = 0$
(6.20)	$A^3E - A^2BD - 2A^2C^2 + 18ADF + 12CD^2 + 3GH = 0$
(6.24)	$A^3F - A^2CD + 4D^3 + H^2 = 0$
(7.6)	$9AP + 12ABI - 3B^2F + 4BCE - CM - 12EJ' + 108FI = 0$
(7.10)	$12ACI + AE^2 - 3BCF + 4C^2E - DM - 9FJ' = 0$
(7.12)	$AQ + 2EL - 3FK = 0$
(7.12)	$AR' - GJ' - 2CO = 0$
(7.12)	$CO - DN - EL = 0$
(7.12)	$ABK - B^2H + BCG - 6CO - 12EL - 18FK + 36HI = 0$
(7.14)	$A^2M - A^2BE - 6ACJ' + 18AEF + 3GK + 12CDE = 0$
(7.16)	$ACK - BCH + C^2G - 6DO - 18FL = 0$
(7.16)	$2ACK + AEG - ABL - 6DO - 3HJ' = 0$
(7.18)	$A^2CE + 2ABCD - 2AC^3 - 3ADJ' - 18CDF - 3GL = 0$

Deg. Order.	
(7.18)	$2A^3I - A^2BF + A^2CE - 2ADJ' + 4D^2E + HK = 0$
(7.22)	$A^2DE - 3A^2CF - ABD^2 + AC^2D + 18D^2F + 3HL = 0$
(8.8)	$AS + 4C^2I - FM - J'^2 = 0$
(8.8)	$4AEI - 4BDI + 4C^2I - 2BEF + 3CE^2 - FM - 3DP = 0$
(8.10)	$AT - 2EO - J'K = 0$
(8.10)	$CQ + EO - 3FN = 0$
(8.10)	$3BCK - BEG - 6CR' + 6EO - 36FN + GM - 3J'K = 0$
(8.10)	$2B^2L - BCK - 3BEG - 18EO + GM - 72IL + 3J'K = 0$
(8.12)	$3A^2P + 4A^2BI - 4AEJ' + 4DE^2 + K^2 = 0$
(8.12)	$2ABCE + ACM - 3AEJ' - 6C^2J' - 18CEF + 3GN = 0$
(8.14)	$ABO - ACN + 3AGI - 9FO - 3J'L = 0$
(8.14)	$AEK - BEH + CEG + 6DQ - 18FO = 0$
(8.14)	$BCL - 2C^2K - CEG + 3DR' - 3J'L = 0$
(8.14)	$ABO + 2ACN - AEK + 2BCL - 4C^2K - 18FO + HM = 0$
(8.16)	$GO - HN - KL = 0$
(8.16)	$4A^2CI - ABCF + 2AC^2E - 3AFJ' - 2CDJ' + GO = 0$
(8.16)	$6A^2CI - 3ABCF + ABDE + 3AC^2E - ADM - 18DEF - 3KL = 0$
(8.20)	$A^2DI - ABDF + AC^2F + 9DF^2 + L^2 = 0$
(8.20)	$AGK - BGH + CG^2 - 6HO - 12L^2 = 0.$

2. By a Compound Syzygy is meant either a simply divisible syzygy, *i. e.* one which is divisible by a power or product of the covariants A, B, ... Z, or else a linear function of simply divisible syzygies. The consideration of the simply divisible syzygies will serve the purpose of the present article.

The linear relations connecting them are as follows:

Deg. Order.

(8.18) $A(CO - DN - EL) - C(AO - DK + EH) + D(AN - CK + EG)$
$+ E(AL + CH - DG) = 0$

(8.22) $A(ACK - BCH + C^2G - 6DO - 18FL) - C(A^2K - ABH + ACG - 12DL + 18FH)$
$+ 6D(AO - 2CL + 3FG) + 18F(AL + CH - DG) = 0$

(8.24) $A(A^2CE + 2ABCD - 2AC^3 - 3ADJ' - 18CDF - 3GL)$
$- C(A^3E - A^2BD - 2A^2C^2 + 18ADF + 12CD^2 + 3GH)$
$+ 3D(A^2J' - A^2BC + 12ACF + 4C^2D + G^2) + 3G(AL + CH - DG) = 0$

(8.28) $A(A^2DE - 3A^2CF - ABD^2 + AC^2D + 18D^2F + 3HL) + 3C(A^3F - A^2CD + 4D^3 + H^2)$
$- D(A^3E - A^2BD - 2A^2C^2 + 18ADF + 12CD^2 + 3GH) - 3H(AL + CH - DG) = 0$

(9.14) $3A(4AEI - 4BDI + 4C^2I - 2BEF + 3CE^2 - FM - 3DP)$
$- C(12ACI + AE^2 - 3BCF + 4C^2E - DM - 9FJ')$
$+ D(9AP + 12ABI - 3B^2F + 4BCE - CM - 12EJ' + 108FI)$
$- 4E(3A^2I - 3ABF + 2ACE + BCD - C^3 - 3DJ' + 27F^2)$
$+ 3F(AM - 2ABE + B^2D - BC^2 - 3CJ' - 36DI + 36EF) = 0$

Deg. Order.

(9.16) $A(CQ+EO-3FN)-C(AQ+2EL-3FK)-E(AO-2CL+3FG)$
$+3F(AN-CK+EG)=0$

(9.16) $A(3BCK-BEG-6CR'+6EO-36FN+GM-3J'K)$
$+A(2B^2L-BCK-3BEG-18EO+GM-72IL+3J'K)$
$-2B^2(AL+CH-DG)+6C(AR'-GJ'-2CO)$
$-2C(ABK-B^2H+BCG-6CO-12EL-18FK+36HI)$
$+12E(AO-2CL+3FG)+36F(AN-CK+EG)$
$-2G(AM-2ABE+B^2D-BC^2-3CJ'-36DI+36EF)$
$+72I(AL+CH-DG)=0$

(9.18) $A(2ABCE+ACM-3AEJ'-6C^2J'-18CEF+3GN)$
$-C(A^2M-A^2BE-6ACJ'+18AEF+12CDE+3GK)$
$+3E(A^2J'-A^2BC+12ACF+4C^2D+G^2)-3G(AN-CK+EG)=0$

(9.20) $A(AEK-BEH+CEG+6DQ-18FO)-6D(AQ+2EL-3FK)$
$-E(A^2K-ABH+ACG-12DL+18FH)+18F(AO-DK+EH)=0$

(9.20) $A(BCL-2C^2K-CEG+3DR'-3J'L)+C(2ACK+AEG-ABL-6DO-3HJ')$
$-3D(AR'-GJ'-2CO)+3J'(AL+CH-DG)=0$

(9.20) $A(ABO-ACN+3AGI-9FO-3J'L)-AB(AO-2CL+3FG)$
$+AC(AN-CK+EG)-BC(AL+CH-DG)$
$-C(ACK-BCH+C^2G-6DO-18FL)$
$+C(2ACK+AEG-ABL-6DO-3HJ')+9F(AO-2CL+3FG)$
$-G(3A^2I-3ABF+2ACE+BCD-C^3-3DJ'+27F^2)$
$+3J'(AL+CH-DG)=0$

(9.20) $A(ABO+2ACN-AEK+2BCL-4C^2K-18FO+HM)-AB(AO-DK+EH)$
$-2AC(AN-CK+EG)-BC(AL+CH-DG)$
$+C(2ACK+AEG-ABL-6DO-3HJ')$
$-D(ABK-B^2H+BCG-6CO-12EL-18FK+36HI)$
$+18F(AO-DK+EH)+E(A^2K-ABH+ACG-12DL+18FH)$
$-H(AM-2ABE+B^2D-BC^2-3CJ'-36DI+36EF)=0$

(9.22) $A(GO-HN-KL)-G(AO-DK+EH)+H(AN-CK+EG)$
$+K(AL+CH-DG)=0$

(9.22) $A(6A^2CI-3ABCF+ABDE+3AC^2E-ADM-18DEF-3KL)$
$-3C(2A^2I-A^2BF+A^2CE-2ADJ'+4D^2E+HK)$
$+D(A^2M-A^2BE-6ACJ'+18AEF+12CDE+3GK)$
$+3K(AL+CH-DG)=0$

(9.22) $3A(4A^2CI-ABCF+2AC^2E-3AFJ'-2CDJ'+GO)$
$-A^2(12ACI+AE^2-3BCF+4C^2E-DM-9FJ')$
$-D(A^2M-A^2BE-6ACJ'+18AEF+3GK+12CDE)$
$+E(A^2E-A^2BD-2A^2C^2+18ADF+12CD^2+3GH)-3G(AO-DK+EH)=0$

Deg. Order.

$$
\begin{aligned}
&(9.22) && 3A(4A^2CI - ABCF + 2AC^2E - 3AFJ' - 2CDJ' + GO) \\
& && -4AC(3A^2I - 3ABF + 2ACE + BCD - C^3 - 3DJ' + 27F^2) \\
& && +2C(A^2CE + 2ABCD - 2AC^3 - 3ADJ' - 18CDF - 3GL) \\
& && +9F(A^2J' - A^2BC + 12ACF + 4C^2D + G^2) - 3G(AO - 2CL + 3GF) = 0
\end{aligned}
$$

$$
\begin{aligned}
&{}^*(9.24) && A^3(ABK - B^2H + BCG - 6CO - 12EL - 18FK + 36HI) \\
& && -AB(A^2K - ABH + ACG - 12DL + 18FH) - 6AC(AO - 2CL + 3FG) \\
& && +12AC(CO - DK + EH) + 12AE(AL + CH - DG) - 12C^2(AL + CH - DG) \\
& && -12D(ACK - BCH + C^2G - 6DO - 18FL) \\
& && +12D(2ACK + AEG - ABL - 6DO - 3HJ') \\
& && +18F(A^2K - ABH + ACG - 12DL + 18FH) \\
& && -12H(3A^2I - 3ABF + 2ACE + BCD - C^3 - 3DJ' + 27F^2) = 0
\end{aligned}
$$

$$
\begin{aligned}
&(9.26) && 3A(A^2DI - ABDF + AC^2F + 9DF^2 + L^2) \\
& && -AD(3A^2I - 3ABF + 2ACE + BCD - C^3 - 3DJ' + 27F^2) \\
& && +C(A^2DE - 3A^2CF - ABD^2 + AC^2D + 18D^2F + 3HL) \\
& && +D(A^2CE + 2ABCD - 2AC^3 - 3ADJ' - 18CDF - 3GL) \\
& && -3L(AL + CH - DG) = 0
\end{aligned}
$$

$$
\begin{aligned}
&(9.26) && A(AGK - BGH + CG^2 - 6HO - 12L^2) - G(A^2K - ABH + ACG - 12DL + 18FH) \\
& && +6H(AO - 2CL + 3FG) + 12L(AL + CH - DG) = 0
\end{aligned}
$$

$$
\begin{aligned}
&(9.26) && 2AD(3A^2I - 3ABF + 2ACE + BCD - C^3 - 3DJ' + 27F^2) \\
& && +2C(A^2DE - 3A^2CF - ABD^2 + AC^2D + 18D^2F + 3HL) \\
& && -3D(2A^2I - A^2BF + A^2CE - 2ADJ' + 4D^2E + HK) \\
& && +3E(A^3F - A^2CD + 4D^3 + H^2) \\
& && -3F(A^3E - A^2BD - 2A^2C^2 + 18ADF + 12CD^2 + 3GH) \\
& && +3H(AO - 2CL + 3FG) - 3H(AO - DK + EH) = 0
\end{aligned}
$$

$$
\begin{aligned}
&(9.26) && A^3(AM - 2ABE + B^2D - BC^2 - 3CJ' - 36DI + 36EF) \\
& && -A^2(A^2M - A^2BE - 6ACJ' + 18AEF + 12CDE + 3GK) \\
& && +AB(A^3E - A^2BD - 2A^2C^2 + 18ADF + 12CD^2 + 3GH) \\
& && -3AC(A^2J' - A^2BC + 12ACF + 4C^2D + G^2) \\
& && +12AD(3A^2I - 3ABF + 2ACE + BCD - C^3 - 3DJ' + 27F^2) \\
& && -12D(A^2CE + 2ABCD - 2AC^3 - 3ADJ' - 18CDF - 3GL) \\
& && -18F(A^3E - A^2BD - 2A^2C^2 + 18ADF + 12CD^2 + 3GH) \\
& && +3G(A^2K - ABH + ACG - 12DL + 18FH) = 0
\end{aligned}
$$

$$
\begin{aligned}
&{}^*(9.28) && A^3(2ACK + AEG - ABL - 6DO - 3HJ') \\
& && -2AC(A^2K - ABH + ACG - 12DL + 18FH) \\
& && -G(A^3E - A^2BD - 2A^2C^2 + 18ADF + 12CD^2 + 3GH) - 12CD(AL + CH - DG) \\
& && +A^2B(AL + CH - DG) + 6AD(AO - 2CL + 3FG) \\
& && +3H(A^2J' - A^2BC + 12ACF + 4C^2D + G^2) = 0
\end{aligned}
$$

---

* Also relations of deg. order (9.24), (9.28), (9.30), and (9.34) obtained by multiplying the relations of deg. 8 by the sextic A.

Deg. Order.

*(9.30) $3A^2(2A^3I - A^2BF + A^2CE - 2ADJ' + 4D^2E + HK)$
$- 2A^3(3A^2I - 3ABF + 2ACE + BCD - C^3 - 3DJ' + 27F^2)$
$- 3AB(A^3F - A^2CD + 4D^3 + H^2)$
$+ AC(A^3E - A^2BD - 2A^2C^2 + 18ADF + 12CD^2 + 3GH)$
$- 12D(A^2DE - 3A^2CF - ABD^2 + AC^2D + 18D^2F + 3HL)$
$+ 54F(A^3F - A^2CD + 4D^3 + H^2)$
$- 3H(A^2K - ABH + ACG - 12DL + 18FH) = 0$

(9.32) $A^3(AO - 2CL + 3FG) - A^3(AO - DK + EH)$
$+ 2A^2C(AL + CH - DG) - AD(A^2K - ABH + ACG - 12DL + 18FH)$
$- 12D^2(AL + CH - DG) - 3G(A^3F - A^2CD + 4D^3 + H^2)$
$+ H(A^3E - A^2BD - 2A^2C^2 + 18ADF + 12CD^2 + 3GH) = 0$

These relations take the place of Prof. Sylvester's syzygants of the second grade, defined as rational integral functions of irreducible ones of the first grade, which vanish when expressed in terms of the covariants (see *American Journal of Mathematics*, Vol. IV, p. 61). Those of the present article are *linear* functions of the syzygants of the first grade, given in the preceding article, the coefficients being covariants.† They are written at full length to exhibit clearly their property of vanishing identically when expressed in terms of the covariants.‡

In the following articles it will be shown that the list is complete.

(II). *Digression on the Generating Function for Syzygants of Binary Quantics.*

3. If, for the quantic of order i, and for its covariants of deg. order (m, n), there be

α asyzygetic covariants,
β syzygies,
γ groundforms,
δ compound covariants,

it is well known that $\alpha + \beta = \gamma + \delta$ (1)

Now suppose that the groundforms are of deg. order $(r.s)$, $(r'.s')$, . . . and let $(1 - a^r x^s)(1 - a^{r'} x^{s'}) \dots$ be denoted by $\Pi(1 - a^r x^s)$, then

$$\frac{1}{\Pi(1 - a^r x^s)} = \Sigma(\gamma + \delta)\, a^m x^n \dots \qquad (2)$$

In this we can by means of (1) replace $\gamma + \delta$ by $\alpha + \beta$, and if, moreover, we

* See the preceding note.

† To make my meaning perfectly clear, such a compound syzygant as $(AL + CH - DG)^2$, should it ever occur, is not considered except as a linear function of $AL(AL + CH - DG)$, $CH(AL + CH - DG)$, and $DG(AL + CH - DG)$.

‡ In this paper the word *syzygant* has the meaning originally given to it by Prof. Sylvester, *loc. cit.*

write $\Sigma a a^m x^n = \phi(a, x)$, $\phi(a, x)$ is the generating function for covariants, and (2) becomes

$$\Sigma \beta a^m x^n = \frac{1}{\Pi(1-a^r x^s)} - \phi(a, x) \qquad (3)$$

This is the generating function for syzygants; its use is limited. by the one assumption made in finding it, to those quantics for which we have a *complete* and *correct* list of groundforms. It matters not how such a list may be found, but the proof of its correctness and completeness must be independent of the now disproved Fundamental Postulate of Tamisage.

4. The principle that the only compound syzygants that need be considered are linear functions of the simply divisible syzygants is of vital importance to the theory, and gives an immediate interpretation to each term in the numerator of the Generating Function for Syzygants. For if Σ denote a ground-syzygant, and A, B, C, . . . be the groundforms; all the syzygants that can be formed from Σ, *i. e.* all the simply divisible syzygants, are found in the expansion of $\dfrac{\Sigma}{(1-A)(1-B)(1-C)\ldots}$. Now the Generating Function for Syzygants can be resolved into a series of fractions of precisely this form, and hence the terms of the numerator will correspond to ground-syzygants.

It must be understood that there are syzygants not only of the first and second grades, but also, as Prof. Sylvester has remarked, of the 3d, 4th, and higher grades.

A Syzygant of the second grade is defined as a *linear* function of the simply divisible syzygants of the first grade, which is identically zero when considered as a function of the groundforms.

A Syzygant of the third grade is defined as a *linear* function of the simply divisible syzygants of the second grade, which is identically zero when considered as a function of the simply divisible syzygants of the first grade.

And generally—a Syzygant of the grade $n + 2$ is a *linear* function of the simply divisible syzygants of the grade $n + 1$, which is identically zero when considered as a function of the simply divisible syzygants of the grade n.

The Σ of the present article may be taken to be a syzygant of any grade, the first article of this paper contains examples of syzygants of the first grade, the second contains examples of syzygants of the second grade, and a simple example of a syzygant of the third grade occurs for the deg. order (11.24) for the binary sextic.

The sextic has in fact a syzygant of the second grade of deg. order (10.18), viz.

$$
\begin{aligned}
(10.18) \qquad & \mathrm{C}\,(\mathrm{AEK} - \mathrm{BEH} + \mathrm{CEG} + 6\mathrm{DQ} - 18\mathrm{FO}) \\
& - 6\mathrm{D}\,(\mathrm{CQ} + \mathrm{EO} - 3\mathrm{FN}) \\
& - \mathrm{E}\,(\mathrm{ACK} - \mathrm{BCH} + \mathrm{C^2G} - 6\mathrm{DO} - 18\mathrm{FL}) \\
& + 18\mathrm{F}\,(\mathrm{CO} - \mathrm{DN} - \mathrm{EL}) = 0.
\end{aligned}
$$

In this, let the syzygants of the first grade, within brackets, be denoted by single letters, so that the syzygant of the second grade is

$$(10.18) \qquad \mathrm{C}\alpha - 6\mathrm{D}\beta - \mathrm{E}\gamma + 18\mathrm{F}\delta = \Gamma \text{ suppose.}$$

Treating the syzygants of the second grade, given in Art. 2, in a similar manner; the syzygants of deg. order (8.18), (8.22) and the first syzygants of deg. order (9.16) and (9.20) respectively give

$$
\begin{aligned}
(8.18) \qquad & \Delta = \mathrm{A}\delta - \mathrm{C}\varepsilon + \mathrm{D}\xi + \mathrm{E}\eta \\
(8.22) \qquad & \Theta = \mathrm{A}\gamma - \mathrm{C}\theta + 6\mathrm{D}\iota + 18\mathrm{F}\eta \\
(9.16) \qquad & \Lambda = \mathrm{A}\beta - \mathrm{C}\varkappa - \mathrm{E}\iota + 3\mathrm{F}\xi \\
(9.20) \qquad & \Omega = \mathrm{A}\alpha - 6\mathrm{D}\varkappa - \mathrm{E}\theta + 18\mathrm{F}\varepsilon.
\end{aligned}
$$

And then the Syzygant of the third grade referred to is

$$
\begin{aligned}
(11.24) \qquad & \mathrm{A}\Gamma - \mathrm{C}\Omega + 6\mathrm{D}\Lambda + \mathrm{E}\Theta - 18\mathrm{F}\Delta \\
& = \mathrm{A}\,(\mathrm{C}\alpha - 6\mathrm{D}\beta - \mathrm{E}\gamma + 18\mathrm{F}\delta) \\
& - \mathrm{C}\,(\mathrm{A}\alpha - 6\mathrm{D}\varkappa - \mathrm{E}\theta + 18\mathrm{F}\varepsilon) \\
& + 6\mathrm{D}\,(\mathrm{A}\beta - \mathrm{C}\varkappa - \mathrm{E}\iota + 3\mathrm{F}\xi) \\
& + \mathrm{E}\,(\mathrm{A}\gamma - \mathrm{C}\theta + 6\mathrm{D}\iota + 18\mathrm{F}\eta) \\
& - 18\mathrm{F}\,(\mathrm{A}\delta - \mathrm{C}\varepsilon + \mathrm{D}\xi + \mathrm{E}\eta) = 0.
\end{aligned}
$$

5. For the Binary Sextic, we have, see Prof. Sylvester's Tables of Generating Functions, etc. (*American Journal of Mathematics*, Vol. II, p. 225),

$$\Pi\,(1 - a^r x^s) = (1 - ax^6) \ldots (1 - a^{15}),$$

where there are 26 factors, one for each groundform. And $\phi\,(a, x) = \dfrac{N}{D}$, where

$$
\begin{aligned}
N = 1 & + a^3(x^2 + x^6 + x^8 + x^{12}) + a^4(x^4 + x^6 + x^{10}) + a^5(x^2 + x^4 + x^8 - x^{16}) \\
& + a^6(x^4 + 2x^6) + a^7(x^2 + x^4 + x^8 - x^{12}) + a^8(x^2 + x^4 + x^6 - x^{14}) \\
& + a^9(x^4 + x^6 - x^{10} - x^{12}) + a^{10}(x^2 + x^4 - x^{12} - x^{14}) + a^{11}(x^4 + x^6 - x^{10} - x^{12}) \\
& + a^{12}(x^2 - x^{10} - x^{12} - x^{14}) + a^{13}(x^4 - x^8 - x^{12} - x^{14}) + a^{14}(-2x^{10} - x^{12}) \\
& + a^{15}(1 - x^8 - x^{12} - x^{14}) + a^{16}(-x^6 - x^{10} - x^{12}) + a^{17}(-x^4 - x^8 - x^{10} - x^{14}) - a^{20}x^{16}.
\end{aligned}
$$

$$D = (1 - a^2)(1 - a^4)(1 - a^6)(1 - a^{10})(1 - a^2x^4)(1 - a^2x^8)(1 - ax^6).$$

Hence the numerator of the Generating Function for Syzygants $\frac{1}{\Pi(1-a^r x^s)}$ $-\phi(a, x)$, is

$$1 - N \left\{ \begin{array}{l} (1-a^3x^2)(1-a^3x^6)(1-a^3x^8)(1-a^3x^{12})(1-a^4x^4)(1-a^4x^6)(1-a^4x^{10}) \\ (1-a^5x^2)(1-a^5x^4)(1-a^5x^8)(1-a^6x^6)^2(1-a^7x^2)(1-a^7x^4)(1-a^8x^2) \\ (1-a^9x^4)(1-a^{10}x^2)(1-a^{12}x^2)(1-a^{15}) \end{array} \right\}$$

viz. N is multiplied by those factors of $\Pi(1-a^r x^s)$ which do not occur in D and the result subtracted from unity.

The result would be an expression of the 139th degree in a, and of the 106th in x, for the numerator. The denominator is of course $\Pi(1-a^r x^s)$.

The numerator for Syzygants of the Binary Sextic is

$$\begin{array}{l} a^5x^{16} + a^6(x^8 + x^{10} + x^{12} + 2x^{14} + x^{16} + x^{18} + x^{20} + x^{24}) \\ + a^7(x^6 + x^{10} + 4x^{12} + x^{14} + 2x^{16} + 2x^{18} + x^{22}) \\ + a^8(2x^8 + 4x^{10} + 2x^{12} + 4x^{14} + 3x^{16} - x^{18} + 2x^{20} - x^{22} - x^{24} - x^{28}) \\ + a^9(x^6 + 4x^8 + 2x^{10} + 5x^{12} + 3x^{14} - 2x^{16} + 2x^{18} - 4x^{20} - 4x^{22} - x^{24} - 4x^{26} - x^{28} - x^{30} - x^{32}) \\ + \ldots\ldots\ldots\ldots\ldots \end{array}$$

The numerator of the Generating Function for Syzygants may be called the Numerator for Syzygants, for two reasons; first, for the sake of brevity, and secondly, since it gives the number of ground-syzygants of any grade. Results obtained from it are, however, liable to correction, whenever ground-syzygants of grade n coexist with ground-syzygants of grade $n+1$ of the same deg. order. Two such cases occur in the present example, viz. there is one ground-syzygant of the second grade and deg. order (9.14) and one of deg. order (9.18), so that the corresponding terms in the Numerator for Syzygants of the Sextic should be written $(4-1)a^9x^{14} + (3-1)a^9x^{18}$. When this is done the positive terms agree with the numbers given in Prof. Sylvester's Table of Syzygies (*American Journal of Mathematics*, Vol. IV, p. 59), except a^9x^6, where it will be shown hereafter that there is an error in the table; and the negative terms agree with the syzygants of the second grade given in Art. 2 of the present paper.

6. For Quantics which have groundforms and syzygies of the same deg. order a complete list of groundforms cannot be obtained by tamisage, and if $\Pi(1-a^r x^s)$ is formed from an incomplete list this will involve a correction of the Numerator for Syzygants.

In fact the preceding article shows that the Numerator for Syzygants is of the form $1-NM$, where N is the numerator of the generating function for covariants, in its representative form, and the multiplier M consists of those factors of $\Pi(1-a^r x^s)$ which do not appear in the denominator of the generating function for covariants. If then a factor $1-a^\rho x^\sigma$ has been omitted in $\Pi(1-a^r x^s)$ it will also have been omitted in M, and the corrected Numerator for Syzygants will be $1-NM(1-a^\rho x^\sigma)$. If more than one groundform has been omitted the correction will be of the same nature.

The correction for those cases in which Ground-Syzygants of grade n exist simultaneously with Ground-Syzygants of grade $n+1$, of the same deg. order, consists in multiplying N by each factor of M successively and separating those terms, as they occur, which in $1-NM$ will correspond to Syzygants of different grades, and thus preventing their combination with each other.

In the case of the Sextic the work may be arranged as follows:

$N=1+$

	+	+	+	+	−	−	−	−
a^3	x^2	x^6	x^8	x^{12}				
a^4	x^4	x^6	x^{10}					
a^5	x^2	x^4	x^8		x^{16}			
a^6	x^4	$2x^6$						
a^7	x^2	x^4	x^8		x^{12}			
a^8	x^2	x^4	x^6		x^{14}			
a^9	x^4	x^6			x^{10}	x^{12}		
a^{10}	x^2	x^4			x^{12}	x^{14}		
a^{11}	x^4	x^6			x^{10}	x^{12}		
a^{12}	x^2				x^{10}	x^{12}	x^{14}	
a^{13}	x^4				x^8	x^{12}	x^{14}	
a^{14}					$2x^{10}$	x^{12}		
a^{15}	1				x^8	x^{12}	x^{14}	
a^{16}					x^6	x^{10}	x^{12}	
a^{17}					x^4	x^8	x^{10}	x^{14}
a^{20}					x^{16}			

$N(1-a^3x^2)=1+$

	+	+	+	−	−	−	−	−	+	+	+	+
a^3	x^6	x^8	x^{12}									
a^4	x^4	x^6	x^{10}									
a^5	x^2	x^4	x^8	x^{16}								
a^6	$2x^6$			x^8	x^{10}	x^{14}						
a^7	x^2	x^4		x^6	$2x^{12}$							
a^8	x^2			x^{10}	x^{14}				x^{18}			
a^9	x^4			$2x^8$	x^{10}	x^{12}						
a^{10}	x^2			x^6	x^{10}	x^{12}	x^{14}		x^{14}			
a^{11}				x^8	x^{10}	x^{12}			x^{16}			
a^{12}	x^2			x^6	x^8	x^{10}	x^{12}	x^{14}	x^{12}	x^{14}		
a^{13}				x^6	x^8	x^{12}	x^{14}		x^{14}	x^{16}		
a^{14}				x^6	x^8	$2x^{10}$	x^{12}		x^{12}	x^{14}		
a^{15}	1			x^4	x^8	x^{12}	x^{14}		x^{12}	x^{14}	x^{16}	
a^{16}				$2x^6$	x^{10}	x^{12}			x^{10}	x^{14}	x^{16}	
a^{17}				x^4	x^8	x^{10}	x^{14}		$2x^{12}$	x^{14}		
a^{18}				x^2					x^{10}	x^{14}	x^{16}	
a^{19}									x^8	x^{12}	x^{14}	
a^{20}				x^{16}					x^6	x^{10}	x^{12}	x^{16}
a^{23}									x^{18}			

and so on until N has been multiplied by all the factors of M. In performing this first multiplication it has been assumed that all the negative terms of N correspond to ground-syzygants, and if this is not so, considerable uncertainty arises; thus if the term $-a^{10}x^{14}$ in N corresponds to a compound syzygant, formed by multiplying a ground-syzygant of deg. order (7.12) by the groundform of deg. order (3.2), it ought to be removed and the two terms $-a^{10}x^{14}$ and $+a^{10}x^{14}$ in $N(1-a^3x^2)$ ought to cancel one another; but if it corresponds to a ground-syzygant these two terms will correspond to a ground-syzygant of the first grade and one of the second grade respectively, and ought to be kept distinct. The same reasoning applies to the deg. orders (12.12), (12.14), (13.14), (14.12), (15.12), (15.14), (16.10), (17.14), and (20.16). A similar assumption is made at each successive multiplication, with the object of making the list of ground-syzygants of the first grade coincide, as closely as possible, with Prof. Sylvester's table, referred to above; and by these means the cases of disagreement have been reduced to three only; viz. there are found, one syzygy of deg. order (9.6) and one of deg. order (11.6), not given in the table, and two of deg. order (15.6) where the table gives three.

The factor $(1-a^3x^2)$ has been chosen for the first multiplication, because all those terms in the positive portion of N which correspond to compound forms can, by its use, be removed at one operation; and no other factor possesses this property. For the Quintic, the only factor possessing this property is $(1-a^5x)$; a circumstance of which Prof. Cayley has taken advantage in the arrangement of the terms in the numerator of his "Real Generating Function" (see his Tenth Memoir on Quantics, Phil. Trans. Part II, 1878, p. 608). In Quantics higher than the Sextic no such factor occurs.

The second, and all the remaining multiplications, are performed in the following manner.

Let $\quad N(1-a^3x^2)=1+\Gamma-\Sigma_1+\Sigma_2,$

then $\quad N(1-a^3x^2)(1-a^rx^s)=1+\Gamma'-\Sigma_1'+\Sigma_2'-\Sigma_3,$

where
$$\Gamma'=\Gamma-a^rx^s$$
$$\Sigma_1'=\Sigma_1+a^rx^s\Gamma$$
$$\Sigma_2'=\Sigma_2+a^rx^s\Sigma_1$$
$$\Sigma_3=a^rx^s\Sigma_2.$$

Here Γ denotes the first positive block, whose terms consist exclusively of ground-types, Σ_1 denotes the negative block and typifies syzygants of the first grade (supposed to be ground-syzygants), and Σ_2 denotes the second positive block and is exclusively typical of syzygants of the second grade, provided those of Σ_1 are *ground*-syzygants, but not otherwise.

7. After the first multiplication, the method of the preceding article cannot be followed with certainty without a preliminary investigation to show precisely what the terms of each block denote (*e. g.* whether *all* those of the second block denote *ground*-syzygants), and hence what terms, if any, will cancel one another. But, in the case of the Sextic, since all the terms of the first block have been, by the first multiplication, made to typify groundforms only; all the terms which, originating in it, are placed in the second block after each multiplication, necessarily correspond to ground-syzygants. For they typify syzygants containing a term which is either the square of one or the product of two groundforms, and these cannot possibly be compound syzygants. Precisely similar reasoning shows that, after any block has been made to contain only terms typical of ground-syzygants of its own grade, all the terms which, originating in it at each multiplication, are transferred to the next block will necessarily typify ground-syzygants of the grade proper to the block to which they have been transferred. In fact no syzygant of grade $n+1$ can possibly be compound which contains, as one of its terms, a compound syzygant of grade n formed by the product of a ground-syzygant of grade n with a single groundform.

The form of the Generating Function, when corrected for cases of coexistence of a groundform and syzygant of the same deg. order, will perhaps place the whole in a clearer light.

Suppose then that, after all the groundforms that can be found by tamisage have been removed into the denominator, the Generating Function has been put in the form

$$\frac{1-\Sigma_1+\Sigma_2-\Sigma_3+\ldots}{\Pi(1-a^r x^s)},$$

the correction for the coexistence of a groundform and syzygant of deg. order $(\rho.\sigma)$ gives it the form

$$\frac{1-(a^\rho x^\sigma+\Sigma_1)+(a^\rho x^\sigma\Sigma_1+\Sigma_2)-(a^\rho x^\sigma\Sigma_2+\Sigma_3)+\ldots}{(1-a^\rho x^\sigma)\,\Pi(1-a^r x^s)}.$$

Now if Σ_1 contains one term which typifies a ground-syzygant (S) and another term which typifies a compound syzygant formed by multiplying S by the groundform $(\rho.\sigma)$, a term of the first block will cancel with one of the second, and this term of $a^\rho x^\sigma \Sigma_1$ has been wrongly placed in the second block; but if Σ_1 contains no such terms, every term of $a^\rho x^\sigma \Sigma_1$ has been rightly placed in the second block. An examination of the syzygants denoted by Σ_1 is therefore necessary to find out which of them, if any, is divisible by the groundform $(\rho.\sigma)$. The syzygants denoted by $a^\rho x^\sigma \Sigma_1$ may all of them be considered as compounds divisible by the groundform $(\rho.\sigma)$; but with this difference, those of them that are used to destroy compound syzygants of the block from which they originate, are considered as compound syzygants simply, but those that are transferred to the next block are considered *as single terms of syzygants of the next higher grade.* Hence when Σ_1 contains only ground-syzygants the compounds $a^\rho x^\sigma \Sigma_1$ will be the products of ground-syzygants with a single groundform, and having been all of them transferred to the next block, will stand (as sample terms) for *ground*-syzygants of the next higher grade. The reasoning is of course unaltered if we write Σ_n for Σ_1 throughout.*

I will conclude this section (which is an improvement on what I called the "Automatic Method of Tamisage" in the J. H. U. Circular for April, 1883), by remarking that $1 - \text{NM}$ (Art. 6) has no factor in common with $\Pi(1 - a^r x^s)$, and consequently the generating function for Syzygants is a fraction *in its lowest terms* whose denominator consists of factors typifying all the groundforms; but its very complicated form for the higher quantics sufficiently indicates the impracticability of finding this generating function by any method which does not involve a previous knowledge of the groundforms.

(III). *Exemplifications and Applications.*

8. It seems proper, at the commencement of this section, to give the leading terms of the covariants J′ and R′, and to state why they have been used instead of J and R.

Copying the leading term of J from Prof. Cayley's tables, and forming that of BC by multiplication, we have

* It was by such reasoning as this that I was led to consider the syzygant of the third grade of deg. order (11.24), given in Art. 4, a sample term of which is the ground-syzygant of deg. order (8.18) multiplied by the single groundform F.

	J, x^4	BC, x^4	J', x^4
a^2ge		$+1$	$+1$
f^2	$+1$		-1
$abdg$		-4	-4
bef	-10	-6	$+4$
c^2g		$+8$	$+8$
cdf	$+4$		-4
ce^2	$+16$	$+15$	-1
d^2e	-12	-10	$+2$
a^0b^2df	$+16$	$+24$	$+8$
b^2e^2	$+9$		-9
bc^2f	-12	-18	-6
$bcde$	-76	-60	$+16$
bd^3	$+48$	$+40$	-8
c^3e	$+48$	$+45$	-3
c^2d^2	-32	-30	$+2$

Here $J' = BC - J$, and it is to be noticed that the part independent of a, or residue, of the source of O, is simply the residue of J' multiplied by b. Such an arrangement, when possible, greatly facilitates the use of the method of residues. Thus (see Prof. Cayley's Tables for the Binary Sextic, *American Journal of Mathematics*, Vol. IV, p. 381 and 382),

$$\text{Res. } G = -2b.\text{Res. } C, \text{ and Res. } O = b.\text{Res. } J',$$

so that the syzygy $2CO + GJ' = AR'$ is immediately verified. An inspection of the syzygies given in Art. 1 of the present paper will show that this is not the only syzygy whose form has been simplified by the use of J' and R'.

The source of R' is given in the following table, together with Prof. Cayley's Q, R, and BK.

R′, x^8
+ 1
− 1
− 4
− 5
+ 9
+ 3
+ 2
− 6
− 3
+ 8
− 4
+ 28
− 27
− 21

It will be seen that $8R' = R + BK - 6Q$. The only apparent advantage in using R′ is that it is somewhat simpler than R. The source of J′ has also a special property which that of J does not possess, viz. operating on the invariant I, of the fourth degree, we have

$$(a\delta_c + 3b\delta_d + 6c\delta_e + 10d\delta_f + 15e\delta_g)\, I = \text{source of } J'.$$

9. In what follows a Syzygant will be denoted by one of its terms placed within $\overline{\underline{\quad\;}}$; thus the four Syzygants, for the Sextic, of deg. order (7.12) may be denoted by $\overline{\underline{AQ}}$, $\overline{\underline{AR'}}$, $\overline{\underline{DN}}$, and $\overline{\underline{HI}}$; but remembering what has been said above (Art. 7), this set of four syzygants is here represented by $\overline{\underline{DN}}$, $\overline{\underline{EL}}$, $\overline{\underline{FK}}$, and $\overline{\underline{GJ'}}$; and then $\overline{\underline{DN}}$ is the syzygant typified by the term $-a^7x^{12}$ in the unmultiplied numerator, and the pair of syzygants corresponding to $-2a^7x^{12}$ in the numerator after the first multiplication is $\overline{\underline{DN}}$, $\overline{\underline{EL}}$. This notation is followed, in the list of ground-syzygants given below, whenever the theory indicates a syzygant containing, as one of its terms, a binary combination of the groundforms of the numerator. When this is not the case the presence of syzygants is, in general, indicated by an asterisk; but it is not certain that all of these are *ground-syzygants* (see Art. 6). The five syzygants $\overline{\underline{DG}}$, $\overline{\underline{DN}}$, $\overline{\underline{DQ}}$, $\overline{\underline{DS}}$ and $\overline{\underline{DT}}$ have been inserted to complete the list as far as degree 9.

(5.16)	(6.8)	(6.10)	(6.12)	(6.14)	(6.16)	(6.18)	(6.20)	(6.24)
$\overline{\underline{DG}}$	$\overline{\underline{EF}}$	$\overline{\underline{EG}}$	$\overline{\underline{F^2}}$	$\overline{\underline{EH}}$	$\overline{\underline{G^2}}$	$\overline{\underline{FH}}$	$\overline{\underline{GH}}$	$\overline{\underline{H^2}}$
				$\overline{\underline{FG}}$				

(7.6)	(7.10)	(7.12)	(7.14)	(7.16)	(7.18)	(7.22)	(8.8)	(8.10)	(8.12)	(8.14)	(8.16)	(8.20)
$\overline{\underline{EJ'}}$	$\overline{\underline{FJ'}}$	$\overline{\underline{DN}}$	$\overline{\underline{GK}}$	$\overline{\underline{FL}}$	$\overline{\underline{GL}}$	$\overline{\underline{HL}}$	$\overline{\underline{FM}}$	$\overline{\underline{EO}}$	$\overline{\underline{GN}}$	$\overline{\underline{DQ}}$	$\overline{\underline{GO}}$	$\overline{\underline{HO}}$
		$\overline{\underline{EL}}$		$\overline{\underline{HJ'}}$	$\overline{\underline{HK}}$		$\overline{\underline{J'^2}}$	$\overline{\underline{FN}}$	$\overline{\underline{K^2}}$	$\overline{\underline{FO}}$	$\overline{\underline{HN}}$	$\overline{\underline{L^2}}$
		$\overline{\underline{FK}}$						$\overline{\underline{GM}}$		$\overline{\underline{HM}}$	$\overline{\underline{KL}}$	
		$\overline{\underline{GJ'}}$						$\overline{\underline{J'K}}$		$\overline{\underline{J'L}}$		

(9.6)	(9.8)	(9.10)	(9.12)	(9.14)	(9.18)	(10.4)	(10.6)	(10.8)	(10.10)	(10.12)	(10.14)	(10.16)
$\overline{\underline{J'M}}$	$\overline{\underline{EQ}}$	$\overline{\underline{DS}}$	$\overline{\underline{DT}}$	$\overline{\underline{GQ}}$	$\overline{\underline{HQ}}$	$\overline{\underline{M^2}}$	$\overline{\underline{ET}}$	$\overline{\underline{FS}}$.	*	*	*	$\overline{\underline{HT}}$
	$\overline{\underline{ER'}}$	$\overline{\underline{KN}}$	$\overline{\underline{FQ}}$	$\overline{\underline{GR'}}$	$\overline{\underline{HR'}}$		$\overline{\underline{MN}}$	$\overline{\underline{N^2}}$	$\overline{\underline{FT}}$	$\overline{\underline{GT}}$	$\overline{\underline{HS}}$	$\overline{\underline{LQ}}$
	$\overline{\underline{J'N}}$		$\overline{\underline{FR'}}$	$\overline{\underline{KO}}$	$\overline{\underline{LO}}$				$\overline{\underline{GS}}$	$\overline{\underline{KQ}}$		$\overline{\underline{LR'}}$
	$\overline{\underline{KM}}$		$\overline{\underline{J'O}}$	$\overline{\underline{LN}}$					$\overline{\underline{J'Q}}$	$\overline{\underline{KR'}}$		$\overline{\underline{O^2}}$
			$\overline{\underline{LM}}$						$\overline{\underline{J'R'}}$	$\overline{\underline{NO}}$		
									$\overline{\underline{MO}}$			

(11.6)	(11.8)	(11.10)	(11.12)	(11.14)	(12.4)	(12.6)	(12.8)	(12.10)	(12.12)	(12.14)	(12.16)
$J'S$	*	*	*	HU	MS	EV	*	*	*	*	HV
	FU	GU	LS	LT		$J'U$	KU	FV	GV		
	$J'T$	KT		OQ		MT	NT	OS	LU		
	KS	NQ		OR'		NS			OT		
	MQ	NR'							Q^2		
	MR'								QR'		
									R'^2		

(13.4)	(13.6)	(13.8)	(13.10)	(13.12)	(13.14)	(14.4)	(14.6)	(14.8)	(14.10)	(14.12)
MU	*	*	GX	*	*	S^2	*	*	*	*
	NU	FX	KV		HX		$J'X$	KX	*	LX
		$J'V$	OU		LV		MV	NV		OV
		QS	QT				ST	QU		
		$R'S$	$R'T$					$R'U$		
								T^2		

(15.4)	(15.6)	(15.8)	(15.10)	(15.12)	(15.14)	(16.4)	(16.6)	(16.8)	(16.10)	(16.12)
EY	NX	*	GY	*	*	U^2	*	KY	*	*
MX	TU	FY	OX		HY		*	QX		LY
SU			QV				$J'Y$	$R'X$		
			$R'V$				SV	TV		

(17.4)	(17.6)	(17.8)	(17.10)	(17.14)	(18.2)	(18.4)	(18.6)	(18.8)	(18.12)
*	NY	*	*	*	EZ	UX	FZ	GZ	HZ
MY	TX		OY					QY	
SX	UV							$R'Y$	
								V^2	

(19.4)	(19.6)	(19.10)	(20.2)	(20.4)	(20.8)	(20.16)	(21.6)
$J'Z$	KZ	LZ	MZ	NZ	OZ	*	QZ
	TY			UY			$R'Z$
	VX			X^2			VY

(22.2)	(22.4)	(23.2)	(24.4)	(25.2)	(27.2)	(30.0)
SZ	TZ	UZ	VZ	XZ	YZ	Z^2
	XY		Y^2			

If we now assume that each asterisk corresponds to a *ground*-syzygant, which is in fact the same as the assumption made in Art. 6, the list gives exactly the same results as Prof. Sylvester's table, except in the three cases already noticed.

10. When certain fundamental syzygants are known, all the rest may be calculated very easily by common algebra. The method will be applied, in this article, to the calculation of the syzygants of degree 9, for the Sextic.

We have, from Art. 1, using the notation of Art. 9,

$$\begin{aligned}
J'\,\overline{\underline{EF}} &= J'(AM - 2ABE + B^2D - BC^2 - 3CJ' - 36DI + 36EF) \\
AB\,\overline{\underline{EJ'}} &= AB(9AP + 12ABI - 3B^2F + 4BCE - CM - 12EJ' + 108FI) \\
B^2\,\overline{\underline{F^2}} &= B^2(3A^2I - 3ABF + 2ACE + BCD - C^3 - 3DJ' + 27F^2) \\
BC\,\overline{\underline{EF}} &= BC(AM - 2ABE + B^2D - BC^2 - 3CJ' - 36DI + 36EF) \\
C\,\overline{\underline{J'^2}} &= C(AS + 4C^2I - FM - J'^2) \\
F\,\overline{\underline{EJ'}} &= F(9AP + 12ABI - 3B^2F + 4BCE - CM - 12EJ' + 108FI) \\
I\,\overline{\underline{F^2}} &= I(3A^2I - 3ABF + 2ACE + BCD - C^3 - 3DJ' + 27F^2)
\end{aligned}$$

Whence,

$$\begin{aligned}
&3J'\,\overline{\underline{EF}} - AB\,\overline{\underline{EJ'}} + B^2\,\overline{\underline{F^2}} - BC\,\overline{\underline{EF}} - 9C\,\overline{\underline{J'^2}} + 9F\,\overline{\underline{EJ'}} - 36I\,\overline{\underline{F^2}} \\
&= 3A(J'M - 3AB^2I - 3ABP - 36AI^2 + 2BEJ' + 36BFI - 24CEI - 3CS + 27FP)
\end{aligned}$$

Now, since the left-hand side of this is a compound syzygant, the right side is one also; and dividing by 3A we have the value of the syzygant (9.6) or $\underline{J'M}$, given in the list, but not in Prof. Sylvester's table. At the same time is found the ground-syzygy of the second grade

$$(10.12)\quad 3J'\,\overline{\underline{EF}} - AB\,\overline{\underline{EJ'}} + B^2\,\overline{\underline{F^2}} - BC\,\overline{\underline{EF}} - 9C\,\overline{\underline{J'^2}} + 9F\,\overline{\underline{EJ'}} - 36I\,\overline{\underline{F^2}} = 3A\,\underline{J'M}.$$

Again if

$$\begin{aligned}
C\,\overline{\underline{EO}} &= C(AT - 2EO - J'K) \\
C\,\overline{\underline{J'K}} &= C(2B^2L - BCK - 3BEG - 18EO + GM - 72IL + 3J'K) \\
E\,\overline{\underline{EL}} &= E(AQ + 2EL - 3FK) \\
E\,\overline{\underline{FK}} &= E(ABK - B^2H + BCG - 6CO - 12EL - 18FK + 36HI) \\
E\,\overline{\underline{GJ'}} &= E(AR' - GJ' - 2CO) \\
G\,\overline{\underline{EJ'}} &= G(9AP + 12ABI - 3B^2F + 4BCE - CM - 12EJ' + 108FI) \\
I\,\overline{\underline{EH}} &= I(AO - DK + EH) \\
I\,\overline{\underline{FG}} &= I(AO - 2CL + 3FG) \\
J'\,\overline{\underline{EG}} &= J'(AN - CK + EG) \\
K\,\overline{\underline{EF}} &= K(AM - 2ABE + B^2D - BC^2 - 3CJ' - 36DI + 36EF) \\
B^2\,\overline{\underline{EH}} &= B^2(AO - DK + EH) \\
B^2\,\overline{\underline{FG}} &= B^2(AO - 2CL + 3FG) \\
BC\,\overline{\underline{EG}} &= BC(AN - CK + EG)
\end{aligned}$$

we have

$$\begin{aligned}
&E\,\overline{\underline{GJ'}} + J'\,\overline{\underline{EG}} - C\,\overline{\underline{EO}} = A(ER' + J'N - CT) \\
&B^2\,\overline{\underline{FG}} - 36I\,\overline{\underline{FG}} - BC\,\overline{\underline{EG}} + 3C\,\overline{\underline{EO}} + C\,\overline{\underline{J'K}} + G\,\overline{\underline{EJ'}} - 12E\,\overline{\underline{GJ'}} \\
&\qquad = A(B^2O - 36IO - BCN + 3CT + 9GP + 12BGI - 12ER') \\
&B^2\,\overline{\underline{EH}} - 36I\,\overline{\underline{EH}} - BC\,\overline{\underline{EG}} - 3C\,\overline{\underline{EO}} + E\,\overline{\underline{FK}} + K\,\overline{\underline{EF}} + 6E\,\overline{\underline{EL}} \\
&\qquad = A(B^2O - 36IO - BCN - 3CT - EBK + KM + 6EQ).
\end{aligned}$$

Here the method only gives three of the four syzygies; the fourth is a fundamental one, introducing the covariant U for the first time, and must be found independently. It is

$$KM + 2E(BK - 3R' - 6Q) = 3AU.$$

There is no other fundamental syzygy of degree 9 but this; all the others are found by a mere repetition of the work given above, and are therefore non-fundamental; to each of them corresponds a ground-syzygant of the second grade of degree 10.

The complete list of syzygies of degree 9 is here given, the corresponding syzygies of the second grade are omitted for the sake of brevity. The names correspond to the list of Art. 9.

Deg. Order.	Name.	
(9.6)	$J'M$	$3AB^2I+3ABP+36AI^2-2BEJ'-36BFI+24CEI+3CS-27FP-J'M=0$
(9.8)	EQ	$B^2O-36IO-BCN-3CT-BEK+KM+6EQ=0$
(9.8)	ER'	$B^2O-36IO-BCN+3CT+9GP+12BGI-12ER'=0$
(9.8)	$J'N$	$CT-ER'-J'N=0$
(9.8)	KM	$3AU-2BEK+6ER'+12EQ-KM=0$
(9.10)	DS	$ABCI+ACP-AIJ'+BFJ'-2CEJ'-E^2F+DS=0$
(9.10)	KN	$12ABCI+ABE^2+9ACP-AEM-6CEJ'-18E^2F+3KN=0$
(9.12)	DT	$BEL-2CEK-E^2G-3J'O+3DT=0$
(9.12)	FQ	$3AIK+BCO-3BFK-C^2N+3CEK-3DT+9FQ=0$
(9.12)	FR'	$2BCO-2C^2N+6CGI-3J'O-9FR'=0$
(9.12)	$J'O$	$6AIK-2BEL-3BFK-12BHI+7CEK+2E^2G-9HP-6J'O=0$
(9.12)	LM	$AEN-6CGI-3J'O+LM=0$
(9.14)	GQ	$2LN-KO-GQ=0$
(9.14)	GR'	$ABCJ'-8AC^2I-AJ'^2+2BC^2F-4C^2E-6CFJ'-GR'=0$
(9.14)	KO	$2A^2EI-4ABDI-ABEF+ACE^2-3ADP+2DEJ'-KO=0$
(9.14)	LN	$A^2EI-ABEF-2AC^2I+3CFJ'+9EF^2-LN=0$
(9.18)	HQ	$AK^2-BHK+CGK-12LO+6HQ=0$
(9.18)	HR'	$BGL-2CGK-EG^2+6LO+3HR'=0$
(9.18)	LO	$2ACDI+ACEF-3DFJ'+LO=0$

11. The arguments in favor of the completeness of the list of formulæ given in this paper may be summarised as follows. The "Numerator for Syzygants," (Art. 5) gives an inferior limit to their number; the corrections of Art. 6 give, on the other hand, a superior limit, and actual calculation shows that this superior limit is attained in every case hitherto considered. It is, however, extremely doubtful whether this will happen for degress from 10 to 17 inclusive, or for the deg. order (20.16); for degrees above 20, and for 18 and 19, the superior and inferior limits coincide, as will be seen by an inspection of the asterisks of Art. 9.

Reduction of Alternating Functions to Alternants.

By Wm. Woolsey Johnson.

Denoting by $\phi(a, bcd \ldots l)$ any function of the n quantities $a, b, c \ldots l$ which is symmetrical with respect to all of the quantities except a, the determinant

$$(1) \qquad \begin{vmatrix} \phi_1(a, bcd \ldots l), & \phi_2(a, bcd \ldots l), \ldots & \phi_n(a, bcd \ldots l) \\ \phi_1(b, acd \ldots l), & \phi_2(b, acd \ldots l), \ldots & \phi_n(b, acd \ldots l) \\ \vdots & \vdots & \vdots \\ \phi_1(l, abc \ldots k), & \phi_2(l, abc \ldots k), \ldots & \phi_n(l, abc \ldots k) \end{vmatrix}$$

is obviously an alternating function of the n quantities.

If each of the functions contains only the leading letter, the determinant takes the form

$$(2) \qquad \begin{vmatrix} f_1(a), & f_2(a), \ldots & f_n(a) \\ f_1(b), & f_2(b), \ldots & f_n(b) \\ \vdots & \vdots & \vdots \\ f_1(l), & f_2(l), \ldots & f_n(l) \end{vmatrix},$$

and is an alternant. The alternant (2) may be represented by means of its principal term

$$(3) \qquad f_1(a).f_2(b).f_3(c) \ldots f_n(l).$$

Suppose now that the principal term of the alternating function (1) can be separated into parts of the form (3); then will the sum of the alternants represented by these partial terms be equal to the given alternating function. To prove this it is only necessary to notice that, since an interchange of two rows in (1) is equivalent to an interchange of the corresponding letters, any term in (1) may be derived from the principal term by a certain transposition of the letters, and in like manner, the corresponding term in each of the alternants may be derived from its principal term by the same transposition of the letters: thus every term in the expansion of (1) is equal to the sum of the corresponding terms in the expansions of the alternants.

If a determinant of the form (1) be expressed by enclosing its principal term in [ ] with periods separating the several elements, the theorem is that the periods may be dispensed with and the symbol treated as an ordinary algebraic quantity. Thus

$$\begin{vmatrix} bcd, & 1, & a, & a^2 \\ cda, & 1, & b, & b^2 \\ dab, & 1, & c, & c^2 \\ abc, & 1, & d, & d^2 \end{vmatrix} = [bcd.1.c.d^2] = [a^0bc^2d^3] = \zeta^{\frac{1}{2}}(a, b, c, d),$$

the difference product of the quantities a, b, c and d.

When the elements of the alternating function are rational and integral, it may thus always be expressed in terms of simple alternants, that is, alternants of powers; and, since a simple alternant vanishes when two of the exponents are equal, it will frequently happen that many of the parts of an alternating function whose elements are polynomials will vanish. For example,

$$\begin{vmatrix} 1, & b^2 + c^2, & a^2 + bc \\ 1, & c^2 + a^2, & b^2 + ca \\ 1, & a^2 + b^2, & c^2 + ab \end{vmatrix} = [a^0b^0c^4] + [a^2b^0c^2] + [abc^2] + [a^3bc^0],$$

all of which except the last vanish; the result is, therefore,

$$[1.c^2 + a^2.c^2 + ab] = [a^3bc^0] = -[a^0bc^3] = -(a + b + c)\,\zeta^{\frac{1}{2}}(a, b, c).$$

2. We need of course only consider alternants of which the lowest exponent is zero; thus, when the alternant is of the third order, we have

$$(2) \qquad a(0, p, q) = \begin{vmatrix} 1, & H_p, & H_q \\ 0, & H_{p-1}, & H_{q-1} \\ 0, & H_{p-2}, & H_{q-2} \end{vmatrix} = H_{p-1}.H_{q-2} - H_{p-2}.H_{q-1};$$

but, even in this the simplest case, the expansion of the result in single symmetric functions is very laborious, the ordinary process producing, when p and q are moderately large, an enormous number of terms which cancel one another. The same is true to a great extent of the process given by Mr. Muir in his Treatise on Determinants, although this process shows that the first term of the result is $\Sigma a^{q-2} b^{p-1}$. The result is, however, readily obtained by means of the formula of reduction established below.

3. We have

$$A(0, p, q) = \begin{vmatrix} 1, & a^p, & a^q \\ 1, & b^p, & b^q \\ 1, & c^p, & c^q \end{vmatrix} = c^q(b^p - a^p) + a^q c^p - b^q(c^p - a^p) - a^q b^p.$$

Assuming $q > p$, by adding $0 = a^{q-p} b^p c^p - a^{q-p} b^p c^p$, this may be written in the form

$$A(0, p, q) = c^q(b^p - a^p) - a^{q-p} c^p (b^p - a^p) - b^q (c^p - a^p) + a^{q-p} b^p (c^p - a^p)$$
$$= c^p (b^p - a^p)(c^{q-p} - a^{q-p}) - b^p (c^p - a^p)(b^{q-p} - a^{q-p});$$

hence $a(0, p, q) = \dfrac{A(0, p, q)}{(b-a)(c-a)(c-b)}$ is the quotient of

$$c^p (b^{p-1} + b^{p-2} a + \ldots + a^{p-1})(c^{q-p-1} + c^{q-p-2} a + \ldots + a^{q-p-1})$$
$$- b^p (c^{p-1} + c^{p-2} a + \ldots + a^{p-1})(b^{q-p-1} + b^{q-p-2} a + \ldots + a^{q-p-1})$$

by $c - b$. Expanding the products, and grouping together similar positive and negative terms, we have, for the value of $(c-b)\, a(0, p, q)$,

$$b^{p-1} c^{p-1} (c^{q-p} - b^{q-p}) + a b^{p-2} c^{p-2} (c^{q-p+1} - b^{q-p+1}) + \ldots$$
$$+ a^{p-2} bc (c^{q-2} - b^{q-2}) + a^{p-1} (c^{q-1} - b^{q-1})$$
$$+ a b^{p-1} c^{p-1} (c^{q-p-1} - b^{q-p}) + a^2 b^{p-2} c^{p-2} (c^{q-p} - b^{q-p}) + \ldots$$
$$+ a^{p-1} bc (c^{q-3} - b^{q-3}) + a^p (c^{q-2} - b^{q-2})$$
$$\vdots \qquad\qquad \vdots \qquad\qquad \vdots \qquad\qquad \vdots$$
$$+ a^{q-p-1} b^{p-1} c^{p-1} (c - b) + a^{q-p} b^{p-2} c^{p-2} (c^2 - b^2) + \ldots$$
$$+ a^{q-3} bc (c^{p-1} - b^{p-1}) + a^{q-2} (c^p - b^p),$$

in which $p(q-p)$ binomials are written in $q-p$ rows and p columns. Replacing the binomials by their quotients by $c - b$, we have the expanded value of $a(0, p, q)$.

Now, if we remove the first row and the last column of the rectangular array of binomials written above, we see that the remaining terms, when divided through by abc, constitute the value of $(c-b)\,a(0,\,p-1,\,q-2)$. Hence
$a(0,p,q)-abc.a(0,p-1,q-2)=$
$b^{p-1}c^{q-2}+b^pc^{q-3}+\ldots+b^{q-2}c^{p-1}+a(b^{p-2}c^{q-2}+b^{p-1}c^{q-3}+\ldots+b^{q-2}c^{p-2})+\ldots$
$\quad+a^{p-1}(c^{q-2}+bc^{q-3}+\ldots+b^{q-2})+a^p(c^{q-3}+bc^{q-4}+\ldots+b^{q-3})+\ldots$
$\quad+a^{q-2}(c^{p-1}+bc^{p-2}+\ldots+b^{p-1})$.

It will be noticed that this is a symmetric function of a, b and c in which every product of the degree $q+p-3$ occurs once, except those in which there is an exponent greater than $q-2$. Denoting this function, which may be called a *curtailed* complete symmetric function, by $H_{q-2,\,p-1}$ (in which the sum of the suffixes indicates the degree), we have the formula of reduction

(3) $$a(0,p,q)=H_{q-2,\,p-1}+abc.a(0,p-1,q-2).$$

4. The formula may also be proved directly by division, as follows: we have $A(0,p,q)=c^q(b^p-a^p)+a^qc^p-$ (the result of interchanging b and c), and, in like manner, $abcA(0,p-1,q-2)=abc^{q-1}(b^{p-1}-a^{p-1})+a^{q-1}bc^p-$ (result of interchanging b and c); hence $A(0,p,q)-abcA(0,p-1,q-2)=c^q(b^p-a^p)-abc^{q-1}(b^{p-1}-a^{p-1})-a^{q-1}c^p(b-a)-$ (result of interchanging b and c).

If, therefore, we divide these terms by $(b-a)(c-a)$ and then subtract the result of interchanging b and c, we shall have the value of
$$(c-b)[a(0,p,q)-abc.a(0,p-1,q-2)].$$
Dividing the terms written above by $b-a$, we have
$c^q(b^{p-1}+ab^{p-2}+\ldots+a^{p-1})-abc^{q-1}(b^{p-2}+ab^{p-3}+\ldots+a^{p-2})-a^{q-1}c^p$
or $b^{p-1}c^{q-1}(c-a)+ab^{p-2}c^{q-1}(c-a)+\ldots$
$$+a^{p-2}bc^{q-1}(c-a)+a^{p-1}c^p(c^{q-p}-a^{q-p});$$
and, dividing this by $c-a$,
$b^{p-1}c^{q-1}+ab^{p-2}c^{q-1}+\ldots+a^{p-2}bc^{q-1}+a^{p-1}c^p(c^{q-p-1}+ac^{q-p-2}+\ldots+a^{q-p-1})$.
Finally, subtracting the result of interchanging b and c, we have
$b^{p-1}c^{p-1}(c^{q-p}-b^{q-p})+ab^{p-2}c^{p-2}(c^{q-p+1}-b^{q-p+1})+\ldots+a^{p-2}bc(c^{q-2}-b^{q-2})$
$\quad+a^{p-1}(c^{q-1}-b^{q-1})+a^p(c^{q-2}-b^{q-2})+\ldots+a^{q-2}(c^p-b^p)$
as before, and dividing by $c-b$, the result is
$$a(0,p,q)-abc.a(0,p-1,q-2)=H_{q-2,\,p-1}.$$

5. When $p=1$, the formula reduces to

(4) $$a(0,1,q)=H_{q-2},$$

which results directly from Jacobi's theorem, equation (2); but, starting from this result, we may give an independent proof of the formula (3) as follows: If we write down all the terms in the complete symmetric function H_{r+s} for three quantities a, b and c in a triangular form a^{r+s}, b^{r+s} and c^{r+s} being the terms at the vertices, it is readily seen that the curtailed function $H_{r,s}$ is obtained by cutting off a small triangle of terms at each vertex, and that the terms in the first of these triangles are the same as those of the expression $a^{r+1}H_{s-1}$. Thus

(5) $$H_{r,s} = H_{r+s} - \Sigma a^{r+1}.H_{s-1},$$

in which the curtailed function is expressed in terms of complete symmetric functions. Putting $r = q - 2$ and $s = p - 1$,

$$H_{q-2,\,p-1} = H_{q+p-3} - \Sigma a^{q-1}.H_{p-2},$$

and, by equation (4),

$$H_{q-2,\,p-1} = \alpha(0,\,1,\,q+p-1) - \Sigma a^{q-1}.\alpha(0,\,1,\,p),$$

or $$\zeta^{\frac{1}{2}}(a,\,b,\,c).H_{q-2,\,p-1} = A(0,\,1,\,q+p-1) - \Sigma a^{q-1}.A(0,\,1,\,p)$$

$$= \begin{vmatrix} 1, & a, & a^{q+p-1} \\ 1, & b, & b^{q+p-1} \\ 1, & c, & c^{q+p-1} \end{vmatrix} - \begin{vmatrix} 1, & a, & a^p(a^{q-1}+b^{q-1}+c^{q-1}) \\ 1, & b, & b^p(a^{q-1}+b^{q-1}+c^{q-1}) \\ 1, & c, & c^p(a^{q-1}+b^{q-1}+c^{q-1}) \end{vmatrix} = - \begin{vmatrix} 1, & a, & a^p(b^{q-1}+c^{q-1}) \\ 1, & b, & b^p(c^{q-1}+a^{q-1}) \\ 1, & c, & c^p(a^{q-1}+b^{q-1}) \end{vmatrix}.$$

But, by the theorem in the preceding paper, this alternating function is equal to

$$-[1.b.a^{q-1}c^p + b^{q-1}c^p] = -[a^{q-1}bc^p + b^q c^p] = -A(q-1,\,1,\,p) - A(0,\,q,\,p).$$

Hence $$\zeta^{\frac{1}{2}}(a,\,b,\,c).H_{q-2,\,p-1} = A(0,\,p,\,q) - A(1,\,p,\,q-1)$$
$$= A(0,\,p,\,q) - abcA(0,\,p-1,\,q-2),$$

or $$\alpha(0,\,p,\,q) = H_{q-2,\,p-1} + abc\alpha(0,\,p-1,\,q-2) \text{ as before.}$$

6. As an example of the use of the formula, let us find the value of $\alpha(0,\,5,\,7)$, that is the quotient of $A(0,\,5,\,7)$ by the difference product of the quantities a, b and c. We have

$$\alpha(0,\,5,\,7) = H_{5,4} + abc\alpha(0,\,4,\,5)$$
$$= H_{5,4} + abcH_{3,3},$$

since $\alpha(0,\,3,\,3)$ vanishes: or, writing out the values in single symmetric functions,

$$\alpha(0,\,5,\,7) = \Sigma a^5b^4 + \Sigma a^5b^3c + \Sigma a^5b^2c^2 + 2\Sigma a^4b^4c + 2\Sigma a^4b^3c^2 + 2\Sigma a^3b^3c^3.$$

Again, to find $\alpha(0,\,3,\,8)$, we have

$$\alpha(0,\,3,\,8) = H_{6,2} + abc\alpha(0,\,2,\,6)$$
$$= H_{6,2} + abcH_{4,1} + a^2b^2c^2\alpha(0,\,1,\,4)$$
$$= H_{6,2} + abcH_{4,1} + a^2b^2c^2H_2;$$

or, in single symmetric functions,

$$\begin{aligned}\alpha(0,3,8) &= \Sigma a^6b^2 + \Sigma a^6bc + \Sigma a^5b^3 + \Sigma a^5b^2c + \Sigma a^4b^4 + \Sigma a^4b^3c + \Sigma a^4b^2c^2 + \Sigma a^3b^3c^2 \\ &\quad + abc\,[\Sigma a^4b + \Sigma a^3b^2 + \Sigma a^3bc + \Sigma a^2b^2c] + a^2b^2c^2\,[\Sigma a^2 + \Sigma ab] \\ &= \Sigma a^6b^2 + \Sigma a^6bc + \Sigma a^5b^3 + 2\Sigma a^5b^2c + \Sigma a^4b^4 + 2\Sigma a^4b^3c + 3\Sigma a^4b^2c^2 + 3\Sigma a^3b^3c^2.\end{aligned}$$

In the final expansion, it is to be noticed that, for a term in which the highest exponent (in this case 6) or the exponent zero occurs, the coefficient is unity; otherwise it is 2, provided the next higher or next lower exponent (5 or 1) occurs; but if not, it is 3, provided the next higher or next lower exponent (4 or 2) occurs, and so on. This is the general rule with, however, the restriction that no coefficient must exceed the number of H's in the development which is *the least of the numbers p and $q-p$.* The restriction takes effect whenever the last H has a suffix greater than 2, as in the first of the examples above.

7. In general

(6) $$\alpha(0,p,q) = H_{q-2,\,p-1} + abcH_{q-4,\,p-2} + a^2b^2c^2H_{q-6,\,p-3} + \text{etc.},$$

the series ending either with an H in which the two suffixes are equal, as in the first example above, or with one in which the second suffix is zero, that is, with a complete symmetric function, as in the second example.

The first of these cases corresponds to the theorem

$$\alpha(0,p+1,p+2) = H_{p,p},$$

a case of the more general theorem

(7) $$\alpha(0,p+1,p+2,\ldots p+n-1) = H_{p,p,\ldots}$$

where $H_{p,p,\ldots}$ denotes the symmetric function of n quantities in which every product of the degree $(n-1)p$ occurs once, except those in which there is an exponent greater than p. This theorem is readily derived from

(8) $$A(0,1,2,\ldots n-2,p+n-1) = \zeta^{\frac{1}{2}}(a,b,c,\ldots l).H_p$$

(which results directly from Jacobi's theorem) by substituting for the n quantities their reciprocals.

8. Equation (5), in which the curtailed symmetric function is expressed in terms of complete symmetric functions, holds for any number of quantities; hence by virtue of equation (8) the process of §5 is applicable to n quantities, the curtailed function being thus in general expressed as the sum of $n-1$ co-factors of alternants. The result is, however, not generally available as a formula of reduction for an alternant: but, in the case of four quantities, if we put $r = s = p$ in equation (5), we have

$$H_{p,p} = H_{2p} - \Sigma a^{p+1}.H_{p-1};$$

whence $\zeta^4(a, b, c, d).H_{p,p} = A(0, 1, 2, 2p+3) - \Sigma a^{p+1}.A(0, 1, 2, p+2)$

$$= \begin{vmatrix} 1, & a, & a^2, & a^{2p+3} \\ 1, & b, & b^2, & b^{2p+3} \\ 1, & c, & c^2, & c^{2p+3} \\ 1, & d, & d^2, & d^{2p+3} \end{vmatrix} - \begin{vmatrix} 1, & a, & a^2, & a^{p+2}(a^{p+1}+b^{p+1}+c^{p+1}+d^{p+1}) \\ 1, & b, & b^2, & b^{p+2}(a^{p+1}+b^{p+1}+c^{p+1}+d^{p+1}) \\ 1, & c, & c^2, & c^{p+2}(a^{p+1}+b^{p+1}+c^{p+1}+d^{p+1}) \\ 1, & d, & d^2, & d^{p+2}(a^{p+1}+b^{p+1}+c^{p+1}+d^{p+1}) \end{vmatrix}$$

$$= -[1.b.c^2.d^{p+2}(a^{p+1}+b^{p+1}+c^{p+1})]$$

$$= -A(p+1, 1, 2, p+2) - A(0, p+2, 2, p+2) - A(0, 1, p+3, p+2).$$

Hence $\zeta^4(a, b, c, d)\, H_{p,p} = A(0, 1, p+2, p+3) - A(1, 2, p+1, p+2)$,

or

(9) $$a(0, 1, p+2, p+3) = H_{p,p} + abcd.a(0, 1, p, p+1),$$

a formula of reduction for an alternant of the form $A(0, 1, p+2, p+3)$. By repeated application of this formula, we have

(10) $$a(0, 1, p+2, p+3) = H_{p,p} + abcdH_{p-2,p-2} + a^2b^2c^2d^2H_{p-4,p-4} + \text{etc.},$$

in which the last term is $(abcd)^{\frac{p}{2}}$ or $(abcd)^{\frac{p-1}{2}}\Sigma ab$, according as p is even or odd. For example,

$$\begin{vmatrix} 1, & a, & a^7, & a^8 \\ 1, & b, & b^7, & b^8 \\ 1, & c, & c^7, & c^8 \\ 1, & d, & d^7, & d^8 \end{vmatrix} \div \begin{vmatrix} 1, & a, & a^2, & a^3 \\ 1, & b, & b^2, & b^3 \\ 1, & c, & c^2, & c^3 \\ 1, & d, & d^2, & d^3 \end{vmatrix} = H_{5,5} + abcdH_{3,3} + a^2b^2c^2d^2H_{1,1}$$

$$= \Sigma a^5b^5 + \Sigma a^5b^4c + \Sigma a^5b^3c^2 + \Sigma a^5b^3cd + \Sigma a^5b^2c^2d + \Sigma a^4b^4c^2 + 2\Sigma a^4b^4cd$$
$$+ \Sigma a^4b^3c^3 + 2\Sigma a^4b^3c^2d + 2\Sigma a^4b^2c^2d^2 + 2\Sigma a^3b^3c^3d + 3\Sigma a^3b^3c^2d^2,$$

in which the coefficients follow the same rule as in §6, but no restriction need be observed with respect to the highest coefficient.

Bibliography of Linear Differential Equations.

By H. B. Nixon and J. C. Fields.

The following list contains the titles of papers on the modern Theory of Linear Differential Equations as it has been developed by Fuchs, Thomé, Frobenius, Hermite, Poincaré and others. The list is probably incomplete, but addenda to it will be published from time to time. Again there may be titles included which strictly should not be, as the authors have not been able to see all of the papers referred to. If any one discovering such titles will kindly communicate with the authors, a proper correction will be made.

André, D. Intégration des équations différentielles linéaires à coefficients quelconques. *C. R.*, lxxxiv, 1018-1020.

——Intégration, sous forme finie, de trois espèces d'équations différentielles linéaires à coefficients quelconques. *C. R.*, lxxxviii, 230-232.

——Intégration, sous forme finie, d'une nouvelle espèce d'équations différentielles linéaires à coefficients variables. *C. R.*, xcii, 121-123. *Résal, J.* (3), vii, 283-288.

——Intégration de trois espèces d'équations différentielles linéaires. *Liouville, J.* (3), vi, 27-49.

——Note sur les développements des puissances de certaines fonctions. *Bull. S. M. F.*, vi, 120-121.

Appell, P. Mémoire sur les équations différentielles linéaires. *Ann. de l'École Normale* (2), 391-424.

——Sur une classe d'équations différentielles linéaires binômes à coefficients algébriques. *Ann. de l'École Normale*, 1883, 9-46.

——Sur les équations différentielles linéaires à une variable indépendante. *C. R.*, xc, 1477-1479.

——Sur la transformation des équations différentielles linéaires. *C. R.*, xci, 211-214.

——Sur les équations différentielles linéaires. *C. R.*, xci, 684-685.

——Sur une classe d'équations différentielles linéaires. *C. R.*, xcii, 61-63.

——Sur une classe d'équations différentielles linéaires à coefficients doublement périodiques. *C. R.*, xcii, 1005-1008.

——Sur des équations différentielles linéaires dont les intégrales vérifient des relations d'une certaine forme. *C. R.*, xciii, 699-702.

——Sur une classe d'équations différentielles linéaires binômes à coefficients algébriques. *C. R.*, xciv, 202-205.

APPELL, M. Sur les fonctions uniformes affectées de coupures et sur une classe d'équations différentielles linéaires. *C. R.*, xcvi, 1018-1020.

AUTONNE, L. Sur les intégrales algébriques des équations différentielles linéaires à coefficients rationnels. *C. R.*, xcvi, 56-58.

BARANIECKI, M. A. Beweis eines Satzes aus der Theorie der hypergeometrischen Functionen. *Par. Denkschr.*, viii (Polish).

BELTRAMI E RAZZABONI. Sopra la memoria del Prof. D. Besso intitolata: Alcune proposizioni sulle equazioni differenziali lineari. *Rom. Acc. L.* (3), vi, 12.

BESSO, D. Sopra una classe d'equazioni del sesto grado resolubili per serie ipergeometriche. *Rom. Acc. L. Mem.*, xiv.

——Sul prodotto di più soluzioni particolari d'un equazione differenziale lineare omogenea e specialmente dell' equazione differenziale del terz' ordine. *Rom. Acc. L. Mem.*, xiv.

——Di alcune proprietà dell' equazione differenziale lineare omogenea del second' ordine e di alcune equazioni algebriche. *Rom. Acc. L. Mem.*, xiv.

——Di alcune proprietà dell' equazione differenziale lineare, non omogenea del second' ordine. *Rom. Acc. L. Mem.*, xiv.

BRIOSCHI, F. La théorie des formes dans l'intégration des équations différentiales linéaires du second ordre. *Clebsch Ann.*, xi, 401-412. *Rend. Ist. Lomb.* (2), x, 48-58.

——Sur les équations différentielles linéaires. *Bull. S. M. F.*, vii, 105-108.

——Sur une équation différentielle du troisième ordre. *Proc. of London M. S.*, xxvii, 126-128.

——Sopra una classe di equazioni differenziali integrabili per funzioni ellittiche. *Acc. Rd. L.* (3), iv, 241-246.

——Sulla origine di talune equazioni differenziali lineari. *Rom. Acc. L.* (3), vi, 42-47.

——Sopra una classe di equazioni differenziali lineari del secondo ordine. *Brioschi Ann.* (2), ix, 11-21.

——Di una proprietà delle equazioni differenziali lineari del secondo ordine. *Brioschi Ann.* (2), x, 1-3.

——Sopra una classe di equazioni differenziali lineari del secondo ordine. *Brioschi Ann.* (2), x, 4-9.

——Sulla generazione di una classe di equazioni differenziali lineari integrabili per funzioni ellittiche. *Brioschi Ann.* (2), x, 74-79.

——Sur l'équation de Lamé. *C. R.*, lxxxv, 1160-1162.

——Sur l'équation de Lamé. *C. R.*, lxxxvii, 313-315.

——Sur une classe d'équations différentielles linéaires. *C. R.*, xci, 317-319.

——Sur quelques équations différentielles linéaires. *C. R.*, xci, 807-809.

——Théoremes relatifs à l'équation de Lamé. *C. R.*, xcii, 325-329.

——Sur la théorie des équations différentielles du second ordre. *C. R.*, xciii, 941-922.

——Sur une application du théorème d'Abel. *C. R.*, xciv, 686-691.

BRIOT ET BOUQUET. Mémoire sur l'intégration des équations différentielles au moyen des fonctions elliptiques. *J. l'Ecole Polytechnique*, xxi.

——Recherches ser les propriétées des fonctions défines par des équations différentielles. *J. l'Ecole Polytechnique*, xxi.

CASORATI, F. Sull' equazione fundamentale nella teoria delle equazioni differenziali lineari. *Rend. Ist. Lomb.* (2), xiii, 176-182.

——Sur la distinction des integrales des équations différentielles linéaires en sous-groupes. *C. R.*, xcii, 175-178, 238-241.

——Sulle equazioni differenziali lineari. *Rom. Acc. L.* (3), vi, 121-124.

CASORATI E BELTRAMI. Relazione. *Rom. Acc. L.* (3), vi, 226.

CATALAN, E., FOLIE, F. Rapports sur ce mémoire. *Bull. de Belg.* (2), xxxviii, 562-566.

CATALAN, E. Rapport sur la note de M. Le Paige. *Bull. de Belg.* (2), lxi, 935-939.

CAYLEY, A. On the Schwarzian derivative and the polyhedral functions. *Trans. of Cambr.*, xiii, 5-68. *Proc. of Cambr.*, iii, 349-351.

——Note on a hypergeometric series. *Quart. J.*, xvi, 268-270.

COMBESCURE, E. Remarques sur les équations différentielles linéaires et du 3me ordre. *C. R.*, lxxxviii, 275-277.

CRAIG, THOMAS. On a Certain Class of Linear Differential Equations. *Am. Jour. of Math.*, Vol. vii, No. 3, 279-288.

DARBOUX, G. Application de la méthode de M. Hermite à l'équation linéaire à coefficients constants avec second membre. *Darboux Bull.* (2), iii, 325-328.

——Sur les systèmes d'équations linéaires à une seule variable indépendante. *C. R.*, xc, 524-526, 596-598.

——Sur une équation linéaire. *C. R.*, xciv, 1645-1648.

——Sur une proposition relative aux équations linéaires. *C. R.*, xciv, 1456-1459.

DILLNER, G. Sur les équations différentielles linéaires simultanees à coefficients rationnels dont la solution dépend de la quadrature d'un même produit algébrique. *C. R.*, xcii, 289-290.

——Sur une propriété des produits des *k* équations différentielles linéaires à coefficients rationnels. *C. R.*, xcii, 290-291.

——Sur une classe d'équations différentielles linéaires. *C. R.*, xci, 687.

ELLIOT, M. Sur une équation linéaire du second ordre à coefficient doublement périodiques. *Acta Math.* (2), 1883, 233-260.

ESCARY. Sur quelques remarques, relatives à l'équation de Lamé. *C. R.*, xci, 40-43, 102-105.

FARKAS, J. Solution d'un système d'équations linéaires. *C. R.*, lxxxvii, 523-526.

——Sur l'application de la theorie des sinus des ordres supérieurs à l'intégrations des équations différentielles linéaires. *C. R.*, xc, 1542-1545.

FLOQUET, G. Sur quelques équations différentielles linéaires. *C. R.*, xci, 880-882.

——Sur la théorie des équations différentielles linéaires. *Ann. de l'École Normale* (2), viii. Suppl. 3-132.

——Sur les équations différentielles linéaires à coefficients périodiques. *Ann. de l'École Normale*, 1883, 47-88.

——Sur les équations différentielles linéaires à coefficients périodiques. *C. R.*, xcii, 1397-1398.

FROBENIUS, G. Ueber die Vertauschung von Argument und Parameter in den Integralen der linearen Differentialgleichungen. *Pr. Berlin.*

——Ueber die regulären Integrale der linearen Differentialgleichungen. *Borchardt J.*, lxxx, 317-333.

——Ueber algebraisch integrirbare lineare Differentialgleichungen. *Borchardt J.*, lxxx, 183-193.

——Vertauschung von Argument und Parameter in den Integralen linearen Differentialgleichungen. *Borchardt J.*, lxxviii, 93-97.

——Ueber die Determinante mehrerer Functionen einer Variabeln. *Borchardt J.*, lxxvii, 245-258.

——Ueber den Begriff der Irreductibilität in der Theorie der linearen Differentialgleichungen. *Borchardt J.*, lxxvi, 236-271.

——Ueber die Integration der linearen Differentialgleichungen durch Reihen. *Borchardt J.*, lxxvi, 214-235.

FUCHS, L. Zur Theorie der linearen Differentialgleichungen mit veränderlichen Coefficienten. *Borchardt J.*, lxvi, 121-160; lxviii, 354-385.

——Bemerkungen zur Abhandlung: "Ueber hypergeometrische Functionen n^{ter} Ordnung." *Borchardt J.*, lxxi, 316; lxxii, 255-262.

——Ueber Relationen, welche für die zwischen je zwei singulären Punkten erstreckten. Integrale der Lösungen linearer Differentialgleichungen stattfinden. *Borchardt J.*, lxxii, 177-214.

——Ueber die Darstellung der Functionen complexer Variabeln, insbesondere der Integrale linearer Differentialgleichungen. *Borchardt J.*, lxxv, 177-223; lxxvi, 175-176.

——Ueber lineare Differentialgleichungen zweiter Ordnung. *Gött. Nachr.*, 1875, 568-581, 612-613. *Borchardt J.*, lxxxi, 97-142.

——Sur quelques propriétés des intégrales des équations différentielles. *Borchardt J.*, lxxxiii, 13-38.

——Ueber die linearen Differentialgleichungen zweiter Ordnung welche algebraische Integrale besitzen. *Borchardt J.*, lxxxv, 1-26.

——Extrait d'une lettre adressée à M. Hermite. *Liouville J.* (3), ii, 158-160.

——Sur les équations différentielles linéaires, qui admettent des intégrales dont les différentielles logarithmiques sont des fonctions doublement périodiques. *Liouville J.* (3), iv, 125-141.

——Sur les équations différentielles linéaires du second ordre. *C. R.*, lxxxii, 1434-1437; lxxxiii, 46-47.

——Ueber eine Classe von Functionen mehrerer Variabeln. *Gött. Nachr.*, 1880, 170-176. *Borchardt J.*, lxxxix, 150-169. *Darboux Bull.* (2), iv, 278-300. *C. R.*, xc, 678-680, 735-736.

——Ueber eine Klasse von Differentialgleichungen, welche durch Abel'sche oder elliptische Functionen integrirbar sind. *Gött. Nachr.*, 1878, 19-33. *Brioschi Ann.* (2), ix, 25-35.

——Sur le développement en séries des intégrales des équations différentielles linéaires. *Brioschi Ann.* (2), iv, 36-49. 1870.

——Ueber Functionen, welche durch Umkehrung der Integrale, von Lösungen der linearen Differentialgleichungen entstehen. *Gött. Nachr.*, 1880, 445-453. *Darboux Bull.* (2), iv, 328-336.

——Ueber lineare homogene Differentialgleichungen, zwischen deren Integralen homogene Relationen höheren als ersten Grades bestehen. *Berl. Ber.*, 1882, 703-710.

——Ueber lineare homogene Differentialgleichungen, zwischen deren Integralen homogene Relationen höheren als ersten Grades bestehen. *Acta Math.* (1), 1882-1883, 321-362.

GEGENBAUER, L. Note über hypergeometrischer Reihen. *Grunert Arch.*, lv, 284-290.

——Beiträge zur Theorie der linearen Differentialgleichungen. *Grunert Arch.*, lv, 258-284.

GOURSAT, E. Mémoire sur les fonctions hypergéométriques d'ordre supérieur. *Ann. de l'École Normale*, 1883, 261-286, 295-430.

——Sur l'équation différentielle linéaire qui admet pour integrale la série hypergéométrique. *Ann. de l'École Normale* (2), x. Suppl. 3-142.

——Sur les intégrales algébriques des équations linéaires. *Darb. Bull.* (2), vi, 120-124.

——Sur un cas de réduction des équations linéaires du quatrième ordre. *C. R.*, c, 233-235.

——Sur une classe d'équations linéaires du quatrième ordre. *C. R.*, xcvii, 31-34.

——Sur les fonctions hypergéométriques d'ordre supérieur. *C. R.*, xcvi, 185-187.

——Sur les fonctions hypergéométriques de deux variables. *C. R.*, xcv, 717-719.

——Sur l'intégration algébrique d'une classe d'équations linéaires. *C. R.*, xcvi, 323-325.

GYLDÉN, H. Application nouvelle de l'équation de Lamé. *C. R.*, xcii, 537-538.

——Sur quelques équations différentielles linéaires du second ordre. *C. R.*, xc, 344-345.

HAAG. Note sur une classe d'équations différentielles. *Bull. S. M. F.*, viii, 80-81.

HALPHEN, G. H. Sur la réduction des équations linéaires aux formes intégrables. *Mémoires présentés par divers sauvants*, lxxviii, No. 1 [Paris (1883)].

——Sur une équation différentielle linéaire du troisème ordre. *Math. Annalen*, xxiv Band, 461-464.

——Sur les invariants des équations différentielles linéaires du quatrième ordre. *Acta Math.* (3), 1883-1884, 325-380.

——Sur une classe d'équations différentielles linéaires. *C. R.*, xcii, 779-782.

——Sur des fonctions qui proviennent de l'équation de Gauss. *C. R.*, xcii, 856-859.

——Sur les multiplicateurs des équations différentielles linéaires. *C. R.*, xcvii, 1408-1411, 1541-1544.

HAMBURGER, M. Bemerkungen über die Form der Integrale der linearen Differentialgleichungen mit veränderlichen Coefficienten. *Borchardt J.*, lxxvi, 113-126.

HAZZIDAKIS, J. N. Ueber eine Eigenschaft der Systeme von linearen homogenen Differentialgleichungen. *Borchardt J.*, xc, 80-82.

HERMITE, CH. Equations differentialcs linéaires. *Darboux Bull.* (2), iii, 311-325.

——Sur l'équation de Lamé. *Brioschi Ann.* (2), ix, 21-24.

——Sur les équations différentielles linéaires du second ordre. *Brioschi Ann.* (2), x, 101-104.

——Lettre à M. Fuchs. *Borchardt J.*, lxxix, 324-338.

——Sur l'intégration de l'équation différentielle de Lamé. *Borchardt J.*, lxxxix, 9-19.

——Sur quelques applications des fonctions elliptiques (seven papers), *C. R.*, lxxxv; (five papers) *C. R.*, lxxxvi; (two papers) *C. R.*, lxxxix; (five papers) *C. R.*, xc; (two papers) *C. R.*, xciii; (five papers) *C. R.*, xciv.

HOSENFELDER, E. Ueber die Integration einer linearen Differentialgleichung n^{ter} Ordnung. *Clebsch Ann.*, iv, 195-212.

HUMBERT, G. Sur l'équation hypergéométrique. *Bull. S. M. F.*, viii, 112-120.

——Sur une formule de M. Hermite. *Bull. S. M. F.*, ix, 42-46.

JORDAN, B. Sur la résolution des équations différentielles linéaires. *C. R.*, lxxiii, 787-791.

JORDAN, C. Sur une application de la théorie des substitutions aux équations différentielles linéaires. *C. R.*, lxxviii, 741-743.

——Sur les équations linéaires du second ordre dont les intégrales sont algébriques. *C. R.*, lxxxii, 605-607; lxxxiii, 1033-1037.

——Mémoire sur les équations différentielles linéaires à intégrale algébrique. *Borchardt J.*, lxxxiv, 89-215.

——Détermination des groupes formés d'un nombre fini de substitutions. *C. R.*, lxxxiv, 1446-1448.

JÜRGENS, E. Zur Theorie der linearen Differentialgleichungen mit veränderlichen Coefficienten. Heidelberg.

——Das Integral $\int_a^\beta \frac{y dz}{x-z}$ und die linearen Differentialgleichungen. *Klein Ann.* xix, 435-461.

——Die Form der Integrale der linearen Differentialgleichungen. *Borchardt J.*, lxxx, 150-168.

KLEIN, F. Ueber lineare Differentialgleichungen. *Erl. Ber.*, 1876.

——Sur les équations différentielles linéaires. *Darboux Bull.* (2), i, 180-184. Uebersetzung derselben Arbeit.

——Ueber lineare Differentialgleichungen. *Clebsch Ann.*, xii, 167-180.

——Ueber lineare Differentialgleichungen. *Clebsch Ann.*, xi, 115-119. Abdruck aus den *Erl. Ber.*, Juni 1876 (vgl. *F. d. M.*, viii, p. 189).

KÖNIGSBERGER, L. Allgemeine Untersuchungen aus der Theorie der Differentialgleichungen. Leipzig: *Teubner.*

——Bemerkung zur Theorie der algebraischen Integrale linearer Differentialgleichungen. *Math. Annalen*, Band 21, 454-456.

——Beziehungen zwischen den Fundamentalintegralen einer linearen homogenen Differentialgleichung zweiter Ordnung. *Math. Annalen*, Band 22, 269-289.

——Ueber die Irreductibilität von Differentialgleichungen. *Gött. Nachr.*, 1881, 222-225.

——Ueber algebraisch-logarithmische Integrale nicht homogener linearer Differentialgleichungen. *Gött. Nachr.*, 1880, 453-455.

——Ueber den Zusammenhang zwischen dem allgemeinen und den particulären Integralen von Differentialgleichungen. *Gött. Nachr.*, 1880, 625-630.

——Ueber algebraische Beziehungen zwischen Integralen verschiedener Differentialgleichungen. *Borchardt J.*, lxxxiv, 284-294.

——Algebraisch-logarithmische Integrale nicht homogener linearer Differentialgleichungen. *Borchardt J.*, xc, 267-281.

——Ueber die Irreductibilität von Differentialgleichungen. *Borchardt J.*, xcii, 291-301.

——Eigenschaften der algebraisch-logarithmischen Integrale linearer nicht homogener Differentialgleichungen. *Borchardt J.*, xciv, 291-311.

——Eigenschaften irreductibler Function. *Borchardt J.*, xcv, 171-196.

——Ueber die Irreductibilität der linearen Differentialgleichungen. *Borchardt J.*, xcvi, 123-151.

LAGUERRE. Remarques sur les équations différentielles linéaires du second ordre. *Bull. S. M. F.*, viii, 35-36.

——Sur les équations différentielles linéaires du troisième ordre. *C. R.*, lxxxviii, 116-119.

——Sur quelques invariants des équations différentielles. *C. R.*, lxxxviii, 224-227.

LEMONNIER, H. Conditions pour que deux équations différentielles linéaires sans second membre aient p solutions communes. Équation qui donne ces solutions. *C. R.*, xcv, 476-479.

MANSION, P. Démonstration de la propriété fondamentale des équations différentielles linéaires. *Bull. de Belg.* (2), xxxviii, 578-591.

MAYER, A. Ueber unbeschränkt integrable System von linearen totalen Differentialgleichungen. *Clebsch Ann.*, v, 448-470.

MITTAG-LEFFLER, G. Integration utaf en klass af lineera differential-equationer. *Öfv. v. Stock.* '79.

——Sur la théorie des équations différentielles linéaires. *C. R.*, xc, 218-221.

——Sur les équations différentielles linéaires à coefficients doublement périodiques. *C. R.*, xc, 299-300.

——Sur les équations différentielles linéaires du second ordre. *C. R.*, xci, 978-980.

MOUTARD. Sur les équations différentielles linéaires du second ordre. *C. R.*, lxxx, 729-733.

PEPIN, TH. Méthode pour obtenir les intégrales algébriques des équations différentielles linéaires du second ordre. *Rom. Acc. P. d. V. L.*, xxxiv, 243-389.

——Sur les équations différentielles du second ordre. *Brioschi Ann.* (2), ix, 1-11.

PEPIN, A. Sur les équations linéaires du second ordre. *C. R.*, lxxxii, 1323-1326.

PFANNENSTIEL, E. Bidrag tell de liniära differentialeqvationernas teori. *Goteborg Handl.*, 1882, 1-50.

PICARD, E. Sur les équations différentielles linéaires à coefficients doublement périodiques. *Borchardt J.*, xc, 281-303.

——Sur la forme des intégrales des équations différentielles du second ordre dans le voisinage de certains points critiques. *C. R.*, lxxxvii, 430-432, 743-745.

——Sur certaines équations différentielles linéaires du second ordre. *C. R.*, xc, 1479-1482.

——Sur les équations linéaires simultanées. *C. R.*, xc, 1065-1067.

——Sur une classe d'équations différentielles linéaires. *C. R.*, xc, 128-131.

——Sur les équations différentielles linéaires à coefficients doublement périodiques. *C. R.*, xc, 293-296.

——Sur une généralisation des fonctions périodiques et sur certaines équations différentielles linéaires. *C. R.*, lxxxix, 140-144.

——Sur les formes des intégrales de certaines équations différentielles linéaires. *C. R.*, xciv, 418-421.

——Sur les groupes de transformation des équations différentielles linéaires. *C. R.*, xcvi, 1131-1134.

POCHHAMMER, L. Notiz über die Herleitung der hypergeometrischen Differentialgleichung. *Borchardt J.*, lxxiii, 85-87.
——Ueber einfach singuläre Punkte linearer Differentialgleichungen. *Borchardt J.*, lxxiii, 69-85.
——Ueber Relationen zwischen den hypergeometrischen Integralen n^{ter} Ordnung. *Borchardt J.*, lxxiii, 135-159.
POINCARÉ, H. Sur les équations différentielles linéaires à intégrales algébriques. *C. R.*, xcii, 698-701.
——Sur les Equations Linéaires aux Différentielles ordinaires et aux Différences finies. *Amer. Jour. of Math.*, Vol. vii, No. 3, 203-259.
——Sur les propriétés des fonctions définées par les équations différentielles. *Jour. de l'Ecole Polyt.*, xxviii, 19-26.
——Sur l'intégration des équations linéaires par le moyen des fonction abéliennes. *C. R.*, xcii, 913-915.
——Sur une classe d'invariants relatifs aux équations linéaires. *C. R.*, xciv, 1402-1405.
——Sur le groupe des équations linéaires. *C. R.*, xcvi, 691-694.
——Sur les groupes des équations linéaires. *C. R.*, xcvi, 1302-1304.
——Sur l'intégration algébrique des équations linéaires. *C. R.*, xcvii, 984-986, 1189-1191.
RICCI, G. Sopra un sistema di due equazioni differenziali lineari. *Battaglini G.*, xv, 135-154.
SAUVAGE, L. Sur les propriétés des fonctions définies par un système d'équations différentielles linéaires et homogènes à une ou plusieurs variables indépendantes. *Ann. de l'Ecole Norm.* (2), xi, 33-78.
SEIFERT, W. Ueber die Integration einer Differentialgleichung. *Diss. Göttingen.*
SEYDLER, A. Bemerkung zur Integration einiger linearen Differentialgleichungen. *Casopis I.*, 195-196 (Bohemian).
Cte DE SPARRE. Sur une équation différentielle linéaire du second ordre. Deux mémoires. *Acta Math.* (3), 1883-1884, 105-139, 289-321.
STARKOF. Sur l'intégration des équations linéaires. *N. C. M. V.*, 225-230.
STARKOFF, A. Zur Frage über die Integration linearer Differentialgleichungen mit veränderlichen Coefficienten. Odessa: *H. Ulrich.*
STICKELBERGER, L. Zur Theorie der linearen Differentialgleichungen. Leipzig: *Teubner.*
TANNERY, J. Sur l'équation différentielle linéaire qui relie au module de la fonction complète de première espèce. *C. R.*, lxxxvi, 811-812.

——Sur quelques propriétés des fonctions complètes de première espèce. *C. R.*, lxxxvi, 950-953.

——Propriétés des intégrales des équations différentielles linéaires à coefficients variables. *Ann. de l'Ecole Norm.* (2), iv, 113-182.

——Sur une équation différentielle linéaire du second ordre. *Ann. de l'Ecole Norm.* (2), viii, 168-184.

TARDY, P. Relazioni tra le radici di alcune equazioni fondamentali determinanti. *Atti della R. Accad. di Torino*, Vol. xix, May, 1884, 835-848.

THOMAE, J. Integration einer linearen Differentialgleichung zweiter Ordnung durch Gauss'sche Reihen. *Schlömilch Z.*, xix, 273-286.

THOMÉ, L. W. Zur Theorie der linearen Differentialgleichungen. *Borchardt J.*, lxxiv, 193-218.

——Zur Theorie der linearen Differentialgleichungen. *Borchardt J.*, lxxv, 265-291; lxxvi, 273-291.

——Zur Theorie der linearen Differentialgleichungen. *Borchardt J.*, lxxviii, 223-245.

——Zur Theorie der linearen Differentialgleichungen. *Borchardt J.*, lxxxi, 1-32.

——Zur Theorie der linearen Differentialgleichungen. *Borchardt J.*, lxxxiii, 89-111.

——Zur Theorie der linearen Differentialgleichungen. *Borchardt J.*, lxxxvii, 222-350.

——Zur Theorie der linearen Differentialgleichungen. *Borchardt J.*, xci, 78-198, 341-346.

——Zur Theorie der linearen Differentialgleichungen. *Borchardt J.*, xcv, 44-98.

——Zur Theorie der linearen Differentialgleichungen. *Borchardt J.*, xcvi, 185-281.

TICHOMANDRITZKY, M. Ueber hypergeometrische Reihen. *Diss. St. Pet.*, 1876 (Russian).

TRUDI, N. Teoria delle equazioni differenziali lineari. *Atti di Napoli*, vi, 71.

VILLARCEAU, Y. Sur l'intégration des équations linéaires. *C. R.*, xci, 13-14.

WINCKLER, A. Integration zweier linearer Differentialgleichungen. *Wien. Ber.*, lxxi.

——Ueber die Integration des linearen Differentialgleichungen zweiter Ordnung. *Wien. Ber.*, lxxv.

——Aeltere und neuere Methoden, lineare Differentialgleichungen durch einfache bestimmte Integrale aufzulösen. Wien: *A. Hölder.*

The Addition-Theorem for Elliptic Functions.

BY WILLIAM E. STORY.

The form of the addition-theorem given below [(33)–(35)] is attributed by Clebsch* to Hermite,† whose note I have not seen, but the same result, presumably obtained by the same method, is given by Bertrand‡ and Koenigsberger§; of the two latter writers Koenigsberger alone investigates the effect of the equality of two or more of the arguments added, and neither considers the validity of the result when a certain intermediate equation (9) has equal roots. For this reason, and because the treatises cited are probably inaccessible to many American students, it seems allowable to present, even in a journal devoted to original research, the whole investigation in a brief but practically complete form.

Let $R(z)$ be a given cubic or quartic polynomial in z; we are concerned with the $2m-1$ (where m is any positive integer) integrals

$$(1) \quad v_1 = \int_{z_1'}^{z_1} \frac{dz}{\sqrt{R(z)}}, \quad v_2 = \int_{z_2'}^{z_2} \frac{dr}{\sqrt{R(z)}}, \quad v_3 = \int_{z_3'}^{z_3} \frac{dr}{\sqrt{R(z)}}, \ldots, \quad v_{2m-1} = \int_{z_{2m-1}'}^{z_{2m-1}} \frac{dz}{\sqrt{R(z)}},$$

whose upper limits $z_1, z_2, z_3, \ldots, z_{2m-1}$, and lower limits $z_1', z_2', z_3', \ldots, z_{2m-1}'$ have any given values, and the sign of $\sqrt{R(z)}$ for any value of z is determined by any convention consistent with continuity. It is to be observed that the number of integrals is odd. Now if $p(z)$ or p is an arbitrary polynomial in z of degree m, and $q(z)$ or q an arbitrary polynomial of degree not exceeding $m-2$, then $p - q\sqrt{R(z)}$ contains $m + 1 + m - 1 = 2m$ arbitrary coefficients, which (*i. e.* whose ratios) may be so taken that

$$(2) \qquad p - q\sqrt{R(z)} = 0$$

---

* Geometrie, I, p. 605, footnote.

† Note sur le calcul différentiel et le calcul intégral, in Lacroix: Calcul diff. et int., 6th ed., Paris, 1862, p. 68.

‡ Calcul intégral, pp. 578-588.

§ Elliptische Functionen, II, pp. 1-17.

for each of the $2m-1$ upper limits of the integrals (1); and if these upper limits are all different, this determination of the relative coefficients of (2) is unique, *i. e.* p and q are determined to a common factor *près*. Then (1) rationalized gives

(3) $$p^2-q^2R(z)=0,$$

a rational equation of the degree $2m$ satisfied by the $2m-1$ given upper limits and therefore by one other value, say z_{2m}, which is thus completely determined by the $2m-1$ given values. Then

(4) $$p^2-q^2R(z)\equiv A(z-z_1)(z-z_2)(z-z_3)\ldots(z-z_{2m}),$$

where A is a constant (depending on the common arbitrary factor of p and q). If the given upper limits are not all different, suppose μ_1 of them equal to z_1, μ_2 equal to $z_2,\ldots,\mu_r$ equal to z_r, so that $\mu_1+\mu_2+\ldots+\mu_r=2m-1$, then the coefficients of p and q can be determined, to a common factor *près*, in only one way, so that

for z_1, $p-q\sqrt{R(z)}$ and its first μ_1-1 derivatives shall vanish,
" z_2, $p-q\sqrt{R(z)}$ " " μ_2-1 " "
. . . .
. . . .
" z_r, $p-q\sqrt{R(z)}$ " " μ_r-1 " " .

Now it is easily seen that if, for any value of z, $p-q\sqrt{R(z)}$ and its first $\mu-1$ derivatives vanish, then also will $p^2-q^2R(z)$ and its first $\mu-1$ derivatives vanish for the same value of z; hence

(5) $$p^2-q^2R(z)\equiv A(z-z_1)^{\mu_1}(z-z_2)^{\mu_2}\ldots(z-z_r)^{\mu_r}(z-z_{2m}),$$

where z_{2m} is a value determined by the $2m-1$ given upper limits, and A is a constant. Similarly if $p_1(z)$ or p_1 is a polynomial of degree m in z, and $q_1(z)$ or q_1 a polynomial in z of degree not exceeding $m-2$, the coefficients of p_1 and q_1 can be determined in one way only, to a constant factor *près*, so that

(6) $$p_1-q_1\sqrt{R(z)}=0$$

for each of the $2m-1$ lower limits of the integral (1), if these lower limits are all different; if any lower limit z' occurs μ times, then p_1 and q_1 are to be so determined that $p_1-q_1\sqrt{R(z)}$ and its first $\mu-1$ derivatives shall vanish for $z=z'$; and the coefficients of p_1 and q_1 so taken determine a value z'_{2m} so connected with the $2m-1$ given lower limits that

(7) $$p_1^2-q_1^2R(z)\equiv B(z-z'_1)(z-z'_2)(z-z'_3)\ldots(z-z'_{2m}),$$

where B is a constant. The value of z_{2m} satisfies (2) as well as (3), and z'_{2m} satisfies

(6), viz. the sign of $\sqrt{R(z)}$ is to be so taken for z_{2m} and z'_{2m} that these equations shall be satisfied. The values z_{2m} and z'_{2m} determine another integral

(8) $$v_{2m}=\int_{z'_{2m}}^{z_{2m}}\frac{dz}{\sqrt{R(z)}},$$

whose relation to the $2m-1$ given integrals we have to investigate.

Let a new variable λ be introduced, and for any given value of λ let $\zeta_1, \zeta_2, \zeta_3, \ldots \zeta_{2m}$ be the $2m$ values of z which satisfy the equation

(8) $$(p+\lambda p_1)-(q+\lambda q_1)\sqrt{R(z)}=0, \text{ or}$$

(9) $$(p+\lambda p_1)^2-(q+\lambda q_1)^2 R(z)=0,$$

so that

(10) $$(p+\lambda p_1)^2-(q+\lambda q_1)^2 R(z)\equiv\psi(z)\equiv C(z-\zeta_1)(z-\zeta_2)(z-\zeta_3)\ldots(z-\zeta_{2m}).$$

If λ varies continuously from 0 to ∞, the roots of (10) vary continuously from $z_1, z_2, z_3, \ldots, z_{2m}$ to $z'_1, z'_2, z'_3, \ldots, z_{2m}$. It is of no consequence if any upper limit does not pass into the lower limit of the same integral by this continuous variation of λ. Since $\zeta_1, \zeta_2, \zeta_3, \ldots \zeta_{2m}$ and C are functions of λ defined by (10), *i. e.* this equation is an identity, we may differentiate it with respect to λ and obtain

(11) $$\left\{\begin{aligned} 2(p+\lambda p_1)p_1-2(q+\lambda q_1)q_1 R(z) &\equiv \frac{2(p+\lambda p_1)}{q+\lambda q_1}\left[(q_1+\lambda q_1)p_1-\frac{(q+\lambda q_1)^2}{p+\lambda p_1}R(z)\right] \\ &\equiv \psi(z)\left[\frac{\frac{\partial\zeta_1}{\partial\lambda}}{\zeta_1-z}+\frac{\frac{\partial\zeta_2}{\partial\lambda}}{\zeta_2-z}+\ldots+\frac{\frac{\partial\zeta_{2m}}{\partial\lambda}}{\zeta_{2m}-z}+\frac{\frac{\partial C}{\partial\lambda}}{C}\right];\end{aligned}\right.$$

but $$(q+\lambda q_1)^2 R(z)\equiv(p+\lambda p_1)^2-\psi(z),$$

$$(q+\lambda q_1)p_1-\frac{(q+\lambda q_1)^2 q_1}{p+\lambda p_1}R(z)\equiv(qp_1-pq_1)+\frac{q_1\psi(z)}{p+\lambda p_1},$$

i. e. (11) may be written

(12) $$2\frac{(p+\lambda p_1)}{q+\lambda q_1}(qp_1-pq_1)+2\frac{q_1\psi(z)}{q+\lambda q_1}\equiv\psi(z)\left[\frac{\frac{\partial\zeta_1}{\partial\lambda}}{\zeta_1-z}+\frac{\frac{\partial\zeta_2}{\partial\lambda}}{\zeta_2-z}+\ldots+\frac{\frac{\partial\zeta_{2m}}{\partial\lambda}}{\zeta_{2m}-z}+\frac{\frac{\partial C}{\partial\lambda}}{C}\right].$$

If α represents any one of the numbers $1, 2, 3, \ldots 2m$,

$$\psi(\zeta_\alpha)=0,\ \left(\frac{\psi(z)}{\zeta_\alpha-z}\right)_{z=\zeta_\alpha}=-\frac{\partial\psi(\zeta_\alpha)}{\partial\zeta_\alpha}=-\Psi(\zeta_\alpha),$$

and by (8) $$\frac{p(\zeta_\alpha)+\lambda p_1(\zeta_\alpha)}{q(\zeta_\alpha)+\lambda q_1(\zeta_\alpha)}=\sqrt{R(\zeta_\alpha)};$$

and (12) gives

$$2\sqrt{R(\zeta_\alpha)}\,[q(\zeta_\alpha)p_1(\zeta_\alpha)-p(\zeta_\alpha)q_1(\zeta_\alpha)]=-\Psi(\zeta_\alpha)\frac{\partial\zeta_\alpha}{\partial\lambda},$$

i. e.

$$(13)\qquad \frac{\frac{\partial\zeta_a}{\partial\lambda}}{\sqrt{R(\zeta_a)}} = -2\frac{q(\zeta_a)p_1(\zeta_a)-p(\zeta_a)q_1(\zeta_a)}{\psi'(\zeta_a)},$$

and hence

$$(14)\qquad \sum_1^{2m}{}_a \frac{\frac{\partial\zeta_a}{\partial\lambda}}{\sqrt{R(\zeta_a)}} = -2\sum_1^{2m}{}_a \frac{q(\zeta_a)p_1(\zeta_a)-p(\zeta_a)q_1(\zeta_a)}{\psi'(\zeta_a)} = 0,$$

by a well-known theorem of rational fractions, since $qp_1 - pq_1$ is of degree not higher than $2m-2$, and $\psi(z)$ is of degree m (see Todhunter's Theory of Equations, p. 325, example 13). If ζ_a is a multiple root of (9), say of order $\mu_a \overline{\gtrless} 2$, then $\frac{\psi(z)}{z-\zeta_a}$ contains $(z-\zeta_a)^{\mu_a-1}$ and (12) shows that qp_1-pq_1 contains $(z-\zeta_a)^{\mu_a-1}$, since $\left(\frac{p+\lambda p_1}{q+\lambda q_1}\right)_{z=\zeta_a} = \sqrt{R(\zeta_a)}$ does not in general vanish. But differentiating (12) $\mu_a - 1$ times and putting $z=\zeta_a$ we obtain

$$2\frac{\partial^{\mu_a-1}}{\partial\zeta_a^{\mu_a-1}}\left[\{q(\zeta_a)p_1(\zeta_a)-p(\zeta_a)q(\zeta_a)\}\sqrt{R(\zeta_a)}\right] = -\frac{\partial^{\mu_a}\psi(\zeta_a)}{\partial\zeta_a^{\mu_a}}\frac{\partial\zeta_a}{\partial\lambda}$$
$$= 2\sqrt{R(\zeta_a)}\frac{\partial^{\mu_a-1}}{\partial\zeta_a^{\mu_a-1}}[q(\zeta_a)p_1(\zeta_a)-p(\zeta_a)q_1(\zeta_a)],$$

i. e.

$$(15)\qquad \frac{\frac{\partial\zeta_a}{\partial\lambda}}{\sqrt{R(\zeta_a)}} = -2\frac{\frac{\partial^{\mu_a-1}}{\partial\zeta_a^{\mu_a-1}}[q(\zeta_a)p_1(\zeta_a)-p(\zeta_a)q_1(\zeta_a)]}{\frac{\partial^{\mu_a}\psi(\zeta_a)}{\partial\zeta_a^{\mu_a}}}$$
$$= -2\frac{\partial^{\mu_a-1}}{\partial\zeta_a^{\mu_a-1}}\left[\frac{q(\zeta_a)p_1(\zeta_a)-p(\zeta_a)q_1(\zeta_a)}{\frac{\partial^{\mu_a}\psi(\zeta_a)}{\partial\zeta_a^{\mu_a}}}\right].$$

Now Jacobi has shown* that if $\phi(x)$ and $\psi(x)$ are polynomials in x such that the degree of $\psi(x)$ exceeds that of $\phi(x)$ by at least one unit, and if a is a multiple root of order μ of $\psi(z)=0$, so that $\psi(x)\equiv(x-a)^\mu\psi_1(x)$, then the μ terms in the development of $\frac{\varphi(x)}{\psi(x)}$ in partial fractions whose denominators have become equal to a are replaced by

$$\frac{\frac{\varphi(a)}{\psi_1(a)}}{(x-a)^\mu} + \frac{\frac{\partial}{\partial a}\left(\frac{\varphi(a)}{\psi_1(a)}\right)}{(x-a)^{\mu-1}} + \frac{\frac{\partial^2}{\partial a^2}\left(\frac{\varphi(a)}{\psi_1(a)}\right)}{2!\,(x-a)^{\mu-2}} + \dots + \frac{\frac{\partial^{\mu-1}}{\partial a^{\mu-1}}\left(\frac{\varphi(a)}{\psi_1(a)}\right)}{(\mu-1)!\,(x-a)}$$
$$= \frac{1}{(\mu-1)!}\frac{\partial^{\mu-1}}{\partial a^{\mu-1}}\left(\frac{\varphi(a)}{(x-a)\psi_1(a)}\right) = \mu\frac{\partial^{\mu-1}}{\partial a^{\mu-1}}\left(\frac{\varphi(a)}{(x-a)\frac{\partial^\mu\psi(a)}{\partial a^\mu}}\right),$$

---

* Disquisitiones analyticae de fractionibus simplicibus. Inaugural dissertation. Werke, herausgegeben von Weierstrass, Vol. III, p. 11.

so that

$$(16) \qquad \frac{\varphi(x)}{\psi(x)} \equiv \sum \mu \frac{\partial^{\mu-1}}{\partial a^{\mu-1}} \left(\frac{\varphi(a)}{(x-a)\dfrac{\partial^{\mu}\psi(a)}{\partial a^{\mu}}} \right),$$

where the summation extends to every root a of $\psi(x) = 0$ and its index of multiplicity μ. Writing $xf(x)$ instead of $\phi(x)$, where $f(x)$ is any polynomial whose degree falls short of that of $\psi(x)$ by at least two units, we obtain from (16)

$$\frac{xf(x)}{\psi(x)} \equiv \sum \mu \frac{\partial^{\mu-1}}{\partial a^{\mu-1}} \left(\frac{af(a)}{(x-a)\dfrac{\partial^{\mu}\psi(a)}{\partial a^{\mu}}} \right),$$

and hence, for $x = 0$,

$$(17) \qquad 0 = \sum \mu \frac{\partial^{\mu-1}}{\partial a^{\mu-1}} \left(\frac{af(a)}{-a\dfrac{\partial^{\mu}\psi(a)}{\partial a^{\mu}}} \right) = -\sum \mu \frac{\partial^{\mu-1}}{\partial a^{\mu-1}} \left(\frac{f(a)}{\dfrac{\partial^{\mu}\psi(a)}{\partial a^{\mu}}} \right),$$

which is the generalization of the formula cited in connection with (14). This formula applied to (15) gives

$$(18) \qquad \sum \mu_a \frac{\dfrac{\partial \zeta_a}{\partial \lambda}}{\sqrt{R(\zeta_a)}} = 0,$$

where the summation extends to every root ζ_a of (9) and its index of multiplicity μ_a. But (18) is only what (14) becomes when μ_a values ζ_a are equal. We have then

$$\sum_1^{2m} {}_a \frac{\dfrac{\partial \zeta_a}{\partial \lambda}}{\sqrt{R(\zeta_a)}} = 0,$$

where the summation extends to $\zeta_1, \zeta_2, \zeta_3, \ldots \zeta_{2m}$, whether these are all different or not. Hence

$$(19) \qquad \sum_1^{2m} {}_a \int_{\infty}^{0} \frac{\dfrac{\partial \zeta_a}{\partial \lambda}}{\sqrt{R(\zeta_a)}} d\lambda = \sum_1^{2m} {}_a \int_{z'_a}^{z_a} \frac{d\zeta_a}{\sqrt{R(\zeta_a)}} = \sum_1^{2m} {}_a v_a = 0;$$

i. e.

$$(20) \qquad v_{2m} = -(v_1 + v_2 + v_3 + \ldots + v_{2m-1}),$$

which is the relation between v_{2m} and the $2m - 1$ given integrals corresponding to the relations above mentioned between z_{2m} and the $2m - 1$ given upper limits and between z'_{2m} and the $2m - 1$ given lower limits. The latter relations may be put into a simpler form (evidently this is only one of $2m$ analogous relations), viz. (4) and (7) give, if the constants A and B be taken equal to unity,

$$p^2(0) - q^2(0)R(0) = z_1 z_2 z_3 \ldots z_{2m}, \quad p_1^2(0) - q_1^2(0)R(0) = z'_1 z'_2 z'_3 \ldots z'_{2m},$$

i. e.

(21) $$z_{2m}=\frac{p^2(0)-q^2(0)\,R(0)}{z_1z_2z_3\ldots z_{2m-1}},\quad z'_{2m}=\frac{p_1^2(0)-q_1^2(0)\,R(0)}{z'_1z'_2z'_3\ldots z'_{2m-1}}.$$

In particular if

(22) $$R(z)\equiv z(1-z)(1-k^2z),\ z=x^2,\ x=\operatorname{sn}(u,k),$$

then

(23) $$\left\{\begin{array}{l}\sqrt{R(z)}\equiv \operatorname{sn}u\operatorname{cn}u\operatorname{dn}u,\ R(0)=0,\ R(1)=0,\ R\left(\frac{1}{k^2}\right)=0,\\ \displaystyle\int_0^z\frac{dz}{\sqrt{R(z)}}=2\int_0^x\frac{dx}{\sqrt{(1-x^2)(1-k^2x^2)}}=2u,\end{array}\right.$$

and if
$$z_1=\operatorname{sn}^2u_1,\ z_2=\operatorname{sn}^2u_2,\ z_3=\operatorname{sn}^2u_3,\ \ldots\ z_{2m}=\operatorname{sn}^2u_{2m},$$
$$z'_1=\operatorname{sn}^2u'_1,\ z'_2=\operatorname{sn}^2u'_2,\ z'_3=\operatorname{sn}^2u'_3,\ \ldots\ z'_{2m}=\operatorname{sn}^2u'_{2m},$$

then

$$\int_{z'_\alpha}^{z_\alpha}\frac{dz}{\sqrt{R(z)}}=2(u_\alpha-u'_\alpha),$$

and (19) and (21) become

(24) $$\sum_1^{2m}{}_\alpha\,u_\alpha=\sum_1^{2m}{}_\alpha\,u'_\alpha,$$

$$z_{2m}=\frac{p^2(0)}{z_1z_2z_3\ldots z_{2m-1}},\quad z'_{2m}=\frac{p_1^2(0)}{z'_1z'_2z'_3\ldots z'_{2m-1}},$$

i. e.

(25) $$\operatorname{sn}u_{2m}=\pm\frac{p(0)}{\operatorname{sn}u_1\operatorname{sn}u_2\operatorname{sn}u_3\ldots\operatorname{sn}u_{2m-1}}.$$

Assuming the same particular form of $R(z)$ and the same values of A and B in (4) and (7), we obtain from the two latter equations

$$p^2(1)-q^2(1)\,R(1)=p^2(1)=(1-z_1)(1-z_2)(1-z_3)\ \ldots\ (1-z_{2m})$$
$$=\operatorname{cn}^2u_1\operatorname{cn}^2u_2\operatorname{cn}^2u_3\ldots\operatorname{cn}^2u_{2m},$$

$$p^2\left(\frac{1}{k^2}\right)-q^2\left(\frac{1}{k^2}\right)R\left(\frac{1}{k^2}\right)=p^2\left(\frac{1}{k^2}\right)=\frac{1}{k^{4m}}(1-k^2z_1)(1-k^2z_2)(1-k^2z_3)\ldots(1-k^2z_{2m})$$
$$=\frac{1}{k^{4m}}\operatorname{dn}^2u_1\operatorname{dn}^2u_2\operatorname{dn}^2u_3\ldots\operatorname{dn}^2u_{2m};$$

and hence follow

(26) $$\operatorname{cn}(u_{2m})=\pm\frac{p(1)}{\operatorname{cn}u_1\operatorname{cn}u_2\operatorname{cn}u_3\ldots\operatorname{cn}u_{2m-1}},$$

(27) $$\operatorname{dn}(u_{2m})=\pm\frac{k^{2m}p\left(\frac{1}{k^2}\right)}{\operatorname{dn}u_1\operatorname{dn}u_2\operatorname{dn}u_3\ldots\operatorname{dn}u_{2m-1}}.$$

It is to be observed that the signs of the right members of (25), (26) and (27) have yet to be determined.

If we assume still further

$$(28)\qquad z_1' = z_2' = z_3' = \ldots = z_{2m-1}' = 0,$$

we have

$$(29)\qquad v_1' = v_2' = v_3' = \ldots = v_{2m-1}' = 0,$$

and, by (7),

$$(30)\qquad p_1^2 - q_1^2 R(z) \equiv z^{2m-1}(z - z_{2m}');$$

now the degree of the lowest term in p_1^2 is even and that of the lowest term in $q_1^2 R(z)$ is odd, so that these terms cannot cancel each other, and the degree of the second (the apparently lowest) term in $z^{2m-1}(z - z_{2m}')$ is $2m - 1$; this second term cannot arise from p_1^2 since it is of odd degree, and it cannot arise from $q_1^2 R(z)$ since the lowest term in $q_1^2 R(z)$ is of degree not higher than $2m - 4 + 1 = 2m - 3$; therefore it does not exist, *i. e.*

$$(31)\qquad z_{2m}' = 0, \quad v_{2m}' = 0,$$

and (24) becomes

$$\sum_1^{2m} u_\alpha = 0,$$

i. e.

$$(32)\qquad u_{2m} = -(u_1 + u_2 + u_3 + \ldots + u_{2m-1}).$$

Equations (25), (26) and (27) may then be written

$$(33)\qquad \operatorname{sn}(u_1 + u_2 + u_3 + \ldots + u_{2m-1}) = \pm \frac{p(0)}{\operatorname{sn} u_1 \operatorname{sn} u_2 \operatorname{sn} u_3 \ldots \operatorname{sn} u_{2m-1}},$$

$$(34)\qquad \operatorname{cn}(u_1 + u_2 + u_3 + \ldots + u_{2m-1}) = \pm \frac{p(1)}{\operatorname{cn} u_1 \operatorname{cn} u_2 \operatorname{cn} u_3 \ldots \operatorname{cn} u_{2m-1}},$$

$$(35)\qquad \operatorname{dn}(u_1 + u_2 + u_3 + \ldots + u_{2m-1}) = \pm \frac{k^{2m} p\left(\frac{1}{k^2}\right)}{\operatorname{dn} u_1 \operatorname{dn} u_2 \operatorname{dn} u_3 \ldots \operatorname{dn} u_{2m-1}},$$

where the coefficients of p are to be determined by the conditions derived from (2), namely

$$(36)\qquad \begin{cases} p(\operatorname{sn}^2 u_1) - q(\operatorname{sn}^2 u_1).\operatorname{sn} u_1 \operatorname{cn} u_1 \operatorname{dn} u_1 = 0, \\ p(\operatorname{sn}^2 u_2) - q(\operatorname{sn}^2 u_2).\operatorname{sn} u_2 \operatorname{cn} u_2 \operatorname{dn} u_2 = 0, \\ p(\operatorname{sn}^2 u_3) - q(\operatorname{sn}^2 u_3).\operatorname{sn} u_3 \operatorname{cn} u_3 \operatorname{dn} u_3 = 0, \\ \cdots\cdots\cdots\cdots\cdots\cdots \\ p(\operatorname{sn}^2 u_{2m-1}) - q(\operatorname{sn}^2 u_{2m-1}).\operatorname{sn} u_{2m-1} \operatorname{cn} u_{2m-1} \operatorname{dn} u_{2m-1} = 0. \end{cases}$$

If we put for convenience

$$\operatorname{sn} u = s,\ \operatorname{cn} u = c,\ \operatorname{dn} u = d,\ \operatorname{sn} u_\alpha = s_\alpha,\ \operatorname{cn} u_\alpha = c_\alpha,\ \operatorname{dn} u_\alpha = d_\alpha,$$

we may write

$$(37)\qquad \begin{cases} p(z) - q(z)\sqrt{R(z)} \equiv s^{2m} + a_1 s^{2m-2} + a_2 s^{2m-4} + \ldots + a_m \\ \qquad\qquad + (b_2 s^{2m-4} + b_3 s^{2m-6} + \ldots + b_m)\, scd, \end{cases}$$

which must then vanish for $s = s_1, s_2, s_3, \ldots, s_{2m-1}$, *i. e.* $a_1, a_2, \ldots, a_m, b_2, \ldots, b_m$ are determined by the linear equations

$$(38)\qquad \begin{cases} 0 = s_1^{2m} + a_1 s_1^{2m-2} + a_2 s_1^{2m-4} + \ldots + a_m + (b_2 s_1^{2m-4} + b_3 s_1^{2m-6} + \ldots + b_m)\, s_1 c_1 d_1, \\ 0 = s_2^{2m} + a_1 s_2^{2m-2} + a_2 s_2^{2m-4} + \ldots + a_m + (b_2 s_2^{2m-4} + b_3 s_2^{2m-6} + \ldots + b_m)\, s_2 c_2 d_2, \\ 0 = s_3^{2m} + a_1 s_3^{2m-2} + a_2 s_3^{2m-4} + \ldots + a_m + (b_2 s_3^{2m-4} + b_3 s_3^{2m-6} + \ldots + b_m)\, s_3 c_3 d_3, \\ \cdots\cdots\cdots\cdots\cdots\cdots \\ 0 = s_{2m-1}^{2m} + a_1 s_{2m-1}^{2m-2} + a_2 s_{2m-1}^{2m-4} + \ldots + a_m \\ \qquad + (b_2 s_{2m-1}^{2m-4} + b_3 s_{2m-1}^{2m-6} + \ldots + b_m)\, s_{2m-1} c_{2m-1} d_{2m-1}, \end{cases}$$

and we have

$$(39)\qquad \begin{cases} p(0) = a_m, \\ p(1) = 1 + a_1 + a_2 + a_3 + \ldots + a_m, \\ p\left(\dfrac{1}{k^2}\right) = \dfrac{1}{k^{2m}}(1 + a_1 k^2 + a_2 k^4 + a_3 k^6 + \ldots + a_m k^{2m}). \end{cases}$$

Write for convenience

$$\begin{vmatrix} s_1^{2m-2}, & s_1^{2m-4}, & \ldots, & s_1^2, & 1, & s_1^{2m-3} c_1 d_1, & s_1^{2m-5} c_1 d_1, & \ldots, & s_1 c_1 d_1 \\ s_2^{2m-2}, & s_2^{2m-4}, & \ldots, & s_2^2, & 1, & s_2^{2m-3} c_2 d_2, & s_2^{2m-5} c_2 d_2, & \ldots, & s_2 c_2 d_2 \\ s_3^{2m-2}, & s_3^{2m-4}, & \ldots, & s_3^2, & 1, & s_3^{2m-3} c_3 d_3, & s_3^{2m-5} c_3 d_3, & \ldots, & s_3 c_3 d_3 \\ \cdots & & & & & & & & \cdots \\ s_{2m-1}^{2m-2}, & s_{2m-1}^{2m-4}, & \ldots, & s_{2m-1}^2, & 1, & s_{2m-1}^{2m-3} c_{2m-1} d_{2m-1}, & s_{2m-1}^{2m-5} c_{2m-1} d_{2m-1}, & \ldots, & s_{2m-1} c_{2m-1} d_{2m-1} \end{vmatrix}$$

$$= \begin{vmatrix} s_\alpha^{2m-2}, s_\alpha^{2m-4}, \ldots, s_\alpha^2, 1, s_\alpha^{2m-3} c_\alpha d_\alpha, s_\alpha^{2m-5} c_\alpha d_\alpha, \ldots, s_\alpha c_\alpha d_\alpha \\ \alpha = 1, 2, 3, \ldots, 2m-1 \end{vmatrix},$$

and similarly for any determinant whose rows differ only in the suffixes involved in them; then (38) and (39) give

$$p(0) \begin{vmatrix} s_\alpha^{2m-2}, s_\alpha^{2m-4}, \ldots, s_\alpha^2, 1, s_\alpha^{2m-3} c_\alpha d_\alpha, s_\alpha^{2m-5} c_\alpha d_\alpha, \ldots, s_\alpha c_\alpha d_\alpha \\ \alpha = 1, 2, 3, \ldots, 2m-1 \end{vmatrix}$$

$$= (-1)^m \begin{vmatrix} s_\alpha^{2m}, s_\alpha^{2m-2}, \ldots, s_\alpha^2, s_\alpha^{2m-3} c_\alpha d_\alpha, s_\alpha^{2m-5} c_\alpha d_\alpha, \ldots, s_\alpha c_\alpha d_\alpha \\ \alpha = 1, 2, 3, \ldots, 2m-1 \end{vmatrix},$$

$$p(1)\begin{vmatrix} s_\alpha^{2m-2}, s_\alpha^{2m-4}, \ldots, s_\alpha^2, 1, s_\alpha^{2m-3}c_\alpha d_\alpha, s_\alpha^{2m-5}c_\alpha d_\alpha, \ldots, s_\alpha c_\alpha d_\alpha \\ \alpha = 1, 2, 3, \ldots, 2m-1 \end{vmatrix}$$

$$= \begin{vmatrix} 1 & , 1 & , \ldots, 1 & , 1, 0 & , 0 & \ldots, 0 \\ s_1^{2m} & , s_1^{2m-2} & , \ldots, s_1^2 & , 1, s_1^{2m-3}c_1d_1 & , s_1^{2m-5}c_1d_1 & , \ldots, s_1c_1d_1 \\ s_2^{2m} & , s_2^{2m-2} & , \ldots, s_2^2 & , 1, s_2^{2m-3}c_2d_2 & , s_2^{2m-5}c_2d_2 & , \ldots, s_2c_2d_2 \\ \cdot & \cdot & \cdot & \cdot & \cdot & \cdot \\ \cdot & \cdot & \cdot & \cdot & \cdot & \cdot \\ s_{2m-1}^{2m}, & s_{2m-1}^{2m-2}, & \ldots, s_{2m-1}^2, & 1, s_{2m-1}^{2m-3}c_{2m-1}d_{2m-1}, & s_{2m-1}^{2m-5}c_{2m-1}d_{2m-1}, & \ldots, s_{2m-1}c_{2m-1}d_{2m-1} \end{vmatrix}$$

$$= \begin{vmatrix} s_\alpha^{2m-2}c_\alpha^2, s_\alpha^{2m-4}c_\alpha^2, \ldots, s_\alpha^2 c_\alpha^2, c_\alpha^2, s_\alpha^{2m-3}c_\alpha d_\alpha, s_\alpha^{2m-5}c_\alpha d_\alpha, \ldots, s_\alpha c_\alpha d_\alpha \\ \alpha = 1, 2, 3, \ldots, 2m-1 \end{vmatrix},$$

$$k^{2m}p\left(\frac{1}{k^2}\right)\begin{vmatrix} s_\alpha^{2m-2}, s_\alpha^{2m-4}, \ldots, s_\alpha^2, 1, s_\alpha^{2m-3}c_\alpha d_\alpha, s_\alpha^{2m-5}c_\alpha d_\alpha, \ldots, s_\alpha c_\alpha d_\alpha \\ \alpha = 1, 2, 3, \ldots, 2m-1 \end{vmatrix}$$

$$= \begin{vmatrix} 1 & , k^2 & , k^4 & , \ldots, k^{2m-2}, k^{2m}, 0 & , 0 & , \ldots, 0 \\ s_1^{2m} & , s_1^{2m-2}, & s_1^{2m-4}, & \ldots, s_1^2 \quad , 1 \quad , s_1^{2m-3}c_1d_1 & , s_1^{2m-5}c_1d_1 & , \ldots, s_1c_1d_1 \\ s_2^{2m} & , s_2^{2m-2}, & s_2^{2m-4}, & \ldots, s_2^2 \quad , 1 \quad , s_2^{2m-3}c_2d_2 & , s_2^{2m-5}c_2d_2 & , \ldots, s_2c_2d_2 \\ \cdot & \cdot & \cdot & \cdot & \cdot & \cdot \\ \cdot & \cdot & \cdot & \cdot & \cdot & \cdot \\ s_{2m-1}^{2m}, & s_{2m-1}^{2m-2}, & s_{2m-1}^{2m-4}, & \ldots, s_{2m-1}^2, 1, s_{2m-1}^{2m-3}c_{2m-1}d_{2m-1}, & s_{2m-1}^{2m-5}c_{2m-1}d_{2m-1}, & \ldots, s_{2m-1}c_{2m-1}d_{2m-1} \end{vmatrix}$$

$$= \begin{vmatrix} s_\alpha^{2m-2}d_\alpha^2, s_\alpha^{2m-4}d_\alpha^2, \ldots, s_\alpha^2 d_\alpha^2, d_\alpha^2, s_\alpha^{2m-3}c_\alpha d_\alpha, s_\alpha^{2m-5}c_\alpha d_\alpha, \ldots, s_\alpha c_\alpha d_\alpha \\ \alpha = 1, 2, 3, \ldots, 2m-1 \end{vmatrix};$$

and (33), (34) and (35) become

(40) $\operatorname{sn}(u_1 + u_2 + u_3 + \ldots + u_{2m-1})$

$$= \pm \frac{\begin{vmatrix} s_\alpha^{2m-1}, s_\alpha^{2m-3}, \ldots, s_\alpha, s_\alpha^{2m-4}c_\alpha d_\alpha, s_\alpha^{2m-6}c_\alpha d_\alpha, \ldots, c_\alpha d_\alpha \\ \alpha = 1, 2, 3, \ldots, 2m-1 \end{vmatrix}}{\begin{vmatrix} s_\alpha^{2m-2}, s_\alpha^{2m-4}, \ldots, 1, s_\alpha^{2m-3}c_\alpha d_\alpha, s_\alpha^{m-5}c_\alpha d_\alpha, \ldots, s_\alpha c_\alpha d_\alpha \\ \alpha = 1, 2, 3, \ldots, 2m-1 \end{vmatrix}},$$

(41) $\operatorname{cn}(u_1 + u_2 + u_3 + \ldots . + u_{2m-1})$

$$= \pm \frac{\begin{vmatrix} s_\alpha^{2m-2}c_\alpha, s_\alpha^{2m-4}c_\alpha, \ldots, c_\alpha, s_\alpha^{2m-3}d_\alpha, s_\alpha^{2m-5}d_\alpha, \ldots, s_\alpha d_\alpha \\ \alpha = 1, 2, 3, \ldots, 2m-1 \end{vmatrix}}{\begin{vmatrix} s_\alpha^{2m-2}, s_\alpha^{2m-4}, \ldots, 1, s_\alpha^{2m-3}c_\alpha d_\alpha, s_\alpha^{2m-5}c_\alpha d_\alpha, \ldots, s_\alpha c_\alpha d_\alpha \\ \alpha = 1, 2, 3, \ldots, 2m-1 \end{vmatrix}},$$

(42) $\operatorname{dn}(u_1 + u_2 + u_3 + \ldots + u_{2m-1})$

$$= \pm \frac{\begin{vmatrix} s_\alpha^{2m-2}d_\alpha, s_\alpha^{2m-4}d_\alpha, \ldots, d_\alpha, s_\alpha^{2m-3}c_\alpha, s_\alpha^{2m-5}c_\alpha, \ldots, s_\alpha c_\alpha \\ \alpha = 1, 2, 3, \ldots, 2m-1 \end{vmatrix}}{\begin{vmatrix} s_\alpha^{2m-2}, s_\alpha^{2m-4}, \ldots, 1, s_\alpha^{2m-3}c_\alpha d_\alpha, s_\alpha^{2m-5}c_\alpha d_\alpha, \ldots, s_\alpha c_\alpha d_\alpha \\ \alpha = 1, 2, 3, \ldots, 2m-1 \end{vmatrix}}.$$

If $\quad u_{2m-1}=0,\ s_{2m-1}=0,\ c_{2m-1}=1,\ d_{2m-1}=1,$

(40), (41) and (42) become, on putting $m-1=n$,

(43) $\operatorname{sn}(u_1+u_2+u_3+\ldots+u_{2n})$

$$=\pm\frac{\left|\begin{array}{l} s_\alpha^{2n},\ s_\alpha^{2n-2},\ \ldots,\ 1,\ s_\alpha^{2n-3}c_\alpha d_\alpha,\ s_\alpha^{2n-5}c_\alpha d_\alpha,\ \ldots,\ s_\alpha c_\alpha d_\alpha \\ \alpha=1,\ 2,\ 3,\ \ldots,\ 2n \end{array}\right|}{\left|\begin{array}{l} s_\alpha^{2n-1},\ s_\alpha^{2n-3},\ \ldots,\ s_\alpha,\ s_\alpha^{2n-2}c_\alpha d_\alpha,\ s_\alpha^{2n-4}c_\alpha d_\alpha,\ \ldots,\ c_\alpha d_\alpha \\ \alpha=1,\ 2,\ 3,\ \ldots,\ 2n \end{array}\right|},$$

(44) $\operatorname{cn}(u_1+u_2+u_3+\ldots,+u_{2n})$

$$=\pm\frac{\left|\begin{array}{l} s_\alpha^{2n-1}c_\alpha,\ s_\alpha^{2n-3}c_\alpha,\ \ldots,\ s_\alpha c_\alpha,\ s_\alpha^{2n-2}d_\alpha,\ s_\alpha^{2n-4}d_\alpha,\ \ldots,\ d_\alpha \\ \alpha=1,\ 2,\ 3,\ \ldots,\ 2n \end{array}\right|}{\left|\begin{array}{l} s_\alpha^{2n-1},\ s_\alpha^{2n-3},\ \ldots,\ s_\alpha,\ s_\alpha^{2n-2}c_\alpha d_\alpha,\ s_\alpha^{2n-4}c_\alpha d_\alpha,\ \ldots,\ c_\alpha d_\alpha \\ \alpha=1,\ 2,\ 3,\ \ldots,\ 2n \end{array}\right|},$$

(45) $\operatorname{dn}(u_1+u_2+u_3+\ldots,+u_{2n})$

$$=\pm\frac{\left|\begin{array}{l} s_\alpha^{2n-1}d_\alpha,\ s_\alpha^{2n-3}d_\alpha,\ \ldots,\ s_\alpha d_\alpha,\ s_\alpha^{2n-2}c_\alpha,\ s_\alpha^{2n-4}c_\alpha,\ \ldots,\ c_\alpha \\ \alpha=1,\ 2,\ 3,\ \ldots,\ 2n \end{array}\right|}{\left|\begin{array}{l} s_\alpha^{2n-1},\ s_\alpha^{2n-3},\ \ldots,\ s_\alpha,\ s_\alpha^{2n-2}c_\alpha d_\alpha,\ s_\alpha^{2n-4}c_\alpha d_\alpha,\ \ldots,\ c_\alpha d_\alpha \\ \alpha=1,\ 2,\ 3,\ \ldots,\ 2n \end{array}\right|},$$

which are then the addition formulae for an even number of arguments.

Formulae (40)–(42) give, with a choice of signs which is at present arbitrary,

(46) $\operatorname{sn}(u_1+u_2+u_3+\ldots+u_{2n+1})$

$$=(-1)^n\frac{\left|\begin{array}{l} s_\alpha^{2n+1},\ s_\alpha^{2n-1},\ \ldots,\ s_\alpha,\ s_\alpha^{2n-2}c_\alpha d_\alpha,\ s_\alpha^{2n-4}c_\alpha d_\alpha,\ \ldots,\ c_\alpha d_\alpha \\ \alpha=1,\ 2,\ 3,\ \ldots,\ 2n+1 \end{array}\right|}{\left|\begin{array}{l} s_\alpha^{2n},\ s_\alpha^{2n-2},\ \ldots,\ 1,\ s_\alpha^{2n-1}c_\alpha d_\alpha,\ s_\alpha^{2n-3}c_\alpha d_\alpha,\ \ldots,\ s_\alpha c_\alpha d_\alpha \\ \alpha=1,\ 2,\ 3,\ \ldots,\ 2n+1 \end{array}\right|},$$

(47) $\operatorname{cn}(u_1+u_2+u_3+\ldots+u_{2n+1})$

$$=\frac{\left|\begin{array}{l} s_\alpha^{2n}c_\alpha,\ s_\alpha^{2n-2}c_\alpha,\ \ldots,\ c_\alpha,\ s_\alpha^{2n-1}d_\alpha,\ s_\alpha^{2n-3}d_\alpha,\ \ldots,\ s_\alpha d_\alpha \\ \alpha=1,\ 2,\ 3,\ \ldots,\ 2n+1 \end{array}\right|}{\left|\begin{array}{l} s_\alpha^{2n},\ s_\alpha^{2n-2},\ \ldots,\ 1,\ s_\alpha^{2n-1}c_\alpha d_\alpha,\ s_\alpha^{2n-3}c_\alpha d_\alpha,\ \ldots,\ s_\alpha c_\alpha d_\alpha \\ \alpha=1,\ 2,\ 3,\ \ldots,\ 2n+1 \end{array}\right|},$$

(48) $\operatorname{dn}(u_1+u_2+u_3+\ldots+u_{2n+1})$

$$=\frac{\left|\begin{array}{l} s_\alpha^{2n}d_\alpha,\ s_\alpha^{2n-2}d_\alpha,\ \ldots,\ d_\alpha,\ s_\alpha^{2n-1}c_\alpha,\ s_\alpha^{2n-3}c_\alpha,\ \ldots,\ s_\alpha c_\alpha \\ \alpha=1,\ 2,\ 3,\ \ldots,\ 2n+1 \end{array}\right|}{\left|\begin{array}{l} s_\alpha^{2n},\ s_\alpha^{2n-2},\ \ldots,\ 1,\ s_\alpha^{2n-1}c_\alpha d_\alpha,\ s_\alpha^{2n-3}c_\alpha d_\alpha,\ \ldots,\ s_\alpha c_\alpha d_\alpha \\ \alpha=1,\ 2,\ 3,\ \ldots,\ 2n+1 \end{array}\right|},$$

from which we get, on putting $u_{2n+1} = 0$ and dividing numerators and denominators by $s_1 s_2 s_3 \ldots s_{2n}$,

(49) $\operatorname{sn}(u_1 + u_2 + u_3 + \ldots + u_{2n})$

$$= \frac{\begin{vmatrix} s_a^{2n}, s_a^{2n-2}, \ldots, 1, s_a^{2n-3}c_a d_a, s_a^{2n-5}c_a d_a, \ldots, s_a c_a d_a \\ a = 1, 2, 3, \ldots, 2n \end{vmatrix}}{\begin{vmatrix} s_a^{2n-1}, s_a^{2n-3}, \ldots, s_a, s_a^{2n-2}c_a d_a, s_a^{2n-4}c_a d_a, \ldots, c_a d_a \\ a = 1, 2, 3, \ldots, 2n \end{vmatrix}},$$

(50) $\operatorname{cn}(u_1 + u_2 + u_3 + \ldots + u_{2n})$

$$= \frac{\begin{vmatrix} s_a^{2n-1}c_a, s_a^{2n-3}c_a, \ldots, s_a c_a, s_a^{2n-2}d_a, s_a^{2n-4}d_a, \ldots, d_a \\ a = 1, 2, 3, \ldots, 2n \end{vmatrix}}{\begin{vmatrix} s_a^{2n-1}, s_a^{2n-3}, \ldots, s_a, s_a^{2n-2}c_a d_a, s_a^{2n-4}c_a d_a, \ldots, c_a d_a \\ a = 1, 2, 3, \ldots, 2n \end{vmatrix}},$$

(51) $\operatorname{dn}(u_1 + u_2 + u_3 + \ldots + u_{2n})$

$$= \frac{\begin{vmatrix} s_a^{2n-1}d_a, s_a^{2n-3}d_a, \ldots, s_a d_a, s_a^{2n-2}c_a, s_a^{2n-4}c_a, \ldots, c_a \\ a = 1, 2, 3, \ldots, 2n \end{vmatrix}}{\begin{vmatrix} s_a^{2n-1}, s_a^{2n-3}, \ldots, s_a, s_a^{2n-2}c_a d_a, s_a^{2n-4}c_a d_a, \ldots, c_a d_a \\ a = 1, 2, 3, \ldots, 2n \end{vmatrix}},$$

and from these again we obtain, on putting $u_{2n} = 0$ and dividing numerators and denominators by $s_1 s_2 s_3 \ldots s_{2n-1}$,

$\operatorname{sn}(u_1 + u_2 + u_3 + \ldots + u_{2n-1})$

$$= (-1)^{n-1} \frac{\begin{vmatrix} s_a^{2n-1}, s_a^{2n-3}, \ldots, s_a, s_a^{2n-4}c_a d_a, s_a^{2n-6}c_a d_a, \ldots, c_a d_a \\ a = 1, 2, 3, \ldots, 2n-1 \end{vmatrix}}{\begin{vmatrix} s_a^{2n-2}, s_a^{2n-4}, \ldots, 1, s_a^{2n-3}c_a d_a, s_a^{2n-5}c_a d_a, \ldots, s_a c_a d_a \\ a = 1, 2, 3, \ldots, 2n-1 \end{vmatrix}},$$

$\operatorname{cn}(u_1 + u_2 + u_3 + \ldots + u_{2n-1}$

$$= \frac{\begin{vmatrix} s_a^{2n-2}c_a, s_a^{2n-4}c_a, \ldots, c_a, s_a^{2n-3}d_a, s_a^{2n-5}d_a, \ldots, s_a d_a \\ a = 1, 2, 3, \ldots, 2n-1 \end{vmatrix}}{\begin{vmatrix} s_a^{2n-2}, s_a^{2n-4}, \ldots, 1, s_a^{2n-3}c_a d_a, s_a^{2n-5}c_a d_a, \ldots, s_a c_a d_a \\ a = 1, 2, 3, \ldots, 2n-1 \end{vmatrix}},$$

$\operatorname{dn}(u_1 + u_2 + u_3 + \ldots + u_{2n-1})$

$$= \frac{\begin{vmatrix} s_a^{2n-2}d_a, s_a^{2n-4}d_a, \ldots, d_a, s_a^{2n-3}c_a, s_a^{2n-5}c_a, \ldots, s_a c_a \\ a = 1, 2, 3, \ldots, 2n-1 \end{vmatrix}}{\begin{vmatrix} s_a^{2n-2}, s_a^{2n-4}, \ldots, 1, s_a^{2n-3}c_a d_a, s_a^{2n-5}c_a d_a, \ldots, s_a c_a d_a \\ a = 1, 2, 3, \ldots, 2n-1 \end{vmatrix}}.$$

From these formulæ it appears that, in passing from the sum of an odd number $2n - 1$ of arguments to the sum of the next even number $2n$ of arguments, the sign in the formula for sn does or does not change according as $n - 1$ is odd or even, but the signs of cn and dn remain unchanged, an odd or even value of $n - 1$; while, in passing from the sum of an even number $2n$ of arguments to the sum of the next odd number $2n + 1$ of arguments, the sign in the formula for sn does or does not change according as n is odd or even, and the signs of cn and dn remain unchanged for an odd or even value of n; *i. e.* in the formula for sn of the sum of an even number of arguments, and in those for cn and dn for the sum of any number of arguments, the sign is invariable, while in the formula for sn of the sum of the successive odd numbers of arguments the sign is alternately $+$ and $-$. But for the sum of two arguments the signs in the formulæ for sn, cn and dn are evidently all $+$, therefore (46)–(51) are correct even to their signs.

It remains only to point out the modifications of the formulæ which are necessary when several arguments are equal, in accordance with the principles above established. It is evident that, if $u_a = u_{a+1} = u_{a+2} = \ldots = u_{a+\mu-1}$, the $\alpha + 1^{\text{th}}$, $\alpha + 2^{\text{th}}$, $\ldots$ $\alpha + \mu - 1^{\text{th}}$ rows of the numerator and denominator of the right member of each of the formulæ (46)–(51) has to be replaced respectively by the 1st, 2nd, $\ldots$, $\mu - 1^{\text{th}}$ derivatives of the αth row of that numerator or denominator. As any common factor of numerator and denominator will disappear from the quotient, it is evident that the derivatives may be taken with respect to u_a instead of s_a, as above, and we know that

$$\frac{\partial s_a}{\partial u_a} = c_a d_a, \quad \frac{\partial c_a}{\partial u_a} = - s_a d_a, \quad \frac{\partial d_a}{\partial u_a} = - k^2 s_a c_a,$$

which enables us to determine the $\alpha + 1^{\text{th}}$, $\alpha + 2^{\text{th}}$, $\ldots$, $\alpha + \mu - 1^{\text{th}}$ row in the case supposed, but the general formulæ seem too complicated to be useful.

BALTIMORE, June 15, 1885.

Note on the Theorem $e^{ix} = \cos x + i \sin x$.

By F. Franklin.

The following geometrical demonstration of this theorem is so obvious that it has probably been given before, but I have been unable to find it. Defining e^z as $\lim_{\omega=\infty}\left(1+\frac{z}{\omega}\right)^\omega$, ω being a positive integer, the geometrical construction for e^{ix} is as follows. From the origin O lay off on the axis of x $OA = 1$, and at A erect a perpendicular $AB = x$; then the point B represents $1 + ix$. Divide AB into ω parts; the first point of division, A_1 say, represents $1 + \frac{ix}{\omega}$. Erect at A_1 a perpendicular A_1A_2, making the triangle OA_1A_2 similar to OAA_1; then A_2 represents $\left(1+\frac{ix}{\omega}\right)^2$.

Constructing in like manner a similar triangle OA_2A_3, and so on, A_ω will represent $\left(1+\frac{ix}{\omega}\right)^\omega$. Now let ω tend to infinity; the limit of the broken line $AA_1A_2 \ldots A_\omega$ is an arc of length x in a circle whose radius is 1; hence the limit of A_ω is a point which represents $\cos x + i \sin x$. Therefore

$$e^{ix} = \cos x + i \sin x.$$

Proof of a Theorem of Tchebycheff's on Definite Integrals.

By F. Franklin.

M. Hermite, in his *Cours professé pendant le 2e Semestre,* 1881–82, states the following theorem as communicated to him by M. Tchebycheff:

Let u and v be two functions of x which between the values $x=0$ and $x=1$ are positive, and both vary in the same sense, so that they shall be continually increasing or continually decreasing; then we shall have the inequality

$$\int_0^1 uvdx > \int_0^1 udx \int_0^1 vdx.$$

But, supposing that one of the functions be increasing and the other decreasing, we shall have, on the contrary,

$$\int_0^1 uvdx < \int_0^1 udx \int_0^1 vdx.$$

M. Hermite proceeds to give a proof of this theorem by M. Picard, which is somewhat indirect and long. The theorem may be proved instantaneously, and no restriction need be placed on the signs of u and v; it admits, too, of an extension to the case of any real limits. The theorem, as thus enlarged, is as follows:

If u and v be two functions of x, both increasing continually or both decreasing continually as x passes from a to b, then

$$(b-a)\int_a^b uvdx > \int_a^b udx \int_a^b vdx;$$

and if one increases throughout while the other decreases throughout,

$$(b-a)\int_a^b uvdx < \int_a^b udx \int_a^b vdx.$$

The proof is as follows. Write $u=f(x)$, $v=\phi(x)$; then

$$\begin{aligned}\int_a^b\int_a^b [f(x)-f(y)][\phi(x)-\phi(y)]\,dxdy &= (b-a)\int_a^b f(x)\,\phi(x)\,dx + (b-a)\int_a^b f(y)\,\phi(y)\,dy \\ &\quad - \int_a^b f(x)\,dx\int_a^b \phi(y)\,dy - \int_a^b \phi(x)\,dx\int_a^b f(y)\,dy \\ &= 2\left\{(b-a)\int_a^b uvdx - \int_a^b udx\int_a^b vdx\right\}.\end{aligned}$$

Now if u and v are always increasing or always decreasing, $f(x) - f(y)$ and $\phi(x) - \phi(y)$ always have like signs, while if one is increasing and the other decreasing they have opposite signs; hence in the former case the double integral above written is necessarily positive, while in the latter case it is necessarily negative. This proves the theorem.

In the same way we may prove the theorem about finite series from which the preceding may at once be deduced, viz.

$$n\Sigma_1^n u_r v_r > \Sigma_1^n u_r \Sigma_1^n v_r \text{ or } n\Sigma_1^n u_r v_r < \Sigma_1^n u_r \Sigma_1^n v_r,$$

according as the u's and v's are both ascending (or both descending) series, or one of them ascending and the other descending.

The above demonstration was suggested by the following very simple method of proving either theorem, which I shall give only as applied to the second; the proof for the first being precisely similar. Consider the rectangular matrices

$$\begin{vmatrix} u_1, & u_2, & u_3, & \ldots & u_n \\ 1, & 1, & 1, & \ldots & 1 \end{vmatrix}, \quad \begin{vmatrix} v_1, & v_2, & v_3, & \ldots & v_n \\ 1, & 1, & 1, & \ldots & 1 \end{vmatrix}.$$

This product is the determinant

$$\begin{vmatrix} \Sigma uv, & \Sigma u \\ \Sigma v, & n \end{vmatrix}, = n\Sigma uv - \Sigma u \Sigma v.$$

But the product of the matrices is equivalent to the sum of the products obtained by multiplying every determinant in the first matrix by the corresponding one in the second, and any such product is $(u_r - u_s)(v_r - v_s)$, which is necessarily positive if the series both ascend or both descend throughout, and necessarily negative if one ascends throughout while the other descends throughout.

It is obvious from the nature of the proof that we may give the theorems a somewhat greater generality, viz., they hold whenever u and v vary throughout the interval in such a way that the greater u is the greater v is, or that the greater u is the smaller v is.* As examples of this somewhat more general theorem, we may note that the inequalities

$$\left(\int_a^b u dx\right)^2 < (b-a)\int_a^b u^2 dx, \quad \int_a^b u dx \int_a^b \frac{dx}{u} > (b-a)^2,$$

hold without any restriction as regards the manner in which u varies.

I add a few particular examples of the theorems, which may not be without interest.

* This, it should be noted, is not the same thing as saying that u and v increase together (or the opposite) as x passes from a to b; the theorem does not necessarily hold when this latter condition is fulfilled.

1. $x\int_0^x \frac{dx}{\sqrt{1-x^2}} < \int_0^x \frac{dx}{\sqrt{1-x}} \int_0^x \frac{dx}{\sqrt{1+x}}$; $\sin^{-1}x < \frac{4}{x}(\sqrt{1+x}+\sqrt{1-x}-\sqrt{1-x^2}-1)$;

$$\frac{\pi}{2} < 4(\sqrt{2}-1),\ \frac{\pi}{12} < 2(\sqrt{6}+\sqrt{2}-\sqrt{3}-2).$$

2. $\left(\int_0^x \frac{dx}{\sqrt{1-x^2}}\right)^2 < x\int_0^x \frac{dx}{1-x^2}$; $(\sin^{-1}x)^2 < \frac{x}{2}\log\frac{1+x}{1-x}$; $\frac{\pi^2}{36} < \frac{1}{4}\log 3$.

3. $\frac{\pi}{2}\int_0^{\frac{\pi}{2}} \sin^2 x dx > \left(\int_0^{\frac{\pi}{2}} \sin x dx\right)^2$; $\frac{\pi^2}{8} > 1$.

4. $\int_1^x \frac{dx}{x}\int_1^x x dx > (x-1)^2$; $\log x > 2\frac{x-1}{x+1}\ (x>1)$; $\log x < 2\frac{x-1}{x+1}\ (x<1)$.

$$\left(\int_1^x \frac{dx}{x}\right)^2 < (x-1)\int_1^x \frac{dx}{x^2};\ \log x < \frac{x-1}{\sqrt{x}},\ (x>1);\ \log x < \frac{x-1}{\sqrt{x}}\ (x<1).$$

Thus $\quad \frac{1}{\sqrt{2}} > \log 2 > \frac{2}{3}$.

5. $\int_0^1 x^{a-1}(1-x)^{b-1}dx < \int_0^1 x^{a-1}dx\int_0^1 (1-x)^{b-1}dx$; $ab\Gamma(a)\Gamma(b) < \Gamma(a+b)$, or $\Gamma(a+1)\Gamma(b+1) < \Gamma(a+b)$.

In particular (a being supposed between 0 and 1) $\Gamma(a)\Gamma(1-a) < \frac{1}{a(1-a)}$, so that $\frac{\pi}{\sin a\pi} < \frac{1}{a(1-a)}$, whence $\sin x > x\left(1-\frac{x}{\pi}\right)$.

6. $K\int_0^K \mathrm{sn}^2 u du > \left(\int_0^K \mathrm{sn}\, u du\right)^2$; $K(K-E) > \left(\log\frac{k'}{1-k}\right)^2$

$$K\int_0^K \mathrm{sn}^2 u du < \int_0^K (1-\mathrm{cn}\, u)du\int_0^K (1+\mathrm{cn}\, u)du;\ K(K-E) < k^2K^2 - \left(\tan^{-1}\frac{k}{k'}\right)^2.$$

The comparison of these inequalities gives superior and inferior limits for E, viz.

$$K - \frac{1}{K}\left(\log\frac{k'}{1-k}\right)^2 > E > k'^2K + \frac{1}{K}\left(\tan^{-1}\frac{k}{k'}\right)^2.$$

7. $x\int_0^x \frac{dx}{\sqrt{1-x^2.1-k^2x^2}} < \int_0^x \frac{dx}{\sqrt{1-x.1-kx}}\int_0^x \frac{dx}{\sqrt{1+x.1+kx}}$

8. $x\int_0^x \frac{dx}{\sqrt{1-x^2.1-k^2x^2}} > \int_0^x \frac{dx}{\sqrt{1-x^2}}\int_0^x \frac{dx}{\sqrt{1-k^2x^2}}$; $\int_0^x \frac{dx}{\sqrt{1-x^2.1-k^2x^2}} > \sin^{-1}x\frac{\sin^{-1}kx}{kx}$,

or $$F(k,\phi) > \frac{\varphi\sin^{-1}(k\sin\varphi)}{k\sin\varphi};\ K > \frac{\pi}{2}\frac{\sin^{-1}k}{k}.$$

9. $\int_0^\phi \sqrt{1-k^2\sin^2\varphi}\,d\varphi\int_0^\phi \frac{d\varphi}{\sqrt{1-k^2\sin^2\varphi}} > \phi^2$; $E(k,\phi)F(k,\phi) > \phi^2$; $EK > \frac{\pi^2}{4}$.

Finally we may notice an obvious geometrical interpretation of the inequality

$$\int_a^b u dx\int_a^b \frac{dx}{u} > (b-a)^2;$$

viz., if the inverse of a closed curve be taken with respect to a circle whose centre is a point within the curve (and such that chords through it cut the curve in only two points), the area of the circle is always less than a geometric mean between the areas of the two curves.

On the Calculation of the Co-factors of Alternants of the Fourth Order.

By Wm. Woolsey Johnson.

1. If we subtract the first row of an alternant of the nth order in which the first column is a column of units, from each of the other rows, the determinant is reduced to the order $n-1$, and each row may be divided by one of the differences involving the first letter. For example,

$$\begin{vmatrix} 1, & a^3, & a^4, & a^5 \\ 1, & b^3, & b^4, & b^5 \\ 1, & c^3, & c^4, & c^5 \\ 1, & d^3, & d^4, & d^5 \end{vmatrix}$$

$$=(b-a)(c-a)(d-a)\begin{vmatrix} b+a, & b^3+b^2a+ba^2+a^3, & b^4+b^3a+b^2a^2+ba^3+a^4 \\ c+a, & c^3+c^2a+ca^2+a^3, & c^4+c^3a+c^2a^2+ca^3+a^4 \\ d+a, & d^3+d^2a+da^2+a^3, & d^4+d^3a+d^2a^2+da^3+a^4 \end{vmatrix}.$$

Subtracting the second column multiplied by a from the third, then the first column multiplied by a^2 from the second, and decomposing the result by columns, the determinant in the right hand member becomes

$$\begin{vmatrix} b+a, & b^3+b^2a, & b^4 \\ c+a, & c^3+c^2a, & c^4 \\ d+a, & d^3+d^2a, & d^4 \end{vmatrix} = a^3A(0,\ 2,\ 4)+aA(0,\ 3,\ 4)+aA(1,\ 2,\ 4)+A(1,\ 3,\ 4),$$

in which the alternants involve the letters b, c and d. These alternants are divisible by $\zeta^{\frac{1}{2}}(b,\ c,\ d)$; hence, since

$$\zeta^{\frac{1}{2}}(a,\ b,\ c,\ d)=(b-a)(c-a)(d-a)\,\zeta^{\frac{1}{2}}(b,\ c,\ d),$$

we have, denoting the co-factor of an alternant by the symbol α,

$$\alpha(0,2,4,5)=a^3\alpha(0,2,4)+a\alpha(0,3,4)+a\alpha(1,2,4)+\alpha(1,3,4).$$

2. It is obvious that the process is general, the numbers of partial columns in the columns of the determinant of the $(n-1)^{\text{th}}$ order being the differences between the consecutive exponents of the original alternant, so that the whole number of alternants of the $(n-1)^{\text{th}}$ order is the product of these differences.

In particular when $n=4$, $\alpha(0,p,q,r)$ is a symmetric function of a, b, c and d of the degree $p+q+r-6$, and we have

$$(1)\qquad \alpha(0,p,q,r)=\sum_u\sum_v\sum_w a^s\alpha(u,\ v,\ w),\quad \begin{cases} u=0,\ 1,\ 2,\ \ldots p-1 \\ v=p,\ p+1,\ \ldots q-1 \\ w=q,\ q+1,\ \ldots r-1 \end{cases};$$

the degree of $\alpha(u, v, w)$ is $u+v+w-3$, and $\theta = p+q+r-(u+v+w)-3$; the greatest value of θ being $r-3$, and its least value zero.

The terms of this sum may be imagined to be arranged in a solid rectangular block, analogous to the plane rectangular array of binomials in §3 of the preceding article. Let us suppose the first symbol, u, of $\alpha(u, v, w)$ to increase downward from zero in the top face to $p-1$ in the bottom face; the second, v, to increase from p on the left to $q-1$ on the right, and the third, w, to increase from q in the front face to $r-1$ in the back face. The first term $a^{r-3}\alpha(0, p, q)$ will thus occupy the upper, left-hand, front corner, and $\alpha(p-1, q-1, r-1)$ will occupy the lower, right-hand, back corner.

3. Now if we remove the terms in the top face, in each of which $u=0$, all the rest may be divided by bcd by subtracting unity from each symbol; and if we also remove the back face, which contains the only term in which $\theta = 0$, all the rest may be divided by a. Dividing the remaining block of terms by $abcd$, the first symbol of the α's will run from zero to $p-2$, the second from $p-1$ to $q-2$, and the third from $q-1$ to $r-3$; the term at the upper, left-hand, front corner will be $a^{r-5}\alpha(0, p-1, q-1)$, and the term at the lower, right-hand, back corner will be $\alpha(p-2, q-2, r-3)$. In other words, the remaining block of terms after division is the value of $\alpha(0, p-1, q-1, r-2)$. Denoting by Q the sum of the terms removed, we have therefore

$$\alpha(0, p, q, r) = Q + abcd . \alpha(0, p-1, q-1, r-2), \tag{2}$$

from which it appears that Q is a symmetric function of a, b, c and d.

4. The function Q consists of the terms situated in the top face and in the back face of the rectangular block. Imagining the top face to be rotated about the common edge into the plane of the back face, the terms constituting Q may be written thus:

$$\begin{aligned} Q = {} & a^{r-3}\alpha(0, p, q) + a^{r-4}\alpha(0, p+1, q) + \ldots + a^{r+p-q-2}\alpha(0, q-1, q) \\ & + a^{r-4}\alpha(0, p, q+1) + a^{r-5}\alpha(0, p+1, q+1) + \ldots + a^{r+p-q-3}\alpha(0, q-1, q+1) \\ & \quad \vdots \qquad\qquad\qquad \vdots \qquad\qquad\qquad \vdots \\ & + a^{q-2}\alpha(0, p, r-1) + a^{q-3}\alpha(0, p+1, r-1) + \ldots + a^{p-1}\alpha(0, q-1, r-1) \\ & + a^{q-3}\alpha(1, p, r-1) + a^{q-4}\alpha(1, p+1, r-1) + \ldots + a^{p-2}\alpha(1, q-1, r-1) \\ & \quad \vdots \qquad\qquad\qquad \vdots \qquad\qquad\qquad \vdots \\ & + a^{q-p-1}\alpha(p-1, p, r-1) + a^{q-p-2}\alpha(p-1, p+1, r-1) \\ & \qquad + \ldots + \alpha(p-1, q-1, r-1), \end{aligned} \tag{3}$$

the first and last symbols remaining constant for each row, and in the columns, the last symbol increasing from q to $r-1$, and then the first symbol increasing from 0 to $p-1$. For example, $\alpha(0, 3, 5, 7) = Q + abcd.\alpha(0, 2, 4, 5)$, where

$$\begin{aligned} Q = {} & a^4\alpha(0, 3, 5) + a^3\alpha(0, 4, 5) \\ & + a^3\alpha(0, 3, 6) + a^2\alpha(0, 4, 6) \\ & + a^2\alpha(1, 3, 6) + a\alpha(1, 4, 6) \\ & + a\alpha(2, 3, 6) + \alpha(2, 4, 6). \end{aligned}$$

Accordingly, by the rule for developing the co-factors of the third order, we have

$$\begin{aligned} Q = {} & a^4[\Sigma b^3c^2 + \Sigma b^3cd + 2\Sigma b^2c^2d] \\ & + a^3[\Sigma b^4c^2 + \Sigma b^4cd + 2\Sigma b^3c^3 + 3\Sigma b^3c^2d + 4\Sigma b^2c^2d^2] \\ & + a^2[\Sigma b^4c^3 + 2\Sigma b^4c^2d + 3\Sigma b^3c^3d + 4\Sigma b^3c^2d^2] \\ & + a\,[\Sigma b^4c^3d + 2\Sigma b^4c^2d^2 + 3\Sigma b^3c^3d^2] \\ & + \Sigma b^4c^3d^2 + 2\Sigma b^3c^3d^3, \end{aligned}$$

which is readily seen to be the symmetric function

$$Q = \Sigma a^4b^3c^2 + \Sigma a^4b^3cd + 2\Sigma a^4b^2c^2d + 2\Sigma a^3b^3c^3 + 3\Sigma a^3b^3c^2d + 4\Sigma a^3b^2c^2d^2.$$

5. It is obvious that, in the example above, the coefficients of those Σ's whose typical terms contain a^4 might have been determined solely by means of the first term in Q, viz., $a^4\alpha(0, 3, 5)$; and in like manner, the coefficients of the remaining Σ's, all of whose typical terms contain a^3, might have been determined from the corresponding terms in the value of Q. Thus, if we first write out the series of simple symmetric functions which enter the expression, beginning with $\Sigma a^4b^3c^2$, then for the purpose of determining the coefficients it is only necessary to consider a portion of Q, viz.,

$$\begin{aligned} Q = {} & a^4\alpha(0, 3, 5) + a^3\alpha(0, 4, 5) \\ & + a^3\alpha(0, 3, 6) + \text{etc.} \\ & + \text{etc.}; \end{aligned}$$

and by means of each term in this reduced expression we may write out a group of coefficients, depending mainly upon the exponents of b and d, in accordance with the rule given in §6 of the preceding article, but with the restriction mentioned therein with respect to the highest admissible coefficient. The process is as follows:

$\Sigma a^4b^3c^2$,	Σa^4b^3cd,	$\Sigma a^4b^2c^2d$,	$\Sigma a^3b^3c^3$,	$\Sigma a^3b^3c^2d$,	$\Sigma a^3b^2c^2d^2$
1	1	2	1	1	1
			1	2	3
1	1	2	2	3	4

$$\therefore\ Q = \Sigma a^4b^3c^2 + \Sigma a^4b^3cd + 2\Sigma a^4b^2c^2d + 2\Sigma a^3b^3c^3 + 3\Sigma a^3b^3c^2d + 4\Sigma a^3b^2c^2d^2.$$

It is to be noticed that in some cases not all the coefficients given by the rule are written down. For example, in the case of the third term of Q, the first term of $a\,(0, 3, 6)$ would be Σb^4c^2, but the term $a^3b^4c^3$ is not a typical term, being in fact a term included in the expression $\Sigma a^4b^3c^3$ whose coefficient, 1, has already been written. We therefore commence with the coefficient belonging to the term Σb^3c^3, which is unity, because the exponent of d is zero.

6. In general, the series of simple symmetric functions to be written begins with $\Sigma a^{r-3}b^{q-2}c^{p-1}$, and the power of a which occurs in the last term of the series is the lowest power which need be retained in the expression for Q. In other words, we may reject all powers of a whose exponents are less than $\frac{1}{4}(p+q+r-6)$.

In determining the coefficients corresponding to the term of Q containing $a\,(u, v, w)$ when u is not zero, it is not necessary to reduce the term to the form $a^0\,(bcd)^u a\,(0, v-u, w-u)$. The coefficient unity occurs when b has its highest exponent, or when d has its lowest exponent; in $a\,(0, v-u, w-u)$ these are $w-u-2$ and zero respectively, which in $a\,(u, v, w)$ become $w-2$ and u respectively, and are still the highest and lowest exponents. So also, these exponents not occurring, the coefficient is 2 if b has its next highest or d its next lowest exponent, and so on; but with the restriction that the highest admissible coefficient is the least of the quantities $w-u-(v-u)$ and $v-u$; that is $w-v$ and $v-u$, the differences of the symbols in $a\,(u, v, w)$. Thus the rule is a simple extension of that given in §6 of the preceding article for the form $a\,(0, p, q)$.

It is obvious, from the mode in which the rule was derived, that the least of the differences of the symbols is always the last coefficient of the group derived from $a\,(u, v, w)$, so that this coefficient may be written at once without reference to the exponents of b and d.

7. By repeated application of formula (2) we should have

$$a\,(0, p, q, r) = Q + abcd\,Q' + a^2b^2c^2d^2\,Q'' + \ldots,$$

but for purposes of calculation it is better to write

$$a\,(0, p, q, r) = Q_1 + Q_2 + \ldots + Q_k, \tag{4}$$

retaining the factor $abcd$ in the expression for Q_2, so that Q_2 denotes the sum of the terms which in the rectangular block considered in §2 are adjacent to the terms included in Q_1. Thus Q_1 consists of those terms of the sum in formula (1) in which $u=0$ or $w=r-1$, Q_2 consists of those among the remaining terms in which $u=1$ or $w=r-2$, and so on. The number, k, of Q's is obviously the least of the numbers p and $r-q$, the height and depth of the rectangular block. In calculating the value of $\alpha(0, p, q, r)$, we need, of course, retain in the expressions for Q_2, Q_3, etc. only the same powers of a that are retained in Q_1. For example, in calculating $\alpha(0, 4, 7, 9)$, $k = 2$, and, since a^4 is the lowest power of a that need be retained, we may write

$$\begin{aligned} Q_1 = & \, a^6\alpha(0, 4, 7) + a^5\alpha(0, 5, 7) + a^4\alpha(0, 6, 7) \\ & + a^5\alpha(0, 4, 8) + a^4\alpha(0, 5, 8) + \text{etc.} \\ & + a^4\alpha(1, 4, 8) + \text{etc.} \\ & + \text{etc.}, \\ Q_2 = & \, a^5\alpha(1, 4, 7) + a^4\alpha(1, 5, 7) + \text{etc.} \\ & + a^4\alpha(2, 4, 7) + \text{etc.} \\ & + \text{etc.} \end{aligned}$$

We may now write the coefficients determined by Q_2 immediately under those determined by Q_1; and, writing for abridgement only the exponents of b, c, and d in each group, the calculation is as follows:

$\Sigma a^6 \ldots$						$\Sigma a^5 \ldots$						$\Sigma a^4 \ldots$	
530,	521,	440,	431,	422,	332	540,	531,	522,	441,	432,	333	442,	433
1	1	1	2	2	3	1	1	1	2	2	2	1	1
						1	2	2	2	3	4	3	3
												2	3
							1	1	1	2	3	2	2
												1	2
1	1	1	2	2	3	2	4	4	5	7	9	9	11

8. If we remove the top line in the expression for Q_1 above, the terms of Q_2 may be derived from the remaining terms by simply transferring a unit from the third symbol of each α to the first symbol, the second symbol and the exponent of a remaining unchanged.

The reason is that we thus pass from the term $a^e\alpha(u, v, w)$ of Q_1 to the term $a^e\alpha(u + 1, v, w - 1)$, which, if it be a term of the rectangular block, is situated therein one place further forward and at the same time one place lower down, and is therefore a term of Q_2. The top line of Q_1 gives rise to no part of Q_2 in this mode of derivation, because it is situated in the front face of the rectangular block; in other words, the third symbol being q in this line, the rule would give an inadmissibly small value of w in formula (1). In like manner, if the whole value of Q_1 were written, the last line, in which the value of u is $p - 1$, would give rise to no part of Q_2 because the rule would give an inadmissibly large value of u in formula (1).

The same rule, of course, applies to the derivation of Q_3 from Q_2, and so on, when $k > 2$. We reject in each case the top line, and we reject the bottom line when the value of u would be too great, that is to say, greater than $p - 1$.

9. The simple relation which exists between the coefficients determined by $\alpha(u, v, w)$ and $\alpha(u+1, v, w-1)$ renders it unnecessary to write the expressions for Q_2, Q_3, etc. For, in $\alpha(u+1, v, w-1)$ the highest exponent of b is a unit less and the lowest exponent of d is a unit greater than the corresponding numbers in $\alpha(u, v, w)$, and moreover, each of the differences between consecutive symbols is less by a unit in the former than in the latter case; hence, whether in any given case the coefficient is determined by the exponent of b or of d, or by the limiting value of the coefficient, in accordance with the rule in §6, it is always a unit less when determined by $\alpha(u+1, v, w-1)$ than when determined by $\alpha(u, v, w)$. For example, all the coefficients determined by Q_2 in the calculation given in §7 might have been thus derived from the corresponding ones determined by Q_1, the upper line of coefficients given by Q_1 producing, for the reason given above, no derived coefficients.

10. If then, after writing the reduced expression for Q_1, we indicate for each line the number of Q's in which it or a corresponding line occurs, we may, as soon as we have written the group of coefficients determined by a term of Q_1, at once write the derived coefficients corresponding to the other Q's.

It follows at once, from the rule to reject the top lines in deriving the successive Q's, that the top line of Q_1 should be marked 1, the next line 2, and so on till we come to the kth line. The subsequent lines are to be marked k unless the rule for rejecting the bottom line in forming the successive Q's applies. Now, since p is the value of v in the first column of the expression for Q, and since we must, as explained in §8, reject the bottom line when the first symbol u, as given by the rule, would be as great as p, it follows that the indicating number of a line cannot exceed the difference between the first two symbols or values of u and v in the first term of the line. This number is therefore to be taken as the indicating number when it is less than k.

11. The calculation of $\alpha(0, 5, 13, 17)$ is given below as an example. The first term is $\Sigma a^{14}b^{11}c^{4}$; the function being of the 29th degree, the lowest power of a to be retained is a^{8}, and the value of k is 4. The simple symmetric functions in the final development are indicated as in the preceding example, §7. It will be noticed that among those commencing with a^{13} none occur in which the exponent of b exceeds 12; and, in general, it follows from the mode in which Q is found that the sum of the exponents of a and b in any term cannot exceed their sum in the first term; in other words, the sum of the two highest exponents cannot exceed $r+q-5$.

CALCULATION OF $\alpha(0, 5, 13, 17)$.

1 | $a^{14}\alpha(0, 5, 13) + a^{13}\alpha(0, 6, 13) + a^{12}\alpha(0, 7, 13) + a^{11}\alpha(0, 8, 13) + a^{10}\alpha(0, 9, 13) + a^{9}\alpha(0, 10, 13) + a^{8}\alpha(0, 11, 13)$

2 | $a^{13}\alpha(0, 5, 14) + a^{12}\alpha(0, 6, 14) + a^{11}\alpha(0, 7, 14) + a^{10}\alpha(0, 8, 14) + a^{9}\alpha(0, 9, 14) + a^{8}\alpha(0, 10, 14)$

3 | $a^{12}\alpha(0, 5, 15) + a^{11}\alpha(0, 6, 15) + a^{10}\alpha(0, 7, 15) + a^{9}\alpha(0, 8, 15) + a^{8}\alpha(0, 9, 15)$

4 | $a^{11}\alpha(0, 5, 16) + a^{10}\alpha(0, 6, 16) + a^{9}\alpha(0, 7, 16) + a^{8}\alpha(0, 8, 16)$

4 | $a^{10}\alpha(1, 5, 16) + a^{9}\alpha(1, 6, 16) + a^{8}\alpha(1, 7, 16)$

3 | $a^{9}\alpha(2, 5, 16) + a^{8}\alpha(2, 6, 16)$

2 | $a^{8}\alpha(3, 5, 16) +$

$\Sigma a^{14}\ldots$	
1140	1
1131	1
1122	1
1050	1
1041	2
1032	2
960	1
951	2
942	3
933	3
870	1
861	2
852	3
843	4
771	2
762	3
753	4
744	5
663	4
654	5
555	5

$\Sigma a^{13}\ldots$				
1240		1		1
1231		1		1
1222		1		1
1150	1	1		2
1141	1	2	1	4
1132	1	2	1	4
1060	1	1		2
1051	2	2	1	5
1042	2	3	2	7
1033	2	3	2	7
970	1	1		2
961	2	2	1	5
952	3	3	2	8
943	3	4	3	10
880	1	1		2
871	2	2	1	5
862	3	3	2	8
853	4	4	3	11
844	4	5	4	13
772	3	3	2	8
763	4	4	3	11
54				
44				
655	55			

$\Sigma a^{12}\ldots$							
1250		1		1			2
1241		1		2	1		4
1232		1		2	1		4
1160	1	1		1			3
1151	1	2	1	2	1		7
1142	1	2	1	3	2	1	10
1133	1	2	1	3	2	1	10
1070	1	1		1			3
1061	2	2	1	2	1		8
1052	2	3	2	3	2	1	13
1043	2	3	2	4	3	2	16
980	1	1		1			3
971	2	2	1	2	1		8
962	3	3	2	3	2	1	14
953	3	4	3	4	3	2	19
944	3	4	3	5	4	3	22
881	2	2	1	2	1		8
872	3	3	2	3	2	1	14
863	4	4	3	4	3	2	20
854	4	5	4	5	4	3	25
773	4	4	3	4	3	2	20
64	5	5	4	5	4	3	26
55	5	6	5	5	4	3	28
65	6	6	5	5	4	3	29

$\Sigma a^{11}\ldots$											
1170	1	1		1			1				4
1161	1	2	1	2	1		2	1			10
1152	1	2	1	3	2	1	3	2	1		16
1143	1	2	1	3	2	1	4	3	2	1	20
1080	1	1		1			1				4
1071	2	2	1	2	1		2	1			11
1062	2	3	2	3	2	1	3	2	1		19
1053	2	3	2	4	3	2	4	3	2	1	26
1044	2	3	2	4	3	2	5	4	3	2	30
990	1	1		1			1				4
981	2	2	1	2	1		2	1			11
972	3	3	2	3	2	1	3	2	1		20
963	3	4	3	4	3	2	4	3	2	1	29
954	3	4	3	5	4	3	5	4	3	2	36
882	3	3	2	3	2	1	3	2	1		20
873	4	4	3	4	3	2	4	3	2	1	30
864	4	5	4	5	4	3	5	4	3	2	39
855	4	5	4	6	5	4	5	4	3	2	42
774	5	5	4	5	4	3	5	4	3	2	40
765	5	6	5	6	5	4	5	4	3	2	45
666	5	7	6	6	5	4	5	4	3	2	47

$\Sigma a^{10}\ldots$

1090	1	1		1			1								4
1081	2	2	1	2	1		2	1			1				12
1072	2	3	2	3	2	1	3	2	1		2	1			22
1063	2	3	2	4	3	2	4	3	2	1	3	2	1		32
1054	2	3	2	4	3	2	5	4	3	2	4	3	2	1	40
991	2	2	1	2	1		2	1			1				12
982	3	3	2	3	2	1	3	2	1		2	1			23
973	3	4	3	4	3	2	4	3	2	1	3	2	1		35
964	3	4	3	5	4	3	5	4	3	2	4	3	2	1	46
955	3	4	3	5	4	3	6	5	4	3	4	3	2	1	50
883	4	4	3	4	3	2	4	3	2	1	3	2	1		36
874	4	5	4	5	4	3	5	4	3	2	4	3	2	1	49
865	4	5	4	6	5	4	6	5	4	3	4	3	2	1	56
775	4	6	5	6	5	4	6	5	4	3	4	3	2	1	58
766	4	6	5	7	6	5	6	5	4	3	4	3	2	1	61

$\Sigma a^{9}\ldots$

992	3	3	2	3	2	1	3	2	1		2	1			1			24
983	3	4	3	4	3	2	4	3	2	1	3	2	1		2	1		38
974	3	4	3	5	4	3	5	4	3	2	4	3	2	1	3	2	1	52
965	3	4	3	5	4	3	6	5	4	3	5	4	3	2	3	2	1	60
884	3	5	4	5	4	3	5	4	3	2	4	3	2	1	3	2	1	54
875	3	5	4	6	5	4	6	5	4	3	5	4	3	2	3	2	1	65
866	3	5	4	6	5	4	7	6	5	4	5	4	3	2	3	2	1	69
776	3	5	4	7	6	5	7	6	5	4	5	4	3	2	3	2	1	72

$\Sigma a^{8}\ldots$

885	2	4	3	6	5	4	6	5	4	3	5	4	3	2	4	3	2	2	1	68
876	2	4	3	6	5	4	7	6	5	4	6	5	4	3	4	3	2	2	1	76
777	2	4	3	6	5	4	8	7	6	5	6	5	4	3	4	3	2	2	1	80

12. Mr. O. H. Mitchell has shown* that the whole number of terms in the co-factor of an alternant is the quotient obtained by dividing the difference product of the exponents by the difference product $\zeta^{\frac{1}{2}}(0, 1, 2 \ldots n-1)$. The number of terms in $\alpha(0, p, q, r)$ is, accordingly,

$$\tfrac{1}{12}\, pqr\,(q-p)(r-p)(r-q).$$

This result may be used as a verification of the calculated coefficients. Thus in the example above the whole number of terms should be 35,360. The single symmetric functions whose coefficients have been obtained represent 24, 12 or 4 terms each, according as the exponents are all different, two alike, or three alike. If then the sums of the coefficients of the terms of these several classes be

* *American Journal of Mathematics*, Vol. IV, p. 341 (December, 1881).

multiplied by 24, 12 and 4 respectively, the sum of the products should be 35,360; and this was found to be the case.

13. The following additional proof of Mr. Mitchell's result may be given: If in $a(p, q, r, \ldots, z)$ we put $a = b = c = \ldots = l = 1$, the result will be equal to the number of terms in question since the value of each term will be unity. Hence this number is the value assumed by the quotient

$$\frac{A(p, q, r, \ldots, z)}{A(0, 1, 2, \ldots, n-1)}$$

under this hypothesis, which causes the quotient to assume the indeterminate form. Now if $\alpha, \beta, \gamma, \ldots, \lambda$ are the logarithms of $a, b, c, \ldots, l$, the alternant $A(p, q, r, \ldots, z)$ is the result of compounding the two arrays

$$\begin{array}{llll} 1, & \alpha, & \frac{\alpha^2}{2!}, & \frac{\alpha^3}{3!}, \ldots \\ 1, & \beta, & \frac{\beta^2}{2!}, & \frac{\beta^3}{3!}, \ldots \\ \vdots & \vdots & \vdots & \vdots \\ 1, & \lambda, & \frac{\lambda^2}{2!}, & \frac{\lambda^3}{3!}, \ldots \end{array} \quad \text{and} \quad \begin{array}{llll} 1, & p, & p^2, & p^3 \ldots \\ 1, & q, & q^2, & q^3 \ldots \\ \vdots & \vdots & \vdots & \vdots \\ 1, & z, & z^2, & z^3 \ldots \end{array}$$

Hence $A(p, q, \ldots, z)$ is the sum of all the products of the determinants which can be formed by selecting any n columns from the first array, each multiplied by the corresponding determinant formed from the second array. The term of lowest degree in $\alpha, \beta, \ldots, \lambda$ is that formed by selecting the first n columns of the first array, and the value of the corresponding determinant formed from the second array is $\zeta^{\frac{1}{2}}(p, q, \ldots, z)$.

When $a = b = \ldots = l = 1$, $\alpha, \beta, \ldots, \lambda$ vanish simultaneously, and the vanishing ratio of any two alternants is that of the terms of lowest degree in $\alpha, \beta, \ldots, \lambda$; that is, the vanishing value of the ratio

$$\frac{A(p, q, \ldots, z)}{A(p', q', \ldots, z')} \text{ is } \frac{\zeta^{\frac{1}{2}}(p, q, \ldots, z)}{\zeta^{\frac{1}{2}}(p', q', \ldots, z')},$$

and in particular the number of terms in $A(p, q, \ldots, z)$ is

$$\frac{\zeta^{\frac{1}{2}}(p, q, \ldots, z)}{\zeta^{\frac{1}{2}}(0, 1, \ldots, n-1)}.$$

Journal of Mathematics.

SIMON NEWCOMB, Editor.
THOMAS CRAIG, Associate Editor.

PUBLISHED UNDER THE AUSPICES OF THE

JOHNS HOPKINS UNIVERSITY.

Πραγμάτων ἔλεγχος οὐ βλεπομένων.

VOLUME VII. NUMBER 1.

BALTIMORE: PRESS OF ISAAC FRIEDENWALD.

AGENTS:

B. WESTERMANN & CO., *New York.*
D. VAN NOSTRAND, *New York.*
E. STEIGER & CO., *New York.*
FERREE & CO., *Philadelphia.*
CUSHINGS & BAILEY, *Baltimore.*
A. WILLIAMS & CO., *Boston.*
TRÜBNER & CO., *London.*
GAUTHIER-VILLARS, *Paris.*
A. ASHER & CO., *Berlin.*
MAYER & MÜLLER, *Berlin.*
ULRICO HOEPLI, *Milan.*
KARL J. TRÜBNER, *Strassburg.*

OCTOBER 1884.

JOHNS HOPKINS UNIVERSITY.

PROGRAMMES IN MATHEMATICS, 1884–85.

GRADUATE COURSES.

PROFESSOR NEWCOMB:

Analytical and Celestial Mechanics.
Twice weekly, through the year.

DR. STORY:

General Introductory Course for Graduates (including Theory of Numbers, Higher Algebra, Higher Plane Curves, Surfaces and Twisted Curves, Quaternions, Calculus of Operations, Probabilities, Partial Differential Equations, Elliptic Functions, and Mechanics.

Five times weekly, through the year.

This course is intended as preparatory for all the more advanced courses, and candidates for the Doctor's degree in Mathematics are expected to take it in the first year of their candidacy, if they have not previously taken it.

Theory of Numbers.
Twice weekly, first half-year.

Higher Algebra.
Twice weekly, second half-year.

Modern Synthetic Geometry.
Three times weekly, first half-year.

Quaternions.
Three times weekly, second half-year.

Mathematical Seminary.
Weekly through the year.

The exercises of this Seminary will consist of original work by the students, under the guidance of the Director, in prescribed subjects. The subjects chosen will be such as promise continuous work for a considerable length of time, the object being to impart the habit of investigation, rather than to reach results. The students will be required to make weekly reports of progress, the results obtained will be discussed, and new lines of research suggested from time to time.

DR. CRAIG:

Theory of Functions (including Elliptic Functions).
Three times weekly, through the year.

Hydrodynamics
Three times weekly, first half-year.

Calculus of Variations.
Twice weekly, first half-year.

Linear Differential Equations.
Twice weekly, second half-year.

In this course it is intended to give an account of the more recent investigations in the Theory of Linear Differential Equations, particular attention being directed to the work of Fuchs, Klein, and Poincaré.

Partial Differential Equations.
Twice weekly, second half-year.

Mathematical Seminary.
Weekly, through the year.

The subjects to which attention will be particularly directed are the Theory of Analytical Functions and Lame's Functions. During the first two or three meetings of the Seminary the Director will occupy the hour, and after that time the students will read dissertations on subjects selected for them by the Director. The work assigned will be divided into three parts: solution of problems, the historical investigation of the above mentioned subjects, and reports on current mathematical journals.

DR. FRANKLIN:

Problems in Mechanics.
Twice weekly, through the year.

Historical Lectures on Mathematical Topics by the Instructors, Fellows, and some of the Graduate Students.
Once in two weeks, through the year.

MATHEMATICAL SOCIETY.

The Mathematical Society, composed of the instructors and advanced students, will meet monthly as heretofore for the presentation and discussion of papers or oral communications.

UNDERGRADUATE COURSES.

FIRST YEAR:

Conic Sections.
Twice weekly, through the year. DR. STORY.

Differential and Integral Calculus.
Three times weekly, through the year.
DR. FRANKLIN.

SECOND YEAR:

Total Differential Equations.
Twice weekly, through the year. DR. CRAIG.

Theory of Equations.
Three times weekly, first half-year.
DR. FRANKLIN.

Solid Analytic Geometry.
Three times weekly, second half-year.
DR. FRANKLIN

Journal of Mathematics.

SIMON NEWCOMB, Editor.
THOMAS CRAIG, Associate Editor.

Published under the Auspices of the

JOHNS HOPKINS UNIVERSITY.

Πραγμάτων ἔλεγχος οὐ βλεπομένων.

Volume VII. Number 2.

BALTIMORE: PRESS OF ISAAC FRIEDENWALD.

AGENTS:

B. Westermann & Co., *New York.*
D. Van Nostrand, *New York.*
E. Steiger & Co., *New York.*
Jansen, McClurg & Co., *Chicago.*
Cushings & Bailey, *Baltimore.*
Trübner & Co., *London.*
A. Hermann, *Paris.*
Gauthier-Villars, *Paris.*
A. Asher & Co., *Berlin.*
Mayer & Muller, *Berlin.*
Ulrico Hoepli, *Milan.*
Karl J. Trübner, *Strassburg.*

January 1885.

CONTENTS.

The subscription price of the Journal is \$5.00 a volume; single numbers \$1.50.

Subscriptions from countries included in the Postal Union may be sent by *international* money order, made payable to *Nicholas Murray.* Drafts, checks, and domestic postal money orders should be made payable to *American Journal of Mathematics.*

N. B.—Persons wishing to dispose of complete sets of Vol. I. or single copies of No. 2, 3, or 4 of that volume, will please communicate with the Editor.

It is *particularly* requested that all communications and subscriptions be addressed as follows:

EDITOR, American Journal of Mathematics,
Johns Hopkins University, BALTIMORE, MD.

BALTIMORE, January 1, 1885.

Journal of Mathematics.

SIMON NEWCOMB, Editor.
THOMAS CRAIG, Associate Editor.

PUBLISHED UNDER THE AUSPICES OF THE

JOHNS HOPKINS UNIVERSITY.

Πραγμάτων ἔλεγχος οὐ βλεπομένων.

VOLUME VII. NUMBER 3.

BALTIMORE: PRESS OF ISAAC FRIEDENWALD.

AGENTS:

B. WESTERMANN & CO., *New York.*
D. VAN NOSTRAND, *New York.*
E. STEIGER & CO., *New York.*
JANSEN, MCCLURG & CO., *Chicago.*
CUSHINGS & BAILEY, *Baltimore.*
CUPPLES, UPHAM & CO., *Boston.*

TRÜBNER & CO., *London.*
A. HERMANN, *Paris.*
GAUTHIER-VILLARS, *Paris.*
MAYER & MÜLLER, *Berlin.*
ULRICO HOEPLI, *Milan.*
KARL J. TRÜBNER, *Strassburg.*

APRIL 1885.

G. MITTAG-LEFFLER.

4to. Vol. VI in progress. \$3 per Volume.

Band I.

Niels H. ABEL's Portrait in Lichtdruck.

APPELL, sur les fonctions uniformes d'un point analytique (x, y). 2 Mémoires.

— développements en série dans une aire limitée par des arcs de cercle.

BOURGUET, note sur les intégrales Eulériennes.

— sur quelques intégrales définies.

FUCHS, über lineare homogene Differentialgleichungen, zwischen deren Integralen homogene Relationen höheren als ersten Grades bestehen.

GOURSAT, sur un théorème de M. Hermite.

GYLDÉN, eine Annäherungsmethode im Probleme der drei Körper.

HERMITE, sur une relation donnée par M. Cayley dans la théorie des fonctions elliptiques.

MALMSTEN, zur Theorie der Leibrenten.

NETTO, zur Theorie der Discriminanten.

PICARD, sur une classe de groupes discontinus de substitutions linéaires et sur les fonctions de deux variables indépendantes restant invariables par ces substitutions.

POINCARÉ, théorie des groupes fuchsiens.

— mémoire sur les fonctions fuchsiennes.

REYE, das Problem der Configurationen.

— die Hexaëder- und die Octaëder-Configurationen $(12_6, 16_3)$.

SCHERING, zur Theorie der quadratischen Reste.

ZEUTHEN, sur un groupe de théorèmes et formules de la géométrie énumérative.

Band II.

APPELL, sur une classe de fonctions de deux variables indépendantes.

BENDIXSON, quelques théorèmes de la théorie des ensembles.

BOURGUET, sur les intégrales eulériennes et quelques autres fonctions uniformes.

— sur la fonction eulérienne.

CANTOR, sur une propriété du système de tous les nombres algébriques réels.

— une contribution à la théorie des ensembles.

— sur les séries trigonométriques.

— extension d'un théorème de la théorie des séries trigonométriques.

— sur les ensembles infinis et linéaires de points I–IV.

— fondements d'une théorie générale des ensembles.

— sur divers théorèmes de la théorie des ensembles de points situés dans un espace continu à n dimensions. Première communication.

CRONE, sur une espèce de courbes symétriques de la sixième classe.

ELLIOT, sur une équation linéaire du second ordre à coefficients doublement périodiques.

GOURSAT, sur une classe de fonctions représentées par des intégrales définies.

HERMITE ET LIPSCHITZ, sur quelques points dans la théorie des nombres.

MELLIN, über die transcendente Function

$$Q(x) = \Gamma(x) - P(x).$$

PICARD, sur des fonctions de deux variables indépendantes analogues aux fonctions modulaires.

POINCARÉ, sur les fonctions de deux variables.

VALENTINER, zur Theorie der Raumcurven.

Band III.

BELTRAMI, sur les couches de niveau électromagnétiques.

HALPHEN, sur les invariants des équations différentielles linéaires du quatrième ordre.

KÖNIGSBERGER, über die einer beliebigen Differentialgleichung erster Ordnung angehörigen selbständigen Transcendenten.

KRAUSE, sur la transformation des fonctions elliptiques.

— sur la transformation des fonctions hyperelliptiques de premier ordre.

KRAUSE, sur le multiplicateur des fonctions hyperelliptiques de premier ordre.

KRAZER UND PRYM, über die Verallgemeinerung der Riemann'schen Thetaformel.

LE PAIGE, sur les surfaces du troisième ordre.

LINDELÖF, une question de rentes viagères.

MELLIN, eine Verallgemeinerung der Gleichung

$$\Gamma(1+x)\Gamma(1-x) = \frac{\pi x}{\sin \pi x},$$

über gewisse durch die Gammafunction ausdrückbare unendliche Producte.

POINCARÉ, mémoire sur les groupes kleinéens.

PRYM, ein neuer Beweis für die Riemann'sche Thetaformel.

Ableitung einer allgemeinen Thetaformel.

Cte DE SPARRE, sur une équation différentielle linéaire du second ordre: Deux mémoires.

STERN, note sur certaines équations différentielles linéaires.

Band IV.

APPELL, sur les fonctions de trois variables réelles satisfaisant à l'équation différentielle $\Delta F = 0$.

BJERKNES, recherches hydrodynamiques. I. Les équations hydrodynamiques et les relations supplémentaires.

CANTOR, de la puissance des ensembles parfaits de points.

DARBOUX, sur l'équation aux dérivées partielles du troisième ordre des systèmes orthogonaux.

GOURSAT, démonstration du théorème de Cauchy.

HERMITE ET FUCHS, sur un développement en fraction continue.

HERMITE ET LIPSCHITZ, sur l'usage des produits infinis dans la théorie des fonctions elliptiques.

KOWALEVSKI, über die Reduction einer bestimmten Klasse Abel'scher Integrale 3ten Ranges auf elliptische Integrale.

LAGUERRE, sur quelques points de la théorie des équations numériques.

MATTHIESSEN, Untersuchungen über die Lage der Brennlinien eines unendlich dünnen Strahlenbündels gegeneinander und gegen einen Hauptstrahl.

MITTAG-LEFFLER, sur la représentation analytique des fonctions monogènes uniformes d'une variable indépendante.

démonstration nouvelle du théorème de Laurent.

POINCARÉ, sur les groupes des équations linéaires.

SCHEEFFER, Beweis des Laurent'schen Satzes.

SONINE, sur la généralisation d'une formule d'Abel.

ZELLER, zu Eulers Recursionsformel für die Divisorensummen.

Band V.

FIEDLER, über die Durchdringung gleichseitiger Rotationshyperboloide von parallelen Axen.

GOURSAT, sur une classe d'intégrales doubles.

HERMITE, sur quelques conséquences arithmétiques des formules de la théorie des fonctions elliptiques.

KREY, einige Anzahlen für Kegelflächen.

LE PAIGE, nouvelles recherches sur les surfaces du troisième ordre.

MALMSTEN, sur la formule sommatoire d'Euler.

PHRAGMÉN, Beweis eines Satzes aus der Mannigfaltigkeitslehre.

PICARD, sur les formes quadratiques ternaires indéfinies à indéterminées conjuguées et sur les fonctions hyperfuchsiennes correspondantes.

POINCARÉ, mémoire sur les fonctions zetafuchsiennes.

SCHEEFFER, allgemeine Untersuchungen über Rectification der Curven.

zur Theorie der stetigen Functionen einer reellen Veränderlichen.

SCHROTTER, Beiträge zur Theorie der elliptischen Functionen.

ZEUTHEN, sur les pentaèdres complets inscrits à une surface cubique.

BIBLIOTHECA MATHEMATICA

CONTENTS.

The subscription price of the Journal is \$5.00 a volume; single numbers \$1.50.

Subscriptions from countries included in the Postal Union may be sent by international money order, made payable to *Nicholas Murray*. Drafts, checks, and domestic postal money orders should be made payable to *American Journal of Mathematics*.

N. B.—Persons wishing to dispose of complete sets of Vol. I. or single copies of No. 2, 3, or 4 of that volume, will please communicate with the Editor.

It is requested that all scientific communications be addressed to the Editor of the American Journal of Mathematics, and all business or financial communications to the Publication Agency, Johns Hopkins University, Baltimore, Md., U. S. A.

Lightning Source UK Ltd.
Milton Keynes UK
UKHW020606110119
335177UK00005B/387/P

9 780266 538837